3.3 Pop Quiz

1. Determine the amplitude, period, and phase shift for
   $y = -5 \sin(2x + 2\pi/3)$.

2. List the coordinates for the five key points for one cycle of
   $y = 3 \sin(2x)$.

3. If $y = \cos(x)$ is shifted $\pi/2$ to the right, reflected in the
   x-axis, and shifted 3 units upward, then what is the equation
   of the curve in its final position?

4. What is the range of $f(x) = -4 \sin(x - 3) + 2$?

5. What is the frequency of the sine wave determined by
   $y = \sin(500\pi x)$, where x is time in minutes?

HISTORICAL NOTES

These brief margin essays
are about some of the
mathematicians who
developed the topics in
this text, providing a
sense of the historical
significance of the
precalculus topics.

Historical Note

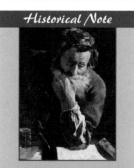

Archimedes of Syracuse (287 B.C.–212
B.C.) was a Greek mathematician,
physicist, engineer, astronomer, and
philosopher. Many consider him one
of the greatest mathematicians in
antiquity. Archimedes seems to have
carried out the first theoretical calculation of π, concluding that $223/71 <
\pi < 22/7$. He used a limiting process
very similar to what you will do if you
try the Concepts of Calculus at the
end of this chapter.

Applications

When you reach for a light switch, your brain controls the angles at your elbow
and your shoulder that put your hand at the location of the switch. An arm on a robot works in much the same manner. In the next example we see how the law of
sines and the law of cosines are used to determine the proper angles at the joints so
that a robot's hand can be moved to a position given in the coordinates of the
workspace.

Example 7 Positioning a robotic arm

A robotic arm with a 0.5-meter segment and a 0.3-meter segment is attached at the origin, as shown in Fig. 3.114. The computer-controlled arm is positioned by rotating
each segment through angles θ_1 and θ_2, as shown in Fig. 3.114. Given that we want to
have the end of the arm at the point (0.7, 0.2), find θ_1 and θ_2 to the nearest tenth of a
degree.

Solution

In the right triangle shown in Fig. 3.114, we have $\tan \omega = 0.2/0.7$ and

$$\omega = \tan^{-1}\left(\frac{0.2}{0.7}\right) \approx 15.95°.$$

CONCEPTS OF CALCULUS

All chapters end
with a discussion of
a particular concept
of calculus, as well
as exercises designed
to illustrate that
concept. This one-
page feature gives you
a preview of important
topics in calculus.

Concepts of Calculus

Area of a circle and π

*The idea of a limiting value to a process was used long before it was formalized in
calculus. Ancient mathematicians discovered the formula for the area of a circle and
the value of π using the formula for the area of a triangle $A = \frac{1}{2}bh$, trigonometry,
and the idea of limits. For the following exercises imagine that you are an ancient
mathematician (with a modern calculator), but you have never heard of π or the formula $A = \pi r^2$.*

Exercises

1. A regular pentagon is inscribed in a circle of radius r as
   shown in the following figure.

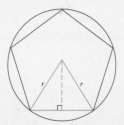

   The pentagon is made up of five isosceles triangles.
   Show that the area of the pentagon is $\frac{5\sin(72°)}{2}r^2$.

2. A regular polygon of n sides is inscribed in a circle of
   radius r. Write the area of the n-gon in terms of r and n.

3. The area of a regular n-gon inscribed in a circle of
   radius r is a constant multiple of r^2. Find the constant
   for a decagon, kilogon, and megagon.

4. What happens to the shape of the inscribed n-gon as n
   increases?

5. What would you use as a formula for the area of a circle
   of radius r? (You have never heard of π.)

6. Find a formula for the perimeter of an n-gon inscribed
   in a circle of radius r.

7. Find a formula for the circumference of a circle of
   radius r. (You have never heard of π.)

8. Suppose that the π-key on your calculator is broken.
   What expression could you use to calculate π accurate
   to nine decimal places?

2ND EDITION

Fundamentals
of Precalculus

Mark Dugopolski

Southeastern Louisiana University

PEARSON

Addison
Wesley

Boston San Francisco New York
London Toronto Sydney Tokyo Singapore Madrid
Mexico City Munich Paris Cape Town Hong Kong Montreal

Publisher: Greg Tobin
Executive Editor: Anne Kelly
Project Editor: Joanne Dill
Editorial Assistant: Leah Goldberg
Senior Managing Editor: Karen Wernholm
Senior Production Supervisor: Peggy McMahon
Design Supervisor: Barbara T. Atkinson
Cover Designer: Barbara T. Atkinson
Cover Photo: Carabiner—Goran Kuzmanovsk/Shutterstock; Rock Climbers—Chris Hill/Shutterstock
Photo Researcher: Beth Anderson
Media Producer: Michelle Small
Software Development: Mary Durnwald and Elizabeth S. DeWerth
Executive Marketing Manager: Becky Anderson
Marketing Coordinator: Bonnie Gill
Senior Author Support/Technology Specialist: Joe Vetere
Rights and Permissions Advisor: Shannon Barbe
Manufacturing Manager: Evelyn Beaton
Text Design, Production Coordination, Illustrations, and Composition: Nesbitt Graphics, Inc.

For permission to use copyrighted material, grateful acknowledgment has been made to the copyright holders on page C-1 in the back of the book, which is hereby made part of this copyright page.

Many of the designations used by manufacturers and sellers to distinguish their products are claimed as trademarks. Where those designations appear in this book, and Pearson Education was aware of a trademark claim, the designations have been printed in initial caps or all caps.

Library of Congress Cataloging-in-Publication Data

Dugopolski, Mark.

   Fundamentals of precalculus / Mark Dugopolski.

      p. cm.

   Includes index.

   ISBN 0-321-50697-9 (student ed.) — ISBN 0-321-53675-4 (AIE)

  1. Functions—Textbooks. I. Title.

QA331.3.D84 2009

515—dc22   2009060059

ISBN-13: 978-0-321-53675-4   ISBN-10: 0-321-53675-4

4 5 6 7 8 9 10—QGT—11 10

CONTENTS

Preface   *ix*

Supplements List   *xiii*

Function Gallery   *xvii*

1 Graphs and Functions

1.1 Real Numbers and Their Properties   *2*
The Real Numbers • Properties of the Real Numbers • Additive Inverses • Relations • Absolute Value • Equations Involving Absolute Value

1.2 Linear and Absolute Value Inequalities   *11*
Interval Notation • Linear Inequalities • Compound Inequalities • Absolute Value Inequalities • Modeling with Inequalities

1.3 Equations and Graphs in Two Variables   *22*
The Cartesian Coordinate System • The Distance Formula • The Midpoint Formula • The Circle • The Line • Using a Graph to Solve an Equation

1.4 Linear Equations in Two Variables   *35*
Slope of a Line • Point-Slope Form • Slope-Intercept Form • Using Slope to Graph a Line • The Three Forms for the Equation of a Line • Parallel Lines • Perpendicular Lines • Applications

1.5 Functions   *45*
The Function Concept • Identifying Functions • Domain and Range • Function Notation • The Average Rate of Change of a Function • Constructing Functions

1.6 Graphs of Relations and Functions   *58*
Graphing Equations • Semicircles • Piecewise Functions • Increasing, Decreasing, and Constant

1.7 Families of Functions, Transformations, and Symmetry   *69*
Translation • Reflection • Stretching and Shrinking • The Linear Family of Functions • Symmetry • Reading Graphs to Solve Inequalities

1.8 Operations with Functions   *83*
Basic Operations with Functions • Composition of Functions • Applications

1.9 Inverse Functions   *92*
One-to-One Functions • Inverse Functions • Inverse Functions Using Function Notation • Graphs of f and f^{-1} • Finding Inverse Functions Mentally

Chapter 1 Highlights   *104*
Chapter 1 Review Exercises   *107*
Concepts of Calculus: Limits   *112*

2 Polynomial and Rational Functions

2.1 Quadratic Functions and Inequalities *114*
Two Forms for Quadratic Functions • Opening, Vertex, and Axis of
Symmetry • Intercepts • Quadratic Inequalities • Applications of
Maximum and Minimum

2.2 Complex Numbers *125*
Definitions • Addition, Subtraction, and Multiplication • Division of
Complex Numbers • Roots of Negative Numbers • Imaginary Solutions to
Quadratic Equations

2.3 Zeros of Polynomial Functions *132*
The Remainder Theorem • Synthetic Division • The Factor Theorem • The
Fundamental Theorem of Algebra • The Rational Zero Theorem

2.4 The Theory of Equations *143*
The Number of Roots of a Polynomial Equation • The Conjugate Pairs
Theorem • Descartes's Rule of Signs

2.5 Miscellaneous Equations *151*
Factoring Higher-Degree Equations • Equations Involving Square Roots •
Equations with Rational Exponents • Equations of Quadratic Type •
Equations Involving Absolute Value

2.6 Graphs of Polynomial Functions *161*
Drawing Good Graphs • Symmetry • Behavior at the x-Intercepts • The
Leading Coefficient Test • Sketching Graphs of Polynomial Functions •
Polynomial Inequalities

2.7 Rational Functions and Inequalities *175*
Rational Functions and Their Domains • Horizontal and Vertical
Asymptotes • Oblique Asymptotes • Sketching Graphs of Rational
Functions • Rational Inequalities • Applications

Chapter 2 Highlights *190*
Chapter 2 Review Exercises *192*
Concepts of Calculus: Instantaneous rate of change *195*

3 Trigonometric Functions

3.1 Angles and Their Measurements *197*
Degree Measure of Angles • Radian Measure of Angles • Arc Length

3.2 The Sine and Cosine Functions *209*
Definition • Sine and Cosine of a Multiple of 45° • Sine and Cosine of
a Multiple of 30° • Sine and Cosine of an Arc • Approximate Values for
Sine and Cosine • The Fundamental Identity • Modeling the Motion of
a Spring

3.3 The Graphs of the Sine and Cosine Functions *217*
The Graph of $y = \sin(x)$ • The Graph of $y = \cos(x)$ • Transformations
of Sine and Cosine • Changing the Period • The General Sine Wave •
Frequency

3.4 The Other Trigonometric Functions and Their Graphs *231*
Definitions • Graph of $y = \tan(x)$ • Graph of $y = \cot(x)$ • Graph of $y = \sec(x)$ • Graph of $y = \csc(x)$

3.5 The Inverse Trigonometric Functions *242*
The Inverse of the Sine Function • The Inverse Cosine Function • Inverses of Tangent, Cotangent, Secant, and Cosecant • Compositions of Functions

3.6 Right Triangle Trigonometry *252*
Trigonometric Ratios • Right Triangles • Solving a Right Triangle • Applications

3.7 Identities *263*
Pythagorean Identities • Odd and Even Identities • Sum and Difference Identities • Cofunction Identities • Double-Angle and Half-Angle Identities • Product and Sum Identities • The Function $y = a \sin x + b \cos x$

3.8 Conditional Trigonometric Equations *274*
Cosine Equations • Sine Equations • Tangent Equations • Equations Involving Multiple Angles • More Complicated Equations • Modeling Projectile Motion

3.9 The Law of Sines and the Law of Cosines *287*
Oblique Triangles • The Law of Sines • The Ambiguous Case (SSA) • The Law of Cosines • Applications

Chapter 3 Highlights *301*
Chapter 3 Review Exercises *306*
Concepts of Calculus: Area of a circle and π *310*

4 Exponential and Logarithmic Functions

4.1 Exponential Functions and Their Applications *312*
The Definition • Domain of Exponential Functions • Graphing Exponential Functions • The Exponential Family of Functions • Exponential Equations • The Compound Interest Model • Continuous Compounding and the Number e

4.2 Logarithmic Functions and Their Applications *325*
The Definition • Graphs of Logarithmic Functions • The Logarithmic Family of Functions • Logarithmic and Exponential Equations • Applications

4.3 Rules of Logarithms *335*
The Inverse Rules • The Logarithm of a Product • The Logarithm of a Quotient • The Logarithm of a Power • Using the Rules • The Base-Change Formula

4.4 More Equations and Applications *345*
Logarithmic Equations • Exponential Equations • Strategy for Solving Equations • Radioactive Dating • Newton's Model for Cooling • Paying off a Loan

Chapter 4 Highlights *355*
Chapter 4 Review Exercises *357*
Concepts of Calculus: Evaluating transcendental functions *360*

5 Conic Sections, Polar Coordinates, and Parametric Equations

5.1 The Parabola *362*
Definition • Developing the Equation • The Standard Equation of a
Parabola • Graphing a Parabola • Parabolas Opening to the Left or Right •
Applications

5.2 The Ellipse and the Circle *371*
Definition of Ellipse • The Equation of the Ellipse • Graphing
an Ellipse Centered at the Origin • Translation of Ellipses • The Circle •
Applications

5.3 The Hyperbola *383*
The Definition • Developing the Equation • Graphing a Hyperbola
Centered at (0, 0) • Hyperbolas Centered at (h, k) • Finding the Equation
of a Hyperbola • Classifying the Conics

5.4 Polar Coordinates *396*
Polar Coordinates • Polar-Rectangular Conversions • Polar Equations •
Converting Equations

5.5 Polar Equations of the Conics *406*
Alternative Definition of the Conics • Polar Equations of the Conics •
Equivalency of the Definitions

5.6 Parametric Equations *411*
Graphs of Parametric Equations • Eliminating the Parameter • Writing
Parametric Equations • Shooting Baskets with Parametric Equations

Chapter 5 Highlights *417*
Chapter 5 Review Exercises *419*
Concepts of Calculus: The reflection property of a parabola *423*

A Appendix: Basic Algebra Review

A.1 Exponents and Radicals *AA-1*
Exponential Expressions • Negative Integral Exponents • Rules of
Exponents • Roots • Rational Exponents • Radical Notation • The
Product and Quotient Rules for Radicals • Simplified Form and
Rationalizing the Denominator • Operations with Radical Expressions

A.2 Polynomials *AA-11*
Definitions • Naming and Evaluating Polynomials • Addition and
Subtraction of Polynomials • Multiplication of Polynomials • Using FOIL •
Special Products • Division of Polynomials

A.3 Factoring Polynomials *AA-17*
Factoring Out the Greatest Common Factor • Factoring by Grouping •
Factoring $ax^2 + bx + c$ • Factoring the Special Products • Factoring the
Difference and Sum of Two Cubes • Factoring Completely

A.4 Rational Expressions *AA-22*
Reducing • Multiplication and Division • Building Up the Denominator •
Addition and Subtraction

Answers to Selected Exercises *A-1*
Credits *C-1*
Index of Applications *I-1*
Index *I-3*

Fundamentals of Precalculus is designed to review the fundamental topics that are necessary for success in calculus. Containing only five chapters, this text may be covered in a one-semester or one-term course, with a minimum of deleting or skipping around. This text contains the rigor essential for building a strong foundation of mathematical skills and concepts, and at the same time supports students' mathematical needs with a number of tools newly developed for this revision. A student who is well acquainted with the material in this text will have the necessary skills, understanding, and insights required to succeed in calculus.

Content Changes for the Second Edition

The purpose of this revision was to make this text easier to teach and learn from, as well as to make it more pedagogically sound. I have extensively updated explanations, examples, exercises, and art in response to comments from the users of the first edition. All changes were made with the goal of giving the student a smoother path to success. Using a pedagogical approach to color, more color has been added to examples to emphasize key computations. The art package has been revamped with new photographs and situational art, and it accompanies a fresh new design that will help students visualize the mathematics being discussed. The notation of limits is now used to describe the behavior of polynomial, rational, exponential, logarithmic, and trigonometric functions. Instructors who want to give their students a good introduction to limits can do so by also covering the Concepts of Calculus at the end of each chapter.

New or Enhanced Features

I have included several new features and revised some existing features. These improvements are designed to make the book easier to use and to support student and instructor success. I believe students and instructors will welcome the following new or enhanced features:

For Students

Pop Quizzes (page 11) Included in every section of the text, Pop Quizzes give instructors and students convenient quizzes that can be used to measure student progress. They are short and can be used in the classroom to confirm comprehension of the basics. The answers appear in the Annotated Instructor's Edition only.

Historical Notes (page 23) Located in the margins throughout the text, these brief essays are designed to connect the topics of precalculus to the mathematicians who first studied them and to give precalculus a human face.

Thinking Outside the Box (page 174) Found throughout the text, these problems are designed to get students (and instructors) to do some mathematics just for fun. Yes, fun. I enjoyed writing and solving these problems and hope that you will, too. The problems can be used for individual or group work, so be creative and try Thinking Outside the Box. The answers are given in the Annotated Instructor's Edition only, and complete solutions can be found in the Instructor's Solutions Manual.

Foreshadowing Calculus (page 121) This feature gives a brief indication of the connection between certain algebraic topics and calculus and identifies topics that will be important later on in calculus.

Concepts of Calculus (page 310) All chapters end with a discussion of a particular concept of calculus, as well as exercises designed to illustrate that concept. This one-page feature gives students a preview of important topics of calculus and may be used as a writing or collaborative learning assignment.

Function Galleries (page 240) Located throughout the text, these function summaries have been improved and redesigned to help students link the visual aspects of various families of functions to the properties of the functions. They are also gathered together at the beginning of the text.

Art Annotations (page 115) have been added to make the mathematical art a better learning tool and easier for students to use.

For Instructors

Annotated Instructor's Edition New to this edition, this text is available with answers beside the exercises for most exercises, plus a complete answer section at the back of the text.

Example Numbers (page 333) Occurring at the end of the direction line for nearly all groups of exercises, the Example Numbers indicate which examples correspond to the exercises. This keying of exercises and examples appears only in the Annotated Instructor's Edition and will make it easier for instructors to assign exercises for homework.

Insider's Guide This new supplement includes resources to help faculty with course preparation and classroom management. Included are helpful teaching tips correlated to each section of the text as well as additional resources for classroom enrichment.

Continuing Features

Chapter Opener Each chapter begins with a discussion of a famous bridge. Modern bridges are among the largest and most artistic structures ever built. Bridges were chosen to begin the chapters because a bridge and this text both have a similar purpose: a bridge is built to link two locations with a faster and more efficient path and this text was written to link a student's past mathematical training with the content of calculus in a fast and efficient manner.

Graphing Calculator Discussions Optional graphing calculator discussions are included in the text. They are clearly marked by graphing calculator icons so that they can be easily skipped if desired. Although use of this technology is optional, students who do not use a graphing calculator can still benefit from the technology discussions. In this text, the graphing calculator is used as a tool to support and enhance the algebraic conclusions, not to make conclusions. Any graphing utility or computer mathematics package can be used in place of a graphing calculator. Due to rapid changes in technology and the differences among various brands, this text does not give specific instructions on how to use the various graphing calculators that are available. However, the Graphing Calculator Manual that accompanies this text provides students with keystroke operations for many of the more popular graphing calculators, using specific examples from this text.

Summaries of important concepts are included to help students clarify ideas that have multiple parts. For example, see the Summary of the Types of Symmetry in Section 2.6 on page 162.

Strategies contain general guidelines for accomplishing tasks. They have been given a more visible design so that students will be more aware of them and use them to sharpen their problem-solving skills. For example, see the Strategy for Solving Exponential and Logarithmic Equations in Section 4.4 on page 348.

Procedures are similar to Strategies but are more specific and more algorithmic. They have also been highlighted in this edition and are designed to enable students to use a step-by-step approach to problem solving. For example, see the Procedure for Finding an Inverse Function by the Switch-and-Solve Method in Section 1.9 on page 96.

For Thought Each exercise set is preceded by a set of ten true/false questions that review the basic concepts in the section, help check student understanding before beginning the exercises, and offer opportunities for writing and/or discussion. The answers to all For Thought exercises are included in the back of the student edition of this text.

Highlights (page 190) This end-of-chapter feature has been completely rewritten so that it is a more effective tool for reviewing the chapter. Each highlight now contains an overview of all of the concepts presented in the chapter along with brief examples to illustrate the concepts.

Chapter Review Exercises These exercises are designed to give students a comprehensive review of the chapter without reference to individual sections and to prepare students for a chapter test.

Basic Algebra Review This four-section review of basic algebra concepts is located in Appendix A. Some or all of this review can be used as a starting point for the course or can be left for individual students to use as needed.

Index of Applications The many applications contained within the text are listed in the Index of Applications that appears at the end of the text. The applications are referenced by page and grouped by subject matter.

Acknowledgments

Thanks to the many professors and students who have used my texts in the past. I am always glad to hear about what you like and don't like about my texts. Thanks to the following reviewers whose comments and suggestions were invaluable in preparing this second edition:

Alison Ahlgren, University of Illinois at Urbana–Champaign
David Akins, El Camino College
Richard Rockwell, Pacific Union College
Michael Montano, Riverside Community College
Alexander Vaninsky, Hostos Community College, CUNY
Denise Widup, University of Wisconsin, Parkside

Thanks to Edgar N. Reyes, Southeastern Louisiana University, for working all of the exercises and writing the Solutions Manuals; Rebecca W. Muller, Southeastern Louisiana University, for writing the Instructor's Testing Manual; Darryl Nester, Bluffton University, for writing the Graphing Calculator Manual; and Abby Tanenbaum for editing the Insider's Guide to Teaching with Fundamentals of Precalculus. I wish to express my thanks to Edgar Reyes and Lauri Semarne for accuracy-checking this text. A special thanks also goes to Nesbitt Graphics, the compositor, for the superb work they did on this book.

Finally, it has been another great experience working with the talented and dedicated Pearson Addison-Wesley team. I am grateful for their assistance, encouragement, and direction throughout this project: Greg Tobin, Anne Kelly, Becky Anderson, Peggy McMahon, Joanne Dill, Barbara Atkinson, Joe Vetere, Michelle Small, Beth Anderson, Leah Goldberg, and Bonnie Gill.

As always, thanks to my wife Cheryl, whose love, encouragement, understanding, support, and patience are invaluable.

Mark Dugopolski
Ponchatoula, Louisiana

SUPPLEMENTS LIST

▪▪▪Student Supplements

Student's Solutions Manual
- By Edgar N. Reyes, *Southeastern Louisiana University*.
- Provides detailed solutions to all odd-numbered text exercises.
- ISBN: 0-321-53662-2 and 978-0-321-53662-4

Graphing Calculator Manual
- By Darryl Nester, *Bluffton University*.
- Provides instructions and keystroke operations for the TI-83/83 Plus, TI-84 Plus, TI-85, TI-86, and TI-89.
- Also contains worked-out examples taken directly from the text.
- ISBN: 0-321-53661-4 and 978-0-321-53661-7

Video Lectures on CD with Optional Captioning
- Videos provide comprehensive coverage of each section and topic in the text in an engaging format on CD that stresses student interaction.
- Complete set of digitized videos, ideal for distance learning or supplemental instruction at home or on campus.
- Videos include optional text captioning
- ISBN: 0-321-53712-2 and 978-0-321-53712-6

A Review of Algebra
- By Heidi Howard, *Florida Community College at Jacksonville*.
- Provides additional support for those students needing further algebra review.
- ISBN: 0-201-77347-3 and 978-0-201-77347-7

▪▪▪Instructor Supplements

Annotated Instructor's Edition
- NEW! Special edition of the text.
- Provides answers in the margins to many text exercises, plus a full answer section at the back of the text.
- ISBN: 0-321-53675-4 and 978-0-321-53675-4

Instructor's Solutions Manual
- By Edgar N. Reyes, *Southeastern Louisiana University*.
- Provides complete solutions to all text exercises, including the For Thought, Linking Concepts, and Concepts of Calculus exercises.
- ISBN: 0-321-53676-2 and 978-0-321-53676-1

Instructor's Testing Manual
- By Rebecca W. Muller, *Southeastern Louisiana University*.
- Contains six alternative forms of tests per chapter (two are multiple choice).
- Answers keys are included.
- ISBN: 0-321-53677-0 and 978-0-321-53677-8

TestGen®
- Enables instructors to build, edit, print, and administer tests.
- Features a computerized bank of questions developed to cover all text objectives.
- Available within MyMathLab® or from the Instructor Resource Center at www.aw-bc.com/irc.

PowerPoint Lecture Presentations with Active Learning Questions
- Features presentations written and designed specifically for this text, including figures and examples from the text.
- Active Learning Questions for use with classroom response systems include Multiple Choice questions to review lecture material.
- Available within MyMathLab® or from the Instructor Resource Center at www.aw-bc.com/irc.

NEW! Insider's Guide to Teaching with Fundamentals of Precalculus
- Provides helpful teaching tips correlated to each section of the text as well as general teaching advice.
- Available within MyMathLab® or from the Instructor Resource Center at www.aw-bc.com/irc.
- ISBN: 0-321-53655-X and 978-0-321-53655-6

Pearson Adjunct Support Center
- Offers consultation on suggested syllabi, helpful tips on using the textbook support package, assistance with content, and advice on classroom strategies
- Available Sunday–Thursday evenings from 5 P.M. to midnight EST: telephone: 1-800-435-4084; e-mail: AdjunctSupport@aw.com; fax: 1-877-262-9774

■■■ MyMathLab®

MyMathLab MyMathLab® is a series of text-specific, easily customizable online courses for Pearson Education's textbooks in mathematics and statistics. Powered by CourseCompass™ (Pearson Education's online teaching and learning environment) and MathXL® (our own online homework, tutorial, and assessment system), MyMathLab gives you the tools you need to deliver all or a portion of your course online, whether your students are in a lab setting or working from home. MyMathLab provides a rich and flexible set of course materials, featuring *free-response* exercises that are algorithmically generated for unlimited practice and mastery. Students can also use online tools, such as video lectures, animations, and a multimedia textbook, to independently improve their understanding and performance. Instructors can use MyMathLab's homework and test managers to select and assign online exercises correlated directly with the textbook, and they can also create and assign their own online exercises and import TestGen tests for added flexibility. MyMathLab's online gradebook—designed specifically for mathematics and statistics—automatically tracks students' homework and test results and gives the instructor control over how to calculate final grades. MyMathLab also includes access to **Pearson's Tutor Center**, which provides students with tutoring via toll-free phone, fax, email, and interactive Web sessions. Instructors can also add offline (paper-and-pencil) grades to the gradebook. MyMathLab is available to qualified adopters. For more information, visit our Web site at www.mymathlab.com or contact your sales representative.

■■■ MathXL®

MathXL MathXL® is a powerful online homework, tutorial, and assessment system that accompanies Pearson Education's textbooks in mathematics or statistics. With MathXL, instructors can create, edit, and assign online homework and tests using algorithmically generated exercises correlated with the textbook at the objective level. They can also create and assign their own online exercises and import TestGen tests for added flexibility. All student work is tracked in MathXL's online gradebook. Students can take chapter tests in MathXL and receive personalized study plans based on their test results. The study plan diagnoses weaknesses and links students directly to tutorial exercises for the objectives they need to study and on which they need to be retested. Students can also access supplemental animations and video clips directly from selected exercises. MathXL is available to qualified adopters. For more information, visit our Web site at www.mathxl.com, or contact your sales representative.

■■■ MathXL® Tutorials on CD

(ISBN: 0-321-53646-0 and 978-0-321-53646-4)

This interactive tutorial CD-ROM provides algorithmically generated practice exercises that are correlated at the objective level to the exercises in the textbook. Every practice exercise is accompanied by an example and a guided solution designed to involve students in the solution process. Selected exercises may also include a video clip to help students visualize concepts. The software provides helpful feedback for incorrect answers and can generate printed summaries of students' progress.

■ ■ ■ Video Lectures on CD with Optional Captioning (ISBN: 0-321-53712-2 and 978-0-321-53712-6)

The video lectures for this text are available on CD-ROM, making it easy and convenient for students to watch the videos from a computer at home or on campus. The videos feature an engaging team of mathematics instructors who present a lecture for each section of the text, covering key definitions, examples, and procedures. The format provides distance-learning students with comprehensive video instruction for each section in the book, but also allows students needing only small amounts of review to watch instruction on a specific skill or procedure. The videos have an optional text captioning window; the captions can be easily turned off or on for individual student needs.

Constant Function

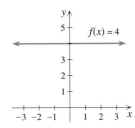

Domain $(-\infty, \infty)$
Range $\{4\}$
Constant on $(-\infty, \infty)$
Symmetric about y-axis

Identity Function

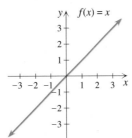

Domain $(-\infty, \infty)$
Range $(-\infty, \infty)$
Increasing on $(-\infty, \infty)$
Symmetric about origin

Linear Function

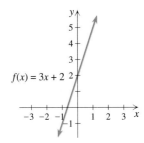

Domain $(-\infty, \infty)$
Range $(-\infty, \infty)$
Increasing on $(-\infty, \infty)$

Absolute-Value Function

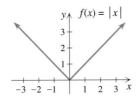

Domain $(-\infty, \infty)$
Range $[0, \infty)$
Increasing on $(0, \infty)$
Decreasing on $(-\infty, 0)$
Symmetric about y-axis

Square Function

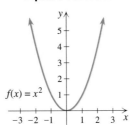

Domain $(-\infty, \infty)$
Range $[0, \infty)$
Increasing on $(0, \infty)$
Decreasing on $(-\infty, 0)$
Symmetric about y-axis

Square-Root Function

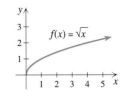

Domain $[0, \infty)$
Range $[0, \infty)$
Increasing on $(0, \infty)$

Cube Function

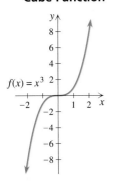

Domain $(-\infty, \infty)$
Range $(-\infty, \infty)$
Increasing on $(-\infty, \infty)$
Symmetric about origin

Cube-Root Function

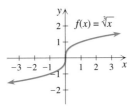

Domain $(-\infty, \infty)$
Range $(-\infty, \infty)$
Increasing on $(-\infty, \infty)$
Symmetric about origin

Greatest Integer Function

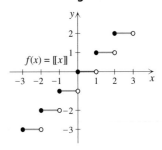

Domain $(-\infty, \infty)$
Range $\{n \mid n$ is an integer$\}$
Constant on $[n, n + 1)$
for every integer n

Function Gallery: Some Inverse Functions

Linear

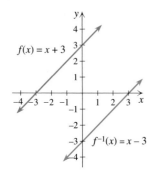

$f(x) = x + 3$

$f^{-1}(x) = x - 3$

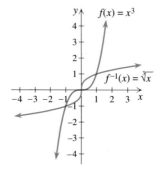

$f(x) = 5x$

$f^{-1}(x) = \dfrac{x}{5}$

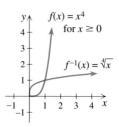

$f^{-1}(x) = \dfrac{x + 3}{2}$

$f(x) = 2x - 3$

Powers and Roots

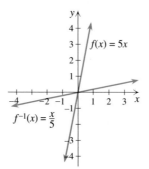

$f(x) = x^2$ for $x \geq 0$

$f^{-1}(x) = \sqrt{x}$

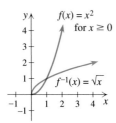

$f(x) = x^3$

$f^{-1}(x) = \sqrt[3]{x}$

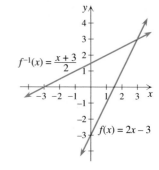

$f(x) = x^4$ for $x \geq 0$

$f^{-1}(x) = \sqrt[4]{x}$

Function Gallery: Polynomial Functions

Linear: $f(x) = mx + b$, domain $(-\infty, \infty)$, range $(-\infty, \infty)$ if $m \neq 0$, slope m, y-intercept $(0, b)$

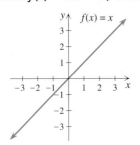

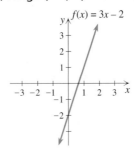

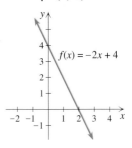

Slope 1, y-intercept $(0, 0)$ Slope 3, y-intercept $(0, -2)$ Slope -2, y-intercept $(0, 4)$

Quadratic: $f(x) = ax^2 + bx + c$ or $f(x) = a(x - h)^2 + k$, domain $(-\infty, \infty)$, vertex (h, k) or $x = -b/(2a)$

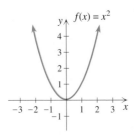

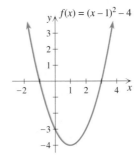

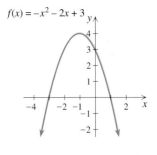

Vertex $(0, 0)$
Range $[0, \infty)$

Vertex $(1, -4)$
Range $[-4, \infty)$

Vertex $(-1, 4)$
Range $(-\infty, 4]$

Cubic: $f(x) = ax^3 + bx^2 + cx + d$, domain $(-\infty, \infty)$, range $(-\infty, \infty)$

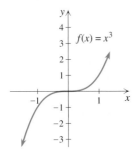

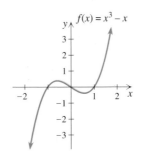

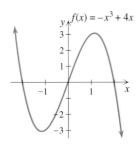

Quartic or Fourth-Degree: $f(x) = ax^4 + bx^3 + cx^2 + dx + e$, domain $(-\infty, \infty)$

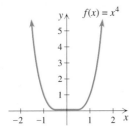

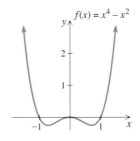

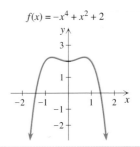

Function Gallery: Some Basic Rational Functions

Horizontal Asymptote *x*-axis and Vertical Asymptote *y*-axis

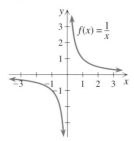

$f(x) = \dfrac{1}{x}$

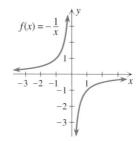

$f(x) = -\dfrac{1}{x}$

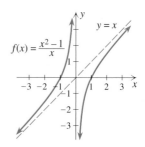

$f(x) = \dfrac{1}{x^2}$

Various Asymptotes

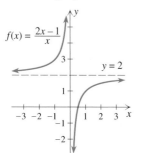

$f(x) = \dfrac{2x-1}{x}$

$y = 2$

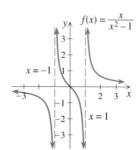

$f(x) = \dfrac{x}{x^2-1}$

$x = -1$

$x = 1$

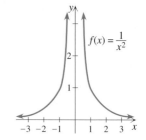

$f(x) = \dfrac{x^2-1}{x}$

$y = x$

Function Gallery: **Exponential and Logarithmic Functions**

Exponential: $f(x) = a^x$, domain $(-\infty, \infty)$, range $(0, \infty)$

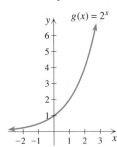

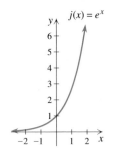

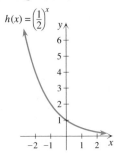

Increasing on $(-\infty, \infty)$
y-intercept $(0, 1)$

Decreasing on $(-\infty, \infty)$
y-intercept $(0, 1)$

Increasing on $(-\infty, \infty)$
y-intercept $(0, 1)$

Logarithmic: $f^{-1}(x) = \log_a(x)$, domain $(0, \infty)$, range $(-\infty, \infty)$

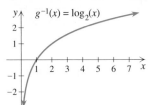

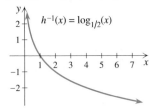

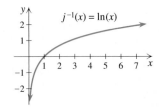

Increasing on $(0, \infty)$
x-intercept $(1, 0)$

Decreasing on $(0, \infty)$
x-intercept $(1, 0)$

Increasing on $(0, \infty)$
x-intercept $(1, 0)$

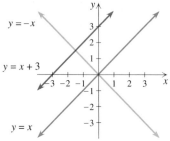

Linear

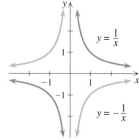

Quadratic

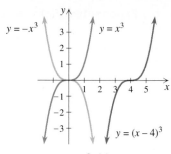

Cubic

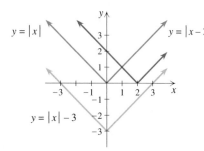

Absolute value

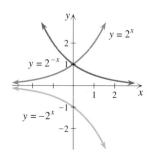

Exponential

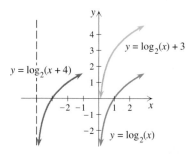

Logarithmic

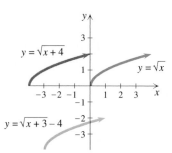

Square root

Reciprocal

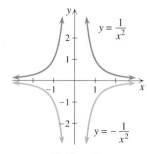

Rational

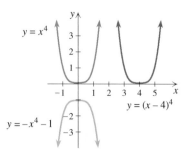

Fourth degree

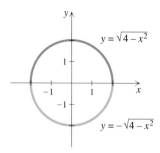

Semicircle

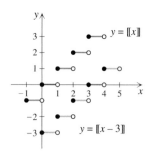

Greatest integer

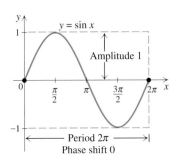

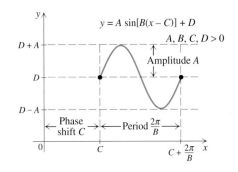

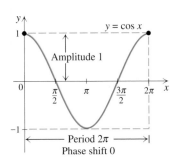

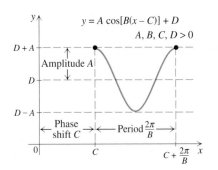

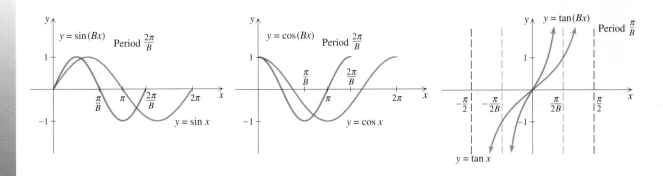

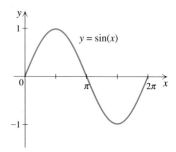

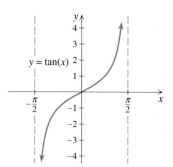

Domain	$(-\infty, \infty)$	$(-\infty, \infty)$	$x \neq \dfrac{\pi}{2} + k\pi$
(*k* any integer)			
Range	$[-1, 1]$	$[-1, 1]$	$(-\infty, \infty)$
Period	2π	2π	π
Fundamental cycle	$[0, 2\pi]$	$[0, 2\pi]$	$\left[-\dfrac{\pi}{2}, \dfrac{\pi}{2}\right]$

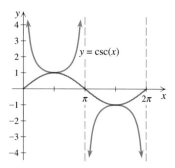

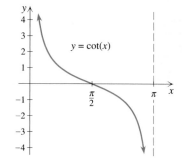

Domain	$x \neq k\pi$	$x \neq \dfrac{\pi}{2} + k\pi$	$x \neq k\pi$
(*k* any integer)			
Range	$(-\infty, -1] \cup [1, \infty)$	$(-\infty, -1] \cup [1, \infty)$	$(-\infty, \infty)$
Period	2π	2π	π
Fundamental cycle	$[0, 2\pi]$	$[0, 2\pi]$	$[0, \pi]$

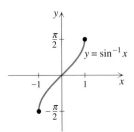

Domain $[-1, 1]$
Range $\left[-\frac{\pi}{2}, \frac{\pi}{2}\right]$

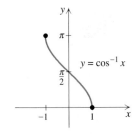

Domain $[-1, 1]$
Range $[0, \pi]$

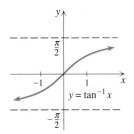

Domain $(-\infty, \infty)$
Range $\left(-\frac{\pi}{2}, \frac{\pi}{2}\right)$

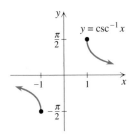

Domain $(-\infty, -1] \cup [1, \infty)$
Range $\left[-\frac{\pi}{2}, 0\right) \cup \left(0, \frac{\pi}{2}\right]$

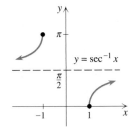

Domain $(-\infty, -1] \cup [1, \infty)$
Range $\left[0, \frac{\pi}{2}\right) \cup \left(\frac{\pi}{2}, \pi\right]$

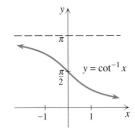

Domain $(-\infty, \infty)$
Range $(0, \pi)$

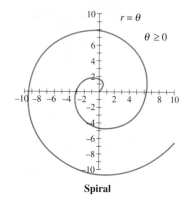

$r = \theta$

$\theta \geq 0$

Spiral

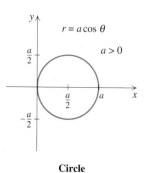

$r = a \cos \theta$

$a > 0$

Circle

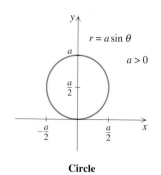

$r = a \sin \theta$

$a > 0$

Circle

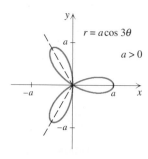

$r = a \cos 3\theta$

$a > 0$

Three-Leaf Rose

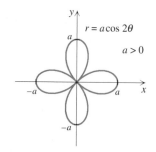

$r = a \cos 2\theta$

$a > 0$

Four-Leaf Rose

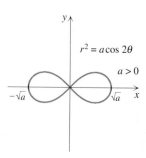

$r^2 = a \cos 2\theta$

$a > 0$

Lemniscate

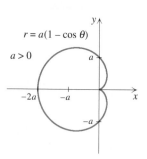

$r = a(1 - \cos \theta)$

$a > 0$

Cardioid

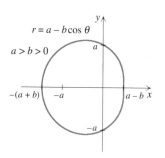

$r = a - b \cos \theta$

$a > b > 0$

Limaçon

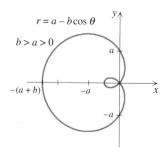

$r = a - b \cos \theta$

$b > a > 0$

Limaçon

CHAPTER 1

Graphs and Functions

The Brooklyn Bridge

Location Manhattan and Brooklyn, New York

Completion Date 1883

Cost $18 million

Length 3460 feet

Longest Single Span 1595 feet

Engineers John Roebling, Washington Roebling

CONSIDERED a brilliant feat of 19th-century engineering, the Brooklyn Bridge was the first suspension bridge to use steel for its cable wire. In 1883 it was the longest suspension bridge in the world. But the bridge was not built without problems. Before construction began, the bridge's chief engineer, John Roebling, died from tetanus. The project was taken over and seen to completion by his son, Washington Roebling. Three years later, Roebling was crippled by an illness that is known today as "the bends." The bedridden engineer watched the bridge's progress for the next 11 years from his apartment with a telescope and passed on orders to workers through his wife, Emily. Today, the Brooklyn Bridge is the second busiest bridge in New York City. One hundred forty-four thousand vehicles cross the bridge every day.

Real Numbers and Their Properties

The numbers that we use the most in calculus are the real numbers. So we begin our preparation for calculus by reviewing the set of real numbers and their properties.

The Real Numbers

A **set** is a collection of objects or **elements.** The set containing the numbers 1, 2, and 3 is written as {1, 2, 3}. To indicate a continuing pattern, we use three dots as in {1, 2, 3, . . .}. The set of real numbers is a collection of many types of numbers. To better understand the real numbers we recall some of the basic subsets of the real numbers:

Subset	Name (symbol)
{1, 2, 3, . . .}	**Counting** or **natural numbers** (N)
{0, 1, 2, 3, . . .}	**Whole numbers** (W)
{. . . , −3, −2, −1, 0, 1, 2, 3, . . .}	**Integers** (Z)

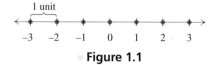

■ **Figure 1.1**

Numbers can be pictured as points on a line, the **number line.** To draw a number line, draw a line and label any convenient point with the number 0. Now choose a convenient length, one **unit,** and use it to locate evenly spaced points as shown in Fig. 1.1. The positive integers are located to the right of zero and the negative integers to the left of zero. The numbers corresponding to the points on the line are called the **coordinates** of the points.

The integers and their ratios form the set of **rational numbers,** Q. The rational numbers also correspond to points on the number line. For example, the rational number 1/2 is found halfway between 0 and 1 on the number line. In set notation, the set of rational numbers is written as

$$\left\{\frac{a}{b} \,\middle|\, a \text{ and } b \text{ are integers with } b \neq 0\right\}.$$

This notation is read "The set of all numbers of the form a/b such that a and b are integers with b not equal to zero." In our set notation we used letters to represent integers. A letter that is used to represent a number is called a **variable.**

There are infinitely many rational numbers located between each pair of consecutive integers, yet there are infinitely many points on the number line that do not correspond to rational numbers. The numbers that correspond to those points are called **irrational** numbers. In decimal notation, the rational numbers are the numbers that are repeating or terminating decimals, and the irrational numbers are the nonrepeating nonterminating decimals. For example, the number 0.595959 . . . is a rational number because the pair 59 repeats indefinitely. By contrast, notice that in the number 5.010010001 . . . , each group of zeros contains one more zero than the previous group. Because no group of digits repeats, 5.010010001 . . . is an irrational number.

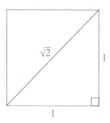

■ **Figure 1.2**

Numbers such as $\sqrt{2}$ or π are also irrational. We can visualize $\sqrt{2}$ as the length of the diagonal of a square whose sides are one unit in length. See Fig. 1.2. In any circle, the ratio of the circumference c to the diameter d is π ($\pi = c/d$).

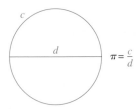

Figure 1.3

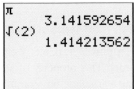

Figure 1.4

Johann Heinrich Lambert (1728–1777) was a German mathematician, physicist, and astronomer. His father was a poor tailor, so Johann had to struggle to gain an education. Lambert studied light intensity and was the first to introduce hyberbolic functions into trigonometry. He proved that π is an irrational number. Lambert also devised theorems regarding conic sections that made the calculation of the orbits of comets simpler.

See Fig. 1.3. It is difficult to see that numbers like $\sqrt{2}$ and π are irrational because their decimal representations are not apparent. However, the irrationality of π was proven in 1767 by Johann Heinrich Lambert, and it can be shown that the square root of any positive integer that is not a perfect square is irrational.

Since a calculator operates with a fixed number of decimal places, it gives us a *rational approximation* for an irrational number such as $\sqrt{2}$ or π. See Fig. 1.4. □

The set of rational numbers, Q, together with the set of irrational numbers, I, is called the set of **real numbers,** R. The following are examples of real numbers:

$$-3, \quad -0.025, \quad 0, \quad \frac{1}{3}, \quad 0.595959\ldots, \quad \sqrt{2}, \quad \pi, \quad 5.010010001\ldots$$

These numbers are **graphed** on a number line in Fig. 1.5.

Figure 1.5

Since there is a one-to-one correspondence between the points of the number line and the real numbers, we often refer to a real number as a point. Figure 1.6 shows how the various subsets of the real numbers are related to one another.

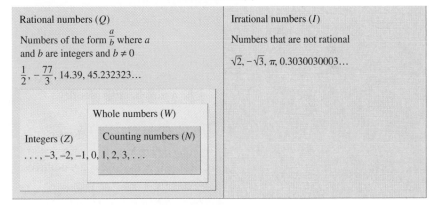

Figure 1.6

To indicate that a number is a member of a set, we write $a \in A$, which is read "a is a member of set A." We write $a \notin A$ for "a is not a member of set A." Set A is a subset of set B ($A \subseteq B$) means that every member of set A is also a member of set B, and A is not a subset of B ($A \nsubseteq B$) means that there is at least one member of A that is not a member of B.

Example **1** **Classifying numbers and sets of numbers**

Determine whether each statement is true or false and explain. See Fig. 1.6.

a. $0 \in R$ **b.** $\pi \in Q$ **c.** $R \subseteq Q$ **d.** $I \nsubseteq Q$ **e.** $\sqrt{5} \in Q$

Solution

a. True, because 0 is a member of the set of whole numbers, a subset of the set of real numbers.

b. False, because π is irrational.

c. False, because every irrational number is a member of R but not Q.

d. True, because the irrational numbers and the rational numbers have no numbers in common.

e. False, because the square root of any integer that is not a perfect square is irrational. ◼

Properties of the Real Numbers

In arithmetic we can observe that $3 + 4 = 4 + 3, 6 + 9 = 9 + 6$, etc. We get the same sum when we add two real numbers in either order. This property of addition of real numbers is the **commutative property.** Using variables, the commutative property of addition is stated as $a + b = b + a$ for any real numbers a and b. There is also a commutative property of multiplication, which is written as $a \cdot b = b \cdot a$ or $ab = ba$. There are many properties concerning the operations of addition and multiplication on the real numbers that are useful in algebra.

Properties of the Real Numbers

For any real numbers a, b, and c:

$a + b$ and ab are real numbers	**Closure property**
$a + b = b + a$ and $ab = ba$	**Commutative properties**
$a + (b + c) = (a + b) + c$ and $a(bc) = (ab)c$	**Associative properties**
$a(b + c) = ab + ac$	**Distributive property**
$0 + a = a$ and $1 \cdot a = a$ (Zero is the **additive identity,** and 1 is the **multiplicative identity.**)	**Identity properties**
$0 \cdot a = 0$	**Multiplication property of zero**
For each real number a, there is a unique real number $-a$ such that $a + (-a) = 0$. ($-a$ is the **additive inverse** of a.)	**Additive inverse property**
For each nonzero real number a, there is a unique real number $1/a$ such that $a \cdot 1/a = 1$. ($1/a$ is the **multiplicative inverse** or **reciprocal** of a.)	**Multiplicative inverse property**

The closure property indicates that the sum and product of any pair of real numbers is a real number. The commutative properties indicate that we can add or multiply in either order and get the same result. Since we can add or multiply only a pair of numbers, the associative properties indicate two different ways to obtain the result when adding or multiplying three numbers. The operations within parentheses are performed first. Because of the commutative property, the distributive property can be used also in the form $(b + c)a = ab + ac$.

Note that the properties stated here involve only addition and multiplication, considered the basic operations of the real numbers. Subtraction and division are defined in terms of addition and multiplication. By definition $a - b = a + (-b)$ and $a \div b = a \cdot 1/b$ for $b \neq 0$. Note that $a - b$ is called the **difference** of a and b and $a \div b$ is called the **quotient** of a and b.

Example 2 Using the properties

Complete each statement using the property named.

a. $a7 = $ _____, commutative
b. $2x + 4 = $ _____, distributive
c. $8($ _____ $) = 1$, multiplicative inverse
d. $\frac{1}{3}(3x) = $ _____, associative

Solution

a. $a7 = 7a$ **b.** $2x + 4 = 2(x + 2)$

c. $8\left(\frac{1}{8}\right) = 1$ **d.** $\frac{1}{3}(3x) = \left(\frac{1}{3} \cdot 3\right)x$ ■

Additive Inverses

The negative sign is used to indicate negative numbers as in -7 (negative seven). If the negative sign precedes a variable as in $-b$ it is read as "additive inverse" or "opposite" because $-b$ could be positive or negative. If b is positive then $-b$ is negative and if b is negative then $-b$ is positive.

Using two "opposite" signs has a cancellation effect. For example, $-(-5) = 5$ and $-(-(-3)) = -3$. Note that the additive inverse of a number can be obtained by multiplying the number by -1. For example, $-1 \cdot 3 = -3$.

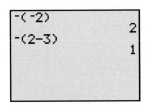

Calculators usually use the negative sign (-) to indicate opposite or negative and the subtraction sign ($-$) for subtraction as shown in Fig. 1.7. □

■ Figure 1.7

We know that $a + b = b + a$ for any real numbers a and b, but is $a - b = b - a$ for any real numbers a and b? In general, $a - b$ is not equal to $b - a$. For example, $7 - 3 = 4$ and $3 - 7 = -4$. So subtraction is not commutative. Since $a - b + b - a = 0$, we can conclude that $a - b$ and $b - a$ are opposites or additive inverses of each other. We summarize these properties of opposites as follows.

Properties of Opposites

For any real numbers a and b:

1. $-1 \cdot a = -a$ (The product of -1 and a is the opposite of a.)
2. $-(-a) = a$ (The opposite of the opposite of a is a.)
3. $-(a - b) = b - a$ (The opposite of $a - b$ is $b - a$.)

Example 3 Using properties of opposites

Use the properties of opposites to complete each equation.

a. $-(-\pi) = $ _____   **b.** $-1(-2) = $ _____   **c.** $-1(x - h) = $ _____

Solution

a. $-(-\pi) = \pi$
b. $-1(-2) = -(-2) = 2$
c. $-1(x - h) = -(x - h) = h - x$ ■

Relations

Symbols such as $<, >, =, \leq$, and $\geq$ are called **relations** because they indicate how numbers are related. We can visualize these relations by using a number line. For example, $\sqrt{2}$ is located to the right of 0 in Fig. 1.5, so $\sqrt{2} > 0$. Since $\sqrt{2}$ is to the left of π in Fig. 1.5, $\sqrt{2} < \pi$. In fact, if a and b are any two real numbers, we say that a is less than b (written $a < b$) provided that a is to the left of b on the number line. We say that a is greater than b (written $a > b$) if a is to the right of b on the number line. We say $a = b$ if a and b correspond to the same point on the number line. The fact that there are only three possibilities for ordering a pair of real numbers is called the **trichotomy property.**

Trichotomy Property

For any two real numbers a and b, exactly one of the following is true: $a < b, a = b$, or $a > b$.

The trichotomy property is very natural to use. For example, if we know that $r = t$ is false, then we can conclude (using the trichotomy property) that either $r > t$ or $r < t$ is true. If we know that $w + 6 > z$ is false, then we can conclude that $w + 6 \leq z$ is true. The following four properties of equality are also very natural to use, and we often use them without even thinking about them.

Properties of Equality

For any real numbers a, b, and c:

1. $a = a$ **Reflexive property**
2. If $a = b$, then $b = a$.   **Symmetric property**
3. If $a = b$ and $b = c$, then $a = c$.   **Transitive property**
4. If $a = b$, then a and b may be substituted   **Substitution property**
   for one another in any expression involving
   a or b.

Absolute Value

The **absolute value** of a (in symbols, $|a|$) can be thought of as the distance from a to 0 on a number line. Since both 3 and -3 are three units from 0 on a number line as shown in Fig. 1.8, $|3| = 3$ and $|-3| = 3$:

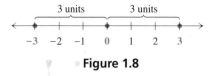

■ **Figure 1.8**

A symbolic definition of absolute value is written as follows.

Definition: Absolute Value

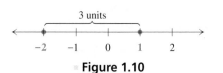

Figure 1.9

For any real number a,

$$|a| = \begin{cases} a & \text{if } a \geq 0 \\ -a & \text{if } a < 0. \end{cases}$$

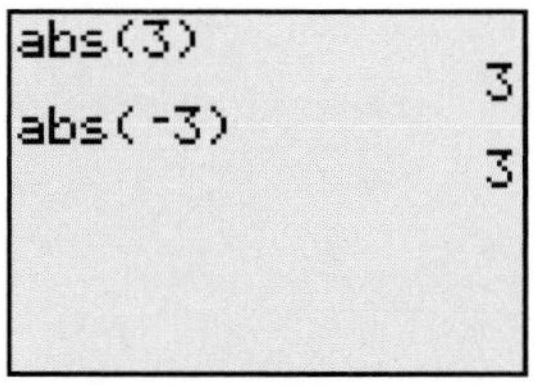

A calculator typically uses **abs** for absolute value as shown in Fig. 1.9. □
The symbolic definition of absolute value indicates that for $a \geq 0$ we use the equation $|a| = a$ (the absolute value of a is just a). For $a < 0$ we use the equation $|a| = -a$ (the absolute value of a is the opposite of a, a positive number).

Example 4 Using the definition of absolute value

Use the symbolic definition of absolute value to simplify each expression.

a. $|5.6|$ **b.** $|0|$ **c.** $|-3|$

Solution

a. Since $5.6 \geq 0$, we use the equation $|a| = a$ to get $|5.6| = 5.6$.
b. Since $0 \geq 0$, we use the equation $|a| = a$ to get $|0| = 0$.
c. Since $-3 < 0$, we use the equation $|a| = -a$ to get $|-3| = -(-3) = 3$. ■

The definition of absolute value guarantees that the absolute value of any number is nonnegative. The definition also implies that additive inverses (or opposites) have the same absolute value. These properties of absolute value and two others are stated as follows.

Properties of Absolute Value

For any real numbers a and b:

1. $|a| \geq 0$ (The absolute value of any number is nonnegative.)
2. $|-a| = |a|$ (Additive inverses have the same absolute value.)
3. $|a \cdot b| = |a| \cdot |b|$ (The absolute value of a product is the product of the absolute values.)
4. $\left|\dfrac{a}{b}\right| = \dfrac{|a|}{|b|}, b \neq 0$ (The absolute value of a quotient is the quotient of the absolute values.)

Absolute value is used in finding the distance between points on a number line. Since 9 lies four units to the right of 5, the distance between 5 and 9 is 4. In symbols, $d(5, 9) = 4$. We can obtain 4 by $9 - 5 = 4$ or $|5 - 9| = 4$. In general, $|a - b|$ gives the distance between a and b for any values of a and b. For example, the distance between -2 and 1 in Fig. 1.10 is three units and

$$d(-2, 1) = |-2 - 1| = |-3| = 3.$$

3 units

$$\begin{array}{ccccc} \bullet & & & \bullet & \\ -2 & -1 & 0 & 1 & 2 \end{array}$$

Figure 1.10

Distance Between Two Points on the Number Line

If a and b are any two points on the number line, then the distance between a and b is $|a - b|$. In symbols, $d(a, b) = |a - b|$.

Note that $d(a, 0) = |a - 0| = |a|$, which is consistent with the definition of absolute value of a as the distance between a and 0 on the number line.

Example **5** Distance between two points on a number line

Find the distance between -3 and 5 on the number line.

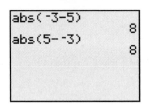

■ **Figure 1.11**

Figure 1.12

Solution

The points corresponding to -3 and 5 are shown on the number line in Fig. 1.11. The distances between these points is found as follows:

$$d(-3, 5) = |-3 - 5| = |-8| = 8$$

Notice that $d(-3, 5) = d(5, -3)$:

$$d(5, -3) = |5 - (-3)| = |8| = 8$$

When you use a calculator to find the absolute value of a difference or a sum, you must use parentheses as shown in Fig. 1.12. ■

Example **6** Absolute value equations for distance

Translate each sentence into an absolute value equation and solve it.

a. The distance between x and 3 is 6 units.
b. The distance between -2 and x is 5 units.

Solution

a. If x is 6 units away from 3, then $|x - 3| = 6$. Since both -3 and 9 are 6 units away from 3, the solution set is $\{-3, 9\}$.
b. If x is 5 units away from -2, then $|x - (-2)| = 5$ or $|x + 2| = 5$. Since both -7 and 3 are 5 units away from -2, the solution set is $\{-7, 3\}$. ■

Equations Involving Absolute Value

Not all absolute value equations are solved as easily as those in Example 6. To solve equations involving absolute value, remember that $|x| = x$ if $x \geq 0$, and $|x| = -x$ if $x < 0$. The absolute value of x is greater than or equal to 0 for any real number x. So an equation such as $|x| = -6$ has no solution. Since a number and its opposite have the same absolute value, $|x| = 4$ is equivalent to $x = 4$ or $x = -4$. The only number that has 0 absolute value is 0. These ideas are summarized as follows.

S U M M A R Y   **Basic Absolute Value Equations**

Absolute value equation	Equivalent statement	Solution set
$|x| = k \ (k > 0)$	$x = -k$ or $x = k$	$\{-k, k\}$
$|x| = 0$	$x = 0$	$\{0\}$
$|x| = k \ (k < 0)$		$\varnothing$

Example **7** Equations involving absolute value

Solve each equation.

a. $|3x| = 12$
b. $2|x + 8| + 6 = 0$

Solution

a. First write an equivalent statement without absolute value symbols.

$$|3x| = 12$$
$$3x = -12 \quad \text{or} \quad 3x = 12$$
$$x = -4 \quad \text{or} \quad x = 4$$

Check that $|3(4)| = 12$ and $|3(-4)| = 12$. The solution set is $\{-4, 4\}$.

b. First isolate $|x + 8|$.

$$2|x + 8| + 6 = 0$$
$$2|x + 8| = -6$$
$$|x + 8| = -3$$

Since the left side of the equation is nonnegative, the solution set is $\varnothing$. ∎

For Thought

True or False? Explain.

1. Zero is the only number that is both rational and irrational.

2. Every real number has a multiplicative inverse.

3. If a is not less than and not equal to 3, then a is greater than 3.

4. If $a \leq w$ and $w \leq z$, then $a < z$.

5. For any real numbers a, b, and c, $a - (b - c) = (a - b) - c$.

6. If a and b are any two real numbers, then the distance between a and b on the number line is $a - b$.

7. Calculators give only rational answers.

8. For any real numbers a and b, the opposite of $a + b$ is $a - b$.

9. The solution set to $|x - 9| = 2$ is $\{7, 11\}$.

10. The solution set to $|x - 5| = -3$ is $\varnothing$.

1.1 Exercises

Match each given statement with its symbolic form and determine whether the statement is true or false. If the statement is false, correct it.

1. The number $\sqrt{2}$ is a real number.

2. The number $\sqrt{3}$ is rational.

3. The number 0 is not an irrational number.

4. The number -6 is not an integer.

5. The set of integers is a subset of the real numbers.

6. The set of irrational numbers is a subset of the rationals.

7. The set of real numbers is not a subset of the rational numbers.

8. The set of natural numbers is not a subset of the whole numbers.

   a. $\sqrt{3} \in Q$ **b.** $-6 \notin Z$ **c.** $R \not\subseteq Q$

   d. $I \subseteq Q$ **e.** $\sqrt{2} \in R$ **f.** $N \not\subseteq W$

   g. $Z \subseteq R$ **h.** $0 \notin I$

Determine which elements of the set $\{-3.5, -\sqrt{2}, -1, 0, 1, \sqrt{3}, 3.14, \pi, 4.3535 \ldots, 5.090090009 \ldots\}$ *are members of the following sets.*

9. Real numbers   **10.** Rational numbers

11. Irrational numbers   **12.** Integers

13. Whole numbers   **14.** Natural numbers

Complete each statement using the property named.

15. $7 + x = $ _____, commutative

16. $5(4y) = $ _____, associative

17. $5(x + 3) = $ _____, distributive

18. $-3(x - 4) = $ _____, distributive

19. $5x + 5 = $ _____, distributive

20. $-5x + 10 = $ _____, distributive

21. $-13 + (4 + x) = $ _____, associative

22. $yx = $ _____, commutative

23. $0.125(\underline{\hspace{1cm}}) = 1$, multiplicative inverse

24. $-3 + (\underline{\hspace{1cm}}) = 0$, additive inverse

Use the properties of opposites to complete each equation.

25. $-\left(-\sqrt{3}\right) = $ _____   **26.** $-1(-6.4) = $ _____

27. $-1(x^2 - y^2) = $ _____   **28.** $-(1 - a^2) = $ _____

Use the symbolic definition of absolute value to simplify each expression.

29. $|7.2|$ **30.** $|0/3|$ **31.** $\left|-\sqrt{5}\right|$ **32.** $|-3/4|$

Find the distance on the number line between each pair of numbers.

33. $8, 13$ **34.** $1, 99$ **35.** $-5, 17$ **36.** $22, -9$

37. $-6, -18$ **38.** $-3, -14$ **39.** $-\dfrac{1}{2}, \dfrac{1}{4}$ **40.** $-\dfrac{1}{2}, -\dfrac{3}{4}$

Translate each sentence into an absolute value equation and solve it.

41. The distance between x and 7 is 3 units.

42. The distance between x and 2 is 7 units.

43. The distance between -1 and x is 3 units.

44. The distance between -4 and x is 8 units.

45. The distance between x and -9 is 4 units.

46. The distance between x and -6 is 5 units.

Solve each absolute value equation.

47. $|x| = 9$ **48.** $|x| = 13.6$

49. $|5x - 4| = 0$ **50.** $|4 - 3x| = 0$

51. $|2x - 3| = 7$ **52.** $|3x + 4| = 12$

53. $2|x + 5| - 10 = 0$ **54.** $6 - 4|x + 3| = -2$

55. $8|3x - 2| = 0$ **56.** $5|6 - 3x| = 0$

57. $2|x| + 7 = 6$ **58.** $5 + 3|x - 4| = 0$

Solve each problem.

59. Graph the numbers $\frac{1}{2}, -\frac{1}{2}, \frac{1}{3}, -\frac{1}{3}, 0, \frac{5}{12}$, and $-\frac{5}{12}$ on a number line. Explain how you decided where to put the numbers. Arrange these same numbers in order from smallest to largest. Explain your method. Did you use a calculator? If so, explain how it could be done without one.

60. Use a calculator to arrange the numbers $\frac{10}{3}, \sqrt{10}, \frac{22}{7}, \pi$, and $\frac{157}{50}$ in order from smallest to largest. Explain what you did to make your decisions on the order of these numbers. Could these numbers be arranged without using a calculator? How do these numbers differ from those in the previous exercise?

Thinking Outside the Box I

Paying Up   A king agreed to pay his gardener one dollar's worth of titanium per day for seven days of work on the castle grounds. The king has a seven-dollar bar of titanium that is segmented so that it can be broken into seven one-dollar pieces, but it is bad luck to break a seven-dollar bar of titanium more than twice. How can the king make two breaks in the bar and pay the gardener exactly one dollar's worth of titanium per day for seven days?

∎ **Figure for Thinking Outside the Box I**

1.1 Pop Quiz

1. Is 0 an irrational number?

2. Simplify $|-2|$.   3. Simplify $-(1 - y)$.

4. Find the distance between -3 and 9.

5. Write an absolute value equation that indicates that the distance between x and -3 is 6.

6. Solve $|2x - 3| = 1$.

7. Solve $|x + 2| = 0$.

8. Solve $|3x - 7| + 6 = 2$.

1.2 Linear and Absolute Value Inequalities

An equation states that two algebraic expressions are equal, while an **inequality** or **simple inequality** is a statement that two algebraic expressions are not equal in a particular way. Inequalities are stated using less than ($<$), less than or equal to ($\leq$), greater than ($>$), or greater than or equal to ($\geq$). In this section we study some basic inequalities.

Interval Notation

The solution set to an inequality is the set of all real numbers for which the inequality is true. The solution set to the inequality $x > 3$ is written $\{x \mid x > 3\}$ and consists of all real numbers to the right of 3 on the number line. This set is also called the **interval** of numbers greater than 3, and it is written in **interval notation** as $(3, \infty)$. The graph of the interval $(3, \infty)$ is shown in Fig. 1.13. A parenthesis is used next to the 3 to indicate that 3 is not in the interval or the solution set. The infinity symbol (∞) is not used as a number, but only to indicate that there is no bound on the numbers greater than 3.

The solution set to $x \leq 4$ is written in set notation as $\{x \mid x \leq 4\}$ and in interval notation as $(-\infty, 4]$. The symbol $-\infty$ means that all numbers to the left of 4 on the number line are in the set and the bracket means that 4 is in the set. The graph of the interval $(-\infty, 4]$ is shown in Fig. 1.14. Intervals that use the infinity symbol are **unbounded** intervals. The following summary lists the different types of unbounded intervals used in interval notation and the graphs of those intervals on a number line. An unbounded interval with an endpoint is **open** if the endpoint is not included in the interval and **closed** if the endpoint is included. Graphs of inequalities are also drawn using an open circle instead of a parenthesis when an endpoint is not included and a solid circle instead of a bracket when an endpoint is included. The advantage of the parenthesis and bracket notation is that it corresponds to the interval notation.

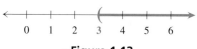
Figure 1.13

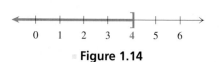
Figure 1.14

SUMMARY	Interval Notation for Unbounded Intervals

Set	Interval notation	Type	Graph
$\{x \mid x > a\}$	(a, ∞)	Open	
$\{x \mid x < a\}$	$(-\infty, a)$	Open	
$\{x \mid x \geq a\}$	$[a, \infty)$	Closed	
$\{x \mid x \leq a\}$	$(-\infty, a]$	Closed	
Real numbers	$(-\infty, \infty)$	Open	

We use a parenthesis when an endpoint of an interval is not included in the solution set and a bracket when an endpoint is included. A bracket is never used next to ∞ because infinity is not a number. On the graphs above, the number lines are shaded, showing that the solutions include all real numbers in the given interval.

Example **1** **Interval notation**

Write an inequality whose solution set is the given interval.

a. $(-\infty, -9)$ **b.** $[0, \infty)$

Solution

a. The interval $(-\infty, -9)$ represents all real numbers less than -9. It is the solution set to $x < -9$.
b. The interval $[0, \infty)$ represents all real numbers greater than or equal to 0. It is the solution set to $x \geq 0$. ■

Linear Inequalities

Replacing the equal sign in the general linear equation $ax + b = 0$ by any of the symbols $<$, $\leq$, $>$, or $\geq$ gives a **linear inequality.** Two inequalities are **equivalent** if they have the same solution set. We solve linear inequalities like we solve linear equations by performing operations on each side to get equivalent inequalities. However, the rules for inequalities are slightly different from the rules for equations.

Adding any real number to both sides of an inequality results in an equivalent inequality. For example, adding 3 to both sides of $-4 < 5$ yields $-1 < 8$, which is true. Adding or subtracting the same number simply moves the original numbers to the right or left along the number line and does not change their order.

The order of two numbers will also be unchanged when they are multiplied or divided by the same positive real number. For example, $10 < 20$, and after dividing both numbers by 10 we have $1 < 2$. Multiplying or dividing two numbers by a negative number will change the order. For example, $4 > 2$, but after multiplying both numbers by -1 we have $-4 < -2$. Likewise, $-10 < 20$, but after dividing both numbers by -10 we have $1 > -2$. *When an inequality is multiplied or divided by a negative number, the direction of the inequality symbol is reversed.* These ideas are stated symbolically in the following box for $<$, but they also hold for $>$, $\leq$, and $\geq$.

Properties of Inequality

If A and B are algebraic expressions and C is a nonzero real number, then the inequality $A < B$ is equivalent to

1. $A \pm C < B \pm C$,
2. $CA < CB$ (for C positive), $CA > CB$ (for C negative),
3. $\dfrac{A}{C} < \dfrac{B}{C}$ (for C positive), $\dfrac{A}{C} > \dfrac{B}{C}$ (for C negative).

Example **2** Solving a linear inequality

Solve $-3x - 9 < 0$. Write the solution set in interval notation and graph it.

Solution

Isolate the variable as is done in solving equations.

$$-3x - 9 < 0$$
$$-3x - 9 + 9 < 0 + 9 \qquad \text{Add 9 to each side.}$$
$$-3x < 9$$
$$x > -3 \qquad \text{Divide each side by } -3, \text{ reversing the inequality.}$$

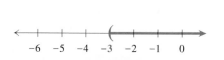

Figure 1.15

The solution set is the interval $(-3, \infty)$ and its graph is shown in Fig. 1.15. Checking the solution to an inequality is generally not as simple as checking an equation, because usually there are infinitely many solutions. We can do a "partial check" by checking one number in $(-3, \infty)$ and one number not in $(-3, \infty)$. For example, $0 > -3$ and $-3(0) - 9 < 0$ is correct, while $-6 < -3$ and $-3(-6) - 9 < 0$ is incorrect. ■

We can also perform operations on each side of an inequality using a variable expression. Addition or subtraction with variable expressions will give equivalent inequalities. However, we must always watch for undefined expressions. *Multiplication and division with a variable expression are usually avoided because we do not know whether the expression is positive or negative.*

Example **3** Solving a linear inequality

Solve $\frac{1}{2}x - 3 \geq \frac{1}{4}x + 2$ and graph the solution set.

Solution

Multiply each side by the LCD to eliminate the fractions.

$$\frac{1}{2}x - 3 \geq \frac{1}{4}x + 2$$

$$4\left(\frac{1}{2}x - 3\right) \geq 4\left(\frac{1}{4}x + 2\right) \quad \text{Multiply each side by 4.}$$

$$2x - 12 \geq x + 8$$

$$x - 12 \geq 8$$

$$x \geq 20$$

Figure 1.16

The solution set is the interval $[20, \infty)$. See Fig. 1.16 for its graph. ■

Compound Inequalities

A **compound inequality** is a sentence containing two simple inequalities connected with "and" or "or." The solution to a compound inequality can be an interval of real numbers that does not involve infinity, a **bounded** interval of real numbers. For example, the solution set to the compound inequality $x \geq 2$ and $x \leq 5$ is the set of real numbers between 2 and 5, inclusive. This inequality is also written as $2 \leq x \leq 5$. Its solution set is $\{x \,|\, 2 \leq x \leq 5\}$, which is written in interval notation as $[2, 5]$. Because $[2, 5]$ contains both of its endpoints, the interval is **closed.** The following summary lists the different types of bounded intervals used in interval notation and the graphs of those intervals on a number line.

S U M M A R Y **Interval Notation for Bounded Intervals**

Set	Interval notation	Type	Graph
$\{x \,\|\, a < x < b\}$	(a, b)	Open	
$\{x \,\|\, a \leq x \leq b\}$	$[a, b]$	Closed	
$\{x \,\|\, a \leq x < b\}$	$[a, b)$	Half open or half closed	
$\{x \,\|\, a < x \leq b\}$	$(a, b]$	Half open or half closed	

The notation $a < x < b$ is used only when x is between a and b, and a is less than b. We do *not* write inequalities such as $5 < x < 3$, $4 > x < 9$, or $2 < x > 8$.

The **intersection** of sets A and B is the set $A \cap B$ (read "A intersect B"), where $x \in A \cap B$ if and only if $x \in A$ and $x \in B$. (The symbol $\in$ means "belongs to.") The **union** of sets A and B is the set $A \cup B$ (read "A union B"), where $x \in A \cup B$ if and only if $x \in A$ or $x \in B$. In solving compounded inequalities it is often necessary to find intersections and unions of intervals.

Example **4** Intersections and unions of intervals

Let $A = (1, 5)$, $B = [3, 7)$, and $C = (6, \infty)$. Write each of the following sets in interval notation.

a. $A \cup B$ **b.** $A \cap B$ **c.** $A \cup C$ **d.** $A \cap C$

Solution

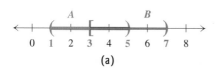

■ **Figure 1.17**

a. Graph both intervals on the number line as shown in Fig. 1.17(a). The union of two intervals is the set of points that are in one, the other, or both intervals. For a union, nothing is omitted. The union consists of all points shaded in the figure. So $A \cup B = (1, 7)$.

b. The intersection of A and B is the set of points that belong to both intervals. The intersection consist of the points that are shaded twice in Fig. 1.17(a). So $A \cap B = [3, 5)$.

c. Graph both intervals on the number line as shown in Fig. 1.17(b). For a union, nothing is omitted. So $A \cup C = (1, 5) \cup (6, \infty)$. Note that $A \cup C$ cannot be written as a single interval.

d. Since there are no points shaded twice in Fig. 1.17(b), $A \cap C = \varnothing$. ■

The solution set to a compound inequality using the connector "or" is the union of the two solution sets, and the solution set to a compound inequality using "and" is the intersection of the two solution sets.

Example **5** Solving compound inequalities

Solve each compound inequality. Write the solution set using interval notation and graph it.

a. $2x - 3 > 5$ and $4 - x \leq 3$ **b.** $4 - 3x < -2$ or $3(x - 2) \leq -6$
c. $-4 \leq 3x - 1 < 5$

Solution

a. $\quad 2x - 3 > 5 \quad$ and $\quad 4 - x \leq 3$

$\qquad\quad 2x > 8 \quad$ and $\qquad -x \leq -1$

$\qquad\qquad x > 4 \quad$ and $\qquad\quad x \geq 1$

Graph $(4, \infty)$ and $[1, \infty)$ on the number line as shown in Fig. 1.18(a). The intersection of the intervals is the set of points that are shaded twice in Fig. 1.18(a). So the intersection is the interval $(4, \infty)$ and $(4, \infty)$ is the solution set to the compound inequality. Its graph is shown in Fig. 1.18(b).

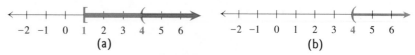

■ **Figure 1.18**

b. $\quad 4 - 3x < -2 \quad$ or $\quad 3(x - 2) \leq -6$

$\qquad\quad -3x < -6 \quad$ or $\qquad x - 2 \leq -2$

$\qquad\qquad x > 2 \quad$ or $\qquad\qquad x \leq 0$

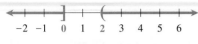

Figure 1.19

Figure 1.20

The union of the intervals $(2, \infty)$ and $(-\infty, 0]$ consists of all points that are shaded in Fig. 1.19. This set cannot be written as a single interval. So the solution set is $(-\infty, 0] \cup (2, \infty)$ and its graph is shown in Fig. 1.19.

c. We could write $-4 \le 3x - 1 < 5$ as the compound inequality $-4 \le 3x - 1$ and $3x - 1 < 5$, and then solve each simple inequality. Since each is solved using the same sequence of steps, we can solve the original inequality without separating it:

$$-4 \le 3x - 1 < 5$$

$$-4 + 1 \le 3x - 1 + 1 < 5 + 1 \qquad \text{Add 1 to each part of the inequality.}$$

$$-3 \le 3x < 6$$

$$\frac{-3}{3} \le \frac{3x}{3} < \frac{6}{3} \qquad\qquad \text{Divide each part by 3.}$$

$$-1 \le x < 2$$

The solution set is the half-open interval $[-1, 2)$, graphed in Fig. 1.20. ∎

It is possible that all real numbers satisfy a compound inequality or no real numbers satisfy a compound inequality.

$\mathcal{Example}$ **6** **All or nothing**

Solve each compound inequality.

a. $3x - 9 \le 9$ or $4 - x \le 3$ **b.** $-\dfrac{2}{3}x < 4$ and $\dfrac{3}{4}x < -6$

Solution

a. Solve each simple inequality and find the union of their solution sets:

$$3x - 9 \le 9 \qquad \text{or} \qquad 4 - x \le 3$$

$$3x \le 18 \qquad \text{or} \qquad -x \le -1$$

$$x \le 6 \qquad \text{or} \qquad x \ge 1$$

The union of $(-\infty, 6]$ and $[1, \infty)$ is the set of all real numbers, $(-\infty, \infty)$.

b. Solve each simple inequality and find the intersection of their solution sets:

$$-\frac{2}{3}x < 4 \qquad\qquad \text{and} \qquad\qquad \frac{3}{4}x < -6$$

$$\left(-\frac{3}{2}\right)\left(-\frac{2}{3}x\right) > \left(-\frac{3}{2}\right)4 \quad \text{and} \quad \left(\frac{4}{3}\right)\left(\frac{3}{4}x\right) < \left(\frac{4}{3}\right)(-6)$$

$$x > -6 \qquad\qquad \text{and} \qquad\qquad x < -8$$

Since $(-6, \infty) \cap (-\infty, -8) = \varnothing$, there is no solution to the compound inequality. ∎

Absolute Value Inequalities

Recall that the absolute value of a number is the number's distance from 0 on the number line. The inequality $|x| < 3$ means that x is less than three units from 0. See Fig. 1.21. The real numbers that are less than three units from 0 are precisely the numbers that satisfy $-3 < x < 3$. So the solution set to $|x| < 3$ is the open interval $(-3, 3)$.

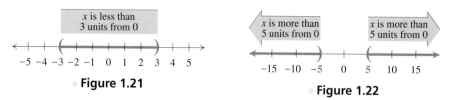

■ **Figure 1.21**

■ **Figure 1.22**

The inequality $|x| > 5$ means that x is more than five units from 0 on the number line, which is equivalent to the compound inequality $x > 5$ or $x < -5$. See Fig. 1.22. So the solution to $|x| > 5$ is the union of two intervals, $(-\infty, -5) \cup (5, \infty)$. These ideas about absolute value inequalities are summarized as follows.

S U M M A R Y   **Basic Absolute Value Inequalities (for $k > 0$)**

Absolute value inequality	Equivalent statement	Solution set in interval notation	Graph of solution set
$|x| > k$	$x < -k$ or $x > k$	$(-\infty, -k) \cup (k, \infty)$	
$|x| \geq k$	$x \leq -k$ or $x \geq k$	$(-\infty, -k] \cup [k, \infty)$	
$|x| < k$	$-k < x < k$	$(-k, k)$	
$|x| \leq k$	$-k \leq x \leq k$	$[-k, k]$	

In the next example we use the rules for basic absolute value inequalities to solve more complicated absolute value inequalities.

Example **7** Absolute value inequalities

Solve each absolute value inequality and graph the solution set.

a. $|3x + 2| < 7$ **b.** $-2|4 - x| \leq -4$ **c.** $|7x - 9| \geq -3$

Solution

a.
$$|3x + 2| < 7$$
$$-7 < 3x + 2 < 7 \qquad \text{Write the equivalent compound inequality.}$$
$$-7 - 2 < 3x + 2 - 2 < 7 - 2 \qquad \text{Subtract 2 from each part.}$$
$$-9 < 3x < 5$$
$$-3 < x < \frac{5}{3} \qquad \text{Divide each part by 3.}$$

■ **Figure 1.23**

The solution set is the open interval $\left(-3, \frac{5}{3}\right)$. The graph is shown in Fig. 1.23.

b. $-2|4 - x| \le -4$ Divide each side by -2, reversing the inequality.

$|4 - x| \ge 2$

$4 - x \le -2$ or $4 - x \ge 2$ Write the equivalent compound inequality.

$-x \le -6$ or $-x \ge -2$

$x \ge 6$ or $x \le 2$ Multiply each side by

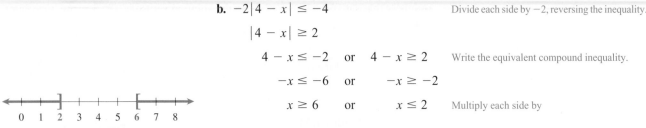

Figure 1.24

Figure 1.25

The solution set is $(-\infty, 2] \cup [6, \infty)$. Its graph is shown in Fig. 1.24. Check 8, 4, and 0 in the original inequality. If $x = 8$, we get $-2|4 - 8| \le -4$, which is true. If $x = 4$, we get $-2|4 - 4| \le -4$, which is false. If $x = 0$, we get $-2|4 - 0| \le -4$, which is true.

c. The expression $|7x - 9|$ has a nonnegative value for every real number x. So the inequality $|7x - 9| \ge -3$ is satisfied by every real number. The solution set is $(-\infty, \infty)$, and its graph is shown in Fig. 1.25. ■

Modeling with Inequalities

Inequalities occur in applications just as equations do. In fact, in real life, equality in anything is usually the exception. The solution to a problem involving an inequality is generally an interval of real numbers. In this case we often ask for the range of values that solve the problem.

Example **8** **An application involving inequality**

Remington scored 74 on his midterm exam in history. If he is to get a B, the average of his midterm and final exam must be between 80 and 89 inclusive. In what range must his final exam score lie for him to get a B in the course?

Solution

Let x represent Remington's final exam score. The average of his midterm and final must satisfy the following inequality:

$$80 \le \frac{74 + x}{2} \le 89$$

$$160 \le 74 + x \le 178$$

$$86 \le x \le 104$$

His final exam score must lie in the interval [86, 104] if he is to get a B. ■

When discussing the error made in a measurement, we may refer to the *absolute error* or the *relative error*. For example, if L is the actual length of an object and x is the length determined by a measurement, then the absolute error is $|x - L|$ and the relative error is $|x - L|/L$.

■ Foreshadowing Calculus

In Example 9 we find values for x that determine the size of the relative error. This same idea is used in calculus in the study of limits.

Example **9** **Application of absolute value inequality**

A technician is testing a scale with a 50 lb block of steel. The scale passes this test if the relative error when weighing this block is less than 0.1%. If x is the reading on the scale, then for what values of x does the scale pass this test?

Solution

If the relative error must be less than 0.1%, then x must satisfy the following inequality:

$$\frac{|x - 50|}{50} < 0.001$$

Solve the inequality for x:

$$|x - 50| < 0.05$$

$$-0.05 < x - 50 < 0.05$$

$$49.95 < x < 50.05$$

So the scale passes the test if it shows a weight in the interval $(49.95, 50.05)$. ▪

Absolute value inequalities like the one in Example 9 are used to measure "closeness" in the study of limits in calculus. For example, $|x - 2| < 0.01$ indicates that x differs from 2 by less than 0.01. So x is really close to 2. Note that $x - 2 < 0.01$ does not mean that x is close to 2 because this inequality is satisfied even if x is -1000.

For Thought

True or False? Explain.

1. The inequality $-3 < x + 6$ is equivalent to $x + 6 > -3$.

2. The inequality $-2x < -6$ is equivalent to $\frac{-2x}{-2} < \frac{-6}{-2}$.

3. The smallest real number that satisfies $x > 12$ is 13.

4. The number -6 satisfies $|x - 6| > -1$.

5. $(-\infty, -3) \cap (-\infty, -2) = (-\infty, -2)$

6. $(5, \infty) \cap (-\infty, -3) = (-3, 5)$

7. All negative numbers satisfy $|x - 2| < 0$.

8. The compound inequality $x < -3$ or $x > 3$ is equivalent to $|x| < -3$.

9. The inequality $|x| + 2 < 5$ is equivalent to $-5 < x + 2 < 5$.

10. The fact that the difference between your age, y, and my age, m, is at most 5 years is written $|y - m| \le 5$.

1.2 Exercises

For each interval write an inequality whose solution set is the interval, and for each inequality write the solution set in interval notation. See the summary of interval notation for unbounded intervals on page 12.

1. $(-\infty, 12)$

2. $(-\infty, -3]$

3. $[-7, \infty)$

4. $(1.2, \infty)$

5. $x \ge -8$

6. $x < 54$

7. $x < \pi/2$

8. $x \ge \sqrt{3}$

Solve each inequality. Write the solution set using interval notation and graph it.

9. $3x - 6 > 9$

10. $2x + 1 < 6$

11. $7 - 5x \le -3$

12. $-1 - 4x \ge 7$

13. $\frac{1}{2}x - 4 < \frac{1}{3}x + 5$

14. $\frac{1}{2} - x > \frac{x}{3} + \frac{1}{4}$

15. $\dfrac{7 - 3x}{2} \geq -3$

16. $\dfrac{5 - x}{3} \leq -2$

17. $\dfrac{2x - 3}{-5} \geq 0$

18. $\dfrac{5 - 3x}{-7} \leq 0$

19. $-2(3x - 2) \geq 4 - x$

20. $-5x \leq 3(x - 9)$

Write as a single interval. See the summary of interval notation for bounded intervals on page 14.

21. $(-3, \infty) \cup (5, \infty)$

22. $(-\infty, 0) \cup (-\infty, 6)$

23. $(3, 5) \cup (-3, \infty)$

24. $(4, 7) \cap (3, \infty)$

25. $(-\infty, -2) \cap (-5, \infty)$

26. $(-3, \infty) \cap (2, \infty)$

27. $(-\infty, -5) \cap (-2, \infty)$

28. $(-\infty, -3) \cup (-7, \infty)$

29. $(-\infty, 4) \cup [4, 5]$

30. $[3, 5] \cup [5, 7]$

Solve each compound inequality. Write the solution set using interval notation and graph it.

31. $5 > 8 - x$ and $1 + 0.5x < 4$

32. $5 - x < 4$ and $0.2x - 5 < 1$

33. $\dfrac{2x - 5}{-2} < 2$ and $\dfrac{2x + 1}{3} > 0$

34. $\dfrac{4 - x}{2} > 1$ and $\dfrac{2x - 7}{-3} < 1$

35. $1 - x < 7 + x$ or $4x + 3 > x$

36. $5 + x > 3 - x$ or $2x - 3 > x$

37. $\dfrac{1}{2}(x + 1) > 3$ or $0 < 7 - x$

38. $\dfrac{1}{2}(x + 6) > 3$ or $4(x - 1) < 3x - 4$

39. $1 - \dfrac{3}{2}x < 4$ and $\dfrac{1}{4}x - 2 \leq -3$

40. $\dfrac{3}{5}x - 1 > 2$ and $5 - \dfrac{2}{5}x \geq 3$

41. $1 < 3x - 5 < 7$

42. $-3 \leq 4x + 9 \leq 17$

43. $-2 \leq 4 - 6x < 22$

44. $-13 < 5 - 9x \leq 41$

Solve each absolute value inequality. Write the solution set using interval notation and graph it. See the summary of basic absolute value inequalities on page 17.

45. $|3x - 1| < 2$

46. $|4x - 3| \leq 5$

47. $|5 - 4x| \leq 1$

48. $|6 - x| < 6$

49. $5 \geq |4 - x|$

50. $3 < |2x - 1|$

51. $|5 - 4x| < 0$

52. $|3x - 7| \geq -5$

53. $|4 - 5x| < -1$

54. $|2 - 9x| \geq 0$

55. $3|x - 2| + 6 > 9$

56. $3|x - 1| + 2 < 8$

57. $\left|\dfrac{x - 3}{2}\right| > 1$

58. $\left|\dfrac{9 - 4x}{2}\right| < 3$

Write an inequality of the form $|x - a| < k$ or of the form $|x - a| > k$ so that the inequality has the given solution set. (Hint: $|x - a| < k$ means that x is less than k units from a and $|x - a| > k$ means that x is more than k units from a on the number line.)

59. $(-5, 5)$

60. $(-2, 2)$

61. $(-\infty, -3) \cup (3, \infty)$

62. $(-\infty, -1) \cup (1, \infty)$

63. $(4, 8)$

64. $(-3, 9)$

65. $(-\infty, 3) \cup (5, \infty)$

66. $(-\infty, -1) \cup (5, \infty)$

For each graph write an absolute value inequality that has the given solution set.

67.

68.

69.

70.

71.

72.

Solve each problem.

73. *Final Exam Score* Lucky scored 65 on his Psychology 101 midterm. If the average of his midterm and final must be between 79 and 90 for a B, then for what range of scores on the final exam would Lucky get a B?

74. *Bringing Up Your Average*   Felix scored 52 and 64 on his first two tests in Sociology 212. What must he get on the third test to get an average for the three tests above 70?

75. *Weighted Average with Whole Numbers*   Ingrid scored 65 on her calculus midterm. If her final exam counts twice as much as her midterm exam, then for what range of scores on her final would she get an average between 79 and 90?

76. *Weighted Average with Fractions*   Elizabeth scored 62, 76, and 80 on three equally weighted tests in French. If the final exam score counts for two-thirds of the grade and the tests count for one-third, then what range of scores on the final exam would give her a final average over 70?

77. *Price Range for a Car*   Yolanda is shopping for a used car in a city where the sales tax is 10% and the title and license fee is $300. If the maximum that she can spend is $8000, then she should look at cars in what price range?

78. *Price of a Burger*   The price of Elaine's favorite Big Salad at the corner restaurant is 10 cents more than the price of Jerry's hamburger. After treating a group of friends to lunch, Jerry is certain that for 10 hamburgers and 5 salads he spent more than $9.14, but not more than $13.19, including tax at 8% and a 50 cent tip. In what price range is a hamburger?

79. *Bicycle Gear Ratio*   The gear ratio r for a bicycle is defined by the formula

$$r = \frac{Nw}{n},$$

where N is the number of teeth on the chainring (by the pedal), n is the number of teeth on the cog (by the wheel), and w is the wheel diameter in inches. The following chart gives uses for the various gear ratios.

Ratio	Use	
$r > 90$	down hill	
$70 < r \le 90$	level	
$50 < r \le 70$	mild hill	
$35 < r \le 50$	long hill	

A bicycle with a 27-in. diameter wheel has 50 teeth on the chainring and 5 cogs with 14, 17, 20, 24, and 29 teeth. Find the gear ratio with each of the five cogs. Does this bicycle have a gear ratio for each of the four types of pedaling described in the table?

80. *Selecting the Cogs*   Use the formula from the previous exercise to answer the following.
   a. If a single-speed 27-in. bicycle has 17 teeth on the cog, then for what numbers of teeth on the chainring will the gear ratio be between 60 and 80?

   b. If a 26-in. bicycle has 40 teeth on the chainring, then for what numbers of teeth on the cog will the gear ratio be between 60 and 75?

81. *Expensive Models*   The prices of the 2006 BMW 745 Li and 760 Li differ by more than $25,000. The list price of the 745 Li is $74,595. Assuming that you do not know which model is more expensive, write an absolute value inequality that describes the price of the 760 Li. What are the possibilities for the price of the 760 Li?

82. *Difference in Prices*   There is less than $800 difference between the price of a $14,200 Ford and a comparable Chevrolet. Express this as an absolute value inequality. What are the possibilities for the price of the Chevrolet?

83. *Controlling Temperature*   Michelle is trying to keep the water temperature in her chemistry experiment at 35°C. For the experiment to work, the relative error for the actual temperature must be less than 1%. Write an absolute value inequality for the actual temperature. Find the interval in which the actual temperature must lie.

84. *Laying Out a Track*   Melvin is attempting to lay out a track for a 100 m race. According to the rules of competition, the relative error of the actual length must be less than 0.5%. Write an absolute value inequality for the actual length. Find the interval in which the actual length must fall.

85. *Acceptable Bearings*   A spherical bearing is to have a circumference of 7.2 cm with an error of no more than 0.1 cm. Use an absolute value inequality to find the acceptable range of values for the diameter of the bearing.

86. *Acceptable Targets*   A manufacturer makes circular targets that have an area of 15 ft². According to competition rules, the area can be in error by no more than 0.5 ft². Use an absolute value inequality to find the acceptable range of values for the radius.

Area: 15 ± 0.5 ft²

■ **Figure for Exercise 86**

Thinking Outside the Box II

One in a Million   If you write the integers from 1 through 1,000,000 inclusive, then how many ones will you write?

1.2 Pop Quiz

Solve each inequality. Write the solution set using interval notation.

1. $x - \sqrt{2} \geq 0$

2. $6 - 2x < 0$

3. $x > 4$ or $x \geq -1$

4. $x - 1 > 5$ and $2x < 18$

5. $|x| > 6$

6. $|x - 1| \leq 2$

1.3 Equations and Graphs in Two Variables

In algebra and calculus we study relationships between two variables. For example, p might represent the price of gasoline and n the number of gallons consumed in a month; r might be the radius of a circle and A its area; x might be the number of toppings on a medium pizza and y the cost of the pizza. We can visualize the relationship between two variables in a two-dimensional coordinate system.

The Cartesian Coordinate System

If x and y are real numbers, then (x, y) is called an **ordered pair** of real numbers. The numbers x and y are the **coordinates** of the ordered pair, with x being the **first coordinate** or **abscissa,** and y being the **second coordinate** or **ordinate.** For example, Table 1.1 shows the number of toppings on a medium pizza and the corresponding cost. The ordered pair $(3, 11)$ indicates that a three-topping pizza costs $11. The order of the numbers matters. In this context, $(11, 3)$ would indicate that an 11-topping pizza costs $3.

To picture ordered pairs of real numbers we use the **rectangular coordinate system** or **Cartesian coordinate system,** named after the French mathematician René Descartes (1596–1650). The Cartesian coordinate system consists of two number lines drawn perpendicular to one another, intersecting at zero on each number line as shown in Fig. 1.26. The point of intersection of the number lines is called the

■ **Table 1.1**

Toppings x	Cost y	
0	$ 5	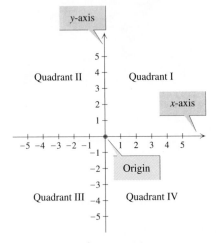
1	7	
2	9	
3	11	
4	13	

Historical Note

René Descartes (1596–1650) was a noted French philosopher, mathematician, and scientist. He has been called the "founder of modern philosophy" and the "father of modern mathematics." He ranks as one of the most important and influential thinkers of modern times.

■ **Figure 1.26**

origin. The horizontal number line is the ***x*-axis** and its positive numbers are to the right of the origin. The vertical number line is the ***y*-axis** and its positive numbers are above the origin. The two number lines divide the plane into four regions called **quadrants,** numbered as shown in Fig. 1.26. The quadrants do not include any points on the axes. We call a plane with a rectangular coordinate system the **coordinate plane** or the ***xy*-plane.**

Just as every real number corresponds to a point on the number line, every ordered pair of real numbers (a, b) corresponds to a point P in the *xy*-plane. For this reason, ordered pairs of numbers are often called **points.** So a and b are the coordinates of (a, b) or the coordinates of the point P. Locating the point P that corresponds to (a, b) in the *xy*-plane is referred to as **plotting** or **graphing** the point, and P is called the *graph* of (a, b). In general, a **graph** is a set of points in the rectangular coordinate system.

■ **Figure 1.27**

Example 1 Plotting points

Plot the points $(3, 5)$, $(4, -5)$, $(-3, 4)$, $(-2, -5)$, and $(0, 2)$ in the *xy*-plane.

Solution

The point $(3, 5)$ is located three units to the right of the origin and five units above the *x*-axis as shown in Fig. 1.27. The point $(4, -5)$ is located four units to the right of the origin and five units below the *x*-axis. The point $(-3, 4)$ is located three units to the left of the origin and four units above the *x*-axis. The point $(-2, -5)$ is located two units to the left of the origin and five units below the *x*-axis. The point $(0, 2)$ is on the *y*-axis because its first coordinate is zero. ■

Note that for points in quadrant I, both coordinates are positive. In quadrant II the first coordinate is negative and the second is positive, while in quadrant III, both coordinates are negative. In quadrant IV the first coordinate is positive and the second is negative.

The Distance Formula

If a and b are real numbers, then the distance between them on the number line is $|a - b|$. Now consider the points $A(x_1, y_1)$ and $B(x_2, y_2)$ shown in Fig. 1.28. Let AB represent the length of line segment $\overline{AB}$. Now $\overline{AB}$ is the hypotenuse of the right triangle in Fig. 1.28. Since A and C lie on a horizontal line, the distance between them

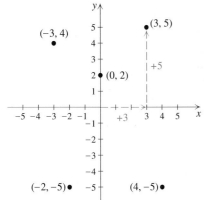

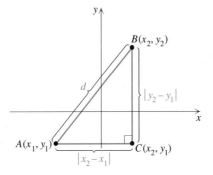

■ **Figure 1.28**

is $|x_2 - x_1|$. Likewise $CB = |y_2 - y_1|$. Since the sum of the squares of the legs of a right triangle is equal to the square of the hypotenuse (the Pythagorean theorem) we have

$$d^2 = |x_2 - x_1|^2 + |y_2 - y_1|^2.$$

Since the distance between two points is a nonnegative real number, we have $d = \sqrt{(x_2 - x_1)^2 + (y_2 - y_1)^2}$. The absolute value symbols are replaced with parentheses, because $|a - b|^2 = (a - b)^2$ for any real numbers a and b.

Theorem:
The Distance Formula

> The distance d between the points (x_1, y_1) and (x_2, y_2) is given by the formula
> $$d = \sqrt{(x_2 - x_1)^2 + (y_2 - y_1)^2}.$$

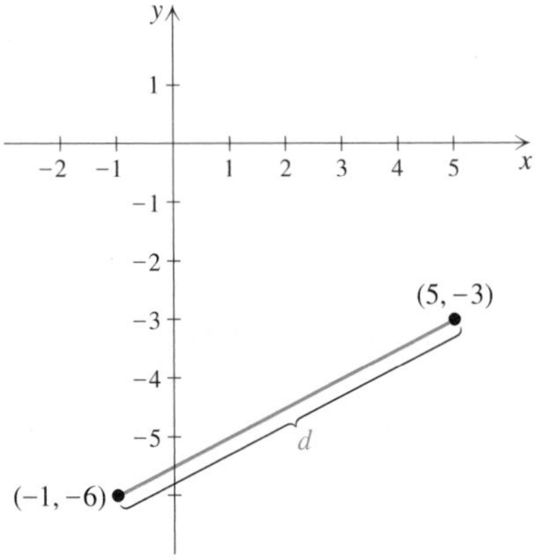

Example **2** **Finding the distance between two points**

Find the distance between $(5, -3)$ and $(-1, -6)$.

Solution

Let $(x_1, y_1) = (5, -3)$ and $(x_2, y_2) = (-1, -6)$. These points are shown on the graph in Fig. 1.29. Substitute these values into the distance formula:

$$
\begin{aligned}
d &= \sqrt{(-1 - 5)^2 + (-6 - (-3))^2} \\
&= \sqrt{(-6)^2 + (-3)^2} \\
&= \sqrt{36 + 9} = \sqrt{45} = 3\sqrt{5}
\end{aligned}
$$

The exact distance between the points is $3\sqrt{5}$.   ■

■ **Figure 1.29**

Note that the distance between two points is the same regardless of which point is chosen as (x_1, y_1) or (x_2, y_2).

The Midpoint Formula

When you average two test scores (by finding their sum and dividing by 2), you are finding a number midway between the two scores. Likewise, the midpoint of the line segment with endpoints -1 and 7 in Fig. 1.30 is $(-1 + 7)/2$ or 3. In general, $(a + b)/2$ is the midpoint of the line segment with endpoints a and b shown in Fig. 1.31.

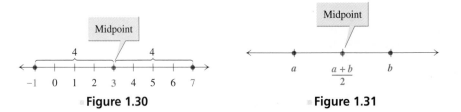

■ **Figure 1.30**   ■ **Figure 1.31**

Theorem: Midpoint on
a Number Line

> If a and b are real numbers, then $\dfrac{a + b}{2}$ is midway between them on the number line.

PROOF The distance between two numbers on the number line is the absolute value of their difference. So

$$\left| a - \frac{a+b}{2} \right| = \left| \frac{2a}{2} - \frac{a+b}{2} \right| = \left| \frac{a-b}{2} \right| = \frac{|a-b|}{2}$$

and
$$\left| b - \frac{a+b}{2} \right| = \left| \frac{2b}{2} - \frac{a+b}{2} \right| = \left| \frac{b-a}{2} \right| = \frac{|b-a|}{2}.$$

Since $|a-b| = |b-a|$, the distances from $\frac{a+b}{2}$ to a and from $\frac{a+b}{2}$ to b are equal. Since $\frac{a+b}{2}$ is equidistant from a and b it must be between a and b. ◼

We can find the midpoint of a line segment in the xy-plane in the same manner.

**Theorem:
The Midpoint Formula**

The midpoint of the line segment with endpoints (x_1, y_1) and (x_2, y_2) is
$$\left(\frac{x_1 + x_2}{2}, \frac{y_1 + y_2}{2} \right).$$

◼ **Figure 1.32**

PROOF Start with a line segment with endpoints $A(x_1, y_1)$ and $B(x_2, y_2)$ as shown in Fig. 1.32. Let M be the midpoint of $\overline{AB}$. Draw horizontal and vertical line segments as shown in Fig. 1.32 forming three right triangles. Since M is the midpoint of $\overline{AB}$ the two small right triangles are congruent. So D is the midpoint of $\overline{AC}$ and E is the midpoint of $\overline{BC}$. Since the midpoint on a number line is found by adding and dividing by two, the x-coordinate of D is $\frac{x_1 + x_2}{2}$ and the y-coordinate of E is $\frac{y_1 + y_2}{2}$. So the midpoint M is $\left(\frac{x_1 + x_2}{2}, \frac{y_1 + y_2}{2} \right)$. ◼

Example **3** **Finding a midpoint**

Find the midpoint of the line segment that has endpoints $(5, -3)$ and $(-1, 6)$.

Solution

Let $(x_1, y_1) = (5, -3)$ and $(x_2, y_2) = (-1, 6)$ in the midpoint formula:

$$\left(\frac{x_1 + x_2}{2}, \frac{y_1 + y_2}{2} \right) = \left(\frac{5 + (-1)}{2}, \frac{-3 + (-6)}{2} \right) = \left(2, -\frac{9}{2} \right)$$

So the midpoint of the line segment is $(2, -9/2)$. ◼

The Circle

An ordered pair is a **solution to** or **satisfies** an equation in two variables if the equation is correct when the variables are replaced by the coordinates of the ordered pair. For example, $(3, 11)$ satisfies $y = 2x + 5$ because $11 = 2(3) + 5$ is correct. The **solution set** to an equation in two variables is the set of all ordered pairs that satisfy the equation. The graph of (the solution set to) an equation is a geometric object that gives us a visual image of an algebraic object. Circles provide a nice example of this relationship between algebra and geometry.

A **circle** is the set of all points in a plane that lie a fixed distance from a given point in the plane. The fixed distance is called the **radius,** and the given point is the **center.** The distance formula can be used to write an equation for the circle shown in Fig. 1.33 with center (h, k) and radius r for $r > 0$. A point (x, y) is on the circle if and only if it satisfies the equation

$$\sqrt{(x - h)^2 + (y - k)^2} = r.$$

Since both sides of $\sqrt{(x - h)^2 + (y - k)^2} = r$ are positive, we can square each side to get the following **standard form** for the equation of a circle.

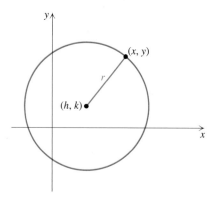

▪ **Figure 1.33**

Theorem: Equation for a Circle in Standard Form

> The equation for a circle with center (h, k) and radius r for $r > 0$ is
> $$(x - h)^2 + (y - k)^2 = r^2.$$
> A circle centered at the origin has equation $x^2 + y^2 = r^2$.

Note that squaring both sides of an equation produces an equivalent equation only when both sides are positive. If we square both sides of $\sqrt{x} = -3$, we get $x = 9$. But $\sqrt{9} \neq -3$ since the square root symbol always represents the nonnegative square root. Raising both sides of an equation to a power is discussed further in Section 2.5.

Example **4** Graphing a circle

Sketch the graph of the equation $(x - 1)^2 + (y + 2)^2 = 9$.

Solution

Since $(x - h)^2 + (y - k)^2 = r^2$ is a circle with center (h, k) and radius r (for $r > 0$), the graph of $(x - 1)^2 + (y + 2)^2 = 9$ is a circle with center $(1, -2)$ and radius 3. You can draw the circle as in Fig. 1.34 with a compass. To draw a circle by hand, locate the points that lie 3 units above, below, right, and left of the center, as shown in Fig. 1.34. Then sketch a circle through these points.

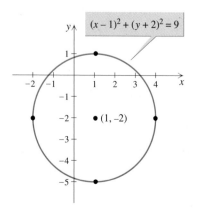

▪ **Figure 1.34**

To support these results with a graphing calculator you must first solve the equation for y:

$$(x - 1)^2 + (y + 2)^2 = 9$$
$$(y + 2)^2 = 9 - (x - 1)^2$$
$$y + 2 = \pm\sqrt{9 - (x - 1)^2}$$
$$y = -2 \pm \sqrt{9 - (x - 1)^2}$$

Now enter y_1 and y_2 as in Fig. 1.35(a). Set the viewing window as in Fig. 1.35(b). The graph in Fig. 1.35(c) supports our previous conclusion. A circle looks round only if the same unit distance is used on both axes. Some calculators automatically draw the graph with the same unit distance on both axes given the correct command (Zsquare on a TI-83).

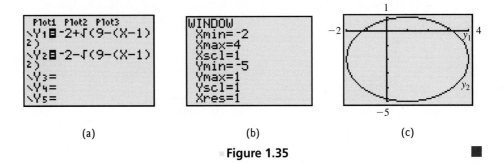

(a)   (b)   (c)

■ **Figure 1.35**   ■

Note that an equation such as $(x - 1)^2 + (y + 2)^2 = -9$ is not satisfied by any pair of real numbers, because the left-hand side is a nonnegative real number, while the right-hand side is negative. The equation $(x - 1)^2 + (y + 2)^2 = 0$ is satisfied only by $(1, -2)$. Since only one point satisfies $(x - 1)^2 + (y + 2)^2 = 0$, its graph is sometimes called a degenerate circle with radius zero. We study circles again later in this text when we study the conic sections.

In the next example we start with a description of a circle and write its equation.

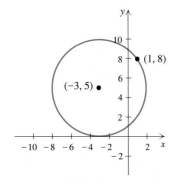

■ **Figure 1.36**

Example **5** **Writing an equation of a circle**

Write the standard equation for the circle with the center $(-3, 5)$ and passing through $(1, 8)$ as shown in Fig. 1.36.

Solution

The radius of this circle is the distance between $(-3, 5)$ and $(1, 8)$:

$$r = \sqrt{(8 - 5)^2 + (1 - (-3))^2} = \sqrt{9 + 16} = 5$$

Now use $h = -3$, $k = 5$, and $r = 5$ in the standard equation of the circle $(x - h)^2 + (y - k)^2 = r^2$.

$$(x - (-3))^2 + (y - 5)^2 = 5^2$$

So the equation of the circle is $(x + 3)^2 + (y - 5)^2 = 25$.   ■

If the equation of a circle is not given in standard form, we can convert it to standard form using a process called **completing the square.** Completing the square means finding the third term of a perfect square trinomial when given the first two. That is, if we start with $x^2 + bx$, then what third term will make a perfect square trinomial? Since

$$\left(x + \frac{b}{2}\right)^2 = x^2 + 2 \cdot \frac{b}{2} \cdot x + \left(\frac{b}{2}\right)^2 = x^2 + bx + \left(\frac{b}{2}\right)^2$$

adding $\left(\frac{b}{2}\right)^2$ to $x^2 + bx$ completes the square. For example, the perfect square trinomial that starts with $x^2 + 6x$ is $x^2 + 6x + 9$. Note that 9 can be found by taking one-half of 6 and squaring.

Rule for Completing the Square of $x^2 + bx + ?$

The last term of a perfect square trinomial (with $a = 1$) is the square of one-half of the coefficient of the middle term. In symbols, the perfect square trinomial whose first two terms are $x^2 + bx$ is

$$x^2 + bx + \left(\frac{b}{2}\right)^2.$$

Example **6** **Changing an equation of a circle to standard form**

Graph the equation $x^2 + 6x + y^2 - 5y = -\frac{1}{4}$.

Solution

Complete the square for both x and y to get the standard form.

$$x^2 + 6x + 9 + y^2 - 5y + \frac{25}{4} = -\frac{1}{4} + 9 + \frac{25}{4} \quad \left(\frac{1}{2} \cdot 6\right)^2 = 9, \left(\frac{1}{2} \cdot 5\right)^2 = \frac{25}{4}$$

$$(x + 3)^2 + \left(y - \frac{5}{2}\right)^2 = 15 \qquad \text{Factor the trinomials on the left side.}$$

The graph is a circle with center $\left(-3, \frac{5}{2}\right)$ and radius $\sqrt{15}$. See Fig. 1.37. ■

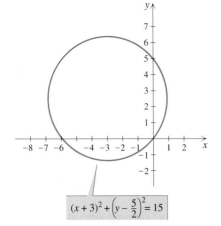

$$(x + 3)^2 + \left(y - \frac{5}{2}\right)^2 = 15$$

■ **Figure 1.37**

Whether an equation of the form $x^2 + y^2 + Ax + By = C$ is a circle depends on the value of C. We can always complete the squares for $x^2 + Ax$ and $y^2 + By$ by adding $(A/2)^2$ and $(B/2)^2$ to each side of the equation. Since a circle must have a positive radius we will have a circle only if $C + (A/2)^2 + (B/2)^2 > 0$.

The Line

For the circle, we started with the geometric definition and developed the algebraic equation. We would like to do the same thing for lines, but geometric definitions of lines are rather vague. Euclid's definition was "length without breadth." A modern geometry textbook states that a line is "a straight set of points extending infinitely in both directions." Another modern textbook defines lines algebraically as the graph of an equation of the form $Ax + By = C$. So we will accept the following theorem without proof.

Theorem: Equation of a Line in Standard Form

If A, B, and C are real numbers, then the graph of the equation

$$Ax + By = C$$

is a straight line, provided A and B are not both zero. Every straight line in the coordinate plane has an equation in the form $Ax + By = C$, the **standard form** for the equation of a line.

An equation of the form $Ax + By = C$ is called a **linear equation in two variables.** The equations,

$$2x + 3y = 5, \quad x = 4, \quad \text{and} \quad y = 5$$

are linear equations in standard form. An equation such as $y = 3x - 1$ that can be rewritten in standard form is also called a linear equation.

There is only one line containing any two distinct points. So to graph a linear equation we simply find two points that satisfy the equation and draw a line through them. We often use the point where the line crosses the x-axis, the **x-intercept,** and the point where the line crosses the y-axis, the **y-intercept.** Since every point on the x-axis has y-coordinate 0, we find the x-intercept by replacing y with 0 and then solving the equation for x. Since every point on the y-axis has x-coordinate 0, we find the y-intercept by replacing x with 0 and then solving for y. If the x- and y-intercepts are both at the origin, then you must find another point that satisfies the equation.

Example **7** Graphing lines and showing the intercepts

Graph each equation. Be sure to find and show the intercepts.

a. $2x - 3y = 9$ **b.** $y = 40 - x$

Solution

a. Since the y-coordinate of the x-intercept is 0, we replace y by 0 in the equation:

$$2x - 3(0) = 9$$
$$2x = 9$$
$$x = 4.5$$

To find the y-intercept, we replace x by 0 in the equation:

$$2(0) - 3y = 9$$
$$-3y = 9$$
$$y = -3$$

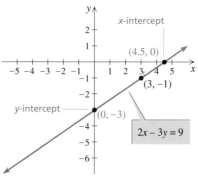

Figure 1.38

The x-intercept is $(4.5, 0)$ and the y-intercept is $(0, -3)$. Locate the intercepts and draw the line as shown in Fig. 1.38. To check, locate a point such as $(3, -1)$, which also satisfies the equation, and see if the line goes through it.

b. If $x = 0$, then $y = 40 - 0 = 40$ and the y-intercept is $(0, 40)$. If $y = 0$, then $0 = 40 - x$ or $x = 40$. The x-intercept is $(40, 0)$. Draw a line through these points as shown in Fig. 1.39 on the next page. Check that $(10, 30)$ and $(20, 20)$ also satisfy $y = 40 - x$ and the line goes through these points.

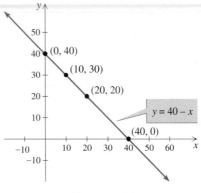

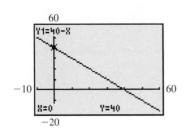

■ **Figure 1.40**

■ **Figure 1.39**

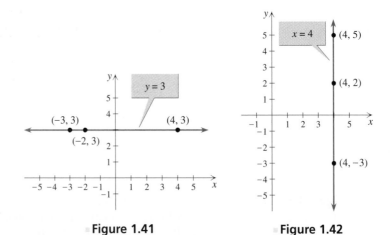

The calculator graph shown in Fig. 1.40 is consistent with the graph in Fig. 1.39. Note that the viewing window is set to show the intercepts. ■

Example **8** Graphing horizontal and vertical lines

Sketch the graph of each equation in the rectangular coordinate system.

a. $y = 3$ **b.** $x = 4$

Solution

a. The equation $y = 3$ is equivalent to $0 \cdot x + y = 3$. Because x is multiplied by 0, we can choose any value for x as long as we choose 3 for y. So ordered pairs such as $(-3, 3)$, $(-2, 3)$, and $(4, 3)$ satisfy the equation $y = 3$. The graph of $y = 3$ is the horizontal line shown in Fig. 1.41.

b. The equation $x = 4$ is equivalent to $x + 0 \cdot y = 4$. Because y is multiplied by 0, we can choose any value for y as long as we choose 4 for x. So ordered pairs such as $(4, -3)$, $(4, 2)$, and $(4, 5)$ satisfy the equation $x = 4$. The graph of $x = 4$ is the vertical line shown in Fig. 1.42.

■ **Figure 1.41**   ■ **Figure 1.42**

Note that you cannot graph $x = 4$ using the Y= key on your calculator. You can graph it on a calculator using polar coordinates or parametric equations (discussed later in this text). ■

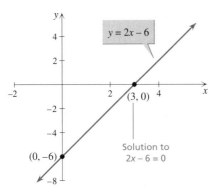

Figure 1.43

In the context of two variables the equation $x = 4$ has infinitely many solutions. Every ordered pair on the vertical line in Fig. 1.42 satisfies $x = 4$. In the context of one variable $x = 4$ has only one solution, 4.

Using a Graph to Solve an Equation

Graphing and solving equations go hand in hand. For example, the graph of $y = 2x - 6$ in Fig. 1.43 has x-intercept $(3, 0)$, because if $x = 3$ then $y = 0$. Of course 3 is also the solution to the corresponding equation $2x - 6 = 0$ (where y is replaced by 0). For this reason, the solution to an equation is also called a **zero** or **root** of the equation. Every x-intercept on a graph provides us a solution to the corresponding equation. However, an x-intercept on a graph may not be easy to identify. In the next example we see how a graphing calculator identifies an x-intercept and thus gives us the approximate solution to an equation.

Example **9** **Using a graph to solve an equation**

Use a graphing calculator to solve $0.55(x - 3.45) + 13.98 = 0$.

Solution

First graph $y = 0.55(x - 3.45) + 13.98$ using a viewing window that shows the x-intercept as in Fig. 1.44(a). Next press ZERO or ROOT on the CALC menu. The calculator can find a zero between a *left bound* and a *right bound*, which you must enter. The calculator also asks you to make a guess. See Fig. 1.44(b). The more accurate the guess, the faster the calculator will find the zero. The solution to the equation rounded to two decimal places is -21.97. See Fig. 1.44(c). As always, consult your caluluator manual if you are having difficulty.

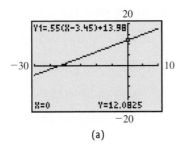
(a)

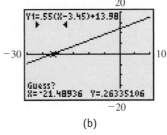
(b)

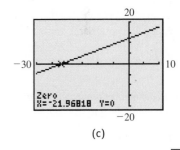
(c)

Figure 1.44

For Thought

True or False? Explain.

1. The point $(2, -3)$ is in quadrant II.

2. The point $(4, 0)$ is in quadrant I.

3. The distance between (a, b) and (c, d) is
$\sqrt{(a - b)^2 + (c - d)^2}$.

4. The equation $3x^2 + y = 5$ is a linear equation.

5. The solution to $7x - 9 = 0$ is the x-coordinate of the x-intercept of $y = 7x - 9$.

6. $\sqrt{7^2 + 9^2} = 7 + 9$

7. The origin lies midway between $(1, 3)$ and $(-1, -3)$.

8. The distance between $(3, -7)$ and $(3, 3)$ is 10.

9. The x-intercept for the graph of $3x - 2y = 7$ is $(7/3, 0)$.

10. The graph of $(x + 2)^2 + (y - 1)^2 = 5$ is a circle centered at $(-2, 1)$ with radius 5.

1.3 Exercises

For each point shown in the xy-plane, write the corresponding ordered pair and name the quadrant in which it lies or the axis on which it lies.

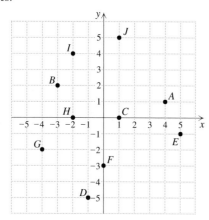

1. A **2.** B **3.** C **4.** D

5. E **6.** F **7.** G **8.** H

9. I **10.** J

For each pair of points find the distance between them and the midpoint of the line segment joining them.

11. $(1, 3), (4, 7)$ **12.** $(-3, -2), (9, 3)$

13. $(-1, -2), (1, 0)$ **14.** $(-1, 0), (1, 2)$

15. $(-1, 1), \left(-1 + 3\sqrt{3}, 4\right)$

16. $\left(1 + \sqrt{2}, -2\right), \left(1 - \sqrt{2}, 2\right)$

17. $(1.2, 4.8), (-3.8, -2.2)$ **18.** $(-2.3, 1.5), (4.7, -7.5)$

19. $(a, 0), (b, 0)$ **20.** $(a, 0), \left(\dfrac{a + b}{2}, 0\right)$

21. $(\pi, 0), (\pi/2, 1)$ **22.** $(0, 0), (\pi/2, 1)$

Determine the center and radius of each circle and sketch the graph.

23. $x^2 + y^2 = 16$ **24.** $x^2 + y^2 = 1$

25. $(x + 6)^2 + y^2 = 36$ **26.** $x^2 + (y - 2)^2 = 16$

27. $y^2 = 25 - (x + 1)^2$ **28.** $x^2 = 9 - (y - 3)^2$

29. $(x - 2)^2 = 8 - (y + 2)^2$

30. $(y + 2)^2 = 20 - (x - 4)^2$

Write the standard equation for each circle.

31. Center at $(0, 0)$ with radius 7

32. Center at $(0, 0)$ with radius 5

33. Center at $(-2, 5)$ with radius $1/2$

34. Center at $(-1, -6)$ with radius $1/3$

35. Center at $(3, 5)$ and passing through the origin

36. Center at $(-3, 9)$ and passing through the origin

37. Center at $(5, -1)$ and passing through $(1, 3)$

38. Center at $(-2, -3)$ and passing through $(2, 5)$

Determine the center and radius of each circle and sketch the graph. See the rule for completing the square on page 28.

39. $x^2 + y^2 = 9$ **40.** $x^2 + y^2 = 100$

41. $x^2 + y^2 + 6y = 0$ **42.** $x^2 + y^2 = 4x$

43. $x^2 - 6x + y^2 - 8y = 0$

44. $x^2 + 10x + y^2 - 8y = -40$

45. $x^2 + y^2 = 4x + 3y$ **46.** $x^2 + y^2 = 5x - 6y$

47. $x^2 + y^2 = \dfrac{x}{2} - \dfrac{y}{3} - \dfrac{1}{16}$ **48.** $x^2 + y^2 = x - y + \dfrac{1}{2}$

Write the standard equation for each of the following circles.

49. a.

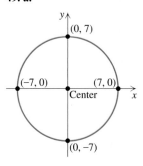

b.

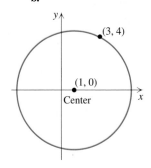

c.

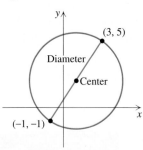

50. a.

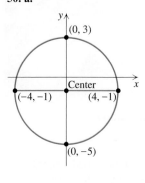

b.

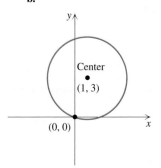

c.

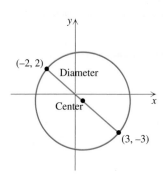

Write the equation of each circle in standard form. The coordinates of the center and the radius for each circle are integers.

51. a.

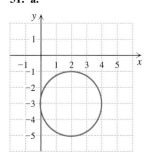

b.

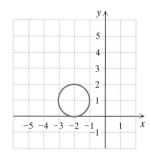

c.

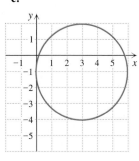

d.

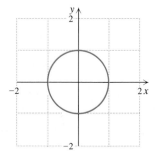

52. a.

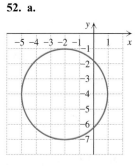

b.

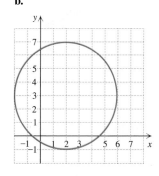

c.

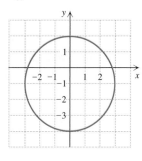

d.

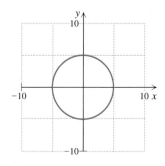

Sketch the graph of each linear equation. Be sure to find and show the x- and y-intercepts.

53. $y = 3x - 4$

54. $y = 5x - 5$

55. $3x - y = 6$

56. $5x - 2y = 10$

57. $x = 3y - 90$

58. $x = 80 - 2y$

59. $\frac{2}{3}y - \frac{1}{2}x = 400$

60. $\frac{1}{2}x - \frac{1}{3}y = 600$

61. $2x + 4y = 0.01$

62. $3x - 5y = 1.5$

63. $0.03x + 0.06y = 150$

64. $0.09x - 0.06y = 54$

Graph each equation in the rectangular coordinate system.

65. $x = 5$

66. $y = -2$

67. $y = 4$

68. $x = -3$

69. $x = -4$

70. $y = 5$

71. $y - 1 = 0$

72. $5 - x = 4$

Find the solution to each equation by reading the accompanying graph.

73. $2.4x - 8.64 = 0$

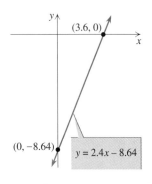

74. $8.84 - 1.3x = 0$

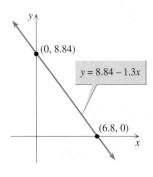

75. $-\frac{3}{7}x + 6 = 0$

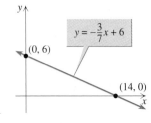

76. $\frac{5}{6}x + 30 = 0$

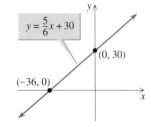

Use a graphing calculator to estimate the solution to each equation to two decimal places. Then find the solution algebraically and compare it with your estimate.

77. $1.2x + 3.4 = 0$

78. $3.2x - 4.5 = 0$

79. $0.03x - 3497 = 0$

80. $0.09x + 2000 = 0$

81. $4.3 - 3.1(2.3x - 9.9) = 0$

82. $9.4x - 4.37(3.5x - 9.76) = 0$

Solve each problem.

83. *First Marriage* The median age at first marriage for women went from 20.8 in 1970 to 25.1 in 2000 as shown in the accompanying figure (U.S. Census Bureau, www.census.gov).
   a. Find the midpoint of the line segment in the figure and interpret your result.

   b. Find the distance between the two points shown in the figure and interpret your result.

■ **Figure for Exercise 83**

84. *Unmarried Couples* The number of unmarried-couple households h (in millions), can be modeled using the equation $h = 0.171n + 2.913$, where n is the number of years since 1990 (U.S. Census Bureau, www.census.gov).
   a. Find and interpret the n-intercept for the line. Does it make sense?

   b. Find and interpret the h-intercept for the line.

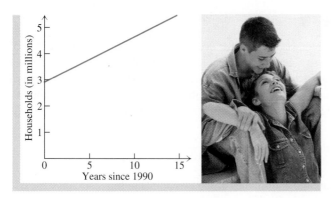

Figure for Exercise 84

85. *Capsize Control* The capsize screening value C is an indicator of a sailboat's suitability for extended offshore sailing. C is determined by the formula

$$C = 4D^{-1/3}B,$$

where D is the displacement of the boat in pounds and B is its beam (or width) in feet. Sketch the graph of this equation for B ranging from 0 to 20 ft assuming that D is fixed at 22,800 lbs. Find C for the Island Packet 40, which has a displacement of 22,800 pounds and a beam of 12 ft 11 in. (Island Packet Yachts, www.ipy.com).

86. *Limiting the Beam* The International Offshore Rules require that the capsize screening value C (from the previous exercise) be less than or equal to 2 for safety. What is the maximum

allowable beam (to the nearest inch) for a boat with a displacement of 22,800 lbs? For a fixed displacement, is a boat more or less likely to capsize as its beam gets larger?

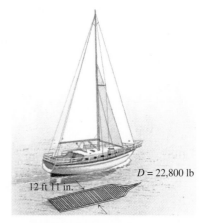

$D = 22{,}800$ lb

12 ft 11 in.

Figure for Exercises 85 and 86

Thinking Outside the Box III

Methodical Mower Eugene is mowing a rectangular lawn that is 300 ft by 400 ft. He starts at one corner and mows a swath of uniform width around the outside edge in a clockwise direction. He continues going clockwise, widening the swath that is mowed and shrinking the rectangular section that is yet to be mowed. When he is half done with the lawn, how wide is the swath?

1.3 Pop Quiz

1. Find the distance between $(-1, 3)$ and $(3, 5)$.

2. Find the center and radius for the circle
$$(x - 3)^2 + (y + 5)^2 = 81.$$

3. Find the center and radius for the circle
$$x^2 + 4x + y^2 - 10y = -28.$$

4. Find the equation of the circle that passes through the origin and has center at $(3, 4)$.

5. Find all intercepts for $2x - 3y = 12$.

6. Which point is on both of the lines $x = 5$ and $y = -1$?

1.4 Linear Equations in Two Variables

In Section 1.3 we graphed lines, including horizontal and vertical lines. We learned that every line has an equation in standard form $Ax + By = C$. In this section we will continue to study lines.

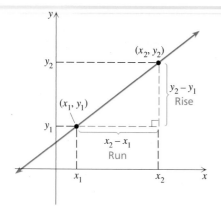

Figure 1.45

Slope of a Line

A road that has a 5% grade rises 5 feet for every horizontal run of 100 feet. A roof that has a 5–12 pitch rises 5 feet for every horizontal run of 12 feet. The grade of a road and the pitch of a roof are measurements of *steepness*. The steepness or *slope* of a line in the *xy*-coordinate system is the ratio of the **rise** (the change in *y*-coordinates) to the **run** (the change in *x*-coordinates) between two points on the line:

$$\text{slope} = \frac{\text{change in } y\text{-coordinates}}{\text{change in } x\text{-coordinates}} = \frac{\text{rise}}{\text{run}}$$

If (x_1, y_1) and (x_2, y_2) are the coordinates of the two points in Fig. 1.45, then the rise is $y_2 - y_1$ and the run is $x_2 - x_1$:

Definition: Slope

> The **slope** of the line through (x_1, y_1) and (x_2, y_2) with $x_1 \neq x_2$ is
>
> $$\frac{y_2 - y_1}{x_2 - x_1}.$$

Note that if (x_1, y_1) and (x_2, y_2) are two points for which $x_1 = x_2$ then the line through them is a vertical line. Since this case is not included in the definition of slope, a vertical line does not have a slope. We also say that the slope of a vertical line is undefined. If we choose two points on a horizontal line then $y_1 = y_2$ and $y_2 - y_1 = 0$. For any horizontal line the rise between two points is 0 and the slope is 0.

Example 1 Finding the slope

In each case find the slope of the line that contains the two given points.

a. $(-3, 4), (-1, -2)$ **b.** $(-3, 7), (5, 7)$ **c.** $(-3, 5), (-3, 8)$

Solution

a. Use $(x_1, y_1) = (-3, 4)$ and $(x_2, y_2) = (-1, -2)$ in the formula:

$$\text{slope} = \frac{y_2 - y_1}{x_2 - x_1} = \frac{-2 - 4}{-1 - (-3)} = \frac{-6}{2} = -3$$

The slope of the line is -3.

b. Use $(x_1, y_1) = (-3, 7)$ and $(x_2, y_2) = (5, 7)$ in the formula:

$$\text{slope} = \frac{y_2 - y_1}{x_2 - x_1} = \frac{7 - 7}{5 - (-3)} = \frac{0}{8} = 0$$

The slope of this horizontal line is 0.

c. The line through $(-3, 5)$ and $(-3, 8)$ is a vertical line and so it does not have a slope. ■

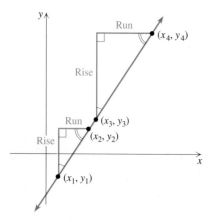

Figure 1.46

The slope of a line is the same number regardless of which two points on the line are used in the calculation of the slope. To understand why, consider the two triangles shown in Fig. 1.46. These triangles have the same shape and are called similar triangles. Because the ratios of corresponding sides of similar triangles are equal, the ratio of rise to run is the same for either triangle.

Point-Slope Form

Suppose that a line through (x_1, y_1) has slope m. Every other point (x, y) on the line must satisfy the equation

$$\frac{y - y_1}{x - x_1} = m$$

because any two points can be used to find the slope. Multiply both sides by $x - x_1$ to get $y - y_1 = m(x - x_1)$, which is the **point-slope form** of the equation of a line.

**Theorem:
Point-Slope Form**

> The equation of the line (in point-slope form) through (x_1, y_1) with slope m is
>
> $$y - y_1 = m(x - x_1).$$

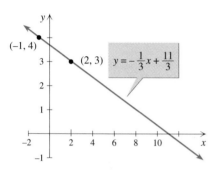

Figure 1.47

In Section 1.3 we started with the equation of a line and graphed the line. Using the point-slope form, we can start with a graph of a line or a description of the line and write the equation for the line.

Example 2 The equation of a line given two points

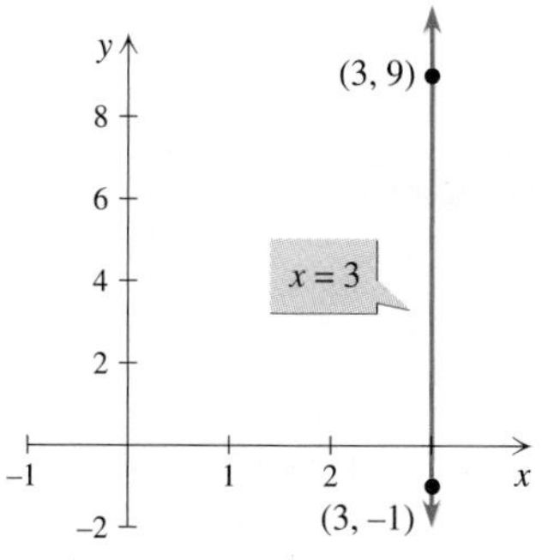

In each case graph the line through the given pair of points. Then find the equation of the line and solve it for y if possible.

a. $(-1, 4), (2, 3)$ **b.** $(2, 5), (-6, 5)$ **c.** $(3, -1), (3, 9)$

Solution

a. Find the slope of the line shown in Fig. 1.47 as follows:

$$m = \frac{y_2 - y_1}{x_2 - x_1} = \frac{3 - 4}{2 - (-1)} = \frac{-1}{3} = -\frac{1}{3}$$

Now use a point, say $(2, 3)$, and $m = -\frac{1}{3}$ in the point-slope form:

$$y - y_1 = m(x - x_1)$$

$$y - 3 = -\frac{1}{3}(x - 2) \qquad \text{The equation in point-slope form}$$

$$y - 3 = -\frac{1}{3}x + \frac{2}{3}$$

$$y = -\frac{1}{3}x + \frac{11}{3} \qquad \text{The equation solved for } y$$

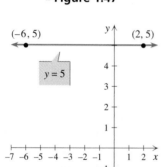

Figure 1.48

b. The slope of the line through $(2, 5)$ and $(-6, 5)$ shown in Fig. 1.48 is 0. The equation of this horizontal line is $y = 5$.

c. The line through $(3, -1)$ and $(3, 9)$ shown in Fig. 1.49 is vertical and it does not have slope. Its equation is $x = 3$. ■

Slope-Intercept Form

The line $y = mx + b$ goes through $(0, b)$ and $(1, m + b)$. Between these two points the rise is m and the run is 1. So the slope is m. Since $(0, b)$ is the y-intercept and m is the slope, $y = mx + b$ is called **slope-intercept form**. Any equation in standard

Figure 1.49

form $Ax + By = C$ can be rewritten in slope-intercept form by solving the equation for y provided $B \neq 0$. If $B = 0$ the line is vertical and has no slope.

Theorem:
Slope-Intercept Form

> The equation of a line (in slope-intercept form) with slope m and y-intercept $(0, b)$ is
>
> $$y = mx + b.$$
>
> Every nonvertical line has an equation in slope-intercept form.

If you know the slope and y-intercept for a line then you can use slope-intercept form to write its equation. For example, the equation of the line through $(0, 9)$ with slope 4 is $y = 4x + 9$. In the next example we use the slope-intercept form to determine the slope and y-intercept for a line.

Example **3** **Find the slope and y-intercept**

Identify the slope and y-intercept for the line $2x - 3y = 6$.

Solution

First solve the equation for y to get it in slope-intercept form:

$$2x - 3y = 6$$
$$-3y = -2x + 6$$
$$y = \frac{2}{3}x - 2$$

So the slope is $\frac{2}{3}$ and the y-intercept is $(0, -2)$. ■

Using Slope to Graph a Line

Slope is the ratio $\frac{\text{rise}}{\text{run}}$ that results from moving from one point to another on a line. A positive rise indicates a motion upward and a negative rise indicates a motion downward. A positive run indicates a motion to the right and a negative run indicates a motion to the left. If you start at any point on a line with slope $\frac{1}{2}$, then moving up 1 unit and 2 units to the right will bring you back to the line. On a line with slope -3 or $\frac{-3}{1}$, moving down 3 units and 1 unit to the right will bring you back to the line.

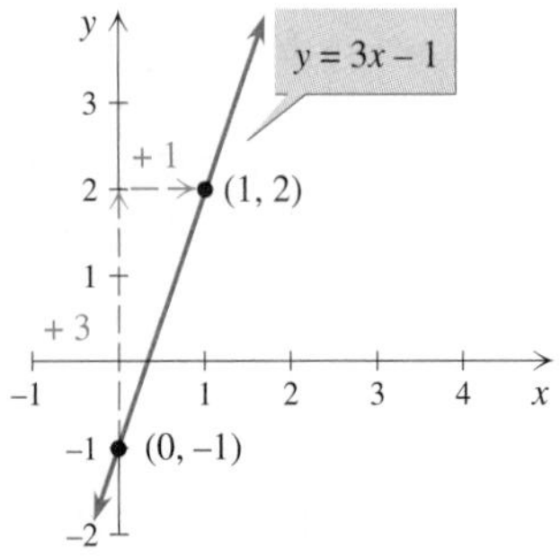

Figure 1.50

Example **4** **Graphing a line using its slope and y-intercept**

Graph each line.

a. $y = 3x - 1$ **b.** $y = -\frac{2}{3}x + 4$

Solution

a. The line $y = 3x - 1$ has y-intercept $(0, -1)$ and slope 3 or $\frac{3}{1}$. Starting at $(0, -1)$ we obtain a second point on the line by moving up 3 units and 1 unit to the right. So the line goes through $(0, -1)$ and $(1, 2)$ as shown in Fig. 1.50.

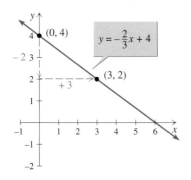

Figure 1.51

b. The line $y = -\frac{2}{3}x + 4$ has y-intercept $(0, 4)$ and slope $-\frac{2}{3}$ or $\frac{-2}{3}$. Starting at $(0, 4)$ we obtain a second point on the line by moving down 2 units and then 3 units to the right. So the line goes through $(0, 4)$ and $(3, 2)$ as shown in Fig. 1.51.   ■

As the x-coordinate increases on a line with positive slope, the y-coordinate increases also. As the x-coordinate increases on a line with negative slope, the y-coordinate decreases. Figure 1.52 shows some lines of the form $y = mx$ with positive slopes and negative slopes. Observe the effect that the slope has on the position of the line.

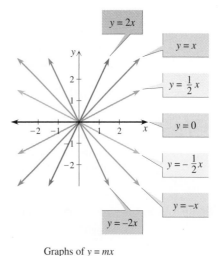

Graphs of $y = mx$

Figure 1.52

The Three Forms for the Equation of a Line

There are three forms for the equation of a line. The following strategy will help you decide when and how to use these forms.

STRATEGY	**Finding the Equation of a Line**

1. Since vertical lines have no slope they cannot be written in slope-intercept form $y = mx + b$ or point-slope form $y - y_1 = m(x - x_1)$. All lines can be described by an equation in standard form $Ax + By = C$.

2. For any constant k, $y = k$ is a horizontal line and $x = k$ is a vertical line.

3. If you know two points on a line, then find the slope.

4. If you know the slope and point on the line, use point-slope form. If the point is the y-intercept, then use slope-intercept form.

5. Final answers are usually written in slope-intercept or standard form. Standard form is often simplified by using only integers for the coefficients.

Example **5** Standard form using integers

Find the equation of the line through $\left(0, \frac{1}{3}\right)$ with slope $\frac{1}{2}$. Write the equation in standard form using only integers.

Solution

Since we know the slope and y-intercept, start with slope-intercept form:

$$y = \frac{1}{2}x + \frac{1}{3} \quad \text{Slope-intercept form}$$

$$-\frac{1}{2}x + y = \frac{1}{3}$$

$$-6\left(-\frac{1}{2}x + y\right) = -6 \cdot \frac{1}{3} \quad \text{Multiply by } -6 \text{ to get integers.}$$

$$3x - 6y = -2 \quad \text{Standard form with integers}$$

Any integral multiple of $3x - 6y = -2$ would also be standard form, but we usually use the smallest possible positive coefficient for x. ■

Parallel Lines

Two lines in a plane are said to be **parallel** if they have no points in common. Any two vertical lines are parallel, and slope can be used to determine whether nonvertical lines are parallel. For example, the lines $y = 3x - 4$ and $y = 3x + 1$ are parallel because their slopes are equal.

Theorem: Parallel Lines

> Two nonvertical lines in the coordinate plane are parallel if and only if their slopes are equal.

PROOF Suppose that $y = m_1x + b_1$ and $y = m_2x + b_2$ are two lines with different y-intercepts ($b_1 \neq b_2$). These lines are parallel if and only if $m_1x + b_1 = m_2x + b_2$ has no solution. This equation has no solution if and only if $m_1 = m_2$. ■

Example **6** **Writing equations of parallel lines**

Find the equation in slope-intercept form of the line through $(1, -4)$ that is parallel to $y = 3x + 2$.

Solution

Since $y = 3x + 2$ has slope 3, any line parallel to it also has slope 3. Write the equation of the line through $(1, -4)$ with slope 3 in point-slope form:

$$y - (-4) = 3(x - 1) \quad \text{Point-slope form}$$

$$y + 4 = 3x - 3$$

$$y = 3x - 7 \quad \text{Slope-intercept form}$$

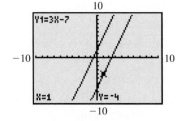

Figure 1.53

The line $y = 3x - 7$ goes through $(1, -4)$ and is parallel to $y = 3x + 2$. The graphs of $y_1 = 3x - 7$ and $y_2 = 3x + 2$ in Fig. 1.53 support the answer. ■

Perpendicular Lines

Two lines are **perpendicular** if they intersect at a right angle. Slope can be used to determine whether lines are perpendicular. For example, lines with slopes such as 2/3 and −3/2 are perpendicular. The slope −3/2 is the opposite of the reciprocal of 2/3. In the following theorem we use the equivalent condition that the product of the slopes of two perpendicular lines is −1, provided they both have slopes.

Theorem:
Perpendicular Lines

Two lines with slopes m_1 and m_2 are perpendicular if and only if $m_1m_2 = -1$.

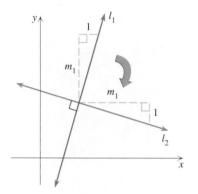

■ **Figure 1.54**

PROOF   The phrase "if and only if" means that there are two statements to prove. First we prove that if l_1 with slope m_1 and l_2 with slope m_2 are perpendicular, then $m_1m_2 = -1$. Assume that $m_1 > 0$. At the intersection of the lines draw a right triangle using a rise of m_1 and a run of 1, as shown in the Fig. 1.54. Rotate l_1 (along with the right triangle) 90 degrees so that it coincides with l_2. Now use the triangle in its new position to determine that $m_2 = \frac{1}{-m_1}$ or $m_1m_2 = -1$.

The second statement to prove is that $m_1m_2 = -1$ or $m_2 = \frac{1}{-m_1}$ implies that the lines are perpendicular. Start with the two intersecting lines and the two congruent right triangles as shown in Fig. 1.54. It takes a rotation of 90 degrees to get the vertical side marked m_1 to coincide with the horizontal side marked m_1. Since the triangles are congruent, rotating 90 degrees makes the triangles coincide and the lines coincide. So the lines are perpendicular.   ■

Example **7** **Writing equations of perpendicular lines**

Find the equation of the line perpendicular to the line $3x - 4y = 8$ and containing the point $(-2, 1)$. Write the answer in slope-intercept form.

Solution

Rewrite $3x - 4y = 8$ in slope-intercept form:

$$-4y = -3x + 8$$

$$y = \frac{3}{4}x - 2 \qquad \text{Slope of this line is 3/4.}$$

Since the product of the slopes of perpendicular lines is −1, the slope of the line that we seek is −4/3. Use the slope −4/3 and the point $(-2, 1)$ in the point-slope form:

$$y - 1 = -\frac{4}{3}(x - (-2))$$

$$y - 1 = -\frac{4}{3}x - \frac{8}{3}$$

$$y = -\frac{4}{3}x - \frac{5}{3}$$

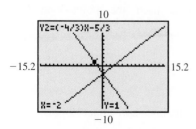

■ **Figure 1.55**

The last equation is the required equation in slope-intercept form. The graphs of these two equations should look perpendicular.

If you graph $y_1 = \frac{3}{4}x - 2$ and $y_2 = -\frac{4}{3}x - \frac{5}{3}$ with a graphing calculator, the graphs will not appear perpendicular in the standard viewing window because each axis has a different unit length. The graphs appear perpendicular in Fig. 1.55 because the unit lengths were made equal with the ZSquare feature of the TI-83.   ■

Applications

Example 8 Finding a linear equation

The monthly cost for a cell phone is $35 for 100 minutes and $45 for 200 minutes. See Fig. 1.56. The cost C is determined from the time t by a linear equation.

a. Find the linear equation.
b. What is the cost for 400 minutes per month?

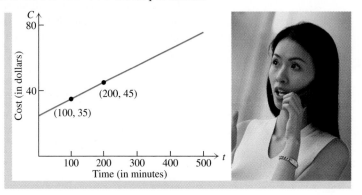

■ **Figure 1.56**

Solution

a. First find the slope:

$$m = \frac{C_2 - C_1}{t_2 - t_1} = \frac{45 - 35}{200 - 100} = \frac{10}{100} = 0.10$$

The slope is $0.10 per minute. Now find b by using $C = 35$, $t = 100$, and $m = 0.10$ in the slope-intercept form $C = mt + b$:

$$35 = 0.10(100) + b$$

$$35 = 10 + b$$

$$25 = b$$

So the formula is $C = 0.10t + 25$.

b. Use $t = 400$ in the formula $C = 0.10t + 25$:

$$C = 0.10(400) + 25 = 65$$

The cost for 400 minutes per month is $65. ■

Note that in Example 8 we could have used the point-slope form $C - C_1 = m(t - t_1)$ to get the formula $C = 0.10t + 25$. Try it.

For Thought

True or False? Explain.

1. The slope of the line through $(2, 2)$ and $(3, 3)$ is $3/2$.

2. The slope of the line through $(-3, 1)$ and $(-3, 5)$ is 0.

3. Any two distinct parallel lines have equal slopes.

4. The graph of $x = 3$ in the coordinate plane is the single point $(3, 0)$.

5. Two lines with slopes m_1 and m_2 are perpendicular if $m_1 = -1/m_2$.

6. Every line in the coordinate plane has an equation in slope-intercept form.

7. The slope of the line $y = 3 - 2x$ is 3.

8. Every line in the coordinate plane has an equation in standard form.

9. The line $y = 3x$ is parallel to the line $y = -3x$.

10. The line $x - 3y = 4$ contains the point $(1, -1)$ and has slope $1/3$.

1.4 Exercises

Find the slope of the line containing each pair of points.

1. $(-2, 3), (4, 5)$

2. $(-1, 2), (3, 6)$

3. $(1, 3), (3, -5)$

4. $(2, -1), (5, -3)$

5. $(5, 2), (-3, 2)$

6. $(0, 0), (5, 0)$

7. $\left(\dfrac{1}{8}, \dfrac{1}{4}\right), \left(\dfrac{1}{4}, \dfrac{1}{2}\right)$

8. $\left(-\dfrac{1}{3}, \dfrac{1}{2}\right), \left(\dfrac{1}{6}, \dfrac{1}{3}\right)$

9. $(5, -1), (5, 3)$

10. $(-7, 2), (-7, -6)$

Find the equation of the line through the given pair of points. Solve it for y if possible.

11. $(-1, -1), (3, 4)$

12. $(-2, 1), (3, 5)$

13. $(-2, 6), (4, -1)$

14. $(-3, 5), (2, 1)$

15. $(3, 5), (-3, 5)$

16. $(-6, 4), (2, 4)$

17. $(4, -3), (4, 12)$

18. $(-5, 6), (-5, 4)$

Write an equation in slope-intercept form for each of the lines shown.

19.

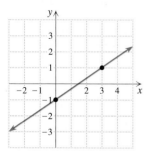

20.

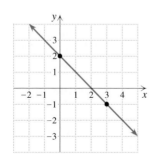

21.

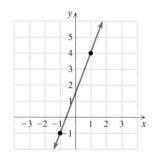

22.

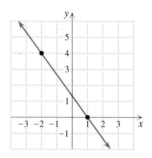

23.

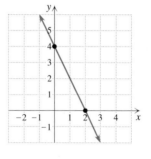

24.

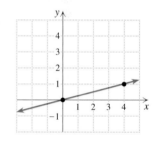

25.

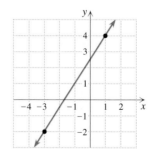

26.

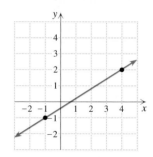

Write each equation in slope-intercept form and identify the slope and y-intercept of the line.

27. $3x - 5y = 10$

28. $2x - 2y = 1$

29. $y - 3 = 2(x - 4)$

30. $y + 5 = -3(x - (-1))$

31. $y + 1 = \dfrac{1}{2}(x - (-3))$

32. $y - 2 = -\dfrac{3}{2}(x + 5)$

33. $y - 4 = 0$

34. $-y + 5 = 0$

35. $y - 0.4 = 0.03(x - 100)$

36. $y + 0.2 = 0.02(x - 3)$

Use the y-intercept and slope to sketch the graph of each equation.

37. $y = \dfrac{1}{2}x - 2$

38. $y = \dfrac{2}{3}x + 1$

39. $y = -3x + 1$

40. $y = -x + 3$

41. $y = -\dfrac{3}{4}x - 1$

42. $y = -\dfrac{3}{2}x$

43. $x - y = 3$

44. $2x - 3y = 6$

45. $y - 5 = 0$

46. $6 - y = 0$

Find the equation of the line through the given pair of points in standard form using only integers. See the strategy for finding the equation of a line on page 39.

47. $(3, 0)$ and $(0, -4)$

48. $(-2, 0)$ and $(0, 3)$

49. $(2, 3)$ and $(-3, -1)$

50. $(4, -1)$ and $(-2, -6)$

51. $(-4, 2)$ and $(-4, 5)$

52. $(-3, 6)$ and $(9, 6)$

Find the slope of each line described.

53. A line parallel to $y = 0.5x - 9$

54. A line parallel to $3x - 9y = 4$

55. A line perpendicular to $3y - 3x = 7$

56. A line perpendicular to $2x - 3y = 8$

57. A line perpendicular to the line $x = 4$

58. A line parallel to $y = 5$

Write an equation in standard form using only integers for each of the lines described. In each case make a sketch.

59. The line with slope 2, going through $(1, -2)$

60. The line with slope -3, going through $(-3, 4)$

61. The line through $(1, 4)$, parallel to $y = -3x$

62. The line through $(-2, 3)$ parallel to $y = \dfrac{1}{2}x + 6$

63. The line perpendicular to $x - 2y = 3$ and containing $(-3, 1)$

64. The line perpendicular to $3x - y = 9$ and containing $(0, 0)$

65. The line perpendicular to $x = 4$ and containing $(2, 5)$

66. The line perpendicular to $y = 9$ and containing $(-1, 3)$

Solve each problem.

67. *Celsius to Fahrenheit Formula*   Fahrenheit temperature F is determined from Celsius temperature C by a linear equation. Water freezes at 0°C or 32°F and boils at 100°C or 212°F. Use the point-slope formula to write F in terms of C. Find the Fahrenheit temperature of an oven that is 150°C.

68. *Cost of Business Cards*   Speedy Printing charges $23 for 200 deluxe business cards and $35 for 500 deluxe business cards. Speedy uses a linear equation to determine the cost from the number of cards. Find the equation and find the cost of 700 deluxe business cards.

69. *Volume Discount*   Mona Kalini gives a bus tour of Honolulu to one person for $49, two people for $48 each, three people for $47 each, etc. Write a linear equation that gives the cost per person c in terms of the number of people n. What is her revenue for a bus load of 40 people?

70. *Ticket Pricing*   At $10 per ticket, the Rhythm Kings will fill all 8000 seats in the Saint Cloud Arena. For every $1 increase in ticket price, 500 tickets will go unsold. Find a linear equation that gives the number of tickets sold n in terms of the ticket price p. Find the revenue if the tickets are $20 each.

71. *Computers and Printers*   An office manager will spend a total of $60,000 on computers at $2000 each and printers at $1500 each. Find a linear equation that gives the number of computers purchased c in terms of the number of printers purchased p. Interpret the slope.

72. *Carpenters and Helpers*   A contractor plans to distribute a total of $2400 in bonus money to nine carpenters and three helpers. The carpenters all get the same amount and the helpers all get the same amount. Find a linear equation that gives the helpers' bonus h in terms of the carpenters' bonus c. Interpret the slope.

Thinking Outside the Box IV

Army of Ants   An army of ants is marching across the kitchen floor. If they form columns with 10 ants in each column, then there are 6 ants left over. If they form columns with 7, 11, or 13 ants in each column, then there are 2 ants left over. What is the smallest number of ants that could be in this army?

1.4 Pop Quiz

1. Find the slope of the line through $(-4, 9)$ and $(5, 6)$.

2. Find the equation of the line through $(3, 4)$ and $(6, 8)$.

3. What is the slope of the line $2x - 7y = 1$?

4. Find the equation of the line with y-intercept $(0, 7)$ that is parallel to $y = 3x - 1$.

5. Find the equation of the line with y-intercept $(0, 8)$ that is perpendicular to $y = \frac{1}{2}x + 4$.

1.5 Functions

From this section to the end of this text we will be studying functions. This important concept is discussed extensively in calculus and also in most other areas of mathematics.

The Function Concept

If you spend $10 on gasoline, then the price per gallon determines the number of gallons that you get. There is a rule: the number of gallons is $10 divided by the price per gallon. The number of hours that you sleep before a test might be related to your grade on the test, but does not determine your grade. There is no rule that will determine your grade from the number of hours of sleep. If the value of a variable y is determined by the value of another variable x, then y **is a function of** x. The phrase "is a function of" means "is determined by." If there is more than one value for y corresponding to a particular x-value, then y is not determined by x and y is not a function of x.

Example **1** Using the phrase "is a function of"

Decide whether a is a function of b, b is a function of a, or neither.

a. Let a represent a positive integer smaller than 100 and b represent the number of divisors of a.

b. Let a represent the age of a United States citizen and b represent the number of days since his/her birth.

c. Let a represent the age of a United States citizen and b represent his/her annual income.

Solution

a. We can determine the number of divisors of any positive integer smaller than 100. So b is a function of a. We cannot determine the integer knowing the number of its divisors, because different integers have the same number of divisors. So a is not a function of b.

b. The number of days since a person's birth certainly determines the age of the person in the usual way. So a is a function of b. However, you cannot determine the number of days since a person's birth from their age. You need more information. For example, the number of days since birth for two 1-year-olds could be 370 and 380 days. So b is not a function of a.

c. We cannot determine the income from the age or the age from the income. We would need more information. Even though age and income are related, the relationship is not strong enough to say that either one is a function of the other. ■

To make the function concept clearer, we now define the noun "function." A *function* is often defined as a rule that assigns each element in one set to a unique element in a second set. Of course understanding this definition depends on knowing the meanings of the words *rule, assigns,* and *unique.* Using the language of ordered pairs we can make the following precise definition in which there are no undefined words.

Definition: Function

> A **function** is a set of ordered pairs in which no two ordered pairs have the same first coordinate and different second coordinates.

If we start with two related variables, we can identify one as the first variable and the other as the second variable and consider the set of ordered pairs containing their corresponding values. If the set of ordered pairs satisfies the function definition, then we say that the second variable is a function of the first. The variable corresponding to the first coordinate is the **independent variable** and the variable corresponding to the second coordinate is the **dependent variable.**

Identifying Functions

A **relation** is a set of ordered pairs. A relation can also be indicated by a verbal description, a graph, a formula or equation, or a table. Not every relation is a function. A function is a special relation. No matter how a relation or function is given there is always an underlying set of ordered pairs.

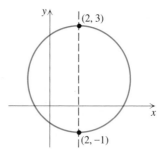

Figure 1.57

When a relation is given by a graph, we can visually check whether there are two ordered pairs with the same first coordinate and different second coordinates. For example, the circle shown in Fig. 1.57 is not the graph of a function, because there are two points on the circle with the same first coordinate. These points lie on a vertical line. In general, if there is a vertical line that crosses a graph more than once, the graph is not the graph of a function. This criterion is known as the **vertical line test.**

Theorem:
The Vertical Line Test

> A graph is the graph of a function if and only if there is no vertical line that crosses the graph more than once.

Every nonvertical line is the graph of a function, because every vertical line crosses a nonvertical line exactly once. Note that the vertical line test makes sense only because we always put the independent variable on the horizontal axis.

Example **2** **Identifying a function from a graph**

Determine which of the graphs shown in Fig. 1.58 are graphs of functions.

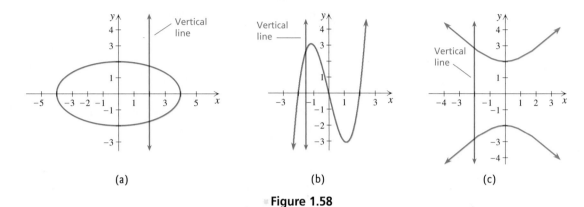

(a) (b) (c)

▪**Figure 1.58**

Solution

For each graph we want to decide whether y is a function of x. That is, can y be uniquely determined from x? In parts (a) and (c) we can draw a vertical line that crosses the graph more than once. So in each case there is an x-coordinate that corresponds to two different y-coordinates. So in parts (a) and (c) we cannot always determine a unique value of y from a given x-coordinate. So y is not a function of x in parts (a) and (c). The graph in Fig. 1.58(b) is the graph of a function because every vertical line appears to cross the graph at most once. So in this case we can determine y for a given x-coordinate and y is a function of x. ◼

A calculator is a virtual function machine. Built-in functions on a calculator are marked with symbols such as $\sqrt{x}, x^2, x!, 10^x, e^x, \ln(x), \sin(x), \cos(x)$, etc. When you provide an x-coordinate and use one of these symbols, the calculator finds the appropriate y-coordinate. The ordered pairs certainly satisfy the definition of function because the calculator will not produce two different second coordinates corresponding to one first coordinate.

In the next example we determine whether a relation is a function when the relation is given as a list of ordered pairs or as a table.

Example **3** **Identifying a function from a list or table**

Determine whether each relation is a function.

a. $\{(1, 3), (2, 3), (4, 9)\}$ **b.** $\{(9, -3), (9, 3), (4, 2), (0, 0)\}$

c.

Quantity	Price each
1–5	$9.40
5–10	$8.75

Solution

a. This set of ordered pairs is a function because no two ordered pairs have the same first coordinate and different second coordinates. The second coordinate is a function of the first coordinate.

b. This set of ordered pairs is not a function because both $(9, 3)$ and $(9, -3)$ are in the set and they have the same first coordinate and different second coordinates. The second coordinate is not a function of the first coordinate.

c. The quantity 5 corresponds to a price of $9.40 and also to a price of $8.75. Assuming that quantity is the first coordinate, the ordered pairs (5, $9.40) and (5, $8.75) both belong to this relation. If you are purchasing items whose price was determined from this table, you would certainly say that something is wrong, the table has a mistake in it, or the price you should pay is not clear. The price is not a function of the quantity purchased. ▪

We have seen and used many functions as formulas. For example, the formula $c = \pi d$ defines a set of ordered pairs in which the first coordinate is the diameter of a circle and the second coordinate is the circumference. Since each diameter corresponds to a unique circumference, the circumference is a function of the diameter. If the set of ordered pairs satisfying an equation is a function, then we say that the equation is a function or the equation defines a function. Other well-known formulas such as $C = \frac{5}{9}(F - 32)$, $A = \pi r^2$, and $V = \frac{4}{3}\pi r^3$ are also functions, but, as we will see in the next example, not every equation defines a function. The variables in the next example and all others in this text represent real numbers unless indicated otherwise.

Example **4** **Identifying a function from an equation**

Determine whether each equation defines y as a function of x.

a. $|y| = x$ **b.** $y = x^2 - 3x + 2$ **c.** $x^2 + y^2 = 1$ **d.** $3x - 4y = 8$

Solution

a. We must determine whether there are any values of x for which there is more than one y-value satisfying the equation. We arbitrarily select a number for x and see. If we select $x = 2$, then the equation is $|y| = 2$, which is satisfied if $y = \pm 2$. So both $(2, 2)$ and $(2, -2)$ satisfy $|y| = x$. So this equation does *not* define y as a function of x.

b. If we select any number for x, then y is calculated by the equation $y = x^2 - 3x + 2$. Since there is only one result when $x^2 - 3x + 2$ is calculated, there is only one y for any given x. So this equation *does* define y as a function of x.

c. Is it possible to pick a number for x for which there is more than one y-value? If we select $x = 0$, then the equation is $0^2 + y^2 = 1$ or $y = \pm 1$. So $(0, 1)$ and $(0, -1)$ both satisfy $x^2 + y^2 = 1$ and this equation does *not* define y as a function of x. Note that $x^2 + y^2 = 1$ is equivalent to $y = \pm\sqrt{1 - x^2}$, which indicates that there are many values for x that would produce two different y-coordinates.

d. The equation $3x - 4y = 8$ is equivalent to $y = \frac{3}{4}x - 2$. Since there is only one result when $\frac{3}{4}x - 2$ is calculated, there is only one y corresponding to any given x. So $3x - 4y = 8$ *does* define y as a function of x. ▪

Domain and Range

A relation is a set of ordered pairs. The **domain** of a relation is the set of all first coordinates of the ordered pairs. The **range** of a relation is the set of all second coordinates of the ordered pairs. The relation

$$\{(19, 2.4), (27, 3.0), (19, 3.6), (22, 2.4), (36, 3.8)\}$$

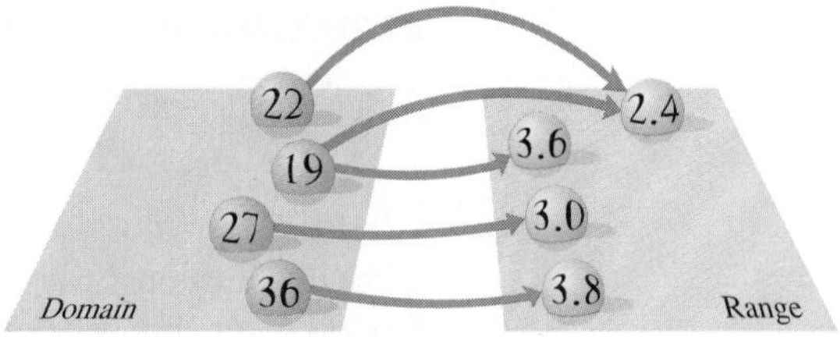

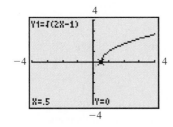

Figure 1.59

shows the ages and grade point averages of five randomly selected students. The domain of this relation is the set of ages, {19, 22, 27, 36}. The range is the set of grade point averages, {2.4, 3.0, 3.6, 3.8}. This relation matches elements of the domain (ages) with elements of the range (grade point averages), as shown in Fig. 1.59. Is this relation a function?

For some relations, all of the ordered pairs are listed, but for others, only an equation is given for determining the ordered pairs. When the domain of the relation is not stated, it is understood that the domain consists of only values of the independent variable that can be used in the expression defining the relation. When we use x and y for the variables, we always assume that x is the independent variable and y is the dependent variable.

Example 5 Determining domain and range

State the domain and range of each relation and whether the relation is a function.

a. $\{(-1, 1), (3, 9), (3, -9)\}$ **b.** $y = \sqrt{2x - 1}$ **c.** $x = |y|$

Solution

a. The domain is the set of first coordinates $\{-1, 3\}$, and the range is the set of second coordinates $\{1, 9, -9\}$. Note that an element of a set is not listed more than once. Since $(3, 9)$ and $(3, -9)$ are in the relation, the relation is not a function.

b. Since y is determined uniquely from x by the formula $y = \sqrt{2x - 1}$, y is a function of x. We are discussing only real numbers. So $\sqrt{2x - 1}$ is a real number provided that $2x - 1 \geq 0$, or $x \geq 1/2$. So the domain of the function is the interval $[1/2, \infty)$. If $2x - 1 \geq 0$ and $y = \sqrt{2x - 1}$, we have $y \geq 0$. So the range of the function is the interval $[0, \infty)$.

The graph of $y = \sqrt{2x - 1}$ shown in Fig. 1.60 supports these answers because the points that are plotted appear to have $x \geq 1/2$ and $y \geq 0$. □

c. The expression $|y|$ is defined for any real number y. So the range is the interval of all real numbers, $(-\infty, \infty)$. Since $|y|$ is nonnegative, the values of x must be nonnegative. So the domain is $[0, \infty)$. Since ordered pairs such as $(2, 2)$ and $(2, -2)$ satisfy $x = |y|$, this equation does not give y as a function of x. ■

Y1=√(2X-1)

X=.5   Y=0

Figure 1.60

In Example 5(b) and (c) we found the domain and range by examining an equation defining a relation. If the relation is a function as in Example 5(b), you can easily draw the graph with a graphing calculator and use it to support your answer. However, to choose an appropriate viewing window you must know the domain and range to begin with. So it is best to use a calculator graph to support your conclusions about domain and range rather than to make conclusions about domain and range. □

Function Notation

A function defined by a set of ordered pairs can be named with a letter. For example,

$$f = \{(2, 5), (3, 8)\}.$$

Since the function f pairs 2 with 5 we write $f(2) = 5$, which is read as "the value of f at 2 is 5" or simply "f of 2 is 5." We also have $f(3) = 8$.

A function defined by an equation can also be named with a letter. For example, the function $y = x^2$ could be named by a new letter, say g. We can then use $g(x)$, read "g of x" as a symbol for the second coordinate when the first coordinate is x. Since y and $g(x)$ are both symbols for the second coordinate we can write $y = g(x)$ and $g(x) = x^2$. Since $3^2 = 9$, the function g pairs 3 with 9 and we write $g(3) = 9$. This notation is called **function notation.**

Example 6 Using function notation

Let $h = \{(1, 4), (6, 0), (7, 9)\}$ and $f(x) = \sqrt{x - 3}$. Find each of the following.

a. $h(7)$ **b.** w, if $h(w) = 0$ **c.** $f(7)$ **d.** x, if $f(x) = 5$

Solution

a. The expression $h(7)$ is the second coordinate when the first coordinate is 7 in the function named h. So $h(7) = 9$.
b. We are looking for a number w for which $h(w) = 0$. That is, the second coordinate is 0 for some unknown first coordinate w. By examining the function h we see that $w = 6$.
c. To find $f(7)$ replace x by 7 in $f(x) = \sqrt{x - 3}$:

$$f(7) = \sqrt{7 - 3} = \sqrt{4} = 2$$

d. To find x for which $f(x) = 5$ we replace $f(x)$ by 5 in $f(x) = \sqrt{x - 3}$:

$$5 = \sqrt{x - 3}$$
$$25 = x - 3$$
$$28 = x$$ ■

Function notation such as $f(x) = 3x + 1$ provides a rule for finding the second coordinate: Multiply the first coordinate (whatever it is) by 3 and then add 1. The x in this notation is called a **dummy variable** because the letter used is unimportant. We could write $f(t) = 3t + 1$,

$$f(\text{first coordinate}) = 3(\text{first coordinate}) + 1,$$

or even $f(\ ) = 3(\ ) + 1$ to convey the same idea. Whatever appears in the parentheses following f must be used in place of x on the other side of the equation.

Example 7 Using function notation with variables

Given that $f(x) = x^2 - 2$ and $g(x) = 2x - 3$, find and simplify each of the following expressions.

a. $f(a)$ **b.** $f(a + 1)$ **c.** $f(x + h) - f(x)$ **d.** $g(x - 2)$ **e.** $g(x + h) - g(x)$

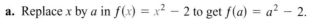

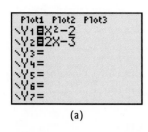

(a)

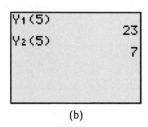

(b)

▪ **Figure 1.61**

Solution

a. Replace x by a in $f(x) = x^2 - 2$ to get $f(a) = a^2 - 2$.

b. $f(a + 1) = (a + 1)^2 - 2$ Replace x by $a + 1$ in $f(x) = x^2 - 2$.

$$= a^2 + 2a - 1$$

c. $f(x + h) - f(x) = (x + h)^2 - x^2$ Replace x with $x + h$ to get $f(x + h)$.

$$= x^2 + 2hx + h^2 - x^2$$

$$= 2hx + h^2$$

d. $g(x - 2) = 2(x - 2) - 3$

$$= 2x - 7$$

e. $g(x + h) - g(x) = 2(x + h) - 3 - (2x - 3)$

$$= 2x + 2h - 3 - 2x + 3$$

$$= 2h$$ ■

A graphing calculator uses subscripts to indicate different functions. For example, if $y_1 = x^2 - 2$ and $y_2 = 2x - 3$, then $y_1(5) = 23$ and $y_2(5) = 7$ as shown in Fig. 1.61(a) and (b). ☐

If a function describes some real application, then a letter that fits the situation is usually used. For example, if watermelons are $3 each, then the cost of x watermelons is given by the function $C(x) = 3x$. The cost of five watermelons is $C(5) = 3 \cdot 5 = \$15$. In trigonometry the abbreviations sin, cos, and tan are used rather than a single letter to name the trigonometric functions. The dependent variables are written as $\sin(x)$, $\cos(x)$, and $\tan(x)$.

The Average Rate of Change of a Function

In Section 1.4 we defined the slope of the line through (x_1, y_1) and (x_2, y_2) as $\frac{y_2 - y_1}{x_2 - x_1}$. We now extend that idea to any function (linear or not).

Definition: Average Rate of Change from x_1 to x_2

If (x_1, y_1) and (x_2, y_2) are two ordered pairs of a function, we define the **average rate of change** of the function as x varies from x_1 to x_2 as

$$\frac{y_2 - y_1}{x_2 - x_1}.$$

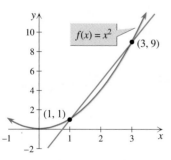

▪ **Figure 1.62**

Note that the x-values can be specified with interval notation. For example, the average rate of change $f(x) = x^2$ on $[1, 3]$ is found as follows:

$$\frac{f(3) - f(1)}{3 - 1} = \frac{9 - 1}{3 - 1} = 4$$

The average rate of change is simply the slope of the line that passes through two points on the graph of the function as shown in Fig. 1.62.

It is not necessary to have a formula for a function to find an average rate of change, as is shown in the next example.

Example **8** **Finding the average rate of change**

The population of California was 29.8 million in 1990 and 33.9 million in 2000 (U.S. Census Bureau, www.census.gov). What was the average rate of change of the population over that time interval?

Solution

The population is a function of the year. The average rate of change of the population is the change in population divided by the change in time:

$$\frac{33.9 - 29.8}{2000 - 1990} = \frac{4.1}{10} = 0.41$$

The average rate of change of the population was 0.41 million people/year or 410,000 people/year. Note that 410,000 people/year is an average and that the population did not actually increase by 410,000 every year. ■

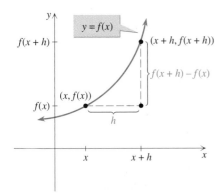

Figure 1.63

The expression $\dfrac{f(x + h) - f(x)}{h}$ is called the **difference quotient.** The difference quotient is the average rate of change between two points that are labeled as shown in Fig. 1.63. In calculus it is often necessary to find and simplify the difference quotient for a function.

Example **9** **Finding a difference quotient**

Find and simplify the difference quotient for each of the following functions.

a. $j(x) = 3x + 2$ **b.** $f(x) = x^2 - 2x$ **c.** $g(x) = \sqrt{x}$ **d.** $y = \dfrac{5}{x}$

Solution

a.
$$\frac{j(x + h) - j(x)}{h} = \frac{3(x + h) + 2 - (3x + 2)}{h}$$

$$= \frac{3x + 3h + 2 - 3x - 2}{h}$$

$$= \frac{3h}{h} = 3$$

■ **Foreshadowing Calculus**

The difference quotient or average rate of change of a function is the rate of change of y over some interval of x-values. In calculus this idea is extended to the instantaneous rate of change, which is the rate of change of y at a single value of x.

b.
$$\frac{f(x + h) - f(x)}{h} = \frac{[(x + h)^2 - 2(x + h)] - (x^2 - 2x)}{h}$$

$$= \frac{x^2 + 2xh + h^2 - 2x - 2h - x^2 + 2x}{h}$$

$$= \frac{2xh + h^2 - 2h}{h}$$

$$= 2x + h - 2$$

c. $\dfrac{g(x + h) - g(x)}{h} = \dfrac{\sqrt{x + h} - \sqrt{x}}{h}$

$$= \dfrac{\left(\sqrt{x + h} - \sqrt{x}\right)\left(\sqrt{x + h} + \sqrt{x}\right)}{h\left(\sqrt{x + h} + \sqrt{x}\right)} \quad \text{Rationalize the numerator.}$$

$$= \dfrac{x + h - x}{h(\sqrt{x + h} + \sqrt{x})}$$

$$= \dfrac{1}{\sqrt{x + h} + \sqrt{x}}$$

d. Use the function notation $f(x) = \frac{5}{x}$ for the function $y = \frac{5}{x}$:

$$\dfrac{f(x + h) - f(x)}{h} = \dfrac{\dfrac{5}{x + h} - \dfrac{5}{x}}{h} = \dfrac{\left(\dfrac{5}{x + h} - \dfrac{5}{x}\right)x(x + h)}{h \cdot x(x + h)}$$

$$= \dfrac{5x - 5(x + h)}{hx(x + h)} = \dfrac{-5h}{hx(x + h)} = \dfrac{-5}{x(x + h)} \qquad ■$$

Note that in Examples 9(b), 9(c), and 9(d) the average rate of change of the function depends on the values of x and h, while in Example 9(a) the average rate of change of the function is constant. In Example 9(c) the expression does not look much different after rationalizing the numerator than it did before. However, we did remove h as a factor of the denominator, and in calculus it is often necessary to perform this step.

Constructing Functions

In the next example we find a formula for, or **construct,** a function relating two variables in a geometric figure.

Example **10** **Constructing a function**

Given that a square has diagonal of length d and side of length s, write the area A as a function of the length of the diagonal.

Solution

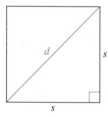

■ **Figure 1.64**

The area of any square is given by $A = s^2$. The diagonal is the hypotenuse of a right triangle as shown in Fig. 1.64. By the Pythagorean theorem, $d^2 = s^2 + s^2, d^2 = 2s^2$, or $s^2 = d^2/2$. Since $A = s^2$ and $s^2 = d^2/2$, we get the formula

$$A = \dfrac{d^2}{2}$$

expressing the area of the square as a function of the length of the diagonal. ■

For Thought

True or False? Explain.

1. Any set of ordered pairs is a function.

2. If $f = \{(1, 1), (2, 4), (3, 9)\}$, then $f(5) = 25$.

3. The domain of $f(x) = 1/x$ is $(-\infty, 0) \cup (0, \infty)$.

4. Each student's exam grade is a function of the student's IQ.

5. If $f(x) = x^2$, then $f(x + h) = x^2 + h$.

6. The domain of $g(x) = |x - 3|$ is $[3, \infty)$.

7. The range of $y = 8 - x^2$ is $(-\infty, 8]$.

8. The equation $x = y^2$ does not define y as a function of x.

9. If $f(t) = \dfrac{t - 2}{t + 2}$, then $f(0) = -1$.

10. The set $\left\{\left(\frac{3}{8}, 8\right), \left(\frac{4}{7}, 7\right), \left(0.16, 6\right), \left(\frac{3}{8}, 5\right)\right\}$ is a function.

1.5 Exercises

For each pair of variables determine whether a is a function of b, b is a function of a, or neither.

1. a is the radius of any U.S. coin and b is its circumference.

2. a is the length of any rectangle with a width of 5 in. and b is its perimeter.

3. a is the length of any piece of U.S. paper currency and b is its denomination.

4. a is the diameter of any U.S. coin and b is its value.

5. a is the universal product code for an item at Wal-Mart and b is its price.

6. a is the final exam score for a student in your class and b is his/her semester grade.

7. a is the time spent studying for the final exam for a student in your class and b is the student's final exam score.

8. a is the age of an adult male and b is his shoe size.

9. a is the height of a car in inches and b is its height in centimeters.

10. a is the cost for mailing a first-class letter and b is its weight.

Use the vertical line test on each graph in Exercises 11–16 to determine whether y is a function of x.

11.

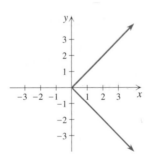

12.

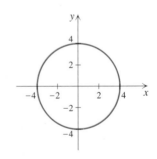

13.

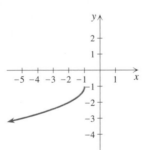

14.

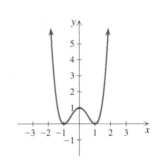

15.

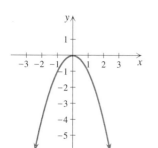

16.

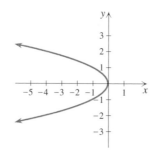

Determine whether each relation is a function.

17. $\{(-1, -1), (2, 2), (3, 3)\}$

18. $\{(0.5, 7), (0, 7), (1, 7), (9, 7)\}$

19. $\{(25, 5), (25, -5), (0, 0)\}$

20. $\{(1, \pi), (30, \pi/2), (60, \pi/4)\}$

21.

x	y
3	6
4	9
3	12

22.

x	y
1	6.98
5	5.98
9	6.98

23.

x	y
-1	1
1	1
-5	1
5	1

24.

x	y
1	1
2	4
3	9
4	16

Determine whether each equation defines y as a function of x.

25. $y = 3x - 8$

26. $y = x^2 - 3x + 7$

27. $x = 3y - 9$

28. $x = y^3$

29. $x^2 = y^2$

30. $y^2 - x^2 = 9$

31. $x = \sqrt{y}$

32. $x = \sqrt[3]{y}$

33. $y + 2 = |x|$

34. $y - 1 = x^2$

35. $x = |2y|$

36. $x = y^2 + 1$

Determine the domain and range of each relation.

37. $\{(-3, 1), (4, 2), (-3, 6), (5, 6)\}$

38. $\{(1, 2), (2, 4), (3, 8), (4, 16)\}$

39. $\{(x, y) | y = 4\}$

40. $\{(x, y) | x = 5\}$

41. $y = |x| + 5$

42. $y = x^2 + 8$

43. $x + 3 = |y|$

44. $x + 2 = \sqrt{y}$

45. $y = \sqrt{x - 4}$

46. $y = \sqrt{5 - x}$

47. $x = -y^2$

48. $x = -|y|$

Let $f = \{(2, 6), (3, 8), (4, 5)\}$ and $g(x) = 3x + 5$. Find the following.

49. $f(2)$

50. $f(4)$

51. $g(2)$

52. $g(4)$

53. x, if $f(x) = 8$

54. x, if $f(x) = 6$

55. x, if $g(x) = 26$

56. x, if $g(x) = -4$

57. $f(4) + g(4)$

58. $f(3) - g(3)$

Let $f(x) = 3x^2 - x$ and $g(x) = 4x - 2$. Find the following.

59. $f(a)$

60. $f(w)$

61. $g(a + 2)$

62. $g(a - 5)$

63. $f(x + 1)$

64. $f(x - 3)$

65. $g(x + h)$

66. $f(x + h)$

67. $f(x + h) - f(x)$

68. $g(x + h) - g(x)$

The following problems involve average rate of change.

69. *Depreciation of a Mustang* If a new Mustang is valued at $16,000 and five years later it is valued at $4000, then what is the average rate of change of its value during those five years?

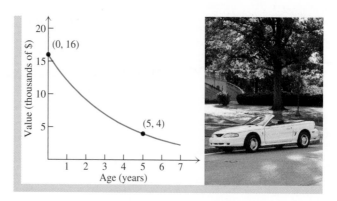

■ **Figure for Exercise 69**

70. *Cost of Gravel*   Wilson's Sand and Gravel will deliver 12 yd³ of gravel for $240, 30 yd³ for $528, and 60 yd³ for $948. What is the average rate of change of the cost as the number of cubic yards varies from 12 to 30? What is the average rate of change as the number of cubic yards varies from 30 to 60?

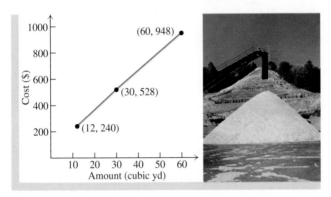

◼ **Figure for Exercise 70**

◼ Foreshadowing Calculus

In calculus we determine what happens to the average rate of change as the length of the interval for the independent variable approaches zero. You can get a hint of that process in the next exercise.

71. *Dropping a Watermelon*   If a comedian drops a watermelon from a height of 64 ft, then its height (in feet) above the ground is given by the function $h(t) = -16t^2 + 64$ where t is time (in seconds). To get an idea of how fast the watermelon is traveling when it hits the ground find the average rate of change of the height on each of the time intervals $[0, 2]$, $[1, 2]$, $[1.9, 2]$, $[1.99, 2]$, and $[1.999, 2]$.

72. *Bungee Jumping*   Billy Joe McCallister jumped off the Tallahatchie Bridge, 70 ft above the water, with a bungee cord tied to his legs. If he was 6 ft above the water 2 sec after jumping, then what was the average rate of change of his altitude as the time varied from 0 to 2 sec?

Find the difference quotient $\dfrac{f(x + h) - f(x)}{h}$ *for each function and simplify it.*

73. $f(x) = 4x$

74. $f(x) = \dfrac{1}{2}x$

75. $f(x) = 3x + 5$

76. $f(x) = -2x + 3$

77. $y = x^2 + x$

78. $y = x^2 - 2x$

79. $y = -x^2 + x - 2$

80. $y = x^2 - x + 3$

81. $g(x) = 3\sqrt{x}$

82. $g(x) = -2\sqrt{x}$

83. $f(x) = \sqrt{x + 2}$

84. $f(x) = \sqrt{\dfrac{x}{2}}$

85. $g(x) = \dfrac{1}{x}$

86. $g(x) = \dfrac{3}{x}$

87. $g(x) = \dfrac{3}{x + 2}$

88. $g(x) = 3 + \dfrac{2}{x - 1}$

Solve each problem.

89. *Constructing Functions*   Consider a square with side of length s, diagonal of length d, perimeter P, and area A.
  a. Write A as a function of s.   **b.** Write s as a function of A.
  c. Write s as a function of d.   **d.** Write d as a function of s.
  e. Write P as a function of s.   **f.** Write s as a function of P.
  g. Write A as a function of P.   **h.** Write d as a function of A.

90. *Constructing Functions*   Consider a circle with area A, circumference C, radius r, and diameter d.
  a. Write A as a function of r.   **b.** Write r as a function of A.
  c. Write C as a function of r.   **d.** Write d as a function of r.
  e. Write d as a function of C.   **f.** Write A as a function of d.
  g. Write d as a function of A.

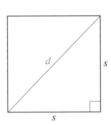

◼ **Figure for Exercise 89**   ◼ **Figure for Exercise 90**

91. *Below Sea Level*   The accompanying table shows the depth below sea level d and the atmospheric pressure A (www.sportsfigures.espn.com). The equation $A(d) = 0.03d + 1$ expresses A as a function of d.
  a. Find the atmospheric pressure for a depth of 100 ft, where nitrogen narcosis begins.

  b. Find the depth at which the pressure is 4.9 atm, the maximum depth for intermediate divers.

Table for Exercise 91

Depth (ft)	Atmospheric Pressure (atm)	
21	1.63	
60	2.8	
100		
	4.9	
200	7.0	
250	8.5	

92. *Computer Spending*   The amount spent online for computers in the year $2000 + n$ can be modeled by the function $C(n) = 0.95n + 5.8$ where n is a whole number and $C(n)$ is billions of dollars.

   a. What does $C(4)$ represent and what is it?

   b. Find the year in which online spending for computers will reach $15 billion?

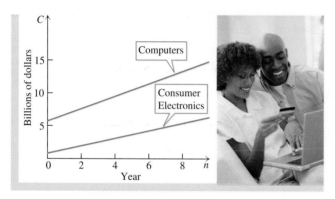

Figure for Exercise 92

93. *Pile of Pipes*   Six pipes, each with radius a, are stacked as shown in the accompanying figure. Construct a function that gives the height h of the pile in terms of a.

Figure for Exercise 93

94. *Angle Bisectors*   The angle bisectors of any triangle meet at a single point. Let a be the length of the hypotenuse of a 30-60-90 triangle and d be the distance from the vertex of the right angle to the point where the angle bisectors meet. Write d as function of a.

Thinking Outside the Box V

Lucky Lucy   Lucy's teacher asked her to evaluate $(20 + 25)^2$. As she was trying to figure out what to do she mumbled, "twenty twenty-five." Her teacher said, "Good, 2025 is correct." Find another pair of two digit whole numbers for which the square of their sum can be found by Lucy's method.

1.5 Pop Quiz

1. Is the radius of a circle a function of its area?

2. Is $\{(2, 4), (1, 8), (2, -4)\}$ a function?

3. Does $x^2 + y^2 = 1$ define y as a function of x?

4. What is the domain of $y = \sqrt{x - 1}$?

5. What is the range of $y = x^2 + 2$?

6. What is $f(2)$ if $f = \{(1, 8), (2, 9)\}$?

7. What is a if $f(a) = 1$ and $f(x) = 2x$?

8. If the cost was $20 in 1998 and $40 in 2008, then what is the average rate of change of the cost for that time period?

9. Find and simplify the difference quotient for $f(x) = x^2 + 3$.

1.6 | Graphs of Relations and Functions

When we graph the set of ordered pairs that satisfy an equation, we are combining algebra with geometry. We saw in Section 1.3 that the graph of any equation of the form $(x - h)^2 + (y - k)^2 = r^2$ is a circle and that the graph of any equation of the form $Ax + By = C$ is a straight line. In this section we will see that graphs of equations have many different geometric shapes.

Graphing Equations

The circle and the line provide nice examples of how algebra and geometry are interrelated. When you see an equation that you recognize as the equation of a circle or a line, sketching a graph is easy to do. Other equations have graphs that are not such familiar shapes. Until we learn to recognize the kinds of graphs that other equations have, we graph other equations by calculating enough ordered pairs to determine the shape of the graph. When you graph equations, try to anticipate what the graph will look like, and after the graph is drawn, pause to reflect on the shape of the graph and the type of equation that produced it. You might wish to look ahead to the Function Gallery on page 79, which shows the basic functions that we will be studying.

△▽ Of course a graphing calculator can speed up this process. Remember that a graphing calculator shows only finitely many points and a graph usually consists of infinitely many points. After looking at the display of a graphing calculator, you must still decide what the entire graph looks like. □

Example **1** The square function

Graph the equation $y = x^2$ and state the domain and range. Determine whether the relation is a function.

Solution

Make a table of ordered pairs that satisfy $y = x^2$:

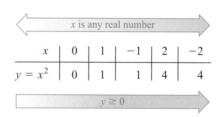

| | x is any real number | | | | |

x	0	1	−1	2	−2
$y = x^2$	0	1	1	4	4

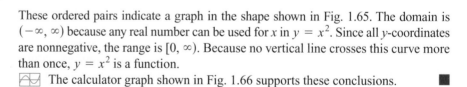

$y \geq 0$

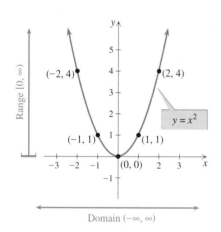

Figure 1.65

These ordered pairs indicate a graph in the shape shown in Fig. 1.65. The domain is $(-\infty, \infty)$ because any real number can be used for x in $y = x^2$. Since all y-coordinates are nonnegative, the range is $[0, \infty)$. Because no vertical line crosses this curve more than once, $y = x^2$ is a function.

△▽ The calculator graph shown in Fig. 1.66 supports these conclusions. ■

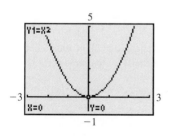

Figure 1.66

The graph of $y = x^2$ is called a **parabola.** We study parabolas in detail in Chapter 2. The graph of the **square-root function** $y = \sqrt{x}$ is half of a parabola.

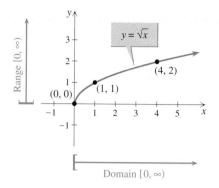

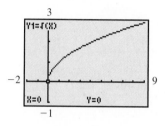

■ **Figure 1.67**

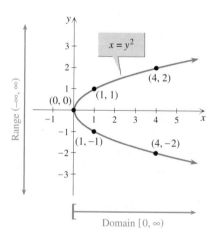

■ **Figure 1.68**

Example **2** **The square-root function**

Graph $y = \sqrt{x}$ and state the domain and range of the relation. Determine whether the relation is a function.

Solution

Make a table listing ordered pairs that satisfy $y = \sqrt{x}$:

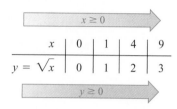

Plotting these ordered pairs suggests the graph shown in Fig. 1.67. The domain of the relation is $[0, \infty)$ and the range is $[0, \infty)$. Because no vertical line can cross this graph more than once, $y = \sqrt{x}$ is a function.

The calculator graph in Fig. 1.68 supports these conclusions. ■

In the next example we graph $x = y^2$ and see that its graph is also a parabola.

Example **3** **A parabola opening to the right**

Graph $x = y^2$ and state the domain and range of the relation. Determine whether the relation is a function.

Solution

Make a table listing ordered pairs that satisfy $x = y^2$. In this case choose y and calculate x:

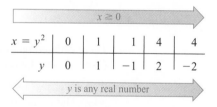

Note that these ordered pairs are the same ones that satisfy $y = x^2$ except that the coordinates are reversed. For this reason the graph of $x = y^2$ in Fig. 1.69 has the same shape as the parabola in Fig. 1.65 and it is also a parabola. The domain of $x = y^2$ is $[0, \infty)$ and the range is $(-\infty, \infty)$. Because we can draw a vertical line that crosses this parabola twice, $x = y^2$ does not define y as a function of x. (Of course, $x = y^2$ does express x as a function of y.) ■

Because $x = y^2$ is equivalent to $y = \pm\sqrt{x}$, the top half of the graph of $x = y^2$ is $y = \sqrt{x}$ and the bottom half is $y = -\sqrt{x}$.

To support these conclusions with a graphing calculator, graph $y_1 = \sqrt{x}$ and $y_2 = -\sqrt{x}$ as shown in Fig. 1.70. □

■ **Figure 1.69**

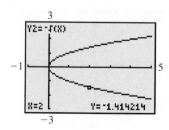

■ **Figure 1.70**

In the next example we graph the **cube function** $y = x^3$ and the **cube-root function** $y = \sqrt[3]{x}$.

Example 4 Cube and cube-root functions

Graph each equation. State the domain and range.

a. $y = x^3$ **b.** $y = \sqrt[3]{x}$

Solution

a. Make a table of ordered pairs that satisfy $y = x^3$:

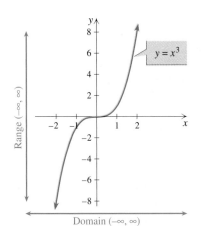

Figure 1.71

		x is any real number			
x	-2	-1	0	1	2
$y = x^3$	-8	-1	0	1	8
		y is any real number			

These ordered pairs indicate a graph in the shape shown in Fig. 1.71. The domain is $(-\infty, \infty)$ and the range is $(-\infty, \infty)$. By the vertical line test, this graph is the graph of a function because no vertical line crosses the curve more than once.

b. Make a table of ordered pairs that satisfy $y = \sqrt[3]{x}$. Note that this table is simply the table for $y = x^3$ with the coordinates reversed.

		x is any real number			
x	-8	-1	0	1	8
$y = \sqrt[3]{x}$	-2	-1	0	1	2
		y is any real number			

These ordered pairs indicate a graph in the shape shown in Fig. 1.72. The domain is $(-\infty, \infty)$ and the range is $(-\infty, \infty)$. Because no vertical line crosses this graph more than once, $y = \sqrt[3]{x}$ is a function.

The calculator graph in Fig. 1.73 supports these conclusions.

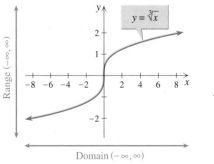

Figure 1.72

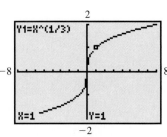

Figure 1.73

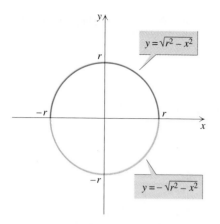

Figure 1.74

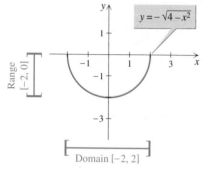

Figure 1.75

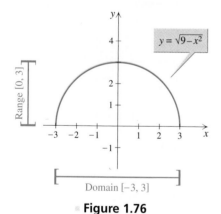

Figure 1.76

Semicircles

The graph of $x^2 + y^2 = r^2$ $(r > 0)$ is a circle centered at the origin of radius r. The circle does not pass the vertical line test, and a circle is not the graph of a function. We can find an equivalent equation by solving for y:

$$x^2 + y^2 = r^2$$
$$y^2 = r^2 - x^2$$
$$y = \pm\sqrt{r^2 - x^2}$$

The equation $y = \sqrt{r^2 - x^2}$ does define y as a function of x. Because y is nonnegative in this equation, the graph is the top semicircle in Fig. 1.74. The top semicircle passes the vertical line test. Likewise, the equation $y = -\sqrt{r^2 - x^2}$ defines y as a function of x, and its graph is the bottom semicircle in Fig. 1.74.

Example 5 Graphing a semicircle

Sketch the graph of each function and state the domain and range of the function.

a. $y = -\sqrt{4 - x^2}$ **b.** $y = \sqrt{9 - x^2}$

Solution

a. Rewrite the equation in the standard form for a circle:

$$y = -\sqrt{4 - x^2}$$
$$y^2 = 4 - x^2 \qquad \text{Square each side.}$$
$$x^2 + y^2 = 4 \qquad \text{Standard form for the equation of a circle.}$$

The graph of $x^2 + y^2 = 4$ is a circle of radius 2 centered at $(0, 0)$. Since y must be negative in $y = -\sqrt{4 - x^2}$, the graph of $y = -\sqrt{4 - x^2}$ is the semicircle shown in Fig. 1.75. We can see from the graph that the domain is $[-2, 2]$ and the range is $[-2, 0]$.

b. Rewrite the equation in the standard form for a circle:

$$y = \sqrt{9 - x^2}$$
$$y^2 = 9 - x^2 \qquad \text{Square each side.}$$
$$x^2 + y^2 = 9$$

The graph of $x^2 + y^2 = 9$ is a circle with center $(0, 0)$ and radius 3. But this equation is not equivalent to the original. The value of y in $y = \sqrt{9 - x^2}$ is nonnegative. So the graph of the original equation is the semicircle shown in Fig. 1.76. We can read the domain $[-3, 3]$ and the range $[0, 3]$ from the graph.   ■

Piecewise Functions

For some functions, different formulas are used in different regions of the domain. Such functions are called **piecewise functions.** The simplest example of such a function is the **absolute value function** $f(x) = |x|$, which can be written as

$$f(x) = \begin{cases} x & \text{for } x \geq 0 \\ -x & \text{for } x < 0. \end{cases}$$

For $x \geq 0$ the equation $f(x) = x$ is used to obtain the second coordinate, and for $x < 0$ the equation $f(x) = -x$ is used. The graph of the absolute value function is shown in the next example.

Example 6   The absolute value function

Graph the equation $y = |x|$ and state the domain and range. Determine whether the relation is a function.

Solution

Make a table of ordered pairs that satisfy $y = |x|$:

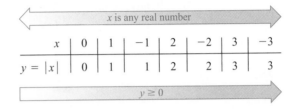

x	0	1	-1	2	-2	3	-3
$y = \lvert x \rvert$	0	1	1	2	2	3	3

These ordered pairs suggest the V-shaped graph of Fig. 1.77. Because no vertical line can cross this graph more than once, $y = |x|$ is a function. The domain is $(-\infty, \infty)$ and the range is $[0, \infty)$.

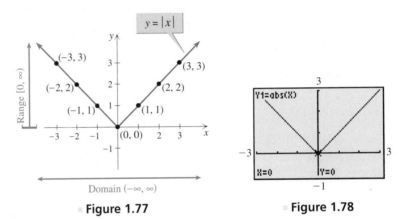

Figure 1.77   **Figure 1.78**

To support these conclusions with a graphing calculator, graph $y_1 = \text{abs}(x)$ as shown in Fig. 1.78. ■

In the next example we graph two more piecewise functions.

Example 7   Graphing a piecewise function

Sketch the graph of each function and state the domain and range.

a. $f(x) = \begin{cases} 1 & \text{for } x < 0 \\ -1 & \text{for } x \geq 0 \end{cases}$ **b.** $f(x) = \begin{cases} x^2 - 4 & \text{for } -2 \leq x \leq 2 \\ x - 2 & \text{for } x > 2 \end{cases}$

Solution

a. For $x < 0$ the graph is the horizontal line $y = 1$. For $x \geq 0$ the graph is the horizontal line $y = -1$. Note that $(0, -1)$ is on the graph shown in Fig. 1.79 but

(0, 1) is not, because when $x = 0$ we have $y = -1$. The domain is the interval $(-\infty, \infty)$ and the range consists of only two numbers, -1 and 1. The range is not an interval. It is written in set notation as $\{-1, 1\}$.

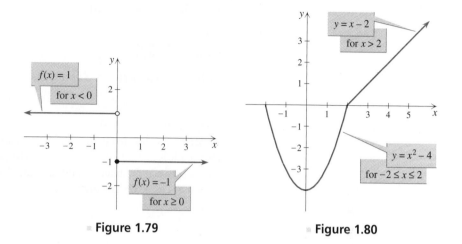

■ **Figure 1.79**   ■ **Figure 1.80**

b. Make a table of ordered pairs using $y = x^2 - 4$ for x between -2 and 2 and $y = x - 2$ for $x > 2$.

x	-2	-1	0	1	2
$y = x^2 - 4$	0	-3	-4	-3	0

x	2.1	3	4	5
$y = x - 2$	0.1	1	2	3

For x in the interval $[-2, 2]$ the graph is a portion of a parabola as shown in Fig. 1.80. For $x > 2$, the graph is a portion of a straight line through $(3, 1)$, $(4, 2)$, and $(5, 3)$. The domain is $[-2, \infty)$, and the range is $[-4, \infty)$.

Figure 1.81(a) shows how to use the inequality symbols from the TEST menu to enter a piecewise function on a calculator. The inequality $x \geq -2$ is not treated as a normal inequality, but instead the calculator gives it a value of 1 when it is satisfied and 0 when it is not satisfied. So, dividing $x^2 - 4$ by the inequalities $x \geq -2$ and $x \leq 2$ will cause $x^2 - 4$ to be graphed only when they are both satisfied and not graphed when they are not both satisfied. The calculator graph in Fig. 1.81(b) is consistent with the conclusions that we have made.

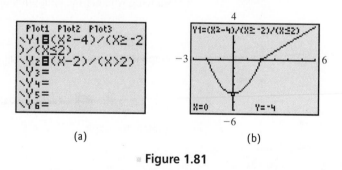

(a)   (b)

■ **Figure 1.81**   ■

Piecewise functions are often found in shipping charges. For example, if the weight in pounds of an order is in the interval $(0, 1]$, the shipping and handling charge is $3. If the weight is in the interval $(1, 2]$, the shipping and handling charge is $4, and so on. The next example is a function that is similar to a shipping and handling charge. This function is referred to as the **greatest integer function** and is written $f(x) = [\![x]\!]$ or $f(x) = \text{int}(x)$. The symbol $[\![x]\!]$ is defined to be the largest integer that is less than or equal to x. For example, $[\![5.01]\!] = 5$, because the greatest integer less than or equal to 5.01 is 5. Likewise, $[\![3.2]\!] = 3$, $[\![-2.2]\!] = -3$, and $[\![7]\!] = 7$.

Example 8 Graphing the greatest integer function

Sketch the graph of $f(x) = [\![x]\!]$ and state the domain and range.

Solution

For any x in the interval $[0, 1)$ the greatest integer less than or equal to x is 0. For any x in $[1, 2)$ the greatest integer less than or equal to x is 1. For any x in $[-1, 0)$ the greatest integer less than or equal to x is -1. The definition of $[\![x]\!]$ causes the function to be constant between the integers and to "jump" at each integer. The graph of $f(x) = [\![x]\!]$ is shown in Fig. 1.82. The domain is $(-\infty, \infty)$, and the range is the set of integers.

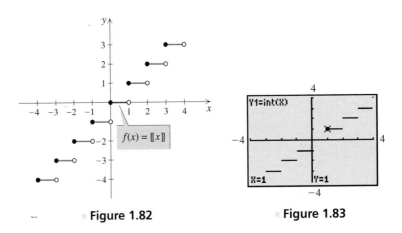

Figure 1.82   **Figure 1.83**

The calculator graph of $y_1 = \text{int}(x)$ looks best in dot mode as in Fig. 1.83, because in connected mode the calculator connects the disjoint pieces of the graph. The calculator graph in Fig. 1.82 supports our conclusion that the graph of this function looks like the one drawn in Fig. 1.82. Note that the calculator is incapable of showing whether the endpoints of the line segments are included. ■

In the next example we vary the form of the greatest integer function, but the graph is still similar to the graph in Fig. 1.82.

Example 9 A variation of the greatest integer function

Sketch the graph of $f(x) = [\![x - 2]\!]$ for $0 \le x \le 5$.

Solution

If $x = 0$, $f(0) = [\![-2]\!] = -2$. If $x = 0.5$, $f(0.5) = [\![-1.5]\!] = -2$. In fact, $f(x) = -2$ for any x in the interval $[0, 1)$. Similarly, $f(x) = -1$ for any x in the interval $[1, 2)$. This pattern continues with $f(x) = 2$ for any x in the interval $[4, 5)$, and $f(x) = 3$ for $x = 5$. The graph of $f(x) = [\![x - 2]\!]$ is shown in Fig. 1.84. ■

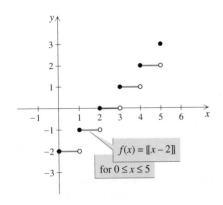

Figure 1.84

Increasing, Decreasing, and Constant

Imagine a point moving from left to right along the graph of a function. If the y-coordinate of the point is getting larger, getting smaller, or staying the same, then the function is said to be **increasing, decreasing,** or **constant,** respectively.

Example **10** Increasing, decreasing, or constant functions

Determine whether each function is increasing, decreasing, or constant by examining its graph.

a.

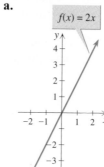

$f(x) = 2x$

b.

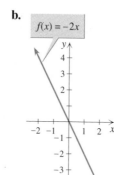

$f(x) = -2x$

c.

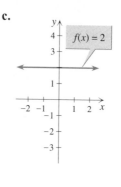

$f(x) = 2$

■ **Foreshadowing Calculus**

Whether a function is increasing, decreasing, or constant is determined analytically in calculus. Here the determination is made graphically.

Solution

If a point is moving left to right along the graph of $f(x) = 2x$, then its y-coordinate is increasing. So $f(x) = 2x$ is an increasing function. If a point is moving left to right along the graph of $f(x) = -2x$, then its y-coordinate is decreasing. So $f(x) = -2x$ is a decreasing function. The function $f(x) = 2$ has a constant y-coordinate and it is a constant function. ■

If the y-coordinates are getting larger, getting smaller, or staying the same when x is in an open interval, then the function is said to be increasing, decreasing, or constant, respectively, on that interval.

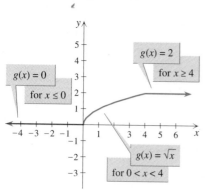

Figure 1.85

Example **11** Increasing, decreasing, or constant on an interval

Sketch the graph of each function and identify any open intervals on which the function is increasing, decreasing, or constant.

a. $f(x) = 4 - x^2$ **b.** $g(x) = \begin{cases} 0 & \text{for} & x \le 0 \\ \sqrt{x} & \text{for} & 0 < x < 4 \\ 2 & \text{for} & x \ge 4 \end{cases}$

Solution

a. The graph of $f(x) = 4 - x^2$ includes the points $(-2, 0)$, $(-1, 3)$, $(0, 4)$, $(1, 3)$, and $(2, 0)$. The graph is shown in Fig. 1.85. The function is increasing on the interval $(-\infty, 0)$ and decreasing on $(0, \infty)$.

b. The graph of g is shown in Fig. 1.86. The function g is constant on the intervals $(-\infty, 0)$ and $(4, \infty)$, and increasing on $(0, 4)$. ■

Figure 1.86

For Thought

True or False? Explain.

1. The range of $y = -x^2$ is $(-\infty, 0]$.

2. The function $y = -\sqrt{x}$ is increasing on $(0, \infty)$.

3. The function $f(x) = \sqrt[3]{x}$ is increasing on $(-\infty, \infty)$.

4. If $f(x) = [\![x + 3]\!]$, then $f(-4.5) = -2$.

5. The range of the function $f(x) = \dfrac{|x|}{x}$ is the interval $(-1, 1)$.

6. The range of $f(x) = [\![x - 1]\!]$ is the set of integers.

7. The only ordered pair that satisfies $(x - 5)^2 + (y + 6)^2 = 0$ is $(5, -6)$.

8. The domain of the function $y = \sqrt{4 - x^2}$ is the interval $[-2, 2]$.

9. The range of the function $y = \sqrt{16 - x^2}$ is the interval $[0, \infty)$.

10. The function $y = \sqrt{4 - x^2}$ is increasing on $(-2, 0)$ and decreasing on $(0, 2)$.

1.6 Exercises

Make a table listing ordered pairs that satisfy each equation. Then graph the equation. Determine the domain and range, and whether y is a function of x.

1. $y = 2x$

2. $x = 2y$

3. $x - y = 0$

4. $x - y = 2$

5. $y = 5$

6. $x = 3$

7. $y = 2x^2$

8. $y = x^2 - 1$

9. $y = 1 - x^2$

10. $y = -1 - x^2$

11. $y = 1 + \sqrt{x}$

12. $y = 2 - \sqrt{x}$

13. $x = y^2 + 1$

14. $x = 1 - y^2$

15. $x = \sqrt{y}$

16. $x - 1 = \sqrt{y}$

17. $y = \sqrt[3]{x} + 1$

18. $y = \sqrt[3]{x} - 2$

19. $x = \sqrt[3]{y}$

20. $x = \sqrt[3]{y} - 1$

21. $y^2 = 1 - x^2$

22. $x^2 + y^2 = 4$

23. $y = \sqrt{1 - x^2}$

24. $y = -\sqrt{25 - x^2}$

25. $y = x^3 + 1$

26. $y = -x^3$

27. $y = 2|x|$

28. $y = |x - 1|$

29. $y = -|x|$

30. $y = -|x + 1|$

31. $x = |y|$

32. $x = |y| + 1$

Make a table listing ordered pairs for each function. Then sketch the graph and state the domain and range.

33. $f(x) = \begin{cases} 2 & \text{for } x < -1 \\ -2 & \text{for } x \geq -1 \end{cases}$

34. $f(x) = \begin{cases} 3 & \text{for } x < 2 \\ 1 & \text{for } x \geq 2 \end{cases}$

35. $f(x) = \begin{cases} x + 1 & \text{for } x > 1 \\ x - 3 & \text{for } x \leq 1 \end{cases}$

36. $f(x) = \begin{cases} 5 - x & \text{for } x \leq 2 \\ x + 1 & \text{for } x > 2 \end{cases}$

37. $f(x) = \begin{cases} \sqrt{x + 2} & \text{for } -2 \leq x \leq 2 \\ 4 - x & \text{for } x > 2 \end{cases}$

38. $f(x) = \begin{cases} \sqrt{x} & \text{for } x \geq 1 \\ -x & \text{for } x < 1 \end{cases}$

39. $f(x) = \begin{cases} \sqrt{-x} & \text{for } x < 0 \\ \sqrt{x} & \text{for } x \geq 0 \end{cases}$

40. $f(x) = \begin{cases} 3 & \text{for } x < 0 \\ 3 + \sqrt{x} & \text{for } x \geq 0 \end{cases}$

41. $f(x) = \begin{cases} x^2 & \text{for } x < -1 \\ -x & \text{for } x \geq -1 \end{cases}$

42. $f(x) = \begin{cases} 4 - x^2 & \text{for } -2 \leq x \leq 2 \\ x - 2 & \text{for } x > 2 \end{cases}$

43. $f(x) = [\![x + 1]\!]$

44. $f(x) = 2[\![x]\!]$

45. $f(x) = [\![x]\!] + 2$ for $0 \leq x < 4$

46. $f(x) = [\![x - 3]\!]$ for $0 < x \leq 5$

From the graph of each function in Exercises 47–54, state the domain, the range, and the intervals on which the function is increasing, decreasing, or constant.

47. a. **b.**

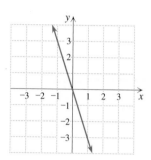
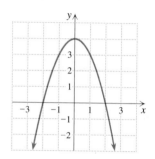

48. a. **b.**

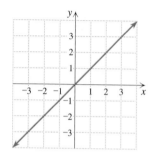
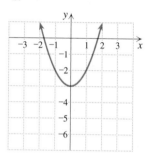

49. a. **b.**

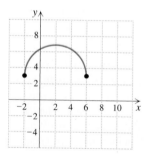
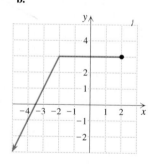

50. a. **b.**

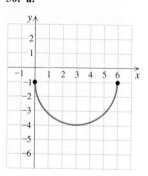
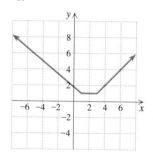

51. a. **b.**

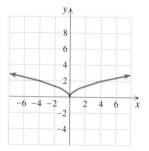
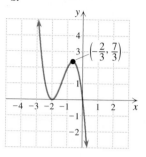

52. a. **b.**

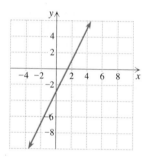
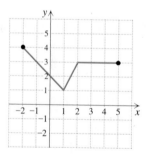

53. a. **b.**

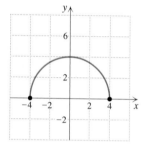
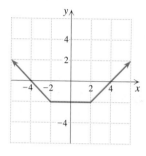

54. a. **b.**

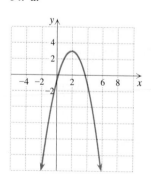
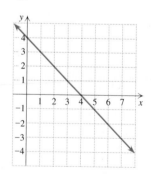

Make a table listing ordered pairs for each function. Then sketch the graph and state the domain and range. Identify any intervals on which f is increasing, decreasing, or constant.

55. $f(x) = 2x + 1$

56. $f(x) = -3x$

57. $f(x) = |x - 1|$

58. $f(x) = |x| + 1$

59. $f(x) = \dfrac{|x|}{x}$

60. $f(x) = \dfrac{2x}{|x|}$

61. $f(x) = \sqrt{9 - x^2}$

62. $f(x) = -\sqrt{1 - x^2}$

63. $f(x) = \begin{cases} x + 1 & \text{for } x \geq 3 \\ x + 2 & \text{for } x < 3 \end{cases}$

64. $f(x) = \begin{cases} \sqrt{-x} & \text{for } x < 0 \\ -\sqrt{x} & \text{for } x \geq 0 \end{cases}$

65. $f(x) = \begin{cases} x + 3 & \text{for } x \leq -2 \\ \sqrt{4 - x^2} & \text{for } -2 < x < 2 \\ -x + 3 & \text{for } x \geq 2 \end{cases}$

66. $f(x) = \begin{cases} 8 + 2x & \text{for } x \leq -2 \\ x^2 & \text{for } -2 < x < 2 \\ 8 - 2x & \text{for } x \geq 2 \end{cases}$

Write a piecewise function for each given graph.

67.

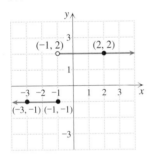

68.

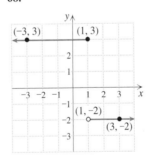

69.

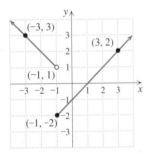

70.

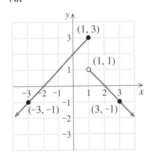

71.

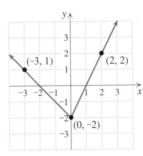

72.

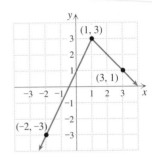

Use the minimum and maximum features of a graphing calculator to find the intervals on which each function is increasing or decreasing. Round approximate answers to two decimal places.

73. $y = 3x^2 - 5x - 4$

74. $y = -6x^2 + 2x - 9$

75. $y = x^3 - 3x$

76. $y = x^4 - 11x^2 + 18$

77. $y = 2x^4 - 12x^2 + 25$

78. $y = x^5 - 13x^3 + 36x$

79. $y = \left| 20 - |x - 50| \right|$

80. $y = x + |x + 30| - |x + 50|$

Solve each problem.

81. *Motor Vehicle Ownership* World motor vehicle ownership in developed countries can be modeled by the function

$$M(t) = \begin{cases} 17.5t + 250 & 0 \leq t \leq 20 \\ 10t + 400 & 20 < t \leq 40 \end{cases}$$

where *t* is the number of years since 1970 and *M(t)* is the number of motor vehicles in millions in the year 1970 + *t* (World Resources Institute, www.wri.org). See the accompanying figure. How many vehicles were there in developed countries in 1988? How many will there be in 2010? What was the average rate of change of motor vehicle ownership from 1984 to 1994?

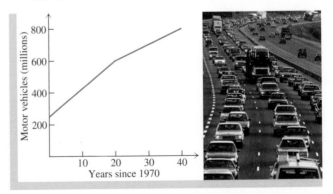

■ **Figure for Exercise 81**

82. *Motor Vehicle Ownership* World motor vehicle ownership in developing countries and Eastern Europe can be modeled by the function

$$M(t) = 6.25t + 50$$

where t is the number of years since 1970 and $M(t)$ is in millions of vehicles (World Resources Institute, www.wri.org). Graph this function. What is the expected average rate of change of motor vehicle ownership from 1990 through 2010? Is motor vehicle ownership expected to grow faster in developed or developing countries in the period 1990 through 2010? (See the previous exercise.)

83. *Filing a Tax Return* An accountant determines the charge for filing a tax return by using the function $C = 50 + 40[\![t]\!]$ for $t > 0$, where C is in dollars and t is in hours. Sketch the graph of this function. For what values of t is the charge over 235?

84. *Shipping Machinery* The cost in dollars of shipping a machine is given by the function $C = 200 + 37[\![w/100]\!]$ for $w > 0$, where w is the weight of the machine in pounds. For which values of w is the cost less than 862?

Thinking Outside the Box VI

Best Fitting Pipe A work crew is digging a pipeline through a frozen wilderness in Alaska. The cross section of the trench is in the shape of the parabola $y = x^2$. The pipe has a circular cross section. If the pipe is too large, then the pipe will not lay on the bottom of the trench.

a. What is the radius of the largest pipe that will lay on the bottom of the trench?

b. If the radius of the pipe is 3 and the trench is in the shape of $y = ax^2$, then what is the largest value of a for which the pipe will lay in the bottom of the trench?

1.6 Pop Quiz

1. Find the domain and range for $y = 1 - \sqrt{x}$.

2. Find the domain and range for $y = \sqrt{9 - x^2}$.

3. Find the range for

$$f(x) = \begin{cases} 2x & \text{for} \quad x \geq 1 \\ 3 - x & \text{for} \quad x < 1 \end{cases}.$$

4. On what interval is $y = x^2$ increasing?

5. On what interval is $y = |x - 3|$ decreasing?

1.7 Families of Functions, Transformations, and Symmetry

If a, h, and k are real numbers with $a \neq 0$, then the graph of $y = af(x - h) + k$ is a **transformation** of the graph of $y = f(x)$. All of the transformations of a function form a **family of functions.** For example, any function of the form $y = a\sqrt{x - h} + k$ is in the square-root family because it is a transformation of $y = \sqrt{x}$. If a transformation changes the shape of a graph then it is a **nonrigid** transformation. Otherwise, it is **rigid.** We will study two types of rigid transformations—*translating* and *reflecting,* and two types of nonrigid transformations—*stretching* and *shrinking.* We start with translating.

Translation

The idea of translation is to move a graph either vertically or horizontally without changing its shape.

Definition: Translation Upward or Downward	If $k > 0$, then the graph of $y = f(x) + k$ is a **translation of k units upward** of the graph of $y = f(x)$. If $k < 0$, then the graph of $y = f(x) + k$ is a **translation of $\|k\|$ units downward** of the graph of $y = f(x)$.

Example **1** Translations upward or downward

Graph the three given functions on the same coordinate plane.

a. $f(x) = \sqrt{x}, g(x) = \sqrt{x} + 3, h(x) = \sqrt{x} - 5$
b. $f(x) = x^2, g(x) = x^2 + 2, h(x) = x^2 - 3$

Solution

a. First sketch $f(x) = \sqrt{x}$ through $(0, 0)$, $(1, 1)$, and $(4, 2)$ as shown in Fig. 1.87. Since $g(x) = \sqrt{x} + 3$ we can add 3 to the y-coordinate of each point to get $(0, 3)$, $(1, 4)$, and $(4, 5)$. Sketch g through these points. Every point on f can be moved up 3 units to obtain a corresponding point on g. We now subtract 5 from the y-coordinates on f to obtain points on h. So h goes through $(0, -5)$, $(1, -4)$, and $(4, -3)$.

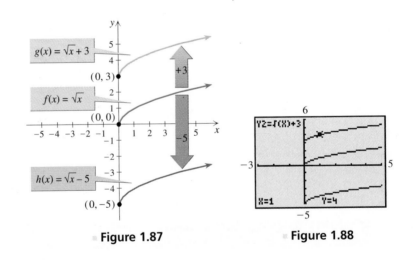

■ **Figure 1.87**   ■ **Figure 1.88**

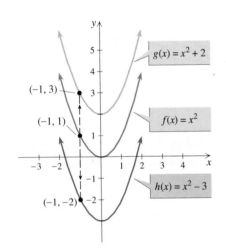

■ **Figure 1.89**

The relationship between f, g, and h can be seen with a graphing calculator in Fig. 1.88. You should experiment with your graphing calculator to see how a change in the formula changes the graph. □

b. First sketch the familiar graph of $f(x) = x^2$ through $(\pm 2, 4)$, $(\pm 1, 1)$, and $(0, 0)$ as shown in Fig. 1.89. Since $g(x) = f(x) + 2$, the graph of g can be obtained by translating the graph of f upward two units. Since $h(x) = f(x) - 3$, the graph of h can be obtained by translating the graph of f downward three units. For example, the point $(-1, 1)$ on the graph of f moves up to $(-1, 3)$ on the graph of g and down to $(-1, -2)$ on the graph of h as shown in Fig. 1.89. ■

Any function of the form $y = a(x - h)^2 + k$ is in the **square family** because it is a transformation of $y = x^2$. The graph of any function in the square family is called a **parabola**.

The graph of $y = f(x) + k$ for $k > 0$ is an upward translation of $y = f(x)$ because the last operation performed is addition of k. To make a graph move horizontally, the first operation performed must be addition or subtraction.

Definition: Translation to the Right or Left

> If $h > 0$, then the graph of $y = f(x - h)$ is a **translation of h units to the right** of the graph of $y = f(x)$. If $h < 0$, then the graph of $y = f(x - h)$ is a **translation of $|h|$ units to the left** of the graph of $y = f(x)$.

Example **2** Translations to the right or left

Graph $f(x) = \sqrt{x}$, $g(x) = \sqrt{x - 3}$, and $h(x) = \sqrt{x + 5}$ on the same coordinate plane.

Solution

First sketch $f(x) = \sqrt{x}$ through $(0, 0)$, $(1, 1)$, and $(4, 2)$ as shown in Fig. 1.90. Since the first operation of g is to subtract 3, we get the corresponding points by adding 3 to each x-coordinate. So g goes through $(3, 0)$, $(4, 1)$, and $(7, 2)$. Since the first operation of h is to add 5, we get corresponding points by subtracting 5 from the x-coordinates. So h goes through $(-5, 0)$, $(-4, 1)$, and $(-1, 2)$.

The calculator graphs of f, g, and h are shown in Fig. 1.91. Be sure to note the difference between $y = \sqrt{(x)} - 3$ and $y = \sqrt{(x - 3)}$ on a calculator.

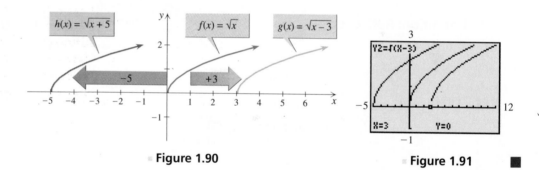

Figure 1.90   **Figure 1.91**   ■

Notice that $y = \sqrt{x - 3}$ lies 3 units to the *right* and $y = \sqrt{x + 5}$ lies 5 units to the *left* of $y = \sqrt{x}$. The next example shows two more horizontal translations.

Example **3** Horizontal translations

Sketch the graph of each function.

a. $f(x) = |x - 1|$ **b.** $f(x) = (x + 3)^2$

Solution

a. The function $f(x) = |x - 1|$ is in the absolute value family and its graph is a translation one unit to the right of $g(x) = |x|$. Calculate a few ordered pairs to get an accurate graph. The points $(0, 1)$, $(1, 0)$, and $(2, 1)$ are on the graph of $f(x) = |x - 1|$ shown in Fig. 1.92 on the next page.

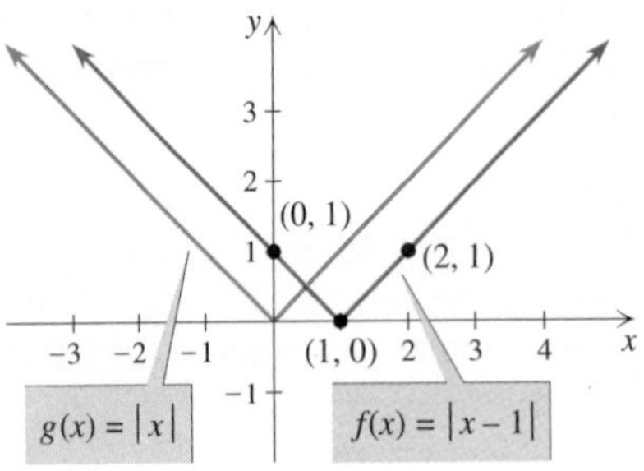

■ **Figure 1.92**

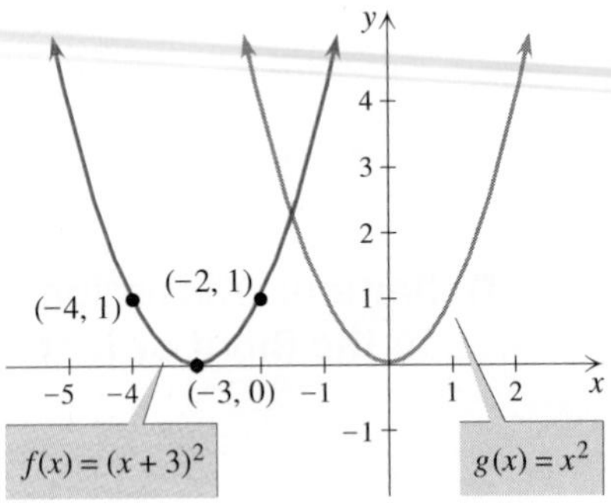

■ **Figure 1.93**

b. The function $f(x) = (x + 3)^2$ is in the square family and its graph is a translation three units to the left of the graph of $g(x) = x^2$. Calculate a few ordered pairs to get an accurate graph. The points $(-3, 0)$, $(-2, 1)$, and $(-4, 1)$ are on the graph shown in Fig. 1.93.   ■

Reflection

The idea in *reflection* is to get a graph that is a mirror image of another graph, where the mirror is placed on the x-axis. To find the mirror image of a point, we simply change the sign of its y-coordinate.

Definition: Reflection

> The graph of $y = -f(x)$ is a **reflection** in the x-axis of the graph of $y = f(x)$.

Example **4** **Graphing using reflection**

Graph each pair of functions on the same coordinate plane.

a. $f(x) = x^2, g(x) = -x^2$
b. $f(x) = x^3, g(x) = -x^3$
c. $f(x) = |x|, g(x) = -|x|$

Solution

a. The graph of $f(x) = x^2$ goes through $(0, 0)$, $(\pm 1, 1)$, and $(\pm 2, 4)$. The graph of $g(x) = -x^2$ goes through $(0, 0)$, $(\pm 1, -1)$, and $(\pm 2, -4)$ as shown in Fig. 1.94.

b. Make a table of ordered pairs for f as follows:

x	-2	-1	0	1	2
$f(x) = x^3$	-8	-1	0	1	8

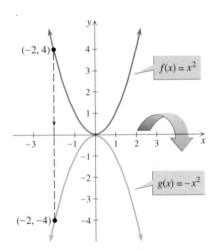

■ **Figure 1.94**

Sketch the graph of f through these ordered pairs as shown in Fig. 1.95. Since $g(x) = -f(x)$, the graph of g can be obtained by reflecting the graph of f in the x-axis. Each point on the graph of f corresponds to a point on the graph of g with the opposite y-coordinate. For example, $(2, 8)$ on the graph of f corresponds to $(2, -8)$ on the graph of g. Both graphs are shown in Fig. 1.95.

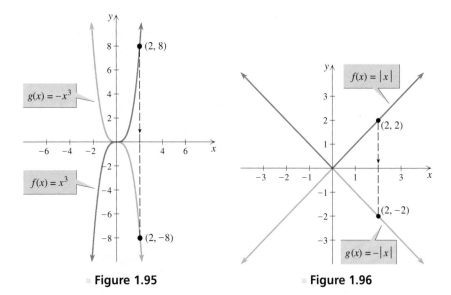

■ **Figure 1.95**   ■ **Figure 1.96**

c. The graph of f is the familiar V-shaped graph of the absolute value function as shown in Fig. 1.96. Since $g(x) = -f(x)$, the graph of g can be obtained by reflecting the graph of f in the x-axis. Each point on the graph of f corresponds to a point on the graph of g with the opposite y-coordinate. For example, $(2, 2)$ on f corresponds to $(2, -2)$ on g. Both graphs are shown in Fig. 1.96. ■

Note that if $y = x^2$ is reflected in the x-axis and then translated one unit upward, the equation for the graph in the final position is $y = -x^2 + 1$. If $y = x^2$ is translated one unit upward and then reflected in the x-axis, the equation for the graph in the final position is $y = -(x^2 + 1)$ or $y = -x^2 - 1$. The order in which the transformations are done can make the final functions different.

Stretching and Shrinking

To *stretch* a graph we multiply the y-coordinates by a number larger than 1. To *shrink* a graph we multiply the y-coordinates by a number between 0 and 1.

Definitions: Stretching and Shrinking

The graph of $y = af(x)$ is obtained from the graph of $y = f(x)$ by

1. **stretching** the graph of $y = f(x)$ by a when $a > 1$, or
2. **shrinking** the graph of $y = f(x)$ by a when $0 < a < 1$.

Example **5** **Graphing using stretching and shrinking**

In each case graph the three functions on the same coordinate plane.

a. $f(x) = \sqrt{x}, g(x) = 2\sqrt{x}, h(x) = \frac{1}{2}\sqrt{x}$

b. $f(x) = x^2, g(x) = 2x^2, h(x) = \frac{1}{2}x^2$

Solution

a. The graph of $f(x) = \sqrt{x}$ goes through $(0, 0)$, $(1, 1)$, and $(4, 2)$ as shown in Fig. 1.97. The graph of g is obtained by stretching the graph of f by a factor of 2. So g goes through $(0, 0)$, $(1, 2)$, and $(4, 4)$. The graph of h is obtained by shrinking the graph of f by a factor of $\frac{1}{2}$. So h goes through $(0, 0)$, $\left(1, \frac{1}{2}\right)$, and $(4, 1)$.

The functions f, g, and h are shown on a graphing calculator in Fig. 1.98. Note how the viewing window affects the shape of the graph. They do not appear as separated on the calculator as they do in Fig. 1.97. □

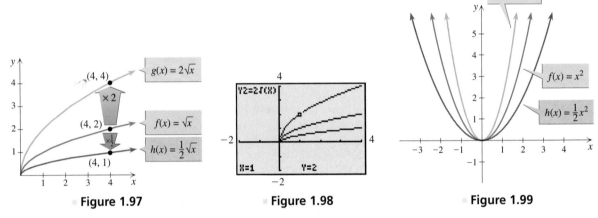

Figure 1.97   Figure 1.98   Figure 1.99

b. The graph of $f(x) = x^2$ is the familiar parabola shown in Fig. 1.99. We stretch it by a factor of 2 to get the graph of g and shrink it by a factor of $\frac{1}{2}$ to get the graph of h. ■

A function involving more than one transformation may be graphed using the following procedure.

PROCEDURE | **Multiple Transformations**

Graph a function involving more than one transformation in the following order:

1. Horizontal translation
2. Stretching or shrinking
3. Reflecting
4. Vertical translation

The function in the next example involves all four of the above transformations.

Example **6** **Graphing using several transformations**

Graph the function $y = 4 - 2\sqrt{x + 1}$.

Solution

First recognize that this function is in the square root family. So its graph is a transformation of the graph of $y = \sqrt{x}$. The graph of $y = \sqrt{x + 1}$ is a *horizontal* translation one unit to the left of the graph of $y = \sqrt{x}$. The graph of $y = 2\sqrt{x + 1}$ is obtained from $y = \sqrt{x + 1}$ by *stretching* it by a factor of 2.

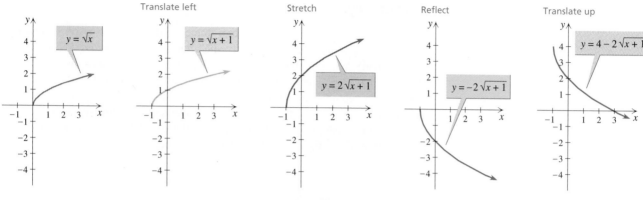

Figure 1.100

Reflect $y = 2\sqrt{x + 1}$ in the *x*-axis to obtain the graph of $y = -2\sqrt{x + 1}$. Finally, the graph of $y = 4 - 2\sqrt{x + 1}$ is a *vertical* translation of $y = -2\sqrt{x + 1}$, four units upward. All of these graphs are shown in Fig. 1.100. ■

The Linear Family of Functions

The function $f(x) = x$ is called the **identity function** because the coordinates in each ordered pair are identical. Its graph is a line through (0, 0) with slope 1. A member of the **linear family,** a **linear function,** is a transformation of the identity function: $f(x) = a(x - h) + k$ where $a \neq 0$. Since *a*, *h*, and *k* are real numbers, we can rewrite this form as a multiple of *x* plus a constant. So a linear function has the form $f(x) = mx + b$, with $m \neq 0$ (the slope-intercept form). If $m = 0$, then the function has the form $f(x) = b$ and it is a **constant function.**

Example **7** Graphing linear functions using transformations

Sketch the graphs of $y = x$, $y = 2x$, $y = -2x$, and $y = -2x - 3$.

Solution

The graph of $y = x$ is a line through (0, 0), (1, 1), and (2, 2). Stretch the graph of $y = x$ by a factor of 2 to get the graph of $y = 2x$. Reflect in the *x*-axis to get the graph of $y = -2x$. Translate downward three units to get the graph of $y = -2x - 3$. See Fig. 1.101.

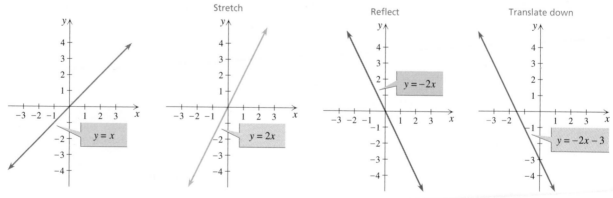

Figure 1.101

Symmetry

The graph of $g(x) = -x^2$ is a reflection in the x-axis of the graph of $f(x) = x^2$. If the paper were folded along the x-axis, the graphs would coincide. See Fig. 1.102. The symmetry that we call reflection occurs between two functions, but the graph of $f(x) = x^2$ has a symmetry within itself. Points such as $(2, 4)$ and $(-2, 4)$ are on the graph and are equidistant from the y-axis. Folding the paper along the y-axis brings all such pairs of points together. See Fig. 1.103. The reason for this symmetry about the y-axis is the fact that $f(-x) = f(x)$ for every value of x. We get the same y-coordinate whether we evaluate the function at a number or at its opposite.

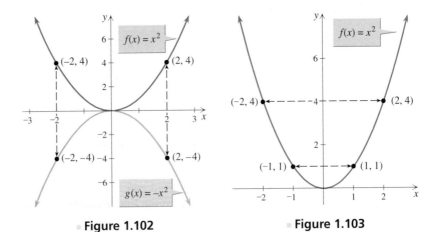

■ **Figure 1.102**   ■ **Figure 1.103**

Definition: Symmetric about the y-Axis

If $f(-x) = f(x)$ for every value of x in the domain of the function f, then f is called an **even function** and its graph is **symmetric about the y-axis.**

Consider the graph of $f(x) = x^3$ shown in Fig. 1.104. On the graph of $f(x) = x^3$ we find pairs of points such as $(2, 8)$ and $(-2, -8)$. The odd exponent in x^3 causes the second coordinate to be negative when the sign of the first coordinate is changed. These points are equidistant from the origin and on opposite sides of the origin. So the symmetry of this graph is about the origin. In this case $f(x)$ and $f(-x)$ are not equal, but $f(-x) = -f(x)$.

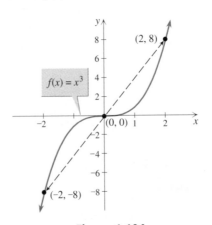

■ **Figure 1.104**

Definition: Symmetric about the Origin

> If $f(-x) = -f(x)$ for every value of x in the domain of the function f, then f is called an **odd function** and its graph is **symmetric about the origin.**

A graph might look like it is symmetric about the y-axis or the origin, but the only way to be sure is to use the definitions of these terms as shown in the following example. Note that an odd power of a negative number is negative and an even power of a negative number is positive. So for any real number x we have $(-x)^n = -x^n$ if n is odd and $(-x)^n = x^n$ if n is even.

Example 8 Determining symmetry in a graph

Discuss the symmetry of the graph of each function.

a. $f(x) = 5x^3 - x$ **b.** $f(x) = |x| + 3$ **c.** $f(x) = x^2 - 3x + 6$

Solution

a. Replace x by $-x$ in the formula for $f(x)$ and simplify:

$$f(-x) = 5(-x)^3 - (-x) = -5x^3 + x$$

Is $f(-x)$ equal to $f(x)$ or the opposite of $f(x)$? Since $-f(x) = -5x^3 + x$, we have $f(-x) = -f(x)$. So f is an odd function and the graph is symmetric about the origin.

b. Since $|-x| = |x|$ for any x, we have $f(-x) = |-x| + 3 = |x| + 3$. Because $f(-x) = f(x)$, the function is even and the graph is symmetric about the y-axis.

c. In this case, $f(-x) = (-x)^2 - 3(-x) + 6 = x^2 + 3x + 6$. So $f(-x) \neq f(x)$, and $f(-x) \neq -f(x)$. This function is neither odd nor even and its graph has neither type of symmetry. ■

Do you see why functions symmetric about the y-axis are called *even* and functions symmetric about the origin are called *odd?* In general, a function defined by a polynomial with even exponents only, such as $f(x) = x^2$ or $f(x) = x^6 - 5x^4 + 2x^2 + 3$, is symmetric about the y-axis. (The constant term 3 has even degree because $3 = 3x^0$.) A function with only odd exponents such as $f(x) = x^3$ or $f(x) = x^5 - 6x^3 + 4x$ is symmetric about the origin. A function containing both even and odd powered terms such as $f(x) = x^2 + 3x$ has neither symmetry. For other types of functions (such as absolute value) you must examine the function more carefully to determine symmetry.

Note that functions cannot have x-axis symmetry. A graph that is symmetric about the x-axis fails the vertical line test. For example, the graph of $x = y^2$ is symmetric about the x-axis, but the graph fails the vertical line test and y is not a function of x.

Reading Graphs to Solve Inequalities

The solution to an inequality in one variable can be read from a graph of an equation in two variables. Now that we have some experience with graphing, we will solve some inequalities by reading graphs.

Example **9** **Using a graph to solve an inequality**

Solve the inequality $(x - 1)^2 - 2 < 0$ by graphing.

Solution

The graph of $y = (x - 1)^2 - 2$ is obtained by translating the graph of $y = x^2$ one unit to the right and two units downward. See Fig. 1.105. To find the x-intercepts we solve $(x - 1)^2 - 2 = 0$:

$$(x - 1)^2 = 2$$
$$x - 1 = \pm\sqrt{2}$$
$$x = 1 \pm \sqrt{2}$$

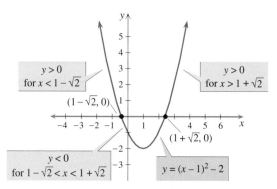

Figure 1.105

The x-intercepts are $\left(1 - \sqrt{2}, 0\right)$ and $\left(1 + \sqrt{2}, 0\right)$. If the y-coordinate of a point on the graph is negative, then the x-coordinate satisfies $(x - 1)^2 - 2 < 0$. So the solution set to $(x - 1)^2 - 2 < 0$ is the open interval $\left(1 - \sqrt{2}, 1 + \sqrt{2}\right)$.

Although a graphing calculator will not find the exact solution to this inequality, you can use TRACE to support the answer and see that y is negative between the x-intercepts. See Fig. 1.106.

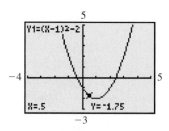

Figure 1.106 ■

Note that the solution set to $(x - 1)^2 - 2 \geq 0$ can also be obtained from the graph in Fig. 1.105. If the y-coordinate of a point on the graph is positive or zero, then the x-coordinate satisfies $(x - 1)^2 - 2 \geq 0$. So the solution set to $(x - 1)^2 - 2 \geq 0$ is $\left(-\infty, 1 - \sqrt{2}\right] \cup \left[1 + \sqrt{2}, \infty\right)$.

Function Gallery: Some Basic Functions and Their Properties

Constant Function

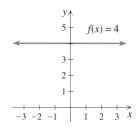

Domain $(-\infty, \infty)$
Range $\{4\}$
Constant on $(-\infty, \infty)$
Symmetric about y-axis

Identity Function

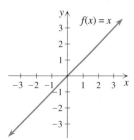

Domain $(-\infty, \infty)$
Range $(-\infty, \infty)$
Increasing on $(-\infty, \infty)$
Symmetric about origin

Linear Function

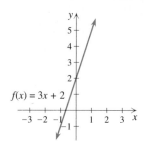

Domain $(-\infty, \infty)$
Range $(-\infty, \infty)$
Increasing on $(-\infty, \infty)$

Absolute-Value Function

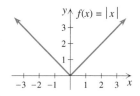

Domain $(-\infty, \infty)$
Range $(0, \infty)$
Increasing on $(0, \infty)$
Decreasing on $(-\infty, 0)$
Symmetric about y-axis

Square Function

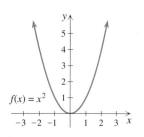

Domain $(-\infty, \infty)$
Range $[0, \infty)$
Increasing on $(0, \infty)$
Decreasing on $(-\infty, 0)$
Symmetric about y-axis

Square-Root Function

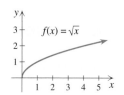

Domain $[0, \infty)$
Range $[0, \infty)$
Increasing on $(0, \infty)$

Cube Function

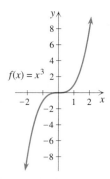

Domain $(-\infty, \infty)$
Range $(-\infty, \infty)$
Increasing on $(-\infty, \infty)$
Symmetric about origin

Cube-Root Function

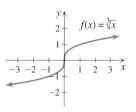

Domain $(-\infty, \infty)$
Range $(-\infty, \infty)$
Increasing on $(-\infty, \infty)$
Symmetric about origin

Greatest Integer Function

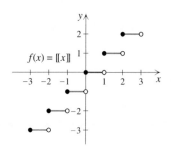

Domain $(-\infty, \infty)$
Range $\{n \mid n \text{ is an integer}\}$
Constant on $[n, n + 1)$
 for every integer n

For Thought

True or False? Explain.

1. The graph of $f(x) = (-x)^4$ is a reflection in the x-axis of the graph of $g(x) = x^4$.

2. The graph of $f(x) = x - 4$ lies four units to the right of the graph of $f(x) = x$.

3. The graph of $y = |x + 2| + 3$ is a translation two units to the right and three units upward of the graph of $y = |x|$.

4. The graph of $f(x) = -3$ is a reflection in the x-axis of the graph of $g(x) = 3$.

5. The functions $y = x^2 + 4x + 1$ and $y = (x + 2)^2 - 3$ have the same graph.

6. The graph of $y = -(x - 3)^2 - 4$ can be obtained by moving $y = x^2$ three units to the right and down four units, and then reflecting in the x-axis.

7. If $f(x) = -x^3 + 2x^2 - 3x + 5$, then $f(-x) = x^3 + 2x^2 + 3x + 5$.

8. The graphs of $f(x) = -\sqrt{x}$ and $g(x) = \sqrt{-x}$ are identical.

9. If $f(x) = x^3 - x$, then $f(-x) = -f(x)$.

10. The solution set to $|x| - 1 \le 0$ is $[-1, 1]$.

1.7 Exercises

Sketch the graphs of each pair of functions on the same coordinate plane.

1. $f(x) = |x|, g(x) = |x| - 4$

2. $f(x) = \sqrt{x}, g(x) = \sqrt{x} + 3$

3. $f(x) = x, g(x) = x + 3$

4. $f(x) = x^2, g(x) = x^2 - 5$

5. $y = x^2, y = (x - 3)^2$

6. $y = |x|, y = |x + 2|$

7. $y = \sqrt{x}, y = \sqrt{x + 9}$

8. $y = x^2, y = (x - 1)^2$

9. $f(x) = \sqrt{x}, g(x) = -\sqrt{x}$

10. $f(x) = x, g(x) = -x$

11. $y = \sqrt{x}, y = 3\sqrt{x}$

12. $y = \sqrt{1 - x^2}, y = 4\sqrt{1 - x^2}$

13. $y = x^2, y = \frac{1}{4}x^2$

14. $y = |x|, y = \frac{1}{3}|x|$

15. $y = \sqrt{4 - x^2}, y = -\sqrt{4 - x^2}$

16. $f(x) = x^2 + 1, g(x) = -(x^2 + 1)$

Match each function in Exercises 17–24 with its graph (a)–(h). See the procedure for multiple transformations on page 74.

17. $y = x^2$

18. $y = (x - 4)^2 + 2$

19. $y = (x + 4)^2 - 2$

20. $y = -2(x - 2)^2$

21. $y = -2(x + 2)^2$

22. $y = -\frac{1}{2}x^2 - 4$

23. $y = \frac{1}{2}(x + 4)^2 + 2$

24. $y = -2(x - 4)^2 - 2$

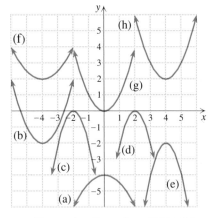

■ **Figure for Exercises 17 to 24**

Write the equation of each graph after the indicated transformation(s).

25. The graph of $y = \sqrt{x}$ is translated two units upward.

26. The graph of $y = \sqrt{x}$ is translated three units downward.

27. The graph of $y = x^2$ is translated five units to the right.

1.7 ▪▪▪ Exercises **81**/

28. The graph of $y = x^2$ is translated seven units to the left.

29. The graph of $y = x^2$ is translated ten units to the right and four units upward.

30. The graph of $y = \sqrt{x}$ is translated five units to the left and twelve units downward.

31. The graph of $y = \sqrt{x}$ is stretched by a factor of 3, translated five units upward, then reflected in the x-axis.

32. The graph of $y = x^2$ is translated thirteen units to the right and six units downward, then reflected in the x-axis.

33. The graph of $y = |x|$ is reflected in the x-axis, stretched by a factor of 3, then translated seven units to the right and nine units upward.

34. The graph of $y = x$ is stretched by a factor of 2, reflected in the x-axis, then translated eight units downward and six units to the left.

Use transformations to graph each function and state the domain and range.

35. $y = (x - 1)^2 + 2$

36. $y = (x + 5)^2 - 4$

37. $y = |x - 1| + 3$

38. $y = |x + 3| - 4$

39. $y = 3x - 40$

40. $y = -4x + 200$

41. $y = \dfrac{1}{2}x - 20$

42. $y = -\dfrac{1}{2}x + 40$

43. $y = -\dfrac{1}{2}|x| + 40$

44. $y = 3|x| - 200$

45. $y = -\dfrac{1}{2}|x + 4|$

46. $y = 3|x - 2|$

47. $y = -\sqrt{x - 3} + 1$

48. $y = -\sqrt{x + 2} - 4$

49. $y = -2\sqrt{x + 3} + 2$

50. $y = -\dfrac{1}{2}\sqrt{x + 2} + 4$

Determine algebraically whether the function is even, odd, or neither. Discuss the symmetry of each function.

51. $f(x) = x^4$

52. $f(x) = x^4 - 2x^2$

53. $f(x) = x^4 - x^3$

54. $f(x) = x^3 - x$

55. $f(x) = (x + 3)^2$

56. $f(x) = (x - 1)^2$

57. $f(x) = |x - 2|$

58. $f(x) = |x| - 9$

59. $f(x) = x$

60. $f(x) = -x$

61. $f(x) = 3x + 2$

62. $f(x) = x - 3$

63. $f(x) = x^3 - 5x + 1$

64. $f(x) = x^6 - x^4 + x^2$

65. $f(x) = 1 + \dfrac{1}{x^2}$

66. $f(x) = (x^2 - 2)^3$

Match each function with its graph (a)–(h).

67. $y = 2 + \sqrt{x}$

68. $y = \sqrt{2 + x}$

69. $y = \sqrt{x^2}$

70. $y = \sqrt{\dfrac{x}{2}}$

71. $y = \dfrac{1}{2}\sqrt{x}$

72. $y = 2 - \sqrt{x - 2}$

73. $y = -2\sqrt{x}$

74. $y = -\sqrt{-x}$

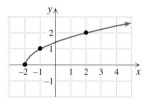

(a)

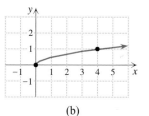

(b)

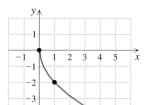

(c)

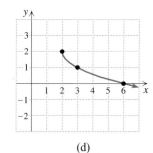

(d)

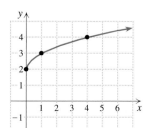

(e)

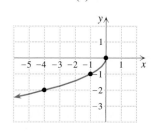

(f)

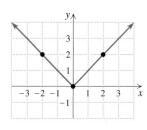

(g)

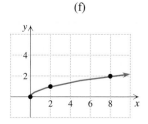

(h)

▪ **Figure for Exercises 67 to 74**

Solve each inequality by reading the corresponding graph.

75. $x^2 - 1 \geq 0$

76. $2x^2 - 3 < 0$

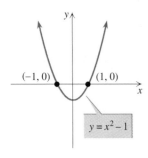

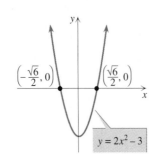

77. $|x - 2| - 3 > 0$

78. $2 - |x + 1| \geq 0$

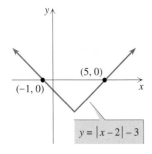

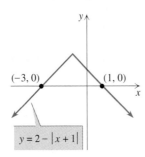

Solve each inequality by graphing an appropriate function. State the solution set using interval notation.

79. $(x - 1)^2 - 9 < 0$

80. $\left(x - \dfrac{1}{2}\right)^2 - \dfrac{9}{4} \geq 0$

81. $5 - \sqrt{x} \geq 0$

82. $\sqrt{x + 3} - 2 \geq 0$

83. $(x - 2)^2 > 3$

84. $(x - 1)^2 < 4$

85. $\sqrt{25 - x^2} > 0$

86. $\sqrt{4 - x^2} \geq 0$

⊞ *Use a graphing calculator to find an approximate solution to each inequality by reading the graph of an appropriate function. Round to two decimal places.*

87. $\sqrt{3}x^2 + \pi x - 9 < 0$

88. $x^3 - 5x^2 + 6x - 1 > 0$

Graph each of the following functions by transforming the given graph of $y = f(x)$.

89. a. $y = 2f(x)$

b. $y = -f(x)$

c. $y = f(x + 1)$

d. $y = f(x - 3)$

e. $y = -3f(x)$

f. $y = f(x + 2) - 1$

g. $y = f(x - 1) + 3$

h. $y = 3f(x - 2) + 1$

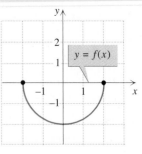

■ **Figure for Exercise 89**

90. a. $y = -f(x)$

b. $y = 2f(x)$

c. $y = -3f(x)$

d. $y = f(x + 2)$

e. $y = f(x - 1)$

f. $y = f(x - 2) + 1$

g. $y = -2f(x + 4)$

h. $y = 2f(x - 3) + 1$

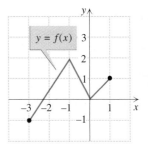

■ **Figure for Exercise 90**

Solve each problem.

91. *Across-the-Board Raise* Each teacher at C. F. Gauss Elementary School is given an across-the-board raise of $2000. Write a function that *transforms* each old salary x into a new salary $N(x)$.

92. *Cost-of-Living Raise* Each registered nurse at Blue Hills Memorial Hospital is first given a 5% cost-of-living raise and then a $3000 merit raise. Write a function that *transforms* each old salary x into a new salary $N(x)$. Does the order in which these raises are given make any difference? Explain.

Thinking Outside the Box VII

Lucky Lucy Ms. Willis asked Lucy to come to the board to find the mean of a pair of one-digit positive integers. Lucy slowly wrote the numbers on the board. While trying to think of what to do next, she rested the chalk between the numbers to make a mark that looked like a decimal point to Ms. Willis. Ms. Willis said "correct" and asked her to find the mean for a pair of two-digit positive integers. Being a quick learner, Lucy again wrote the numbers on the board, rested the chalk between the numbers, and again Ms. Willis said "correct." Lucy had to demonstrate her ability to find the mean for a pair of three-digit and a pair of four-digit positive integers before Ms. Willis was satisfied that she understood the concept. What four pairs of integers did Ms. Willis give to Lucy? Explain why Lucy's method will not work for any other pairs of one-, two-, three-, or four-digit positive integers.

1.7 Pop Quiz

1. What is the equation of the curve $y = \sqrt{x}$ after it is translated 8 units upward?

2. What is the equation of the curve $y = x^2$ after it is translated 9 units to the right?

3. What is the equation of the curve $y = x^3$ after it is reflected in the x-axis?

4. Find the domain and range for $y = -2\sqrt{x - 1} + 5$.

5. If the curve $y = x^2$ is translated 6 units to the right, stretched by a factor of 3, reflected in the x-axis, and translated 4 units upward, then what is the equation of the curve in its final position?

6. Is $y = \sqrt{4 - x^2}$ even, odd, or neither?

1.8 Operations with Functions

In Sections 1.6 and 1.7 we studied the graphs of functions to see the relationships between types of functions and their graphs. In this section we will study various ways in which two or more functions can be combined to make new functions. The emphasis here will be on formulas that define functions.

Basic Operations with Functions

A college student is hired to deliver new telephone books and collect the old ones for recycling. She is paid $6 per hour plus $0.30 for each old phone book she collects. Her salary for a 40-hour week is a function of the number of phone books collected. If x represents the number of phone books collected in one week, then the function $S(x) = 0.30x + 240$ gives her salary in dollars. However, she must use her own car for this job. She figures that her car expenses average $0.20 per phone book collected plus a fixed cost of $20 per week for insurance. We can write her expenses as a function of the number of phone books collected, $E(x) = 0.20x + 20$. Her profit for one week is her salary minus her expenses:

$$P(x) = S(x) - E(x)$$

$$= 0.30x + 240 - (0.20x + 20)$$

$$= 0.10x + 220$$

By subtracting, we get $P(x) = 0.10x + 220$. Her weekly profit is written as a function of the number of phone books collected. In this example we obtained a new function by subtracting two functions. In general, there are four basic arithmetic operations defined for functions.

Definition: Sum, Difference, Product, and Quotient Functions

For two functions f and g, the **sum, difference, product,** and **quotient functions,** functions $f + g, f - g, f \cdot g$, and f/g, respectively, are defined as follows:

$$(f + g)(x) = f(x) + g(x)$$

$$(f - g)(x) = f(x) - g(x)$$

$$(f \cdot g)(x) = f(x) \cdot g(x)$$

$$(f/g)(x) = f(x)/g(x) \qquad \text{provided that } g(x) \neq 0.$$

Example **1** **Evaluating functions**

Let $f(x) = 3\sqrt{x} - 2$ and $g(x) = x^2 + 5$. Find and simplify each expression.

    a. $(f + g)(4)$ **b.** $(f - g)(x)$ **c.** $(f \cdot g)(0)$ **d.** $\left(\dfrac{f}{g}\right)(9)$

Solution

  a. $(f + g)(4) = f(4) + g(4) = 3\sqrt{4} - 2 + 4^2 + 5 = 25$

  b. $(f - g)(x) = f(x) - g(x) = 3\sqrt{x} - 2 - (x^2 + 5) = 3\sqrt{x} - x^2 - 7$

  c. $(f \cdot g)(0) = f(0) \cdot g(0) = (3\sqrt{0} - 2)(0^2 + 5) = (-2)(5) = -10$

  d. $\left(\dfrac{f}{g}\right)(9) = \dfrac{f(9)}{g(9)} = \dfrac{3\sqrt{9} - 2}{9^2 + 5} = \dfrac{7}{86}$

Parts (a), (c), and (d) can be checked with a graphing calculator as shown in Fig. 1.107(a) and (b).

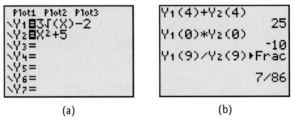

        (a)   (b)

▪ **Figure 1.107** ▪

Think of $f + g, f - g, f \cdot g$, and f/g as generic names for the sum, difference, product, and quotient of the functions f and g. If any of these functions has a particular meaning, as in the phone book example, we can use a new letter to identify it. The domain of $f + g, f - g, f \cdot g$, or f/g is the intersection of the domain of f with the domain of g. Of course, we exclude from the domain of f/g any number for which $g(x) = 0$.

Example **2** **The sum, product, and quotient functions**

Let $f = \{(1, 3), (2, 8), (3, 6), (5, 9)\}$ and $g = \{(1, 6), (2, 11), (3, 0), (4, 1)\}$. Find $f + g, f \cdot g$, and f/g. State the domain of each function.

Solution

The domain of $f + g$ and $f \cdot g$ is $\{1, 2, 3\}$ because that is the intersection of the domains of f and g. The ordered pair $(1, 9)$ belongs to $f + g$ because

$$(f + g)(1) = f(1) + g(1) = 3 + 6 = 9.$$

The ordered pair $(2, 19)$ belongs to $f + g$ because $(f + g)(2) = 19$. The pair $(3, 6)$ also belongs to $f + g$. So

$$f + g = \{(1, 9), (2, 19), (3, 6)\}.$$

Since $(f \cdot g)(1) = f(1) \cdot g(1) = 3 \cdot 6 = 18,$ the pair $(1, 18)$ belongs to $f \cdot g$. Likewise, $(2, 88)$ and $(3, 0)$ also belong to $f \cdot g$. So

$$f \cdot g = \{(1, 18), (2, 88), (3, 0)\}.$$

The domain of f/g is $\{1, 2\}$ because $g(3) = 0$. So

$$\frac{f}{g} = \left\{\left(1, \frac{1}{2}\right), \left(2, \frac{8}{11}\right)\right\}.$$ ◼

In Example 2 the functions are given as sets of ordered pairs, and the results of performing operations with these functions are sets of ordered pairs. In the next example the sets of ordered pairs are defined by means of equations, so the result of performing operations with these functions will be new equations that determine the ordered pairs of the function.

Example **3** **The sum, quotient, product, and difference functions**

Let $f(x) = \sqrt{x}, g(x) = 3x + 1,$ and $h(x) = x - 1.$ Find each function and state its domain.

a. $f + g$ **b.** $\dfrac{g}{f}$ **c.** $g \cdot h$ **d.** $g - h$

Solution

a. Since the domain of f is $[0, \infty)$ and the domain of g is $(-\infty, \infty)$, the domain of $f + g$ is $[0, \infty)$. Since $(f + g)(x) = f(x) + g(x) = \sqrt{x} + 3x + 1,$ the equation defining the function $f + g$ is

$$(f + g)(x) = \sqrt{x} + 3x + 1.$$

b. The number 0 is not in the domain of g/f because $f(0) = 0.$ So the domain of g/f is $(0, \infty).$ The equation defining g/f is

$$\left(\frac{g}{f}\right)(x) = \frac{3x + 1}{\sqrt{x}}.$$

c. The domain of both g and h is $(-\infty, \infty).$ So the domain of $g \cdot h$ is $(-\infty, \infty).$ Since $(3x + 1)(x - 1) = 3x^2 - 2x - 1,$ the equation defining the function $g \cdot h$ is

$$(g \cdot h)(x) = 3x^2 - 2x - 1.$$

d. The domain of both g and h is $(-\infty, \infty).$ So the domain of $g - h$ is $(-\infty, \infty).$ Since $(3x + 1) - (x - 1) = 2x + 2,$ the equation defining $g - h$ is

$$(g - h)(x) = 2x + 2.$$ ◼

Composition of Functions

It is often the case that the output of one function is the input for another function. For example, the number of hamburgers purchased at $1.49 each determines the subtotal. The subtotal is then used to determine the total (including sales tax). So the number of hamburgers actually determines the total and that function is called the *composition* of the other two functions. The composition of functions is defined using function notation as follows.

Definition: Composition of Functions

If f and g are two functions, the **composition** of f and g, written $f \circ g$, is defined by the equation

$$(f \circ g)(x) = f(g(x)),$$

provided that $g(x)$ is in the domain of f. The composition of g and f, written $g \circ f$, is defined by

$$(g \circ f)(x) = g(f(x)),$$

provided that $f(x)$ is in the domain of g.

Note that $f \circ g$ is not the same function as $f \cdot g$, the product of f and g.

For the composition $f \circ g$ to be defined at x, $g(x)$ must be in the domain of f. So the domain of $f \circ g$ is the set of all values of x in the domain of g for which $g(x)$ is in the domain of f. The diagram shown in Fig. 1.108 will help you to understand the composition of functions.

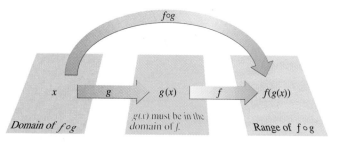

Figure 1.108

Example **4** **Composition of functions defined by sets**

Let $g = \{(1, 4), (2, 5), (3, 6)\}$ and $f = \{(3, 8), (4, 9), (5, 10)\}$. Find $f \circ g$.

Solution

Since $g(1) = 4$ and $f(4) = 9$, $(f \circ g)(1) = 9$. So the ordered pair $(1, 9)$ is in $f \circ g$. Since $g(2) = 5$ and $f(5) = 10$, $(f \circ g)(2) = 10$. So $(2, 10)$ is in $f \circ g$. Now $g(3) = 6$, but 6 is not in the domain of f. So there are only two ordered pairs in $f \circ g$:

$$f \circ g = \{(1, 9), (2, 10)\}$$

In the next example we find specific values of compositions that are defined by equations.

Example **5** **Evaluating compositions defined by equations**

Let $f(x) = \sqrt{x}$, $g(x) = 2x - 1$, and $h(x) = x^2$. Find the value of each expression.

a. $(f \circ g)(5)$ **b.** $(g \circ f)(5)$ **c.** $(h \circ g \circ f)(9)$

Solution

a. $(f \circ g)(5) = f(g(5))$ Definition of composition

$\quad\quad\quad\quad\quad = f(9) \quad\quad g(5) = 2 \cdot 5 - 1 = 9$

$\quad\quad\quad\quad\quad = \sqrt{9}$

$\quad\quad\quad\quad\quad = 3$

b. $(g \circ f)(5) = g(f(5)) = g(\sqrt{5}) = 2\sqrt{5} - 1$

c. $(h \circ g \circ f)(9) = h(g(f(9))) = h(g(3)) = h(5) = 5^2 = 25$

You can check these answers with a graphing calculator as shown in Fig. 1.109(a) and (b). Enter the functions using the Y = key, then go back to the home screen to evaluate. The symbols Y_1, Y_2, and Y_3 are found in the variables menu (VARS) on a TI-83.

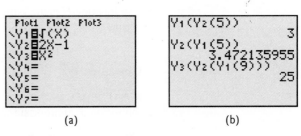

(a)   (b)

■ **Figure 1.109** ■

In Example 4, the domain of g is $\{1, 2, 3\}$ while the domain of $f \circ g$ is $\{1, 2\}$. To find the domain of $f \circ g$, we remove from the domain of g any number x such that $g(x)$ is not in the domain of f. In the next example we construct compositions of functions defined by equations and determine their domains.

Example **6** **Composition of functions defined by equations**

Let $f(x) = \sqrt{x}$, $g(x) = 2x - 1$, and $h(x) = x^2$. Find each composition function and state its domain.

a. $f \circ g$ **b.** $g \circ f$ **c.** $h \circ g$

Solution

a. In the composition $f \circ g$, $g(x)$ must be in the domain of f. The domain of f is $[0, \infty)$. If $g(x)$ is in $[0, \infty)$, then $2x - 1 \geq 0$, or $x \geq \frac{1}{2}$. So the domain of $f \circ g$ is $\left[\frac{1}{2}, \infty\right)$. Since

$$(f \circ g)(x) = f(g(x)) = f(2x - 1) = \sqrt{2x - 1},$$

the function $f \circ g$ is defined by the equation $(f \circ g)(x) = \sqrt{2x - 1}$, for $x \geq \frac{1}{2}$.

b. Since the domain of g is $(-\infty, \infty)$, $f(x)$ is certainly in the domain of g. So the domain of $g \circ f$ is the same as the domain of f, $[0, \infty)$. Since

$$(g \circ f)(x) = g(f(x)) = g(\sqrt{x}) = 2\sqrt{x} - 1,$$

the function $g \circ f$ is defined by the equation $(g \circ f)(x) = 2\sqrt{x} - 1$. Note that $g \circ f$ is generally not equal to $f \circ g$, but in Section 1.9 we will study special types of functions for which they are equal.

c. Since the domain of h is $(-\infty, \infty)$, $g(x)$ is certainly in the domain of h. So the domain of $h \circ g$ is the same as the domain of g, $(-\infty, \infty)$. Since

$$(h \circ g)(x) = h(g(x)) = h(2x - 1) = (2x - 1)^2 = 4x^2 - 4x + 1,$$

the function $h \circ g$ is defined by $(h \circ g)(x) = 4x^2 - 4x + 1$.   ■

Some complicated functions can be thought of as a composition of simpler functions. For example, if $H(x) = (x + 3)^2$, we start with x, add 3 to x, then square the result. These two operations can be accomplished by composition, using $f(x) = x + 3$ followed by $g(x) = x^2$:

$$(g \circ f)(x) = g(f(x)) = g(x + 3) = (x + 3)^2$$

So the function H is the same as the composition of g and f, $H = g \circ f$. Notice that $(f \circ g)(x) = x^2 + 3$ and it is not the same as $H(x)$.

Example **7** **Writing a function as a composition**

Let $f(x) = \sqrt{x}$, $g(x) = x - 3$, and $h(x) = 2x$. Write each given function as a composition of appropriate functions chosen from f, g, and h.

a. $F(x) = \sqrt{x - 3}$ **b.** $G(x) = x - 6$ **c.** $H(x) = 2\sqrt{x} - 3$

Solution

a. The function F consists of subtracting 3 from x and finding the square root of that result. These two operations can be accomplished by composition, using g followed by f. So $F = f \circ g$. Check this answer as follows:

$$(f \circ g)(x) = f(g(x)) = f(x - 3) = \sqrt{x - 3} = F(x)$$

b. Subtracting 6 from x can be accomplished by subtracting 3 from x and then subtracting 3 from that result, so $G = g \circ g$. Check as follows:

$$(g \circ g)(x) = g(g(x)) = g(x - 3) = (x - 3) - 3 = x - 6 = G(x)$$

c. For the function H, find the square root of x, then multiply by 2, and finally subtract 3. These three operations can be accomplished by composition, using f, then h, and then g. So $H = g \circ h \circ f$. Check as follows:

$$(g \circ h \circ f)(x) = g(h(f(x))) = g(h(\sqrt{x})) = g(2\sqrt{x}) = 2\sqrt{x} - 3 = H(x)$$ ■

■ **Foreshadowing Calculus**

One of the big topics in calculus is the instantaneous rate of change of a function. To find the instantaneous rate of change, we often view a complicated function as a composition of simpler functions as is done in Example 7.

Applications

In applied situations, functions are often defined with formulas rather than function notation. In this case, composition can be simply a matter of substitution.

Example **8** **Composition with formulas**

The radius of a circle is a function of the diameter ($r = d/2$) and the area is a function of the radius ($A = \pi r^2$). Construct a formula that expresses the area as a function of the diameter.

Solution

The formula for A as a function of d is obtained by substituting $d/2$ for r:

$$A = \pi r^2 = \pi \left(\frac{d}{2}\right)^2 = \pi \frac{d^2}{4}$$

The function $A = \pi d^2/4$ is the composition of $r = d/2$ and $A = \pi r^2$. ▪

In the next example we find a composition using function notation for the salary of the phone book collector mentioned at the beginning of this section.

Example **9** **Composition with function notation**

A student's salary (in dollars) for collecting x phone books is given by $S(x) = 0.30x + 240$. The amount of withholding (for taxes) is given by $W(x) = 0.20x$, where x is the salary. Express the withholding as a function of the number of phone books collected.

Solution

Note that x represents the number of phone books in $S(x) = 0.30x + 240$ and x represents salary in $W(x) = 0.20x$. So we can replace the salary x in $W(x)$ with $S(x)$ or $0.30x + 240$, which also represents salary:

$$W(S(x)) = W(0.30x + 240) = 0.20(0.30x + 240) = 0.06x + 48$$

Use a new letter to name this function, say T. Then $T(x) = 0.06x + 48$ gives the amount of withholding (for taxes) as a function of x, where x is the number of phone books. ▪

For Thought

True or False? Explain.

1. If $f = \{(2, 4)\}$ and $g = \{(1, 5)\}$, then $f + g = \{(3, 9)\}$.

2. If $f = \{(1, 6), (9, 5)\}$ and $g = \{(1, 3), (9, 0)\}$, then $f/g = \{(1, 2)\}$.

3. If $f = \{(1, 6), (9, 5)\}$ and $g = \{(1, 3), (9, 0)\}$, then $f \cdot g = \{(1, 18), (9, 0)\}$.

4. If $f(x) = x + 2$ and $g(x) = x - 3$, then $(f \cdot g)(5) = 14$.

5. If $s = P/4$ and $A = s^2$, then A is a function of P.

6. If $f(3) = 19$ and $g(19) = 99$, then $(g \circ f)(3) = 99$.

7. If $f(x) = \sqrt{x}$ and $g(x) = x - 2$, then $(f \circ g)(x) = \sqrt{x} - 2$.

8. If $f(x) = 5x$ and $g(x) = x/5$, then $(f \circ g)(x) = (g \circ f)(x) = x$.

9. If $F(x) = (x - 9)^2$, $g(x) = x^2$, and $h(x) = x - 9$, then $F = h \circ g$.

10. If $f(x) = \sqrt{x}$ and $g(x) = x - 2$, then the domain of $f \circ g$ is $[2, \infty)$.

1.8 Exercises

Let $f(x) = x - 3$ and $g(x) = x^2 - x$. Find and simplify each expression.

1. $(f + g)(2)$ **2.** $(g + f)(3)$ **3.** $(f - g)(-2)$

4. $(g - f)(-6)$ **5.** $(f \cdot g)(-1)$ **6.** $(g \cdot f)(0)$

7. $(f/g)(4)$ **8.** $(g/f)(4)$ **9.** $(f + g)(a)$

10. $(f - g)(b)$ **11.** $(f \cdot g)(a)$ **12.** $(f/g)(b)$

Let $f = \{(-3, 1), (0, 4), (2, 0)\}, g = \{(-3, 2), (1, 2), (2, 6), (4, 0)\}$, and $h = \{(2, 4), (1, 0)\}$. Find each function and state the domain of each function.

13. $f + g$ **14.** $f + h$ **15.** $f - g$ **16.** $f - h$

17. $f \cdot g$ **18.** $f \cdot h$ **19.** g/f **20.** f/g

Let $f(x) = \sqrt{x}, g(x) = x - 4$, and $h(x) = \dfrac{1}{x - 2}$. Find an equation defining each function and state the domain of the function.

21. $f + g$ **22.** $f + h$ **23.** $f - h$ **24.** $h - g$

25. $g \cdot h$ **26.** $f \cdot h$ **27.** g/f **28.** f/g

Let $f = \{(-3, 1), (0, 4), (2, 0)\}, g = \{(-3, 2), (1, 2), (2, 6), (4, 0)\}$, and $h = \{(2, 4), (1, 0)\}$. Find each function.

29. $f \circ g$ **30.** $g \circ f$ **31.** $f \circ h$

32. $h \circ f$ **33.** $h \circ g$ **34.** $g \circ h$

Let $f(x) = 3x - 1, g(x) = x^2 + 1$, and $h(x) = \dfrac{x + 1}{3}$. Evaluate each expression. Round approximate answers to three decimal places.

35. $f(g(-1))$ **36.** $g(f(-1))$ **37.** $(f \circ h)(5)$

38. $(h \circ f)(-7)$ **39.** $(f \circ g)(4.39)$ **40.** $(g \circ h)(-9.87)$

41. $(g \circ h \circ f)(2)$ **42.** $(h \circ f \circ g)(3)$ **43.** $(f \circ g \circ h)(2)$

44. $(h \circ g \circ f)(0)$ **45.** $(f \circ h)(a)$ **46.** $(h \circ f)(w)$

47. $(f \circ g)(t)$ **48.** $(g \circ f)(m)$

Let $f(x) = x - 2, g(x) = \sqrt{x}$, and $h(x) = \frac{1}{x}$. Find an equation defining each function and state the domain of the function.

49. $f \circ g$ **50.** $g \circ f$ **51.** $f \circ h$

52. $h \circ f$ **53.** $h \circ g$ **54.** $g \circ h$

55. $f \circ f$ **56.** $g \circ g$ **57.** $h \circ g \circ f$

58. $f \circ g \circ h$ **59.** $h \circ f \circ g$ **60.** $g \circ h \circ f$

Let $f(x) = |x|, g(x) = x - 7$, and $h(x) = x^2$. Write each of the following functions as a composition of functions chosen from f, g, and h.

61. $F(x) = x^2 - 7$ **62.** $G(x) = |x| - 7$

63. $H(x) = (x - 7)^2$ **64.** $M(x) = |x - 7|$

65. $N(x) = (|x| - 7)^2$ **66.** $R(x) = |x^2 - 7|$

67. $P(x) = |x - 7| - 7$ **68.** $Q(x) = (x^2 - 7)^2$

69. $S(x) = x - 14$ **70.** $T(x) = x^4$

Use the two given functions to write y as a function of x.

71. $y = 2a - 3, a = 3x + 1$

72. $y = -4d - 1, d = -3x - 2$

73. $y = w^2 - 2, w = x + 3$

74. $y = 3t^2 - 3, t = x - 1$

75. $y = 3m - 1, m = \dfrac{x + 1}{3}$ **76.** $y = 2z + 5, z = \dfrac{1}{2}x - \dfrac{5}{2}$

Define $y_1 = \sqrt{x + 1}$ and $y_2 = 3x - 4$ on your graphing calculator. For each function y_3, defined in terms of y_1 and y_2, determine the domain and range of y_3 from its graph on your calculator and explain what each graph illustrates.

77. $y_3 = y_1 + y_2$ **78.** $y_3 = 3y_1 - 4$

79. $y_3 = \sqrt{y_2 + 1}$ **80.** $y_3 = \sqrt{y_1 + 1}$

Define $y_1 = \sqrt[3]{x}, y_2 = \sqrt{x}$, and $y_3 = x + 4$. For each function y_4, determine the domain and range of y_4 from its graph on your calculator and explain what each graph illustrates.

81. $y_4 = y_1 + y_2 + y_3$ **82.** $y_4 = \sqrt{y_1 + 4}$

Solve each problem.

83. *Profitable Business*   Charles buys factory reconditioned hedge trimmers for \$40 each and sells them on Ebay for \$68 each. He has a fixed cost of \$200 per month. If x is the number of hedge trimmers he sells per month, then his revenue and cost (in dollars) are given by $R(x) = 68x$ and $C(x) = 40x + 200$. Find a formula for the function $P(x) = R(x) - C(x)$. For what values of x is his profit positive?

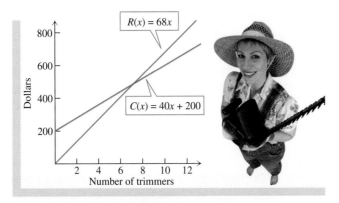

▪ **Figure for Exercise 83**

84. *Profit*   The revenue in dollars that a company receives for installing x alarm systems per month is given by $R(x) = 3000x - 20x^2$, while the cost in dollars is given by $C(x) = 600x + 4000$. The function $P(x) = R(x) - C(x)$ gives the profit for installing x alarm systems per month. Find $P(x)$ and simplify it.

85. Write the area A of a square with a side of length s as a function of its diagonal d.

86. Write the perimeter of a square P as a function of the area A.

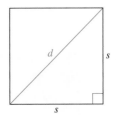

▪ **Figure for Exercises 85 and 86**

87. *Hamburgers*   If hamburgers are \$1.20 each, then $C(x) = 1.20x$ gives the pre-tax cost in dollars for x hamburgers. If sales tax is 5%, then $T(x) = 1.05x$ gives the total cost when the pre-tax cost is x dollars. Write the total cost as a function of the number of hamburgers.

88. *Laying Sod*   Southern Sod will deliver and install 20 pallets of St. Augustine sod for \$2200 or 30 pallets for \$3200 not including tax.
  a. Write the cost (not including tax) as a linear function of x, where x is the number of pallets.

  b. Write a function that gives the total cost (including tax at 9%) as a function of x, where x is the cost (not including tax).

  c. Find the function that gives the total cost as a function of x, where x is the number of pallets.

89. *Displacement-Length Ratio*   The displacement-length ratio D indicates whether a sailboat is relatively heavy or relatively light:

$$D = (d \div 2240) \div x$$

where d is the displacement in pounds and

$$x = (L \div 100)^3$$

where L is the length at the waterline in feet (*Sailing*, www.sailing.com). Assuming that the displacement is 26,000 pounds, write D as a function of L and simplify it.

90. *Sail Area-Displacement Ratio*   The sail area-displacement ratio S measures the sail power available to drive a sailboat:

$$S = A \div y$$

where A is the sail area in square feet and

$$y = (d \div 64)^{2/3}$$

where d is the displacement in pounds. Assuming that the sail area is 6500 square feet, write S as a function of d and simplify it.

91. *Area of a Window*   A window is in the shape of a square with a side of length s, with a semicircle of diameter s adjoining the top of the square. Write the total area of the window W as a function of s.

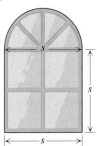

▪ **Figure for Exercises 91 and 92**

92. *Area of a Window*   Using the window of Exercise 91, write the area of the square A as a function of the area of the semicircle S.

Thinking Outside the Box VIII

Whole Number Expression   What is the largest whole number N that cannot be expressed as $N = 3x + 11y$ where x and y are whole numbers?

1.8 Pop Quiz

1. Write the area of a circle as a function of its diameter.

Let $f(x) = x^2$ and $g(x) = x - 2$. Find and simplify.

2. $(f + g)(3)$ 3. $(f \cdot g)(4)$ 4. $(f \circ g)(5)$

Let $m = \{(1, 3), (4, 8)\}$ and $n = \{(3, 5), (4, 9)\}$. Find each function.

5. $m + n$ 6. $n \circ m$

Let $h(x) = x^2$ and $j(x) = \sqrt{x + 2}$. Find the domain of each function.

7. $h + j$ 8. $h \circ j$ 9. $j \circ h$

1.9 Inverse Functions

It is possible for one function to undo what another function does. For example, squaring undoes the operation of taking a square root. The composition of two such functions is the identity function. In this section we explore this idea in detail.

One-to-One Functions

Consider a medium pizza that costs $5 plus $2 per topping. Table 1.2 shows the ordered pairs of the function that determines the cost. Note that for every number of toppings there is a unique cost and for every cost there is a unique number of toppings. There is a **one-to-one correspondence** between the domain and range of this function and the function is a *one-to-one function*. For a function that is not one-to-one, consider a Wendy's menu. Every item corresponds to a unique price, but the price $0.99 corresponds to many different items.

Table 1.2

Toppings x	Cost y	
0	$ 5	
1	7	
2	9	
3	11	
4	13	

Definition:
One-to-One Function

If a function has no two ordered pairs with different first coordinates and the same second coordinate, then the function is called **one-to-one.**

Example **1** One-to-one with ordered pairs

Determine whether each function is one-to-one.

a. $\{(1, 3), (2, 5), (3, 4), (4, 9), (5, 0)\}$
b. $\{(4, 16), (-4, 16), (2, 4), (-2, 4), (5, 25)\}$
c. $\{(3, 0.99), (5, 1.99), (7, 2.99), (8, 3.99), (9, 0.99)\}$

Solution

a. This function is one-to-one because no two ordered pairs have different first co-ordinates and the same second coordinate.

b. This function is not one-to-one because the ordered pairs $(4, 16)$ and $(-4, 16)$ have different first coordinates and the same second coordinate.

c. This function is not one-to-one because the ordered pairs $(3, 0.99)$ and $(9, 0.99)$ have different first coordinates and the same second coordinate. ■

If a function is given as a short list of ordered pairs, then it is easy to determine whether the function is one-to-one. A graph of a function can also be used to determine whether a function is one-to-one using the **horizontal line test.**

Horizontal Line Test

> If each horizontal line crosses the graph of a function at no more than one point, then the function is one-to-one.

The graph of a one-to-one function never has the same y-coordinate for two different x-coordinates on the graph. So if it is possible to draw a horizontal line that crosses the graph of a function two or more times, then the function is not one-to-one.

Example **2** **The horizontal line test**

Use the horizontal line test to determine whether the functions shown in Fig. 1.110 are one-to-one.

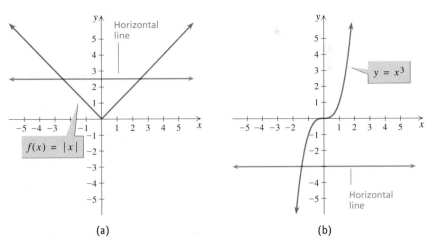

Figure 1.110

Solution

The function $f(x) = |x|$ is not one-to-one because it is possible to draw a horizontal line that crosses the graph twice as shown in Fig. 1.110(a). The function $y = x^3$ in Fig. 1.110(b) is one-to-one because it appears to be impossible to draw a horizontal line that crosses the graph more than once. ■

The horizontal line test explains the visual difference between the graph of a one-to-one function and the graph of a function that is not one-to-one. Because no graph of a function is perfectly accurate, conclusions made from a graph alone may not be correct. For example, from the graph of $y = x^3 - 0.01x$ shown in Fig. 1.111(a) we would conclude that the function is one-to-one. However, another view of the same function in Fig. 1.111(b) shows that the function is not one-to-one.

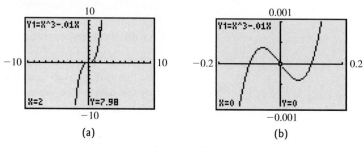

(a)   (b)

■ **Figure 1.111**

Using the equation for a function we can often determine conclusively whether the function is one-to-one. A function f is not one-to-one if it is possible to find two different numbers x_1 and x_2 such that $f(x_1) = f(x_2)$. For example, for the function $f(x) = x^2$, it is possible to find two different numbers, 2 and -2, such that $f(2) = f(-2)$. So $f(x) = x^2$ is not one-to-one. To prove that a function f is one-to-one, we must show that $f(x_1) = f(x_2)$ implies that $x_1 = x_2$.

Example **3** **Using the definition of one-to-one**

Determine whether each function is one-to-one.

a. $f(x) = \dfrac{2x + 1}{x - 3}$ **b.** $g(x) = |x|$

Solution

a. If $f(x_1) = f(x_2)$, then we have the following equation:

$$\frac{2x_1 + 1}{x_1 - 3} = \frac{2x_2 + 1}{x_2 - 3}$$

$$(2x_1 + 1)(x_2 - 3) = (2x_2 + 1)(x_1 - 3) \qquad \text{Multiply by the LCD.}$$

$$2x_1x_2 + x_2 - 6x_1 - 3 = 2x_2x_1 + x_1 - 6x_2 - 3$$

$$7x_2 = 7x_1$$

$$x_2 = x_1$$

Since $f(x_1) = f(x_2)$ implies that $x_1 = x_2$, $f(x)$ is a one-to-one function.

b. If $g(x_1) = g(x_2)$, then $|x_1| = |x_2|$. But this does not imply that $x_1 = x_2$, because two different numbers can have the same absolute value. For example $|3| = |-3|$ but $3 \neq -3$. So g is not one-to-one. ■

Inverse Functions

Consider again the function given in Table 1.2, which determines the cost of a pizza from the number of toppings. Because that function is one-to-one we can make a table in which the number of toppings is determined from the cost as shown in Table 1.3. Of course we could just read Table 1.2 backwards, but we make a new table to emphasize that there is a new function under discussion. Table 1.3 is the *inverse function* for the function in Table 1.2.

Table 1.3

Cost x	Toppings y	
$ 5	0	
7	1	
9	2	
11	3	
13	4	

A function is a set of ordered pairs in which no two ordered pairs have the same first coordinates and different second coordinates. If we interchange the x- and y-coordinates in each ordered pair of a function, as in Tables 1.2 and 1.3, the resulting set of ordered pairs might or might not be a function. If the original function is one-to-one, then the set obtained by interchanging the coordinates in each ordered pair is a function, the inverse function. If a function is one-to-one, then it has an inverse function or it is **invertible.**

**Definition:
Inverse Function**

> The **inverse** of a one-to-one function f is the function f^{-1} (read "f inverse"), where the ordered pairs of f^{-1} are obtained by interchanging the coordinates in each ordered pair of f.

In this notation, the number -1 in f^{-1} does not represent a negative exponent. It is merely a symbol for denoting the inverse function.

Example 4 Finding an inverse function

For each function, determine whether it is invertible. If it is invertible, then find the inverse.

a. $f = \{(-2, 3), (4, 5), (2, 3)\}$ **b.** $g = \{(3, 1), (5, 2), (7, 4), (9, 8)\}$

Solution

a. This function f is *not* one-to-one because of the ordered pairs $(-2, 3)$ and $(2, 3)$. So f is not invertible.

b. The function g is one-to-one, and so g is invertible. The inverse of g is the function $g^{-1} = \{(1, 3), (2, 5), (4, 7), (8, 9)\}$.   ■

Example 5 Using inverse function notation

Let $f = \{(1, 3), (2, 4), (5, 7)\}$. Find f^{-1}, $f^{-1}(3)$, and $(f^{-1} \circ f)(1)$.

Solution

Interchange the x- and y-coordinates of each ordered pair of f to find f^{-1}:

$$f^{-1} = \{(3, 1), (4, 2), (7, 5)\}$$

To find the value of $f^{-1}(3)$, notice that $f^{-1}(3)$ is the second coordinate when the first coordinate is 3 in the function f^{-1}. So $f^{-1}(3) = 1$. To find the composition, use the definition of composition of functions:

$$(f^{-1} \circ f)(1) = f^{-1}(f(1)) = f^{-1}(3) = 1$$ ■

Since the coordinates in the ordered pairs are interchanged, the domain of f^{-1} is the range of f, and the range of f^{-1} is the domain of f. If f^{-1} is the inverse function of f, then certainly f is the inverse of f^{-1}. The functions f and f^{-1} are inverses of each other.

Inverse Functions Using Function Notation

The function $f(x) = 2x + 5$ gives the cost of a pizza where \$5 is the basic cost and x is the number of toppings at \$2 each. If we assume that x could be any real number, this function defines the set

$$f = \{(x, y)\,|\,y = 2x + 5\}.$$

Since the coordinates of each ordered pair are interchanged for f^{-1}, the ordered pairs of f^{-1} must satisfy $x = 2y + 5$:

$$f^{-1} = \{(x, y)\,|\,x = 2y + 5\}.$$

Since $x = 2y + 5$ is equivalent to $y = \dfrac{x - 5}{2}$,

$$f^{-1} = \left\{(x, y)\,\Big|\,y = \frac{x - 5}{2}\right\}.$$

The function f^{-1} is described in function notation as $f^{-1}(x) = \dfrac{x - 5}{2}$. The function f^{-1} gives the number of toppings as a function of cost.

A one-to-one function f pairs members of the domain of f with members of the range of f, and its inverse function f^{-1} exactly reverses those pairings as shown in Fig. 1.112. The above example suggests the following steps for finding an inverse function using function notation.

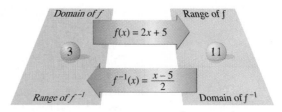

Figure 1.112

PROCEDURE   **Finding $f^{-1}(x)$ by the Switch-and-Solve Method**

To find the inverse of a one-to-one function given in function notation:

1. Replace $f(x)$ by y.
2. Interchange x and y.
3. Solve the equation for y.
4. Replace y by $f^{-1}(x)$.
5. Check that the domain of f is the range of f^{-1} and the range of f is the domain of f^{-1}.

Example 6 The switch-and-solve method

Find the inverse of each function.

a. $f(x) = 4x - 1$ **b.** $f(x) = \dfrac{2x + 1}{x - 3}$

Solution

a. The graph of $f(x) = 4x - 1$ is a line with slope 4. By the horizontal line test the function is one-to-one and invertible. Replace $f(x)$ with y to get $y = 4x - 1$. Next, interchange x and y to get $x = 4y - 1$. Now solve for y:

$$x = 4y - 1$$

$$x + 1 = 4y$$

$$\frac{x + 1}{4} = y$$

Replace y by $f^{-1}(x)$ to get $f^{-1}(x) = \dfrac{x + 1}{4}$. The domain of f is $(-\infty, \infty)$ and that is the range of f^{-1}. The range of f is $(-\infty, \infty)$ and that is the domain of f^{-1}.

b. In Example 3(a) we showed that f is a one-to-one function. So we can find f^{-1} by interchanging x and y and solving for y:

$$y = \frac{2x + 1}{x - 3} \qquad \text{Replace } f(x) \text{ by } y.$$

$$x = \frac{2y + 1}{y - 3} \qquad \text{Interchange } x \text{ and } y.$$

$$x(y - 3) = 2y + 1 \quad \text{Solve for } y.$$

$$xy - 3x = 2y + 1$$

$$xy - 2y = 3x + 1$$

$$y(x - 2) = 3x + 1$$

$$y = \frac{3x + 1}{x - 2}$$

Replace y by $f^{-1}(x)$ to get $f^{-1}(x) = \dfrac{3x + 1}{x - 2}$. The domain of f is all real numbers except 3. The range of f^{-1} is the set of all real numbers except 3. We exclude 3 because $\dfrac{3x + 1}{x - 2} = 3$ has no solution. Check that the range of f is equal to the domain of f^{-1}. ■

If a function f is defined by a short list of ordered pairs, then we simply write all of the pairs in reverse to find f^{-1}. If two functions are defined by formulas, it may not be obvious whether the ordered pairs of one are the reverse of the ordered pairs of the other. However, we can use the compositions of the functions to make the determination as stated in the following theorem.

Theorem: Verifying Whether *f* and *g* Are Inverses

The functions f and g are inverses of each other if and only if

1. $g(f(x)) = x$ for every x in the domain of f and
2. $f(g(x)) = x$ for every x in the domain of g.

Example **7** **Using composition to verify inverse functions**

Determine whether the functions $f(x) = x^3 - 1$ and $g(x) = \sqrt[3]{x + 1}$ are inverse functions.

Solution

Find $f(g(x))$ and $g(f(x))$:

$$f(g(x)) = f\left(\sqrt[3]{x + 1}\right) = \left(\sqrt[3]{x + 1}\right)^3 - 1 = x + 1 - 1 = x$$
$$g(f(x)) = g(x^3 - 1) = \sqrt[3]{x^3 - 1 + 1} = \sqrt[3]{x^3} = x$$

Since $f(g(x)) = x$ is true for any real number (the domain of g) and $g(f(x)) = x$ is true for any real number (the domain of f), the functions f and g are inverses of each other. ▪

Only one-to-one functions are invertible. However, sometimes it is possible to restrict the domain of a function that is not one-to-one so that it is one-to-one on the restricted domain. For example, $f(x) = x^2$ for x in $(-\infty, \infty)$ is not one-to-one and not invertible. But $f(x) = x^2$ for $x \geq 0$ is one-to-one and invertible. The inverse of $f(x) = x^2$ for $x \geq 0$ is the square-root function $f^{-1}(x) = \sqrt{x}$.

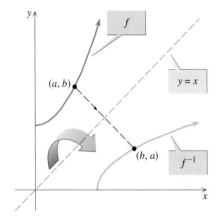

Figure 1.113

Graphs of f and f^{-1}

If a point (a, b) is on the graph of an invertible function f, then (b, a) is on the graph of f^{-1}. See Fig. 1.113. Since the points (a, b) and (b, a) are symmetric with respect to the line $y = x$, the graph of f^{-1} is a reflection of f with respect to the line $y = x$. This reflection property of inverse functions gives us a way to visualize the graphs of inverse functions.

Example **8** **Graphing a function and its inverse**

Find the inverse of the function $f(x) = \sqrt{x - 1}$ and graph both f and f^{-1} on the same coordinate axes.

Solution

If $\sqrt{x_1 - 1} = \sqrt{x_2 - 1}$, then $x_1 = x_2$. So f is one-to-one. Since the range of $f(x) = \sqrt{x - 1}$ is $[0, \infty)$, f^{-1} must be a function with domain $[0, \infty)$. To find a formula for the inverse, interchange x and y in $y = \sqrt{x - 1}$ to get $x = \sqrt{y - 1}$. Solving this equation for y gives $y = x^2 + 1$. So $f^{-1}(x) = x^2 + 1$ for $x \geq 0$. The graph of f is a translation one unit to the right of the graph of $y = \sqrt{x}$, and the graph of f^{-1} is a translation one unit upward of the graph of $y = x^2$ for $x \geq 0$. Both graphs are shown in Fig. 1.114 along with the line $y = x$. Notice that the graph of f^{-1} is a reflection of the graph of f about the line $y = x$, the domain of f is the range of f^{-1}, and the range of f is the domain of f^{-1}. ▪

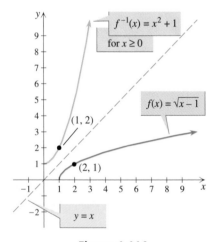

Figure 1.114

Finding Inverse Functions Mentally

It is no surprise that the inverse of $f(x) = x^2$ for $x \geq 0$ is the function $f^{-1}(x) = \sqrt{x}$. For nonnegative numbers, taking the square root undoes what squaring does. It is also no surprise that the inverse of $f(x) = 3x$ is $f^{-1}(x) = x/3$ or that the inverse of $f(x) = x + 9$ is $f^{-1}(x) = x - 9$. If an invertible function involves a single operation, it is usually easy to write the inverse function because for most operations there is an inverse operation. See Table 1.4. If an invertible function involves more than one operation, we find the inverse function by applying the inverse operations in the opposite order from the order in which they appear in the original function.

▪ Table 1.4

Function	Inverse
$f(x) = 2x$	$f^{-1}(x) = x/2$
$f(x) = x - 5$	$f^{-1}(x) = x + 5$
$f(x) = \sqrt{x}$	$f^{-1}(x) = x^2 \ (x \geq 0)$
$f(x) = x^3$	$f^{-1}(x) = \sqrt[3]{x}$
$f(x) = 1/x$	$f^{-1}(x) = 1/x$
$f(x) = -x$	$f^{-1}(x) = -x$

Example **9** **Finding inverses mentally**

Find the inverse of each function mentally.

a. $f(x) = 2x + 1$ **b.** $g(x) = \dfrac{x^3 + 5}{2}$

Solution

a. The function $f(x) = 2x + 1$ is a composition of multiplying x by 2 and then adding 1. So the inverse function is a composition of subtracting 1 and then dividing by 2: $f^{-1}(x) = (x - 1)/2$. See Fig. 1.115.

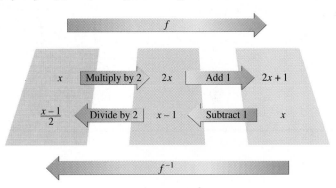

▪ Figure 1.115

b. The function $g(x) = (x^3 + 5)/2$ is a composition of cubing x, adding 5, and dividing by 2. The inverse is a composition of multiplying by 2, subtracting 5, and then taking the cube root: $g^{-1}(x) = \sqrt[3]{2x - 5}$. See Fig. 1.116.

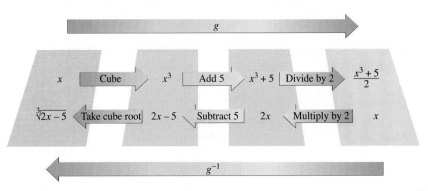

▪ Figure 1.116

Function Gallery: **Some Inverse Functions**

Linear

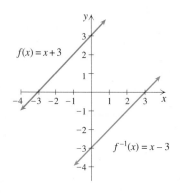

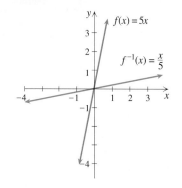

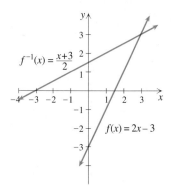

Powers and Roots

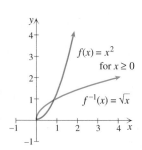

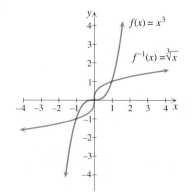

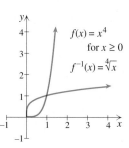

For Thought

True or False? Explain.

1. The inverse of the function $\{(2, 3), (5, 5)\}$ is $\{(5, 2), (5, 3)\}$.

2. The function $f(x) = -2$ is an invertible function.

3. If $g(x) = x^2$, then $g^{-1}(x) = \sqrt{x}$.

4. The only functions that are invertible are the one-to-one functions.

5. Every function has an inverse function.

6. The function $f(x) = x^4$ is invertible.

7. If $f(x) = 3\sqrt{x-2}$, then $f^{-1}(x) = \dfrac{(x+2)^2}{3}$ for $x \geq 0$.

8. If $f(x) = |x - 3|$, then $f^{-1}(x) = |x| + 3$.

9. According to the horizontal line test, $y = |x|$ is one-to-one.

10. The function $g(x) = -x$ is the inverse of the function $f(x) = -x$.

1.9 Exercises

Determine whether each function is one-to-one.

1. $\{(3, 3), (5, 5), (6, 6), (9, 9)\}$

2. $\{(3, 4), (5, 6), (7, 8), (9, 10), (11, 15)\}$

3. $\{(-1, 1), (1, 1), (-2, 4), (2, 4)\}$

4. $\{(3, 2), (5, 2), (7, 2)\}$

5. $\{(1, 99), (2, 98), (3, 97), (4, 96), (5, 99)\}$

6. $\{(-1, 9), (-2, 8), (-3, 7), (1, 9), (2, 8), (3, 7)\}$

Use the horizontal line test to determine whether each function is one-to-one.

7. $f(x) = x^2 - 3x$

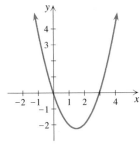

8. $g(x) = |x - 2| + 1$

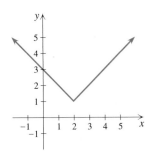

9. $y = \sqrt[3]{x} + 2$

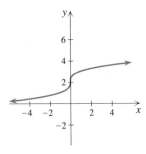

10. $y = (x - 2)^3$

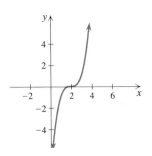

11. $y = x^3 - x$

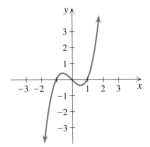

12. $y = \dfrac{1}{x}$

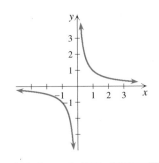

Determine whether each function is one-to-one.

13. $f(x) = 2x - 3$

14. $h(x) = 4x - 9$

15. $q(x) = \dfrac{1 - x}{x - 5}$

16. $g(x) = \dfrac{x + 2}{x - 3}$

17. $p(x) = |x + 1|$

18. $r(x) = 2|x - 1|$

19. $w(x) = x^2 + 3$

20. $v(x) = 2x^2 - 1$

21. $k(x) = \sqrt[3]{x + 9}$

22. $t(x) = \sqrt{x + 3}$

Determine whether each function is invertible. If it is invertible, find the inverse.

23. $\{(9, 3), (2, 2)\}$

24. $\{(4, 5), (5, 6)\}$

25. $\{(-1, 0), (1, 0), (5, 0)\}$

26. $\{(1, 2), (5, 2), (6, 7)\}$

27. $\{(3, 3), (2, 2), (4, 4), (7, 7)\}$

28. $\{(1, 1), (2, 4), (4, 16), (7, 49)\}$

29. $\{(1, 1), (2, 2), (4.5, 2), (5, 5)\}$

30. $\{(0, 2), (2, 0), (1, 0), (4, 6)\}$

Determine whether each function is invertible and explain your answer.

31. The function that pairs the universal product code of an item at Sears with a price.

32. The function that pairs the number of days since your birth with your age in years.

33. The function that pairs the length of a VCR tape in feet with the playing time in minutes.

34. The function that pairs the speed of your car in miles per hour with the speed in kilometers per hour.

35. The function that pairs the number of days of a hotel stay with the total cost for the stay.

36. The function that pairs the number of days that a deposit of $100 earns interest at 6% compounded daily with the amount of interest.

For each function f, find f^{-1}, $f^{-1}(5)$, and $(f^{-1} \circ f)(2)$.

37. $f = \{(2, 1), (3, 5)\}$

38. $f = \{(-1, 5), (0, 0), (2, 6)\}$

39. $f = \{(-3, -3), (0, 5), (2, -7)\}$

40. $f = \{(3.2, 5), (2, 1.99)\}$

Determine whether each function is invertible by inspecting its graph on a graphing calculator.

41. $f(x) = (x + 0.01)(x + 0.02)(x + 0.03)$

42. $f(x) = x^3 - 0.6x^2 + 0.11x - 0.006$

43. $f(x) = |x - 2| - |5 - x|$

44. $f(x) = \sqrt[3]{0.1x + 3} + \sqrt[3]{-0.1x}$

Find the inverse of each function using the procedure for the switch-and-solve method on page 96.

45. $f(x) = 3x - 7$

46. $f(x) = -2x + 5$

47. $f(x) = 2 + \sqrt{x - 3}$

48. $f(x) = \sqrt{3x - 1}$

49. $f(x) = -x - 9$

50. $f(x) = -x + 3$

51. $f(x) = \dfrac{x + 3}{x - 5}$

52. $f(x) = \dfrac{2x - 1}{x - 6}$

53. $f(x) = -\dfrac{1}{x}$

54. $f(x) = x$

55. $f(x) = \sqrt[3]{x - 9} + 5$

56. $f(x) = \sqrt[3]{\dfrac{x}{2}} + 5$

57. $f(x) = (x - 2)^2$ for $x \geq 2$

58. $f(x) = x^2$ for $x \leq 0$

In each case find $f(g(x))$ and $g(f(x))$. Then determine whether g and f are inverse functions.

59. $f(x) = 4x + 4, g(x) = 0.25x - 1$

60. $f(x) = 20 - 5x, g(x) = -0.2x + 4$

61. $f(x) = x^2 + 1, g(x) = \sqrt{x - 1}$

62. $f(x) = \sqrt[4]{x}, g(x) = x^4$

63. $f(x) = \dfrac{1}{x} + 3, g(x) = \dfrac{1}{x - 3}$

64. $f(x) = 4 - \dfrac{1}{x}, g(x) = \dfrac{1}{4 - x}$

65. $f(x) = \sqrt[3]{\dfrac{x - 2}{5}}, g(x) = 5x^3 + 2$

66. $f(x) = x^3 - 27, g(x) = \sqrt[3]{x} + 3$

For each exercise, graph the three functions on the same screen of a graphing calculator. Enter these functions as shown without simplifying any expressions. Explain what these exercises illustrate.

67. $y_1 = \sqrt[3]{x} - 1, y_2 = (x + 1)^3, y_3 = \left(\sqrt[3]{x} - 1 + 1\right)^3$

68. $y_1 = (2x - 1)^{1/3}, y_2 = (x^3 + 1)/2,$
$y_3 = (2((x^3 + 1)/2) - 1)^{1/3}$

Determine whether each pair of functions f and g are inverses of each other.

69.

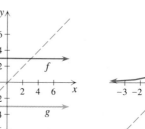

70.

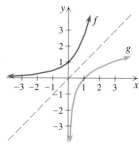

71.

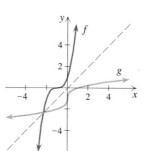

72.

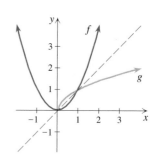

For each function f, sketch the graph of f^{-1}.

73.

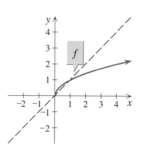

74.

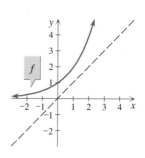

75.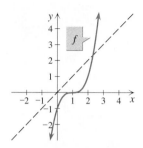

76.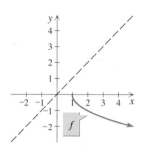

Find the inverse of each function and graph both f and f⁻¹ on the same coordinate plane.

77. $f(x) = 3x + 2$

78. $f(x) = -x - 8$

79. $f(x) = x^2 - 4$ for $x \geq 0$

80. $f(x) = 1 - x^2$ for $x \geq 0$

81. $f(x) = x^3$

82. $f(x) = -x^3$

83. $f(x) = \sqrt{x} - 3$

84. $f(x) = \sqrt{x - 3}$

Find the inverse of each function mentally.

85. a. $f(x) = 5x$

b. $f(x) = x - 88$

c. $f(x) = 3x - 7$

d. $f(x) = 4 - 3x$

e. $f(x) = \dfrac{1}{2}x - 9$

f. $f(x) = -x$

g. $f(x) = \sqrt[3]{x} - 9$

h. $f(x) = 3x^3 - 7$

86. a. $f(x) = \dfrac{x}{2}$

b. $f(x) = x + 99$

c. $f(x) = 5x + 1$

d. $f(x) = 5 - 2x$

e. $f(x) = \dfrac{x}{3} + 6$

f. $f(x) = \dfrac{1}{x}$

g. $f(x) = \sqrt[3]{x - 9}$

h. $f(x) = -x^3 + 4$

Solve each problem.

87. *Price of a Car*   The tax on a new car is 8% of the purchase price P. Express the total cost C as a function of the purchase price. Express the purchase price P as a function of the total cost C.

88. *Volume of a Cube*   Express the volume of a cube $V(x)$ as a function of the length of a side x. Express the length of a side of a cube $S(x)$ as a function of the volume x.

89. *Rowers and Speed*   The world record times in the 2000-m race are a function of the number of rowers as shown in the accompanying figure (www.cbs.sportsline.com). If r is the number of rowers and t is the time in minutes, then the formula $t = -0.39r + 7.89$ models this relationship. Is this function invertible? Find the inverse. If the time for a 2000-m race was 5.55 min, then how many rowers were probably in the boat?

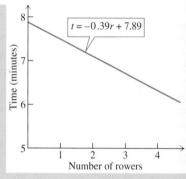

Figure for Exercise 89

90. *Temperature*   The function $C = \dfrac{5}{9}(F - 32)$ expresses the Celsius temperature as a function of the Fahrenheit temperature. Find the inverse function. What is it used for?

91. *Landing Speed*   The function $V = \sqrt{1.496w}$ expresses the landing speed V (in feet per second) as a function of the gross weight (in pounds) of the Piper Cheyenne aircraft. Find the inverse function. Use the inverse function to find the gross weight for a Piper Cheyenne for which the proper landing speed is 115 ft/sec.

92. *Poiseuille's Law*   Under certain conditions, the velocity V of blood in a vessel at distance r from the center of the vessel is given by $V = 500(5.625 \times 10^{-5} - r^2)$ where $0 \leq r \leq 7.5 \times 10^{-3}$. Write r as a function of V.

93. *Depreciation Rate*   The depreciation rate r for a $50,000 new car is given by the function $r = 1 - \left(\dfrac{V}{50{,}000}\right)^{1/5}$, where V is the value of the car when it is five years old.
a. What is the depreciation rate for a $50,000 BMW that is worth $28,000 after 5 years?

b. Write V as a function of r.

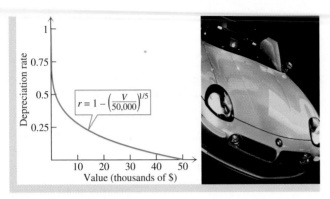

▪ **Figure for Exercise 93**

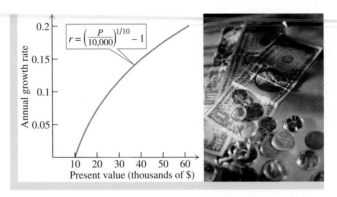

▪ **Figure for Exercise 94**

94. *Annual Growth Rate* One measurement of the quality of a mutual fund is its average annual growth rate over the last 10 years. The function $r = \left(\frac{P}{10,000}\right)^{1/10} - 1$ expresses the average annual growth rate r as a function of the present value P of an investment of $10,000 made 10 years ago. An investment of $10,000 in Fidelity's Contrafund in 1995 was worth $36,555 in 2005 (Fidelity Investments, www.fidelity.com). What was the annual growth rate for that period? Write P as a function of r.

Thinking Outside the Box IX

Costly Computers A school district purchased x computers at y dollars each for a total of $640,000. Both x and y are whole numbers and neither is a multiple of 10. Find the absolute value of the difference between x and y.

1.9 Pop Quiz

1. Is the function $\{(1, 3), (4, 5), (2, 3)\}$ invertible?

2. If $f = \{(5, 4), (3, 6), (2, 5)\}$, then what is $f^{-1}(5)$?

3. If $f(x) = 2x$, then what is $f^{-1}(8)$?

4. Is $f(x) = x^4$ a one-to-one function?

5. If $f(x) = 2x - 1$, then what is $f^{-1}(x)$?

6. If $g(x) = \sqrt[3]{x + 1} - 4$, then what is $g^{-1}(x)$?

7. Find $(h \circ j)(x)$ if $h(x) = x^3 - 5$ and $j(x) = \sqrt[3]{x + 5}$.

▪▪▪Highlights

1.1 Real Numbers and Their Properties

Rationals and Irrationals	Every real number is either rational (a ratio of integers) or irrational (not a ratio of integers).	Rational: $1, 2/3, -44.7$ Irrational: $\sqrt{2}, \sqrt{7}, \pi$
Commutative Property	Addition: $a + b = b + a$ Multiplication: $ab = ba$	$3 + 4 = 4 + 3$ $5 \cdot 7 = 7 \cdot 5$
Associative Property	Addition: $a + (b + c) = (a + b) + c$ Multiplication: $a(bc) = (ab)c$	$3 + (4 + 5) = (3 + 4) + 5$ $3(4 \cdot 5) = (3 \cdot 4)5$
Distributive	$a(b + c) = ab + ac$	$5(6 + 7) = 5 \cdot 6 + 5 \cdot 7$

Absolute Value	A number's distance from zero on the number line If $a \geq 0$, then $	a	= a$. If $a < 0$, then $	a	= -a$.	$	5	= 5,	-5	= 5,	0	= 0$
Distance Between Two Real Numbers	The distance between a and b on a number line is $	a - b	$.	Distance between -5 and 4 is $	-5 - 4	$ or 9.
Absolute Value Equations	If $k > 0$, then $	x	= k \Leftrightarrow x = k$ or $x = -k$. If $k < 0$, then $	x	= k$ has no solution. If $k = 0$, then $	x	= k \Leftrightarrow x = 0$.	$	x	= 5 \Leftrightarrow x = \pm 5$ $	x + 2	= -1$ has no solution. $	x - 3	= 0 \Leftrightarrow x = 3$

1.2 Linear and Absolute Value Inequalities

| **Linear Inequalities** | The inequality symbol is reversed if the inequality is multiplied or divided by a negative number. | $4 - 2x > 10$
$-2x > 6$
$x < -3$ |
| **Absolute Value Inequalities** | $|x| > k \ (k > 0) \Leftrightarrow x > k$ or $x < -k$
$|x| < k \ (k > 0) \Leftrightarrow -k < x < k$
$|x| \leq 0 \Leftrightarrow x = 0$
$|x| \geq 0 \Leftrightarrow x$ is any real number | $|y| > 1 \Leftrightarrow y > 1$ or $y < -1$
$|z| < 2 \Leftrightarrow -2 < z < 2$
$|2b - 5| \leq 0 \Leftrightarrow 2b - 5 = 0$
All real numbers satisfy $|3s - 7| \geq 0$. |

1.3 Equations and Graphs in Two Variables

Distance Formula	Distance between (x_1, y_1) and (x_2, y_2) is $\sqrt{(x_2 - x_1)^2 + (y_2 - y_1)^2}$.	For $(1, 2)$ and $(4, -2)$, $\sqrt{(4 - 1)^2 + (-2 - 2)^2} = 5$.
Midpoint Formula	The midpoint of the line segment with endpoints (x_1, y_1) and (x_2, y_2) is $\left(\frac{x_1 + x_2}{2}, \frac{y_1 + y_2}{2}\right)$.	For $(0, -4)$ and $(6, 2)$, $\left(\frac{0 + 6}{2}, \frac{-4 + 2}{2}\right) = (3, -1)$.
Equation of a Circle	The graph of $(x - h)^2 + (y - k)^2 = r^2 (r > 0)$ is a circle with center (h, k) and radius r.	Circle: $(x - 1)^2 + (y + 2)^2 = 9$ Center $(1, -2)$, radius 3
Linear Equation: Standard Form	$Ax + By = C$ where A and B are not both zero, $x = h$ is a vertical line, $y = k$ is a horizontal line.	$2x + 3y = 6$ is a line. $x = 5$ is a vertical line. $y = 7$ is a horizontal line.

1.4 Linear Equations in Two Variables

Slope Formula	The slope of a line through (x_1, y_1) and (x_2, y_2) is $(y_2 - y_1)/(x_2 - x_1)$ provided $x_1 \neq x_2$.	$(1, 2), (3, -6)$ slope $\frac{-6 - 2}{3 - 1} = -4$
Slope-Intercept Form	$y = mx + b$, slope m, y-intercept $(0, b)$ y is a linear function of x.	$y = 2x + 5$, slope 2, y-intercept $(0, 5)$
Point-Slope Form	The line through (x_1, y_1) with slope m is $y - y_1 = m(x - x_1)$.	Point $(-3, 2)$, $m = 5$ $y - 2 = 5(x - (-3))$
Parallel Lines	Two nonvertical lines are parallel if and only if their slopes are equal.	$y = 7x - 1$ and $y = 7x + 4$ are parallel.
Perpendicular Lines	Two lines with slopes m_1 and m_2 are perpendicular if and only if $m_1 m_2 = -1$.	$y = \frac{1}{2}x + 4$ and $y = -2x - 3$ are perpendicular.

1.5 Functions

Relation	Any set of ordered pairs	$\{(1, 5), (1, 3), (3, 5)\}$
Function	A relation in which no two ordered pairs have the same first coordinate and different second coordinates	$\{(1, 5), (2, 3), (3, 5)\}$
Vertical Line Test	If no vertical line crosses a graph more than once then the graph is a function.	Function Not function
Average Rate of Change	The slope of the line through $(a, f(a))$ and $(b, f(b))$: $\dfrac{f(b) - f(a)}{b - a}$	Average rate of change of $f(x) = x^2$ on $[2, 9]$ is $\dfrac{9^2 - 2^2}{9 - 2}$.
Difference Quotient	Average rate of change of f on $[x, x + h]$: $\dfrac{f(x + h) - f(x)}{h}$	$f(x) = x^2$, difference quotient $= \dfrac{(x + h)^2 - x^2}{h} = 2x + h$

1.6 Graphs of Relations and Functions

Graph of a Relation	An illustration of all ordered pairs of a relation	Graph of $y = x + 1$ shows all ordered pairs in this function.
Circle	A circle is not the graph of a function	Since $(0, \pm 2)$ satisfies $x^2 + y^2 = 4$, it is not a function.
Increasing, Decreasing, or Constant	When going from left to right, a function is increasing if its graph is rising, decreasing if its graph is falling, constant if it is staying the same.	$y = \lvert x \rvert$ is increasing on $(0, \infty)$ and decreasing on $(-\infty, 0)$ $y = 5$ is constant on $(-\infty, \infty)$

1.7 Families of Functions, Transformations, and Symmetry

Transformations of $y = f(x)$	Horizontal: $y = f(x - h)$ Vertical: $y = f(x) + k$ Stretching: $y = af(x)$ for $a > 1$ Shrinking: $y = af(x)$ for $0 < a < 1$ Reflection: $y = -f(x)$	$y = (x - 4)^2$ $y = x^2 + 9$ $y = 3x^2$ $y = 0.5x^2$ $y = -x^2$
Family of Functions	All functions of the form $f(x) = af(x - h) + k$ $(a \neq 0)$ for a given function $y = f(x)$.	The square root family: $y = a\sqrt{x - h} + k$
Even Function	Graph is symmetric with respect to y-axis. $f(-x) = f(x)$	$f(x) = x^2, g(x) = \lvert x \rvert$
Odd Function	Graph is symmetric about the origin. $f(-x) = -f(x)$	$f(x) = x, g(x) = x^3$
Inequalities	$f(x) > 0$ is satisfied on all intervals where the graph of $y = f(x)$ is above the x-axis $f(x) < 0$ is satisfied on all intervals where the graph of $y = f(x)$ is below the x-axis	$4 - x^2 > 0$ on $(-2, 2)$ $4 - x^2 < 0$ on $(-\infty, -2) \cup (2, \infty)$

1.8 Operations with Functions

Sum	$(f + g)(x) = f(x) + g(x)$	$f(x) = x^2 - 4, g(x) = x + 2$ $(f + g)(x) = x^2 + x - 2$
Difference	$(f - g)(x) = f(x) - g(x)$	$(f - g)(x) = x^2 - x - 6$
Product	$(f \cdot g)(x) = f(x) \cdot g(x)$	$(f \cdot g)(x) = x^3 + 2x^2 - 4x - 8$
Quotient	$(f/g)(x) = f(x)/g(x)$	$(f/g)(x) = x - 2$
Composition	$(f \circ g)(x) = f(g(x))$	$(f \circ g)(x) = (x + 2)^2 - 4$ $(g \circ f)(x) = x^2 - 2$

1.9 Inverse Functions

One-to-One Function	A function that has no two ordered pairs with different first coordinates and the same second coordinates	$\{(1, 2), (3, 5), (6, 9)\}$ $g(x) = x + 3$ $f(x) = x^2$ is not one-to-one.
Inverse Function	A one-to-one function has an inverse. The inverse function has the same ordered pairs, with the coordinates reversed.	$f = \{(1, 2), (3, 5), (6, 9)\}$ $f^{-1} = \{(2, 1), (5, 3), (9, 6)\}$ $g(x) = x + 3, g^{-1}(x) = x - 3$
Horizontal Line Test	If there is a horizontal line that crosses the graph of f more than once, then f is not invertible.	$y = 4$ crosses $f(x) = x^2$ twice, so f is not invertible.
Graph of f^{-1}	Reflect the graph of f about the line $y = x$ to get the graph of f^{-1}.	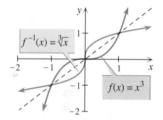

▪▪▪ Chapter 1 Review Exercises

Determine whether each statement is true or false and explain your answer.

1. Every real number is a rational number.

2. Zero is neither rational nor irrational.

3. There are no negative integers.

4. Every repeating decimal number is a rational number.

5. The terminating decimal numbers are irrational numbers.

6. The number $\sqrt{289}$ is a rational number.

7. Zero is a natural number.

8. The multiplicative inverse of 8 is 0.125.

9. The reciprocal of 0.333 is 3.

10. The real number π is irrational.

11. The additive inverse of 0.5 is 0.

12. The distributive property is used in adding like terms.

Solve each equation.

13. $|3q - 4| = 2$

14. $|2v - 1| = 3$

15. $|2h - 3| = 0$

16. $4|x - 3| = 0$

17. $|5 - x| = -1$

18. $|3y - 1| = -2$

Solve each inequality. State the solution set using interval notation and graph the solution set.

19. $4x - 1 > 3x + 2$

20. $6(x - 3) < 5(x + 4)$

21. $5 - 2x > -3$

22. $7 - x > -6$

23. $\dfrac{1}{2}x - \dfrac{1}{3} > x + 2$

24. $0.06x + 1000 > x + 60$

25. $-2 < \dfrac{x - 3}{2} \le 5$

26. $-1 \le \dfrac{3 - 2x}{4} < 3$

27. $3 - 4x < 1$ and $5 + 3x < 8$

28. $-3x < 6$ and $2x + 1 > -1$

29. $-2x < 8$ or $3x > -3$

30. $1 - x < 6$ or $-5 + x < 1$

31. $|x - 3| > 2$

32. $|4 - x| \le 3$

33. $|2x - 7| \le 0$

34. $|6 - 5x| < 0$

35. $|7 - 3x| > -4$

36. $|4 - 3x| \ge 1$

Sketch the graph of each equation. For the circles, state the center and the radius. For the lines state the intercepts.

37. $x^2 + y^2 = 25$

38. $(x - 2)^2 + y^2 = 1$

39. $x^2 + 4x + y^2 = 0$

40. $x^2 - 6x = 2y - y^2 - 1$

41. $x + y = 25$

42. $2x - y = 40$

43. $y = 3x - 4$

44. $y = -\dfrac{1}{2}x + 4$

45. $x = 5$

46. $y = 6$

Solve each problem.

47. Find the exact distance between $(-3, 1)$ and $(2, 4)$.

48. Find the midpoint of the line segment with endpoints $(-1, 1)$ and $(1, 0)$.

49. Write in standard form the equation of the circle that has center $(-3, 5)$ and radius $\sqrt{3}$.

50. Find the center and radius for the circle
$x^2 + y^2 = x - 2y + 1$.

51. Find the x- and y-intercepts for the graph of $3x - 4y = 12$.

52. What is the y-intercept for the graph of $y = 5$?

53. Find the slope of the line that goes through $(3, -6)$ and $(-1, 2)$.

54. Find the slope of the line $3x - 4y = 9$.

55. Find the equation (in slope-intercept form) for the line through $(-2, 3)$ and $(5, -1)$.

56. Find the equation (in standard form using only integers) for the line through $(-1, -3)$ and $(2, -1)$.

57. Find the equation (in standard form using only integers) for the line through $(2, -4)$ that is perpendicular to $3x + y = -5$.

58. Find the equation (in slope-intercept form) for the line through $(2, -5)$ that is parallel to $2x - 3y = 5$.

Graph each set of ordered pairs. State the domain and range of each relation. Determine whether each set is a function.

59. $\{(0, 0), (1, 1), (-2, -2)\}$

60. $\{(0, -3), (1, -1), (2, 1), (2, 3)\}$

61. $\{(x, y) | y = 3 - x\}$

62. $\{(x, y) | 2x + y = 5\}$

63. $\{(x, y) | x = 2\}$

64. $\{(x, y) | y = 3\}$

65. $\{(x, y) | x^2 + y^2 = 0.01\}$

66. $\{(x, y) | x^2 + y^2 = 2x - 4y\}$

67. $\{(x, y) | x = y^2 + 1\}$

68. $\{(x, y) | y = |x - 2|\}$

69. $\{(x, y) | y = \sqrt{x} - 3\}$

70. $\{(x, y) | x = \sqrt{y}\}$

Let $f(x) = x^2 + 3$ and $g(x) = 2x - 7$. Find and simplify each expression.

71. $f(-3)$

72. $g(3)$

73. $g(12)$

74. $f(-1)$

75. x, if $f(x) = 19$

76. x, if $g(x) = 9$

77. $(g \circ f)(-3)$

78. $(f \circ g)(3)$

79. $(f + g)(2)$

80. $(f - g)(-2)$

81. $(f \cdot g)(-1)$

82. $(f/g)(4)$

83. $f(g(2))$

84. $g(f(-2))$

85. $(f \circ g)(x)$

86. $(g \circ f)(x)$

87. $(f \circ f)(x)$

88. $(g \circ g)(x)$

89. $f(a + 1)$

90. $g(a + 2)$

91. $\dfrac{f(3 + h) - f(3)}{h}$

92. $\dfrac{g(5 + h) - g(5)}{h}$

93. $\dfrac{f(x + h) - f(x)}{h}$

94. $\dfrac{g(x + h) - g(x)}{h}$

95. $g\left(\dfrac{x + 7}{2}\right)$

96. $f\left(\sqrt{x - 3}\right)$

97. $g^{-1}(x)$

98. $g^{-1}(-3)$

Use transformations to graph each pair of functions on the same coordinate plane.

99. $f(x) = \sqrt{x}, g(x) = 2\sqrt{x + 3}$

100. $f(x) = \sqrt{x}, g(x) = -2\sqrt{x} + 3$

101. $f(x) = |x|, g(x) = -2|x + 2| + 4$

102. $f(x) = |x|, g(x) = \dfrac{1}{2}|x - 1| - 3$

103. $f(x) = x^2, g(x) = \dfrac{1}{2}(x - 2)^2 + 1$

104. $f(x) = x^2, g(x) = -2x^2 + 4$

For each exercise, graph the function by transforming the given graph of $y = f(x)$.

105. $y = 2f(x - 2) + 1$

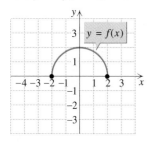

106. $y = 2f(x + 3) - 1$

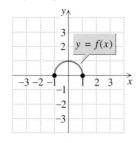

107. $y = -f(x + 1) - 3$

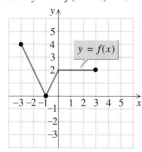

108. $y = -f(x - 1) + 2$

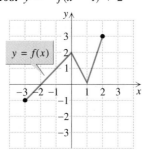

109. $y = -2f(x + 2)$

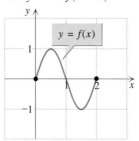

110. $y = -3f(x) + 1$

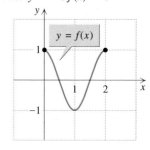

111. $y = -2f(x) + 3$

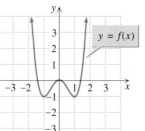

112. $y = 4f(x - 1) + 3$

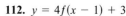
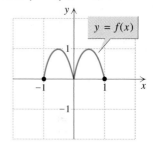

Let $f(x) = \sqrt[3]{x}$, $g(x) = x - 4$, $h(x) = x/3$, and $j(x) = x^2$. Write each function as a composition of the appropriate functions chosen from f, g, h, and j.

113. $F(x) = \sqrt[3]{x - 4}$

114. $G(x) = -4 + \sqrt[3]{x}$

115. $H(x) = \sqrt[3]{\dfrac{x^2 - 4}{3}}$

116. $M(x) = \left(\dfrac{x - 4}{3}\right)^2$

117. $N(x) = \dfrac{1}{3}x^{2/3}$

118. $P(x) = x^2 - 16$

119. $R(x) = \dfrac{x^2}{3} - 4$

120. $Q(x) = x^2 - 8x + 16$

Find the difference quotient for each function and simplify it.

121. $f(x) = -5x + 9$

122. $f(x) = \sqrt{x - 7}$

123. $f(x) = \dfrac{1}{2x}$

124. $f(x) = -5x^2 + x$

Sketch the graph of each function and state its domain and range. Determine the intervals on which the function is increasing, decreasing, or constant.

125. $f(x) = \sqrt{100 - x^2}$

126. $f(x) = -\sqrt{7 - x^2}$

127. $f(x) = \begin{cases} -x^2 & \text{for } x \le 0 \\ x^2 & \text{for } x > 0 \end{cases}$

128. $f(x) = \begin{cases} x^2 & \text{for } x \le 0 \\ x & \text{for } 0 < x \le 4 \end{cases}$

129. $f(x) = \begin{cases} -x - 4 & \text{for } x \le -2 \\ -|x| & \text{for } -2 < x < 2 \\ x - 4 & \text{for } x \ge 2 \end{cases}$

130. $f(x) = \begin{cases} (x+1)^2 - 1 & \text{for } x \le -1 \\ -1 & \text{for } -1 < x < 1 \\ (x-1)^2 - 1 & \text{for } x \ge 1 \end{cases}$

Each of the following graphs is from the absolute-value family. Construct a function for each graph and state the domain and range of the function.

131.

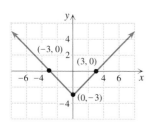

132.

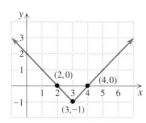

133.

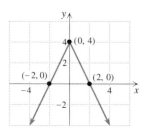

134.

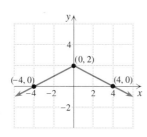

135.

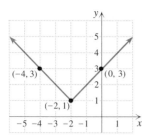

136.

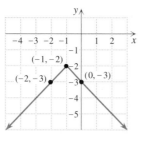

Determine whether the graph of each function is symmetric with respect to the y-axis or the origin.

137. $f(x) = x^4 - 9x^2$

138. $f(x) = |x| - 99$

139. $f(x) = -x^3 - 5x$

140. $f(x) = \dfrac{15}{x}$

141. $f(x) = -x + 1$

142. $f(x) = |x - 1|$

143. $f(x) = \sqrt{x^2}$

144. $f(x) = \sqrt{16 - x^2}$

Graph each pair of functions on the same coordinate plane. What is the relationship between the functions in each pair?

145. $f(x) = \sqrt{x + 3}, g(x) = x^2 - 3$ for $x \ge 0$

146. $f(x) = (x - 2)^3, g(x) = \sqrt[3]{x} + 2$

147. $f(x) = 2x - 4, g(x) = \dfrac{1}{2}x + 2$

148. $f(x) = -\dfrac{1}{2}x + 4, g(x) = -2x + 8$

Determine whether each function is invertible. If the function is invertible, then find the inverse function, and state the domain and range of the inverse function.

149. $\{(\pi, 0), (\pi/2, 1), (2\pi/3, 0)\}$

150. $\{(-2, 1/3), (-3, 1/4), (-4, 1/5)\}$

151. $f(x) = 3x - 21$

152. $f(x) = 3|x|$

153. $y = \sqrt{9 - x^2}$

154. $y = 7 - x$

155. $f(x) = \sqrt{x - 9}$

156. $f(x) = \sqrt{x} - 9$

157. $f(x) = \dfrac{x - 7}{x + 5}$

158. $f(x) = \dfrac{2x - 3}{5 - x}$

159. $f(x) = x^2 + 1$ for $x \le 0$

160. $f(x) = (x + 3)^4$ for $x \ge 0$

Solve each problem.

161. *Turning a Profit* Mary Beth buys roses for $1.20 each and sells them for $2 each at an outdoor market where she rents space for $40 per day. She buys and sells x roses per day. Construct her daily cost, revenue, and profit functions. How many roses must she sell to make a profit?

162. *Tin Can* The surface area S and volume V for a can with a top and bottom are given by

$$S = 2\pi r^2 + 2\pi rh \quad \text{and} \quad V = \pi r^2 h,$$

where r is the radius and h is the height. Suppose that a can has a volume of 1 ft^3.

a. Write its height as a function of its radius.

b. Write its radius as a function of its height.

c. Write its surface area as a function of its radius.

163. *Dropping the Ball*   A ball is dropped from a height of 64 feet. Its height h (in feet) is a function of time t (in seconds), where $h = -16t^2 + 64$ for t in the interval $[0, 2]$. Find the inverse of this function and state the domain of the inverse function.

164. *Sales Tax Function*   If S is the subtotal in dollars of your groceries before a 5% sales tax, then the function $T = 1.05S$ gives the total cost in dollars including tax. Write S as a function of T.

165. Write the diameter of a circle, d, as a function of its area, A.

166. *Circle Inscribed in a Square*   A cylindrical pipe with an outer radius r must fit snugly through a square hole in a wall. Write the area of the square as a function of the radius of the pipe.

167. *Load on a Spring*   When a load of 5 lb is placed on a spring, its length is 6 in., and when a load of 9 lb is placed on the spring, its length is 8 in. What is the average rate of change of the length of the spring as the load varies from 5 lb to 9 lb?

168. *Changing Speed of a Dragster*   Suppose that 2 sec after starting, a dragster is traveling 40 mph, and 5 sec after starting, the dragster is traveling 130 mph. What is the average rate of change of the speed of the dragster over the time interval from 2 sec to 5 sec? What are the units for this measurement?

Thinking Outside the Box X

Nines   Nine people applying for credit at the Highway 99 Loan Company listed nine different incomes each containing a different number of digits. Each of the nine incomes was a whole number of dollars and the maximum income was a nine-digit number. The loan officer found that the arithmetic mean of the nine incomes was $123,456,789. What are the nine incomes?

Concepts of Calculus

Limits

In algebra we can evaluate algebraic expressions for any acceptable value of the variable. The following tables show values of $x^2 + 5$ for certain values of x.

x	2.9	2.99	2.999
$x^2 + 5$	13.41	13.9401	13.994001

x	3.1	3.01	3.001
$x^2 + 5$	14.61	14.0601	14.006001

In calculus we look for trends. What happens to the value of $x^2 + 5$ as x gets closer and closer to 3? From the tables we see that the closer x is to 3, the closer $x^2 + 5$ is to 14. We say that the limit of $x^2 + 5$ as x approaches 3 is 14, and abbreviate this statement as $\lim_{x\to 3}(x^2 + 5) = 14$ or $\lim_{x\to 3}x^2 + 5 = 14$.

Note that we get 14 if we evaluate $3^2 + 5$, but that is not the idea of limits. We are looking for the trend as x approaches but never actually reaches a number. In fact, we often let x approach a number for which the expression cannot be evaluated.

Exercises

1. a. Fill in the second row of each table.

x	1.9	1.99	1.999
$5x - 4$			

x	2.1	2.01	2.001
$5x - 4$			

  b. Can you evaluate $5x - 4$ for $x = 2$?

  c. What is $\lim_{x\to 2}(5x - 4)$?

2. a. Fill in the second row of each table.

x	0.6	0.66	0.666
$\dfrac{24x^2 - 25x + 6}{3x - 2}$			

x	0.7	0.67	0.667
$\dfrac{24x^2 - 25x + 6}{3x - 2}$			

  b. Can you evaluate $\dfrac{24x^2 - 25x + 6}{3x - 2}$ for $x = 2/3$?

  c. What is $\lim_{x\to 2/3}\dfrac{24x^2 - 25x + 6}{3x - 2}$?

3. a. Fill in the second row of each table.

x	0.01	0.0001	0.000001
$(1 + |x|)^{1/|x|}$			

x	-0.01	-0.0001	-0.00001
$(1 + |x|)^{1/|x|}$			

  b. Can you evaluate $(1 + |x|)^{1/|x|}$ for $x = 0$?

  c. What is $\lim_{x\to 0}(1 + |x|)^{1/|x|}$?

4. a. Use a calculator in radian mode to fill in the second row of each table.

x	0.1	0.001	0.0001
$\dfrac{\sin(x)}{x}$			

x	-0.1	-0.001	-0.0001
$\dfrac{\sin(x)}{x}$			

  b. Can you evaluate $\dfrac{\sin(x)}{x}$ for $x = 0$?

  c. What is $\lim_{x\to 0}\dfrac{\sin(x)}{x}$?

2

Polynomial and Rational Functions

The Golden Gate Bridge

Location   San Francisco and Sausalito, California

Completion Date   1937

Cost $27 million

Length   8981 feet

Longest Single Span   4200 feet

Engineer   Joseph B. Strauss

O ften called the most spectacular bridge in the world, the Golden Gate Bridge is located less than 8 miles from the epicenter of the most catastrophic earthquake in history. The bridge is also subject to brutal Pacific winds, tide, and fog. When it was first conceived, building the Golden Gate Bridge seemed like an impossible task. But engineer Joseph Strauss was willing to gamble that his bridge could endure all of these destructive forces. Strauss used more than one million tons of concrete to build the massive blocks that grip the bridge's supporting cables. He completed the bridge only five months after the promised date and $1.3 million under budget. For his efforts Strauss received $1 million and a lifetime bridge pass.

2.1 Quadratic Functions and Inequalities

A **polynomial function** is defined by a polynomial. A **quadratic function** is defined by a quadratic or second-degree polynomial. So a quadratic function has the form $f(x) = ax^2 + bx + c$ where $a \neq 0$. We studied functions of the form $f(x) = a(x - h)^2 + k$ in Section 1.7 (though we did not call them quadratic functions there). Once we see that these forms are equivalent, we can use what we learned in Section 1.7 in our study of quadratic functions here.

Two Forms for Quadratic Functions

It is easy to convert $f(x) = a(x - h)^2 + k$ to the form $f(x) = ax^2 + bx + c$ by squaring the binomial. To convert in the other direction we use completing the square from Section 1.3. Recall that the perfect square trinomial whose first two terms are $x^2 + bx$ is $x^2 + bx + \left(\frac{b}{2}\right)^2$. So to get the last term you *take one-half of the coefficient of the middle term and square it.*

When completing the square in Section 1.3, we added the last term of the perfect square trinomial to both sides of the equation. Here we want to keep $f(x)$ on the left side. To complete the square and change only the right side, we add and subtract on the right side as shown in Example 1(a). Another complication here is that the coefficient of x^2 is not always 1. To overcome this problem, you must factor before completing the square as shown in Example 1(b).

Example **1** Completing the square for a quadratic function

Rewrite each function in the form $f(x) = a(x - h)^2 + k$.

a. $f(x) = x^2 + 6x$ **b.** $f(x) = 2x^2 - 20x + 3$

Solution

a. One-half of 6 is 3 and 3^2 is 9. Adding and subtracting 9 on the right side of $f(x) = x^2 + 6x$ does not change the function.

$$f(x) = x^2 + 6x$$
$$= x^2 + 6x + 9 - 9 \qquad 9 = \left(\frac{1}{2} \cdot 6\right)^2$$
$$= (x + 3)^2 - 9 \qquad \text{Factor.}$$

b. First factor 2 out of the first two terms, because the leading coefficient in $(x - h)^2$ is 1.

$$f(x) = 2x^2 - 20x + 3$$
$$= 2(x^2 - 10x) + 3$$
$$= 2(x^2 - 10x + 25 - 25) + 3 \qquad 25 = \left(\frac{1}{2} \cdot 10\right)^2$$
$$= 2(x^2 - 10x + 25) - 50 + 3 \qquad \text{Remove } -25 \text{ from the parentheses.}$$
$$= 2(x - 5)^2 - 47$$

Because of the 2 preceding the parentheses, the second 25 was doubled when it was removed from the parentheses. ■

In the next example we complete the square and graph a quadratic function.

Example **2** Graphing a quadratic function

Rewrite $f(x) = -2x^2 - 4x + 3$ in the form $f(x) = a(x - h)^2 + k$ and sketch its graph.

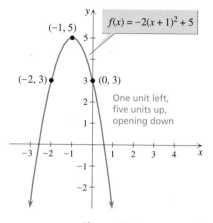

Figure 2.1

Solution

Start by completing the square:

$$f(x) = -2(x^2 + 2x) + 3 \qquad \text{Factor out } -2 \text{ from the first two terms.}$$

$$= -2(x^2 + 2x + 1 - 1) + 3 \qquad \text{Complete the square for } x^2 + 2x.$$

$$= -2(x^2 + 2x + 1) + 2 + 3 \qquad \text{Remove } -1 \text{ from the parentheses.}$$

$$= -2(x + 1)^2 + 5$$

The function is now in the form $f(x) = a(x - h)^2 + k$. The number 1 indicates that the graph of $f(x) = x^2$ is translated one unit to the left. Since $a = -2$, the graph is stretched by a factor of 2 and reflected below the x-axis. Finally, the graph is translated five units upward. The graph shown in Fig. 2.1 includes the points $(0, 3)$, $(-1, 5)$, and $(-2, 3)$. ■

We can use completing the square as in Example 2 to prove the following theorem.

Theorem:
Quadratic Functions

> The graph of any quadratic function is a transformation of the graph of $f(x) = x^2$.

PROOF Since $f(x) = a(x - h)^2 + k$ is a transformation of $f(x) = x^2$, we will show that $f(x) = ax^2 + bx + c$ can be written in that form:

$$f(x) = ax^2 + bx + c$$

$$= a\left(x^2 + \frac{b}{a}x\right) + c \qquad \text{Factor } a \text{ out of the first two terms.}$$

To complete the square for $x^2 + \frac{b}{a}x$, add and subtract $\frac{b^2}{4a^2}$ inside the parentheses:

$$f(x) = a\left(x^2 + \frac{b}{a}x + \frac{b^2}{4a^2} - \frac{b^2}{4a^2}\right) + c$$

$$= a\left(x^2 + \frac{b}{a}x + \frac{b^2}{4a^2}\right) - \frac{b^2}{4a} + c \qquad \text{Remove } -\frac{b^2}{4a^2} \text{ from the parentheses.}$$

$$= a\left(x + \frac{b}{2a}\right)^2 + \frac{4ac - b^2}{4a} \qquad \text{Factor and get a common denominator.}$$

$$= a(x - h)^2 + k. \qquad \text{Let } h = -\frac{b}{2a} \text{ and } k = \frac{4ac - b^2}{4a}.$$

So the graph of any quadratic function is a transformation of $f(x) = x^2$. ▪

In Section 1.7 we stated that any transformation of $f(x) = x^2$ is called a parabola. Therefore, the graph of any quadratic function is a parabola.

The technique of completing the square that was used in the preceding proof is also used to prove the well-known *quadratic formula.* In fact, if you replace $f(x)$ with 0 and solve

$$0 = a\left(x + \frac{b}{2a}\right)^2 + \frac{4ac - b^2}{4a}$$

for x, you get the quadratic formula. You should complete the details.

Theorem:
The Quadratic Formula

> The solutions to $ax^2 + bx + c = 0$ for $a \neq 0$ are given by
>
> $$x = \frac{-b \pm \sqrt{b^2 - 4ac}}{2a}.$$

Opening, Vertex, and Axis of Symmetry

If $a > 0$, the graph of $f(x) = a(x - h)^2 + k$ **opens upward;** if $a < 0$, the graph **opens downward** as shown in Fig. 2.2. Notice that h determines the amount of horizontal translation and k determines the amount of vertical translation of the graph of $f(x) = x^2$. Because of the translations, the point $(0, 0)$ on the graph of $f(x) = x^2$ moves to the point (h, k) on the graph of $f(x) = a(x - h)^2 + k$. Since $(0, 0)$ is the lowest point on the graph of $f(x) = x^2$, the lowest point on any parabola that opens upward is (h, k). Since $(0, 0)$ is the highest point on the graph of $f(x) = -x^2$, the highest point on any parabola that opens downward is (h, k). The point (h, k) is called the **vertex** of the parabola.

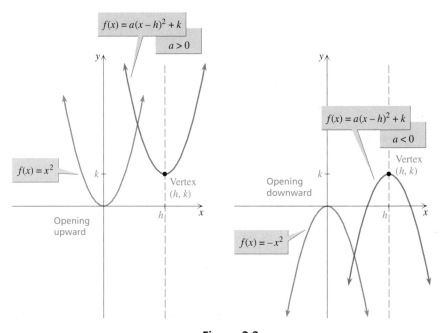

▪ **Figure 2.2**

You can use a graphing calculator to experiment with different values for a, h, and k. Two possibilities are shown in Fig. 2.3(a) and (b). The graphing calculator allows you to see results quickly and to use values that you would not use otherwise. □

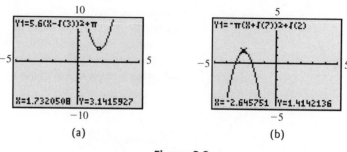

(a)   (b)

■ **Figure 2.3**

We can also find the vertex of the parabola when the function is written in the form $f(x) = ax^2 + bx + c$. The x-coordinate of the vertex is $-b/(2a)$ because we used $h = -b/(2a)$ when we obtained $y = a(x - h)^2 + k$ from $y = ax^2 + bx + c$. The y-coordinate is found by substitution.

SUMMARY   **Vertex of a Parabola**

1. For a quadratic function in the form $f(x) = a(x - h)^2 + k$ the vertex of the parabola is (h, k).

2. For the form $f(x) = ax^2 + bx + c$, the x-coordinate of the vertex is $\frac{-b}{2a}$. The y-coordinate is $f\left(\frac{-b}{2a}\right)$.

Example **3** **Finding the vertex**

Find the vertex of the graph of $f(x) = -2x^2 - 4x + 3$.

Solution

Use $x = -b/(2a)$ to find the x-coordinate of the vertex:

$$x = \frac{-b}{2a} = \frac{-(-4)}{2(-2)} = -1$$

Now find $f(-1)$:

$$f(-1) = -2(-1)^2 - 4(-1) + 3 = 5$$

The vertex is $(-1, 5)$. In Example 2, $f(x) = -2x^2 - 4x + 3$ was rewritten as $f(x) = -2(x + 1)^2 + 5$. In this form we see immediately that the vertex is $(-1, 5)$. The graph in Fig. 2.4 supports these conclusions. ■

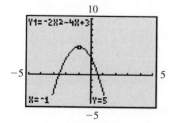

■ **Figure 2.4**

The domain of every quadratic function $f(x) = a(x - h)^2 + k$ is the set of real numbers, $(-\infty, \infty)$. The range of a quadratic function is determined from the second

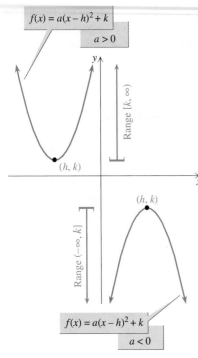

Figure 2.5

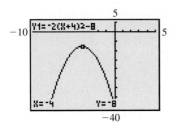

Figure 2.6

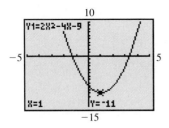

Figure 2.7

coordinate of the vertex. If $a > 0$, the range is $[k, \infty)$ and k is called the **minimum value of the function.** The function is decreasing on $(-\infty, h)$ and increasing on (h, ∞). See Fig. 2.5. If $a < 0$, the range is $(-\infty, k]$ and k is called the **maximum value of the function.** The function is increasing on $(-\infty, h)$ and decreasing on (h, ∞).

The graph of $f(x) = x^2$ is symmetric about the y-axis, which runs vertically through the vertex of the parabola. Since this symmetry is preserved in transformations, the graph of any quadratic function is symmetric about the vertical line through its vertex. The vertical line $x = -b/(2a)$ is called the **axis of symmetry** for the graph of $f(x) = ax^2 + bx + c$. Identifying these characteristics of a parabola before drawing a graph makes graphing easier and more accurate.

Example **4** **Identifying the characteristics of a parabola**

For each parabola, determine whether the parabola opens upward or downward, and find the vertex, axis of symmetry, and range of the function. Find the maximum or minimum value of the function and the intervals on which the function is increasing or decreasing.

a. $y = -2(x + 4)^2 - 8$ **b.** $y = 2x^2 - 4x - 9$

Solution

a. Since $a = -2$, the parabola opens downward. In $y = a(x - h)^2 + k$, the vertex is (h, k). So the vertex is $(-4, -8)$. The axis of symmetry is the vertical line through the vertex, $x = -4$. Since the parabola opens downward from $(-4, -8)$, the range of the function is $(-\infty, -8]$. The maximum value of the function is -8, and the function is increasing on $(-\infty, -4)$ and decreasing on $(-4, \infty)$.

◁◁ The graph in Fig. 2.6 supports these results. □

b. Since $a = 2$, the parabola opens upward. In the form $y = ax^2 + bx + c$, the x-coordinate of the vertex is $x = -b/(2a)$. In this case,

$$x = \frac{-b}{2a} = \frac{-(-4)}{2(2)} = 1.$$

Use $x = 1$ to find $y = 2(1)^2 - 4(1) - 9 = -11$. So the vertex is $(1, -11)$, and the axis of symmetry is the vertical line $x = 1$. Since the parabola opens upward, the vertex is the lowest point and the range is $[-11, \infty)$. The function is decreasing on $(-\infty, 1)$ and increasing on $(1, \infty)$. The minimum value of the function is -11.

◁◁ The graph in Fig. 2.7 supports these results. ■

Intercepts

The x-intercepts and the y-intercept are important points on the graph of a parabola. The x-intercepts are used in solving quadratic inequalities and the y-intercept is the starting point on the graph of a function whose domain is the nonnegative real numbers. The y-intercept is easily found by letting $x = 0$. The x-intercepts are found by letting $y = 0$ and solving the resulting quadratic equation.

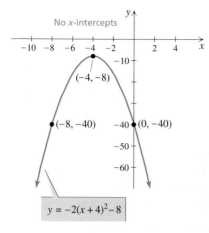

No x-intercepts

$(-4, -8)$

$(-8, -40)$ $(0, -40)$

$y = -2(x + 4)^2 - 8$

Figure 2.8

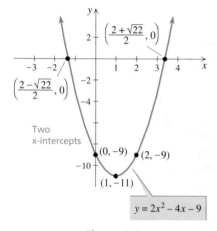

$\left(\dfrac{2 + \sqrt{22}}{2}, 0\right)$

$\left(\dfrac{2 - \sqrt{22}}{2}, 0\right)$

Two
x-intercepts

$(0, -9)$ $(2, -9)$

$(1, -11)$

$y = 2x^2 - 4x - 9$

Figure 2.9

$x - 3$

x

Figure 2.10

Example **5** **Finding the intercepts**

Find the y-intercept and the x-intercepts for each parabola and sketch the graph of each parabola.

a. $y = -2(x + 4)^2 - 8$ **b.** $y = 2x^2 - 4x - 9$

Solution

a. If $x = 0, y = -2(0 + 4)^2 - 8 = -40$. The y-intercept is $(0, -40)$. Because the graph is symmetric about the line $x = -4$, the point $(-8, -40)$ is also on the graph. Since the parabola opens downward from $(-4, -8)$, which is below the x-axis, there are no x-intercepts. If we try to solve $-2(x + 4)^2 - 8 = 0$ to find the x-intercepts, we get $(x + 4)^2 = -4$, which has no real solution. The graph is shown in Fig. 2.8.

b. If $x = 0, y = 2(0)^2 - 4(0) - 9 = -9$. The y-intercept is $(0, -9)$. From Example 4(b), the vertex is $(1, -11)$. Because the graph is symmetric about the line $x = 1$, the point $(2, -9)$ is also on the graph. The x-intercepts are found by solving $2x^2 - 4x - 9 = 0$:

$$x = \frac{4 \pm \sqrt{(-4)^2 - 4(2)(-9)}}{2(2)} = \frac{4 \pm \sqrt{88}}{4} = \frac{2 \pm \sqrt{22}}{2}$$

The x-intercepts are $\left(\dfrac{2 + \sqrt{22}}{2}, 0\right)$ and $\left(\dfrac{2 - \sqrt{22}}{2}, 0\right)$. See Fig. 2.9. ■

Note that if $y = ax^2 + bx + c$ has x-intercepts, they can always be found by using the quadratic formula. The x-coordinates of the x-intercepts are

$$x = \frac{-b \pm \sqrt{b^2 - 4ac}}{2a} = \frac{-b}{2a} \pm \frac{\sqrt{b^2 - 4ac}}{2a}.$$

Note how the axis of symmetry, $x = -b/(2a)$, appears in this formula. The x-intercepts are on opposite sides of the graph and are equidistant from the axis of symmetry.

Quadratic Inequalities

A **sign graph** is a number line that shows where the value of an expression is positive, negative, or zero. For example, the expression $x - 3$ has a positive value if $x > 3$ and a negative value if $x < 3$. If $x = 3$, then the value of $x - 3$ is zero. This information is shown on the sign graph in Fig. 2.10. We can use sign graphs and the rules for multiplying signed numbers to solve **quadratic inequalities**—inequalities that involve quadratic polynomials.

Example **6** **Solving a quadratic inequality using a sign graph of the factors**

Solve each inequality. Write the solution set in interval notation and graph it.

a. $x^2 - x > 6$ **b.** $x^2 - x \leq 6$

Solution

a. Write the inequality as $x^2 - x - 6 > 0$ and factor to get

$$(x - 3)(x + 2) > 0.$$

Note that $x + 2 > 0$ if $x > -2$ and $x + 2 < 0$ if $x < -2$. If $x = -2$, then $x + 2 = 0$. Combine this information with the sign graph for $x - 3$ in Fig. 2.10 to get Fig. 2.11, which shows the signs of both factors.

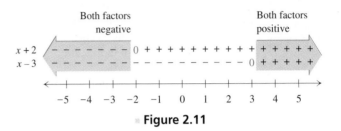

Figure 2.11

The product $(x - 3)(x + 2)$ is positive only where both factors are positive or both factors are negative. So the solution set is $(-\infty, -2) \cup (3, \infty)$ and its graph is in Fig. 2.12.

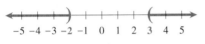

Figure 2.12

b. The inequality $x^2 - x \leq 6$ is equivalent to $(x - 3)(x + 2) \leq 0$. The product $(x - 3)(x + 2)$ is negative when x is between -2 and 3, because that is when the factors have opposite signs on the sign graph in Fig. 2.11. Because $(x - 3)(x + 2) = 0$ if $x = 3$ or $x = -2$, these points are included in the solution set $[-2, 3]$. The graph is shown in Fig. 2.13.

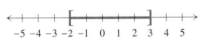

Figure 2.13

The graph of $y = x^2 - x - 6$ in Fig. 2.14 is above the x-axis (y-coordinate greater than 0) when x is in $(-\infty, -2) \cup (3, \infty)$ and on or below the x-axis (y-coordinate less than or equal to 0) when x is in $[-2, 3]$. ■

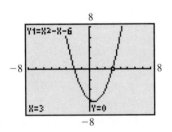

Figure 2.14

Making a sign graph of the factors as in Example 6 shows how the signs of the linear factors determine the solution to a quadratic inequality. Of course, that method works only if you can factor the quadratic polynomial. The **test-point method** works on all quadratic polynomials. For this method, we find the roots of the quadratic polynomial (using the quadratic formula if necessary) and then make a sign graph for the quadratic polynomial itself, not the factors. The signs of the quadratic polynomial are determined by testing numbers in the intervals determined by the roots.

Example **7** **Solving a quadratic inequality using the test-point method**

Solve $2x^2 - 4x - 9 < 0$. Write the solution set in interval notation.

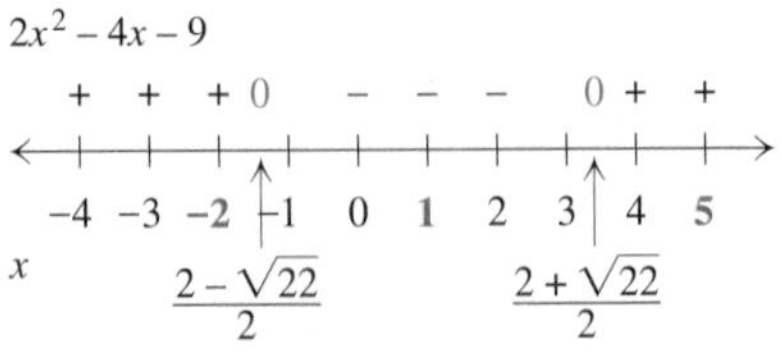

Solution

The roots to $2x^2 - 4x - 9 = 0$ were found to be $\dfrac{2 \pm \sqrt{22}}{2}$ in Example 5(b). Since $\dfrac{2 - \sqrt{22}}{2} \approx -1.3$ and $\dfrac{2 + \sqrt{22}}{2} \approx 3.3$, we make the number line as in Fig. 2.15. Select a convenient test point in each of the three intervals determined by the roots.

Figure 2.15

Our selections -2, 1, and 5 are shown in red on the number line. Now evaluate $2x^2 - 4x - 9$ for each test point:

$$2(-2)^2 - 4(-2) - 9 = 7 \quad \text{Positive}$$
$$2(1)^2 - 4(1) - 9 = -11 \quad \text{Negative}$$
$$2(5)^2 - 4(5) - 9 = 21 \quad \text{Positive}$$

The signs of these results are shown on the number line in Fig. 2.15. Since $2x^2 - 4x - 9 < 0$ between the roots, the solution set is

$$\left(\frac{2 - \sqrt{22}}{2}, \frac{2 + \sqrt{22}}{2} \right).$$

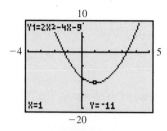

Figure 2.16

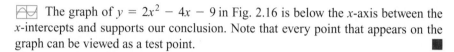 The graph of $y = 2x^2 - 4x - 9$ in Fig. 2.16 is below the x-axis between the x-intercepts and supports our conclusion. Note that every point that appears on the graph can be viewed as a test point. ■

If there are no real solutions to the quadratic equation that corresponds to a quadratic inequality, then the solution set to the inequality is either the empty set or all real numbers. For example, $x^2 + 4 = 0$ has no real solutions. Use any number as a test point to find that the solution set to $x^2 + 4 > 0$ is $(-\infty, \infty)$ and the solution set to $x^2 + 4 < 0$ is the empty set $\varnothing$.

Applications of Maximum and Minimum

If one variable is a quadratic function of another, then the maximum or minimum value of the dependent variable occurs at the vertex of the graph of the quadratic function. A nice application of this idea occurs in modeling projectile motion. The formula $h(t) = -16t^2 + v_0 t + h_0$ is used to find the height in feet at time t in seconds for a projectile that is launched in a vertical direction with initial velocity of v_0 feet per second from an initial height of h_0 feet. Since the height is a quadratic function of time, the maximum height occurs at the vertex of the parabola.

■ **Foreshadowing Calculus**

In algebra we can maximize or minimize a quadratic function because we can easily find the vertex of a parabola. However, max/min problems involving other functions are usually solved using techniques of calculus.

Example **8** **Finding maximum height of a projectile**

A ball is tossed straight upward with an initial velocity of 80 feet per second from a rooftop that is 12 feet above ground level. The height of the ball in feet at time t in seconds is given by $h(t) = -16t^2 + 80t + 12$. Find the maximum height above ground level for the ball.

Solution

Since the height is a quadratic function of t with a negative leading coefficient, the height has a maximum value at the vertex of the parabola. Use $-b/(2a)$ to find the t-coordinate of the vertex. Since $a = -16$ and $b = 80$

$$\frac{-b}{2a} = \frac{-80}{2(-16)} = 2.5.$$

So the ball reaches its maximum height at time $t = 2.5$ seconds. Now

$$h(2.5) = -16(2.5)^2 + 80(2.5) + 12 = 112.$$

So the maximum height of the ball is 112 feet above the ground. ■

In Example 8, the quadratic function was given. In the next example we need to write the quadratic function and then find its maximum value.

■ **Figure 2.17**

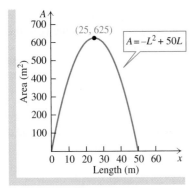

■ **Figure 2.18**

Example **9** **Maximizing area of a rectangle**

If 100 m of fencing will be used to fence a rectangular region, then what dimensions for the rectangle will maximize the area of the region?

Solution

Since the 100 m of fencing forms the perimeter of a rectangle as shown in Fig. 2.17, we have $2L + 2W = 100$, where W is its width and L is its length. Dividing by 2 we get $L + W = 50$ or $W = 50 - L$. Since $A = LW$ for a rectangle, by substituting we get

$$A = LW = L(50 - L) = -L^2 + 50L.$$

So the area is a quadratic function of the length. The graph of this function is the parabola in Fig. 2.18. Since the parabola opens downward, the maximum value of A occurs when

$$L = \frac{-b}{2a} = \frac{-50}{2(-1)} = 25.$$

If $L = 25$ then $W = 25$ also, since $W = 50 - L$. So the length should be 25 meters and the width 25 meters to get the maximum area. The rectangle that gives the maximum area is actually a square with an area of 625 m^2. ■

For Thought

True or False? Explain.

1. The domain and range of a quadratic function are $(-\infty, \infty)$.

2. The vertex of the graph of $y = 2(x - 3)^2 - 1$ is $(3, 1)$.

3. The graph of $y = -3(x + 2)^2 - 9$ has no x-intercepts.

4. The maximum value of y in the function $y = -4(x - 1)^2 + 9$ is 9.

5. For $y = 3x^2 - 6x + 7$, the value of y is at its minimum when $x = 1$.

6. The graph of $f(x) = 9x^2 + 12x + 4$ has one x-intercept and one y-intercept.

7. The graph of every quadratic function has exactly one y-intercept.

8. The inequality $\pi(x - \sqrt{3})^2 + \pi/2 \le 0$ has no solution.

9. The maximum area of a rectangle with fixed perimeter p is $p^2/16$.

10. The function $f(x) = (x - 3)^2$ is increasing on the interval $[-3, \infty)$.

2.1 Exercises

Write each quadratic function in the form $y = a(x - h)^2 + k$ and sketch its graph.

1. $y = x^2 + 4x$

2. $y = x^2 - 6x$

3. $y = x^2 - 3x$

4. $y = x^2 + 5x$

5. $y = 2x^2 - 12x + 22$

6. $y = 3x^2 - 12x + 1$

7. $y = -3x^2 + 6x - 3$

8. $y = -2x^2 - 4x + 8$

9. $y = x^2 + 3x + \dfrac{5}{2}$

10. $y = x^2 - x + 1$

11. $y = -2x^2 + 3x - 1$

12. $y = 3x^2 + 4x + 2$

Find the vertex of the graph of each quadratic function. See the summary on finding the vertex on page 117.

13. $f(x) = 3x^2 - 12x + 1$

14. $f(x) = -2x^2 - 8x + 9$

15. $f(x) = -3(x - 4)^2 + 1$

16. $f(x) = \dfrac{1}{2}(x + 6)^2 - \dfrac{1}{4}$

17. $y = -\dfrac{1}{2}x^2 - \dfrac{1}{3}x$

18. $y = \dfrac{1}{4}x^2 + \dfrac{1}{2}x - 1$

From the graph of each parabola, determine whether the parabola opens upward or downward, and find the vertex, axis of symmetry, and range of the function. Find the maximum or minimum value of the function and the intervals on which the function is increasing or decreasing.

19.

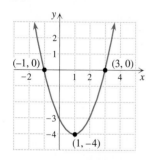

20.

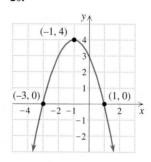

Find the range of each quadratic function and the maximum or minimum value of the function. Identify the intervals on which each function is increasing or decreasing.

21. $f(x) = 3 - x^2$

22. $f(x) = 5 - x^2$

23. $y = (x - 1)^2 - 1$

24. $y = (x + 3)^2 + 4$

25. $y = x^2 + 8x - 2$

26. $y = x^2 - 2x - 3$

27. $y = \dfrac{1}{2}(x - 3)^2 + 4$

28. $y = -\dfrac{1}{3}(x + 6)^2 + 37$

29. $f(x) = -2x^2 + 6x + 9$

30. $f(x) = -3x^2 - 9x + 4$

31. $y = -\dfrac{3}{4}\left(x - \dfrac{1}{2}\right)^2 + 9$

32. $y = \dfrac{3}{2}\left(x - \dfrac{1}{3}\right)^2 - 6$

Identify the vertex, axis of symmetry, y-intercept, x-intercepts, and opening of each parabola, then sketch the graph.

33. $y = x^2 - 3$

34. $y = 8 - x^2$

35. $y = x^2 - x$

36. $y = 2x - x^2$

37. $f(x) = x^2 + 6x + 9$

38. $f(x) = x^2 - 6x$

39. $f(x) = (x - 3)^2 - 4$

40. $f(x) = (x + 1)^2 - 9$

41. $y = -3(x - 2)^2 + 12$

42. $y = -2(x + 3)^2 + 8$

43. $y = -2x^2 + 4x + 1$

44. $y = -x^2 + 2x - 6$

Solve each inequality by making a sign graph of the factors. State the solution set in interval notation and graph it.

45. $2x^2 - x - 3 < 0$

46. $3x^2 - 4x - 4 \le 0$

47. $2x + 15 < x^2$

48. $5x - x^2 < 4$

49. $w^2 - 4w - 12 \ge 0$

50. $y^2 + 8y + 15 \le 0$

51. $t^2 \le 16$

52. $36 \le h^2$

53. $a^2 + 6a + 9 \le 0$

54. $c^2 + 4 \le 4c$

55. $4z^2 - 12z + 9 > 0$

56. $9s^2 + 6s + 1 \ge 0$

Solve each inequality by using the test-point method. State the solution set in interval notation and graph it.

57. $x^2 - 4x + 2 < 0$

58. $x^2 - 4x + 1 \le 0$

59. $x^2 - 9 > 1$

60. $6 < x^2 - 1$

61. $y^2 + 18 > 10y$

62. $y^2 + 3 \ge 6y$

63. $p^2 + 9 > 0$

64. $-5 - s^2 < 0$

65. $a^2 + 20 \le 8a$

66. $6t \le t^2 + 25$

67. $-2w^2 + 5w < 6$

68. $-3z^2 - 5 > 2z$

Identify the solution set to each quadratic inequality by inspecting the graphs of $y = x^2 - 2x - 3$ *and* $y = -x^2 - 2x + 3$ *as shown.*

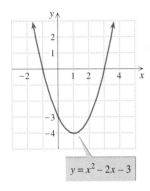

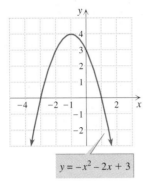

69. $x^2 - 2x - 3 \geq 0$

70. $x^2 - 2x - 3 < 0$

71. $-x^2 - 2x + 3 > 0$

72. $-x^2 - 2x + 3 \leq 0$

73. $x^2 + 2x \leq 3$

74. $x^2 \leq 2x + 3$

The next two exercises incorporate many concepts of quadratics.

75. Let $f(x) = x^2 - 3x - 10$.
   a. Solve $f(x) = 0$.   **b.** Solve $f(x) = -10$.

   c. Solve $f(x) > 0$.   **d.** Solve $f(x) \leq 0$.

   e. Write f in the form $f(x) = a(x - h)^2 + k$ and describe the graph of f as a transformation of the graph of $y = x^2$.

   f. Graph f and state the domain, range, and the maximum or minimum y-coordinate on the graph.

   g. What is the relationship between the graph of f and the answers to parts (c) and (d)?

   h. Find the intercepts, axis of symmetry, vertex, opening, and intervals on which f is increasing or decreasing.

76. Repeat parts (a) through (h) from the previous exercise for $f(x) = -x^2 + 2x + 1$.

Solve each problem.

77. *Maximum Height of a Football*   If a football is kicked straight up with an initial velocity of 128 ft/sec from a height of 5 ft, then its height above the earth is a function of time given by $h(t) = -16t^2 + 128t + 5$. What is the maximum height reached by this ball?

78. *Maximum Height of a Ball*   If a juggler can toss a ball into the air at a velocity of 64 ft/sec from a height of 6 ft, then what is the maximum height reached by the ball?

79. *Shooting an Arrow*   If an archer shoots an arrow straight upward with an initial velocity of 160 ft/sec from a height of 8 ft, then its height above the ground in feet at time t in seconds is given by the function

$$h(t) = -16t^2 + 160t + 8.$$

   a. What is the maximum height reached by the arrow?

   b. How long does it take for the arrow to reach the ground?

80. *Rocket Propelled Grenade*   If a soldier in basic training fires a rocket propelled grenade (RPG) straight up from ground level with an initial velocity of 256 ft/sec, then its height above the ground in feet at time t in seconds is given by the function

$$h(t) = -16t^2 + 256t.$$

   a. What is the maximum height reached by the RPG?

   b. How long does it take for the RPG to reach the ground?

81. *Maximum Area*   Shondra wants to enclose a rectangular garden with 200 yards of fencing. What dimensions for the garden will maximize its area.

82. *Mirror Mirror*   Chantel wants to make a rectangular frame for a mirror using 10 feet of frame molding. What dimensions will maximize the area of the mirror assuming that there is no waste?

83. *Twin Kennels*   Martin plans to construct a rectangular kennel for two dogs using 120 feet of chain-link fencing. He plans to fence all four sides and down the middle to keep the dogs separate. What overall dimensions will maximize the total area fenced?

84. *Cross Fenced*   Kim wants to construct rectangular pens for four animals with 400 feet of fencing. To get four separate pens she will fence a large rectangle and then fence through the middle of the rectangle parallel to the length and parallel to the width. What overall dimensions will maximize the total area of the pens?

85. *Big Barn*   Mike wants to enclose a rectangular area for his rabbits alongside his large barn using 30 feet of fencing. What dimensions will maximize the area fenced if the barn is used for one side of the rectangle?

86. *Maximum Area*   Kevin wants to enclose a rectangular garden using 14 eight-ft railroad ties, which he cannot cut. What are the dimensions of the rectangle that maximize the area enclosed?

87. *Maximizing Revenue*   Mona Kalini gives a walking tour of Honolulu to one person for $49. To increase her business, she advertised at the National Orthodontist Convention that she would lower the price by $1 per person for each additional person, up to 49 people.

a. Write the price per person p as a function of the number of people n.

b. Write her revenue as a function of the number of people on the tour.

c. What is the maximum revenue for her tour?

88. *Concert Tickets*   At $10 per ticket, Willie Williams and the Wranglers will fill all 8000 seats in the Assembly Center. The manager knows that for every $1 increase in the price, 500 tickets will go unsold.

   a. Write the number of tickets sold n as a function of ticket price p.

b. Write the total revenue as a function of the ticket price.

c. What ticket price will maximize the revenue?

Thinking Outside the Box XI

Overlapping Region   A right triangle with sides of length 3, 4, and 5 is drawn so that the endpoints of the side of length 5 are $(0, 0)$ and $(5, 0)$. A square with sides of length 1 is drawn so that its center is the vertex of the right angle and its sides are parallel to the x- and y-axes. What is the area of the region where the square and the triangle overlap?

2.1 Pop Quiz

1. Write $y = 2x^2 + 16x - 1$ in the form $y = a(x - h)^2 + k$.

2. Find the vertex of the graph of $y = 3(x + 4)^2 + 8$.

3. Find the range of $f(x) = -x^2 - 4x + 9$.

4. Find the minimum y-value for $y = x^2 - 3x$.

5. Find the x-intercepts and axis of symmetry for
$y = x^2 - 2x - 8$.

6. Solve the inequality $x^2 + 4x < 0$.

2.2 Complex Numbers

Our system of numbers developed as the need arose. Numbers were first used for counting. As society advanced, the rational numbers were formed to express fractional parts and ratios. Negative numbers were invented to express losses or debts. When it was discovered that the exact size of some very real objects could not be expressed with rational numbers, the irrational numbers were added to the system, forming the set of real numbers. Later still, there was a need for another expansion to the number system. In this section we study that expansion, the set of complex numbers.

Definitions

There are no even roots of negative numbers in the set of real numbers. So, the real numbers are inadequate or incomplete in this regard. The imaginary numbers were invented to complete the set of real numbers. Using real and imaginary numbers, every nonzero real number has *two* square roots, *three* cube roots, *four* fourth roots, and so on. (Actually finding all of the roots of any real number is done in trigonometry.)

   The imaginary numbers are based on the solution of the equation $x^2 = -1$. Since no real number solves this equation, a solution is called an *imaginary number*. The number i is defined to be a solution to this equation: i is a number whose square is -1.

Definition: The Number i	The number i is defined by

$$i^2 = -1.$$

We may also write $i = \sqrt{-1}$.

A complex number is formed as a real number plus a real multiple of i.

Definition: Complex Numbers	The set of **complex numbers** is the set of all numbers of the form $a + bi$, where a and b are real numbers.

In $a + bi$, a is called the **real part** and b is called the **imaginary part.** Two complex numbers $a + bi$ and $c + di$ are **equal** if and only if their real parts are equal ($a = c$) and their imaginary parts are equal ($b = d$). If $b = 0$, then $a + bi$ is a **real number.** If $b \neq 0$, then $a + bi$ is an **imaginary number.**

The form $a + bi$ is the **standard form** of a complex number, but for convenience we use a few variations of that form. If either the real or imaginary part of a complex number is 0, then that part is omitted. For example,

$$0 + 3i = 3i, \qquad 2 + 0i = 2, \qquad \text{and} \qquad 0 + 0i = 0.$$

If b is a radical, then i is usually written before b. For example, we write $2 + i\sqrt{3}$ rather than $2 + \sqrt{3}i$, which could be confused with $2 + \sqrt{3i}$. If b is negative, a subtraction symbol can be used to separate the real and imaginary parts as in $3 + (-2)i = 3 - 2i$. A complex number with fractions, such as $\frac{1}{3} - \frac{2}{3}i$, may be written as $\frac{1 - 2i}{3}$.

Example 1 Standard form of a complex number

Determine whether each complex number is real or imaginary and write it in the standard form $a + bi$.

a. $3i$ **b.** 87 **c.** $4 - 5i$ **d.** 0 **e.** $\dfrac{1 + \pi i}{2}$

Solution

a. The complex number $3i$ is imaginary, and $3i = 0 + 3i$.
b. The complex number 87 is a real number, and $87 = 87 + 0i$.
c. The complex number $4 - 5i$ is imaginary, and $4 - 5i = 4 + (-5)i$.
d. The complex number 0 is real, and $0 = 0 + 0i$.
e. The complex number $\dfrac{1 + \pi i}{2}$ is imaginary, and $\dfrac{1 + \pi i}{2} = \dfrac{1}{2} + \dfrac{\pi}{2}i$. ■

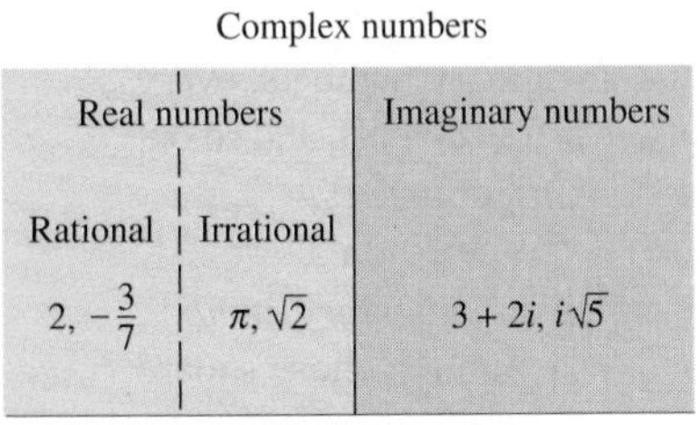

Complex numbers

Real numbers		Imaginary numbers
Rational	Irrational	
$2, -\frac{3}{7}$	$\pi, \sqrt{2}$	$3 + 2i, i\sqrt{5}$

■ **Figure 2.19**

The real numbers can be classified as rational or irrational. The complex numbers can be classified as real or imaginary. The relationship between these sets of numbers is shown in Fig. 2.19.

Addition, Subtraction, and Multiplication

Now that we have defined complex numbers, we define the operations of arithmetic with them.

Definition: Addition, Subtraction, and Multiplication

If $a + bi$ and $c + di$ are complex numbers, we define their sum, difference, and product as follows.

$$(a + bi) + (c + di) = (a + c) + (b + d)i$$

$$(a + bi) - (c + di) = (a - c) + (b - d)i$$

$$(a + bi)(c + di) = (ac - bd) + (bc + ad)i$$

It is not necessary to memorize these definitions, because the results can be obtained by performing the operations as if the complex numbers were binomials with i being a variable, replacing i^2 with -1 wherever it occurs.

Example **2** Operations with complex numbers

Perform the indicated operations with the complex numbers.

a. $(-2 + 3i) + (-4 - 9i)$ **b.** $(-1 - 5i) - (3 - 2i)$ **c.** $2i(3 + i)$
d. $(3i)^2$ **e.** $(-3i)^2$ **f.** $(5 - 2i)(5 + 2i)$

Solution

a. $(-2 + 3i) + (-4 - 9i) = -6 - 6i$
b. $(-1 - 5i) - (3 - 2i) = -1 - 5i - 3 + 2i = -4 - 3i$
c. $2i(3 + i) = 6i + 2i^2 = 6i + 2(-1) = -2 + 6i$
d. $(3i)^2 = 3^2 i^2 = 9(-1) = -9$
e. $(-3i)^2 = (-3)^2 i^2 = 9(-1) = -9$
f. $(5 - 2i)(5 + 2i) = 25 - 4i^2 = 25 - 4(-1) = 29$

Check these results with a calculator that handles complex numbers, as in Fig. 2.20. ■

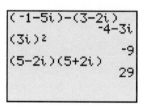

Figure 2.20

We can find whole-number powers of i by using the definition of multiplication. Since $i^1 = i$ and $i^2 = -1$, we have

$$i^3 = i^1 \cdot i^2 = i(-1) = -i \qquad \text{and} \qquad i^4 = i^1 \cdot i^3 = i(-i) = -i^2 = 1.$$

The first eight powers of i are listed here.

$$i^1 = i \qquad i^2 = -1 \qquad i^3 = -i \qquad i^4 = 1$$

$$i^5 = i \qquad i^6 = -1 \qquad i^7 = -i \qquad i^8 = 1$$

This list could be continued in this pattern, but any other whole-number power of i can be obtained from knowing the first four powers. We can simplify a power of i by using the fact that $i^4 = 1$ and $(i^4)^n = 1$ for any integer n.

Example **3** Simplifying a power of i

Simplify.

a. i^{83} **b.** i^{-46}

Solution

a. Divide 83 by 4 and write $83 = 4 \cdot 20 + 3$. So

$$i^{83} = (i^4)^{20} \cdot i^3 = 1^{20} \cdot i^3 = 1 \cdot i^3 = -i.$$

b. Since $-46 = 4(-12) + 2$, we have

$$i^{-46} = (i^4)^{-12} \cdot i^2 = 1^{-12} \cdot i^2 = 1(-1) = -1.$$ ■

Division of Complex Numbers

The complex numbers $a + bi$ and $a - bi$ are called **complex conjugates** of each other.

Example **4** **Complex conjugates**

Find the product of the given complex number and its conjugate.

a. $3 - i$ **b.** $4 + 2i$ **c.** $-i$

Solution

a. The conjugate of $3 - i$ is $3 + i$, and $(3 - i)(3 + i) = 9 - i^2 = 10$.
b. The conjugate of $4 + 2i$ is $4 - 2i$, and $(4 + 2i)(4 - 2i) = 16 - 4i^2 = 20$.
c. The conjugate of $-i$ is i, and $-i \cdot i = -i^2 = 1$. ■

In general we have the following theorem about complex conjugates.

Theorem:
Complex Conjugates

> If a and b are real numbers, then the product of $a + bi$ and its conjugate $a - bi$ is the real number $a^2 + b^2$. In symbols,
>
> $$(a + bi)(a - bi) = a^2 + b^2.$$

We use the theorem about complex conjugates to divide imaginary numbers, in a process that is similar to rationalizing a denominator.

Example **5** **Dividing imaginary numbers**

Write each quotient in the form $a + bi$.

a. $\dfrac{8 - i}{2 + i}$ **b.** $\dfrac{1}{5 - 4i}$ **c.** $\dfrac{3 - 2i}{i}$

Solution

a. Multiply the numerator and denominator by $2 - i$, the conjugate of $2 + i$:

$$\frac{8 - i}{2 + i} = \frac{(8 - i)(2 - i)}{(2 + i)(2 - i)} = \frac{16 - 10i + i^2}{4 - i^2} = \frac{15 - 10i}{5} = 3 - 2i$$

Check division using multiplication: $(3 - 2i)(2 + i) = 8 - i$.

b. $\dfrac{1}{5 - 4i} = \dfrac{1(5 + 4i)}{(5 - 4i)(5 + 4i)} = \dfrac{5 + 4i}{25 + 16} = \dfrac{5 + 4i}{41} = \dfrac{5}{41} + \dfrac{4}{41}i$

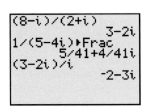

Figure 2.21

Check: $\left(\dfrac{5}{41} + \dfrac{4}{41}i\right)(5 - 4i) = \dfrac{25}{41} + \dfrac{20}{41}i - \dfrac{20}{41}i - \dfrac{16}{41}i^2$

$$= \dfrac{25}{41} + \dfrac{16}{41} = 1.$$

You can also check with a calculator that handles complex numbers as in Fig. 2.21. ▢

c. $\dfrac{3 - 2i}{i} = \dfrac{(3 - 2i)(-i)}{i(-i)} = \dfrac{-3i + 2i^2}{-i^2} = \dfrac{-2 - 3i}{1} = -2 - 3i$

Check: $(-2 - 3i)(i) = -3i^2 - 2i = 3 - 2i.$ ∎

Roots of Negative Numbers

In Examples 2(d) and 2(e), we saw that both $(3i)^2 = -9$ and $(-3i)^2 = -9$. This means that in the complex number system there are two square roots of -9, $3i$ and $-3i$. For any positive real number b, we have $\left(i\sqrt{b}\right)^2 = -b$ and $\left(-i\sqrt{b}\right)^2 = -b$. So there are two square roots of $-b$, $i\sqrt{b}$ and $-i\sqrt{b}$. We call $i\sqrt{b}$ the **principal square root** of $-b$ and make the following definition.

Definition: Square Root of a Negative Number

For any positive real number b, $\sqrt{-b} = i\sqrt{b}.$

In the real number system, $\sqrt{-2}$ and $\sqrt{-8}$ are undefined, but in the complex number system they are defined as $\sqrt{-2} = i\sqrt{2}$ and $\sqrt{-8} = i\sqrt{8}$. Even though we now have meaning for a symbol such as $\sqrt{-2}$, *all operations with complex numbers must be performed after converting to the a + bi form.* If we perform operations with roots of negative numbers using properties of the real numbers, we can get contradictory results:

$$\sqrt{-2} \cdot \sqrt{-8} = \sqrt{(-2)(-8)} = \sqrt{16} = 4 \qquad \text{Incorrect.}$$

$$i\sqrt{2} \cdot i\sqrt{8} = i^2 \cdot \sqrt{16} = -4 \qquad \text{Correct.}$$

The product rule $\sqrt{a} \cdot \sqrt{b} = \sqrt{ab}$ is used *only* for nonnegative numbers a and b.

Example **6** **Square roots of negative numbers**

Write each expression in the form $a + bi$, where a and b are real numbers.

a. $\sqrt{-8} + \sqrt{-18}$ **b.** $\dfrac{-4 + \sqrt{-50}}{4}$ **c.** $\sqrt{-27}\left(\sqrt{9} - \sqrt{-2}\right)$

Solution

The first step in each case is to replace the square roots of negative numbers by expressions with i.

a. $\sqrt{-8} + \sqrt{-18} = i\sqrt{8} + i\sqrt{18} = 2i\sqrt{2} + 3i\sqrt{2}$

$$= 5i\sqrt{2}$$

b. $\dfrac{-4 + \sqrt{-50}}{4} = \dfrac{-4 + i\sqrt{50}}{4} = \dfrac{-4 + 5i\sqrt{2}}{4}$

$$= -1 + \frac{5}{4}i\sqrt{2}$$

c. $\sqrt{-27}\left(\sqrt{9} - \sqrt{-2}\right) = 3i\sqrt{3}\left(3 - i\sqrt{2}\right) = 9i\sqrt{3} - 3i^2\sqrt{6}$

$$= 3\sqrt{6} + 9i\sqrt{3}$$

■

Imaginary Solutions to Quadratic Equations

The quadratic formula $x = \dfrac{-b \pm \sqrt{b^2 - 4ac}}{2a}$ is used to solve the quadratic equation $ax^2 + bx + c = 0$. If the discriminant $b^2 - 4ac$ is negative, then there are no real solutions to the equation. However, square roots of negative numbers exist in the complex number system. So if $b^2 - 4ac < 0$, the equation has two imaginary solutions.

Example **7** Imaginary solutions to a quadratic equation

Find the imaginary solutions to $x^2 - 6x + 11 = 0$.

Solution

Use $a = 1$, $b = -6$, and $c = 11$ in the quadratic formula:

$$x = \frac{-(-6) \pm \sqrt{(-6)^2 - 4(1)(11)}}{2(1)}$$

$$= \frac{6 \pm \sqrt{-8}}{2} = \frac{6 \pm 2i\sqrt{2}}{2} = 3 \pm i\sqrt{2}$$

Check by evaluating $x^2 - 6x + 11$ with $x = 3 + i\sqrt{2}$ as follows:

$$\left(3 + i\sqrt{2}\right)^2 - 6\left(3 + i\sqrt{2}\right) + 11 = 9 + 6i\sqrt{2} + \left(i\sqrt{2}\right)^2 - 18 - 6i\sqrt{2} + 11$$

$$= 9 + 6i\sqrt{2} - 2 - 18 - 6i\sqrt{2} + 11$$

$$= 0$$

■ **Figure 2.22**

The reader should check $3 - i\sqrt{2}$. The imaginary solutions are $3 - i\sqrt{2}$ and $3 + i\sqrt{2}$.

The solutions can be checked with a calculator as in Fig. 2.22. ■

For Thought

True or False? Explain.

1. The multiplicative inverse of i is $-i$.

2. The conjugate of i is $-i$.

3. The set of complex numbers is a subset of the set of real numbers.

4. $\left(\sqrt{3} - i\sqrt{2}\right)\left(\sqrt{3} + i\sqrt{2}\right) = 5$

5. $(2 + 5i)(2 + 5i) = 4 + 25$

6. $5 - \sqrt{-9} = 5 - 9i$

7. If $P(x) = x^2 + 9$, then $P(3i) = 0$.

8. The imaginary number $-3i$ is a root to the equation $x^2 + 9 = 0$.

9. $i^4 = 1$

10. $i^{18} = 1$

2.2 Exercises

Determine whether each complex number is real or imaginary and write it in the standard form $a + bi$.

1. $6i$

2. $-3i + \sqrt{6}$

3. $\dfrac{1 + i}{3}$

4. -72

5. $\sqrt{7}$

6. $-i\sqrt{5}$

7. $\dfrac{\pi}{2}$

8. 0

Perform the indicated operations and write your answers in the form $a + bi$, where a and b are real numbers.

9. $(3 - 3i) + (4 + 5i)$

10. $(-3 + 2i) + (5 - 6i)$

11. $(1 - i) - (3 + 2i)$

12. $(6 - 7i) - (3 - 4i)$

13. $-6i(3 - 2i)$

14. $-3i(5 + 2i)$

15. $(2 - 3i)(4 + 6i)$

16. $(3 - i)(5 - 2i)$

17. $(5 - 2i)(5 + 2i)$

18. $(4 + 3i)(4 - 3i)$

19. $\left(\sqrt{3} - i\right)\left(\sqrt{3} + i\right)$

20. $\left(\sqrt{2} + i\sqrt{3}\right)\left(\sqrt{2} - i\sqrt{3}\right)$

21. $(3 + 4i)^2$

22. $(-6 - 2i)^2$

23. $\left(\sqrt{5} - 2i\right)^2$

24. $\left(\sqrt{6} + i\sqrt{3}\right)^2$

25. i^{17}

26. i^{24}

27. i^{98}

28. i^{19}

29. i^{-4}

30. i^{-13}

31. i^{-1}

32. i^{-27}

Find the product of the given complex number and its conjugate.

33. $3 - 9i$

34. $4 + 3i$

35. $\dfrac{1}{2} + 2i$

36. $\dfrac{1}{3} - i$

37. i

38. $-i\sqrt{5}$

39. $3 - i\sqrt{3}$

40. $\dfrac{5}{2} + i\dfrac{\sqrt{2}}{2}$

Write each quotient in the form $a + bi$.

41. $\dfrac{1}{2 - i}$

42. $\dfrac{1}{5 + 2i}$

43. $\dfrac{-3i}{1 - i}$

44. $\dfrac{3i}{-2 + i}$

45. $\dfrac{-2 + 6i}{2}$

46. $\dfrac{-6 - 9i}{-3}$

47. $\dfrac{-3 + 3i}{i}$

48. $\dfrac{-2 - 4i}{-i}$

49. $\dfrac{1 - i}{3 + 2i}$

50. $\dfrac{4 + 2i}{2 - 3i}$

Write each expression in the form $a + bi$, where a and b are real numbers.

51. $\sqrt{-4} - \sqrt{-9}$

52. $\sqrt{-16} + \sqrt{-25}$

53. $\sqrt{-4} - \sqrt{16}$

54. $\sqrt{-3} \cdot \sqrt{-3}$

55. $\left(\sqrt{-6}\right)^2$

56. $\left(\sqrt{-5}\right)^3$

57. $\sqrt{-2} \cdot \sqrt{-50}$

58. $\dfrac{-6 + \sqrt{-3}}{3}$

59. $\dfrac{-2 + \sqrt{-20}}{2}$

60. $\dfrac{9 - \sqrt{-18}}{-6}$

61. $-3 + \sqrt{3^2 - 4(1)(5)}$

62. $1 - \sqrt{(-1)^2 - 4(1)(1)}$

63. $\sqrt{-8}\left(\sqrt{-2} + \sqrt{8}\right)$

64. $\sqrt{-6}\left(\sqrt{2} - \sqrt{-3}\right)$

Evaluate the expression $\dfrac{-b + \sqrt{b^2 - 4ac}}{2a}$ for each choice of a, b, and c.

65. $a = 1, b = 2, c = 5$

66. $a = 5, b = -4, c = 1$

67. $a = 2, b = 4, c = 3$

68. $a = 2, b = -4, c = 5$

Evaluate the expression $\dfrac{-b - \sqrt{b^2 - 4ac}}{2a}$ for each choice of a, b, and c.

69. $a = 1, b = 6, c = 17$

70. $a = 1, b = -12, c = 84$

71. $a = -2, b = 6, c = 6$

72. $a = 3, b = 6, c = 8$

Find the imaginary solutions to each quadratic equation and check your answers.

73. $x^2 + 1 = 0$ **74.** $x^2 + 9 = 0$

75. $x^2 + 8 = 0$ **76.** $x^2 + 27 = 0$

77. $2x^2 + 1 = 0$ **78.** $3x^2 + 2 = 0$

79. $x^2 - 2x + 2 = 0$ **80.** $x^2 - 4x + 5 = 0$

81. $x^2 - 4x + 13 = 0$ **82.** $x^2 - 2x + 5 = 0$

83. $x^2 - 2x + 4 = 0$ **84.** $x^2 - 4x + 9 = 0$

85. $-2x^2 + 2x = 5$ **86.** $12x - 5 = 9x^2$

87. $4x^2 - 8x + 7 = 0$ **88.** $9x^2 - 6x + 4 = 0$

Thinking Outside the Box XII

Summing Reciprocals   There is only one way to write 1 as a sum of the reciprocals of three different positive integers:

$$\frac{1}{2} + \frac{1}{3} + \frac{1}{6} = 1$$

Find all possible ways to write 1 as a sum of the reciprocals of four different positive integers.

2.2   Pop Quiz

1. Find the sum of $3 + 2i$ and $4 - i$.

2. Find the product of $4 - 3i$ and $2 + i$.

3. Find the product of $2 - 3i$ and its conjugate.

4. Write $\dfrac{5}{2 - 3i}$ in the form $a + bi$.

5. Find i^{27}.

6. What are the two square roots of -16?

2.3   Zeros of Polynomial Functions

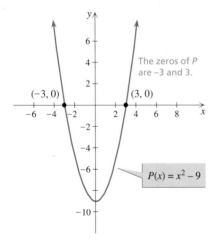

Figure 2.23

We have studied linear and quadratic functions extensively. In this section we will study general polynomial functions.

The Remainder Theorem

If $y = P(x)$ is a polynomial function, then a value of x that satisfies $P(x) = 0$ is called a **zero** of the polynomial function or a zero of the polynomial. For example, 3 and -3 are zeros of the function $P(x) = x^2 - 9$, because $P(3) = 0$ and $P(-3) = 0$. Note that the zeros of $P(x) = x^2 - 9$ are the same as the solutions to the equation $x^2 - 9 = 0$. The real zeros of a polynomial function appear on the graph of the function as the x-coordinates of the x-intercepts. The x-intercepts of the graph of $P(x) = x^2 - 9$ shown in Fig. 2.23 are $(-3, 0)$ and $(3, 0)$.

For polynomial functions of degree 2 or less, the zeros can be found by solving quadratic or linear equations. Our goal in this section is to find all of the zeros of a polynomial function when possible. For polynomials of degree higher than 2, the difficulty of this task ranges from easy to impossible, but we have some theorems to assist us. The remainder theorem relates evaluating a polynomial to division of polynomials.

The Remainder Theorem

If R is the remainder when a polynomial $P(x)$ is divided by $x - c$, then $R = P(c)$.

PROOF   Let $Q(x)$ be the quotient and R be the remainder when $P(x)$ is divided by $x - c$. Since the dividend is equal to the divisor times the quotient plus the remainder, we have

$$P(x) = (x - c)Q(x) + R.$$

This statement is true for any value of x, and so it is also true for $x = c$:

$$P(c) = (c - c)Q(c) + R$$

$$= 0 \cdot Q(c) + R$$

$$= R$$

So $P(c)$ is equal to the remainder when $P(x)$ is divided by $x - c$.   ■

To illustrate the remainder theorem, we will now use long division to evaluate a polynomial. Long division is discussed in Section A.2 of the Appendix.

Example **1** **Using the remainder theorem to evaluate a polynomial**

Use the remainder theorem to find $P(3)$ if $P(x) = 2x^3 - 5x^2 + 4x - 6$.

Solution

By the remainder theorem $P(3)$ is the remainder when $P(x)$ is divided by $x - 3$:

$$
\begin{array}{r}
2x^2 + x + 7 \\
x - 3 \overline{)2x^3 - 5x^2 + 4x - 6} \\
\underline{2x^3 - 6x^2} \\
x^2 + 4x \\
\underline{x^2 - 3x} \\
7x - 6 \\
\underline{7x - 21} \\
15
\end{array}
$$

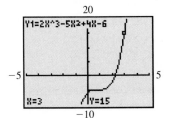

Figure 2.24

The remainder is 15 and therefore $P(3) = 15$. We can check by finding $P(3)$ in the usual manner:

$$P(3) = 2 \cdot 3^3 - 5 \cdot 3^2 + 4 \cdot 3 - 6 = 54 - 45 + 12 - 6 = 15$$

You can check this with a graphing calculator as shown in Fig. 2.24.   ■

Synthetic Division

In Example 1 we found $P(3) = 15$ in two different ways. Certainly, evaluating $P(x)$ for $x = 3$ in the usual manner is faster than dividing $P(x)$ by $x - 3$ using the ordinary method of dividing polynomials. However, there is a faster method, called **synthetic division,** for dividing by $x - 3$. Compare the two methods side by side, both showing $2x^3 - 5x^2 + 4x - 6$ divided by $x - 3$:

Ordinary Division of Polynomials   *Synthetic Division*

$$
\begin{array}{r}
2x^2 + x + 7 \quad \leftarrow \text{Quotient} \\
x - 3 \overline{)2x^3 - 5x^2 + 4x - 6} \\
\underline{2x^3 - 6x^2} \\
x^2 + 4x \\
\underline{x^2 - 3x} \\
7x - 6 \\
\underline{7x - 21} \\
15 \quad \leftarrow \text{Remainder}
\end{array}
$$

$$
\begin{array}{c|rrrr}
3 & 2 & -5 & 4 & -6 \\
  &   & 6 & 3 & 21 \\
\hline
  & 2 & 1 & 7 & 15 \quad \leftarrow \text{Remainder}
\end{array}
$$

Quotient

Synthetic division certainly looks easier than ordinary division, and in general it is faster than evaluating the polynomial by substitution. Synthetic division is used as a quick means of dividing a polynomial by a binomial of the form $x - c$.

In synthetic division we write just the necessary parts of the ordinary division. Instead of writing $2x^3 - 5x^2 + 4x - 6$, write the coefficients 2, -5, 4, and -6. For $x - 3$, write only the 3. The bottom row in synthetic division gives the coefficients of the quotient and the remainder.

To actually perform the synthetic division, start with the following arrangement of coefficients:

$$
\begin{array}{c|rrrr}
3 & 2 & -5 & 4 & -6 \\
  &   &   &   &   \\
\hline
\end{array}
$$

Bring down the first coefficient, 2. Multiply 2 by 3 and write the answer beneath -5. Then add:

$$
\begin{array}{c|rrrr}
3 & 2 & -5 & 4 & -6 \\
  & \downarrow & 6 & & \\
\hline
  & 2 & 1 & &
\end{array}
$$

Multiply → 2   1
                ↑
               Add

Using 3 rather than -3 when dividing by $x - 3$ allows us to multiply and add rather than multiply and subtract as in ordinary division. Now multiply 1 by 3 and write the answer beneath 4. Then add. Repeat the multiply-and-add step for the remaining column:

$$
\begin{array}{c|rrrr}
3 & 2 & -5 & 4 & -6 \\
  &   & 6 & 3 & 21 \\
\hline
  & 2 & 1 & 7 & 15
\end{array}
$$

To perform this arithmetic on a graphing calculator, start with the leading coefficient 2 as the answer. Then repeatedly multiply the answer by 3 and add the next coefficient as shown in Fig. 2.25. □

The quotient is $2x^2 + x + 7$, and the remainder is 15. Since the divisor in synthetic division is of the form $x - c$, the degree of the quotient is always one less than the degree of the dividend.

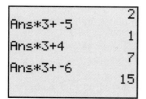

```
                        2
Ans*3+ -5
                        1
Ans*3+4
                        7
Ans*3+ -6
                       15
```

■ **Figure 2.25**

Example **2** **Synthetic division**

Use synthetic division to find the quotient and remainder when $x^4 - 14x^2 + 5x - 9$ is divided by $x + 4$.

Solution

Since $x + 4 = x - (-4)$, we use -4 in the synthetic division. Use $1, 0, -14, 5$, and -9 as the coefficients of the polynomial. We use 0 for the coefficient of the missing x^3-term, as we would in ordinary division of polynomials.

$$
\begin{array}{r|rrrrr}
-4 & 1 & 0 & -14 & 5 & -9 \\
   &   & -4 & 16 & -8 & 12 \\
\hline
\text{Multiply} \rightarrow 1 & & -4 & 2 & -3 & 3 \\
   & & \uparrow & & & \\
   & & \text{Add} & & &
\end{array}
$$

The quotient is $x^3 - 4x^2 + 2x - 3$ and the remainder is 3. ■

To find the value of a polynomial $P(x)$ for $x = c$, we can divide $P(x)$ by $x - c$ or we can substitute c for x and compute. Using long division to divide by $x - c$ is not an efficient way to evaluate a polynomial. However, if we use synthetic division to divide by $x - c$ we can actually evaluate some polynomials using fewer arithmetic operations than we use in the substitution method. Note that in part (a) of the next example, synthetic division takes more steps than substitution, but in part (b) synthetic division takes fewer steps.

Example **3** **Using synthetic division to evaluate a polynomial**

Let $f(x) = x^3$ and $g(x) = x^3 - 3x^2 + 5x - 12$. Use synthetic division to find the following function values.

a. $f(-2)$ **b.** $g(4)$

Solution

a. To find $f(-2)$, divide the polynomial x^3 by $x - (-2)$ or $x + 2$ using synthetic division. Write x^3 as $x^3 + 0x^2 + 0x + 0$, and use $1, 0, 0$, and 0 as the coefficients. We use a zero for each power of x below x^3.

$$
\begin{array}{r|rrrr}
-2 & 1 & 0 & 0 & 0 \\
   &   & -2 & 4 & -8 \\
\hline
   & 1 & -2 & 4 & -8
\end{array}
$$

The remainder is -8, so $f(-2) = -8$. To check, find $f(-2) = (-2)^3 = -8$.

b. To find $g(4)$, use synthetic division to divide $x^3 - 3x^2 + 5x - 12$ by $x - 4$:

$$
\begin{array}{r|rrrr}
4 & 1 & -3 & 5 & -12 \\
  &   & 4 & 4 & 36 \\
\hline
  & 1 & 1 & 9 & 24
\end{array}
$$

The remainder is 24, so $g(4) = 24$. Check this answer by finding $g(4) = 4^3 - 3(4^2) + 5(4) - 12 = 24$. ■

The Factor Theorem

Consider the polynomial function $P(x) = x^2 - x - 6$. We can find the zeros of the function by solving $x^2 - x - 6 = 0$ by factoring:

$$(x - 3)(x + 2) = 0$$

$$x - 3 = 0 \quad \text{or} \quad x + 2 = 0$$

$$x = 3 \quad \text{or} \quad x = -2$$

Both 3 and -2 are zeros of the function $P(x) = x^2 - x - 6$. Note how each factor of the polynomial corresponds to a zero of the function. This example suggests the following theorem.

The Factor Theorem

> The number c is a zero of the polynomial function $y = P(x)$ if and only if $x - c$ is a factor of the polynomial $P(x)$.

PROOF   If c is a zero of the polynomial function $y = P(x)$, then $P(c) = 0$. If $P(x)$ is divided by $x - c$, we get a quotient $Q(x)$ and a remainder R such that

$$P(x) = (x - c)Q(x) + R.$$

By the remainder theorem, $R = P(c)$. Since $P(c) = 0$, we have $R = 0$ and $P(x) = (x - c)Q(x)$, which proves that $x - c$ is a factor of $P(x)$.

If $x - c$ is a factor of $P(x)$, then there is a polynomial $Q(x)$ such that $P(x) = (x - c)Q(x)$. Letting $x = c$ yields $P(c) = (c - c)Q(c) = 0$. So c is a zero of $P(x)$. These two arguments establish the truth of the factor theorem.   ■

Synthetic division can be used in conjunction with the factor theorem. If the remainder of dividing $P(x)$ by $x - c$ is 0, then $P(c) = 0$ and c is a zero of the polynomial function. By the factor theorem, $x - c$ is a factor of $P(x)$.

Example **4** **Using the factor theorem to factor a polynomial**

Determine whether $x + 4$ is a factor of the polynomial $P(x) = x^3 - 13x + 12$. If it is a factor, then factor $P(x)$ completely.

Solution

By the factor theorem, $x + 4$ is a factor of $P(x)$ if and only if $P(-4) = 0$. We can find $P(-4)$ using synthetic division:

$$
\begin{array}{r|rrrr}
-4 & 1 & 0 & -13 & 12 \\
   &   & -4 & 16 & -12 \\
\hline
   & 1 & -4 & 3 & 0
\end{array}
$$

Since $P(-4)$ is equal to the remainder, $P(-4) = 0$ and -4 is a zero of $P(x)$. By the factor theorem, $x + 4$ is a factor of $P(x)$. Since the other factor is the quotient from the synthetic division, $P(x) = (x + 4)(x^2 - 4x + 3)$. Factor the quadratic polynomial to get $P(x) = (x + 4)(x - 1)(x - 3)$.

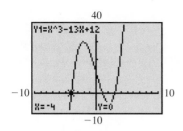

■ Figure 2.26

You can check this result by examining the calculator graph of $y = x^3 - 13x + 12$ shown in Fig. 2.26. The graph appears to cross the x-axis at -4, 1, and 3, supporting the conclusion that $P(x) = (x + 4)(x - 1)(x - 3)$. ■

The Fundamental Theorem of Algebra

Whether a number is a zero of a polynomial function can be determined by synthetic division. But does every polynomial function have a zero? This question was answered in the affirmative by Carl F. Gauss when he proved the fundamental theorem of algebra in his doctoral thesis in 1799 at the age of 22.

The Fundamental Theorem of Algebra

> If $y = P(x)$ is a polynomial function of positive degree, then $y = P(x)$ has at least one zero in the set of complex numbers.

Gauss also proved the n-root theorem of Section 2.4 that says that the number of zeros of a polynomial (or polynomial function) of degree n is at most n. For example, a fifth-degree polynomial function has at least one zero and at most five.

Note that the zeros guaranteed by Gauss are in the set of complex numbers. So the zero might be real or imaginary. The theorem applies only to polynomial functions of degree 1 or more, because a polynomial function of zero degree such as $P(x) = 7$ has no zeros. A polynomial function of degree 1, $f(x) = ax + b$, has exactly one zero which is found by solving $ax + b = 0$. A polynomial function of degree 2, $f(x) = ax^2 + bx + c$, has one or two zeros that can be found by the quadratic formula. For higher degree polynomials the situation is not as simple. The fundamental theorem tells us that a polynomial function has at least one zero but not how to find it. For this purpose we have some other theorems.

The Rational Zero Theorem

Zeros or roots that are rational numbers, the **rational zeros,** are generally the easiest to find. The polynomial function $f(x) = 6x^2 - x - 35$ has two rational zeros that can be found as follows:

$$6x^2 - x - 35 = 0$$
$$(2x - 5)(3x + 7) = 0 \quad \text{Factor.}$$
$$x = \frac{5}{2} \quad \text{or} \quad x = -\frac{7}{3}$$

Note that in $5/2$, 5 is a factor of -35 (the constant term) and 2 is a factor of 6 (the leading coefficient). For the zero $-7/3$, -7 is a factor of -35 and 3 is a factor of 6. Of course, these observations are not surprising, because we used these facts to factor the quadratic polynomial in the first place. Note that there are a lot of other factors of -35 and 6 for which the ratio is *not* a zero of this function. This example illustrates the rational zero theorem, which is also called the rational root theorem.

Historical Note

Carl Friedrich Gauss (1777–1855) was a German mathematician and scientist of profound genius who contributed significantly to many fields, including number theory, analysis, differential geometry, geodesy, magnetism, astronomy, and optics. Gauss had a remarkable influence in many fields of mathematics and science and is ranked as one of history's most influential mathematicians.

The Rational Zero Theorem

If $f(x) = a_n x^n + a_{n-1} x^{n-1} + a_{n-2} x^{n-2} + \cdots + a_1 x + a_0$ is a polynomial function with integral coefficients ($a_n \neq 0$ and $a_0 \neq 0$) and p/q (in lowest terms) is a rational zero of $f(x)$, then p is a factor of the constant term a_0 and q is a factor of the leading coefficient a_n.

PROOF   If p/q is a zero of $f(x)$, then $f(p/q) = 0$:

$$a_n \left(\frac{p}{q}\right)^n + a_{n-1}\left(\frac{p}{q}\right)^{n-1} + a_{n-2}\left(\frac{p}{q}\right)^{n-2} + \cdots + a_1 \frac{p}{q} + a_0 = 0$$

Subtract a_0 from each side and multiply by q^n to get the following equation:

$$a_n p^n + a_{n-1} p^{n-1} q + a_{n-2} p^{n-2} q^2 + \cdots + a_1 p q^{n-1} = -a_0 q^n$$

Since the coefficients are integers and p and q are integers, both sides of this equation are integers. Since p is a factor of the left side p must be a factor of the right side, which is the same integer. Since p/q is in lowest terms, p is not a factor of q. So p must be a factor of a_0.

To prove that q is a factor of a_n, rearrange the last equation as follows:

$$a_{n-1} p^{n-1} q + a_{n-2} p^{n-2} q^2 + \cdots + a_1 p q^{n-1} + a_0 q^n = -a_n p^n.$$

Now q is a factor of a_n by the same argument used previously.   ▪

The rational zero theorem does not identify exactly which rational numbers are zeros of a function; it only gives *possibilities* for the rational zeros.

Example **5** **Using the rational zero theorem**

Find all possible rational zeros for each polynomial function.

a. $f(x) = 2x^3 - 3x^2 - 11x + 6$ **b.** $g(x) = 3x^3 - 8x^2 - 8x + 8$

Solution

a. If the rational number p/q is a zero of $f(x)$, then p is a factor of 6 and q is a factor of 2. The positive factors of 6 are 1, 2, 3, and 6. The positive factors of 2 are 1 and 2. Take each factor of 6 and divide by 1 to get $1/1$, $2/1$, $3/1$, and $6/1$. Take each factor of 6 and divide by 2 to get $1/2$, $2/2$, $3/2$, and $6/2$. Simplify the ratios, eliminate duplications, and put in the negative factors to get

$$\pm 1, \quad \pm 2, \quad \pm 3, \quad \pm 6, \quad \pm\frac{1}{2}, \quad \text{and} \quad \pm\frac{3}{2}$$

as the possible rational zeros to the function $f(x)$.

b. If the rational number p/q is a zero of $g(x)$, then p is a factor of 8 and q is a factor of 3. The factors of 8 are 1, 2, 4, and 8. The factors of 3 are 1 and 3. If we take all possible ratios of a factor of 8 over a factor of 3, we get

$$\pm 1, \quad \pm 2, \quad \pm 4, \quad \pm 8, \quad \pm\frac{1}{3}, \quad \pm\frac{2}{3}, \quad \pm\frac{4}{3}, \quad \text{and} \quad \pm\frac{8}{3}$$

as the possible rational zeros of the function $g(x)$.   ■

Our goal is to find all of the zeros to a polynomial function. The zeros to a polynomial function might be rational, irrational, or imaginary. We can determine the rational zeros by simply evaluating the polynomial function for every number in the list of possible rational zeros. If the list is long, looking at a graph of the function can speed up the process. We will use synthetic division to evaluate the polynomial, because synthetic division gives the quotient polynomial as well as the value of the polynomial.

Example **6** Finding all zeros of a polynomial function

Find all of the real and imaginary zeros for each polynomial function of Example 5.

a. $f(x) = 2x^3 - 3x^2 - 11x + 6$ **b.** $g(x) = 3x^3 - 8x^2 - 8x + 8$

Solution

a. The possible rational zeros of $f(x)$ are listed in Example 5(a). Use synthetic division to check each possible zero to see whether it is actually a zero. Try 1 first.

$$
\begin{array}{r|rrrr}
1 & 2 & -3 & -11 & 6 \\
  &   & 2 & -1 & -12 \\
\hline
  & 2 & -1 & -12 & -6
\end{array}
$$

Since the remainder is -6, 1 is not a zero of the function. Keep on trying numbers from the list of possible rational zeros. To save space, we will not show any more failures. So try $1/2$ next.

$$
\begin{array}{r|rrrr}
\dfrac{1}{2} & 2 & -3 & -11 & 6 \\
  &   & 1 & -1 & -6 \\
\hline
  & 2 & -2 & -12 & 0
\end{array}
$$

Since the remainder in the synthetic division is 0, $1/2$ is a zero of $f(x)$. By the factor theorem, $x - 1/2$ is a factor of the polynomial. The quotient is the other factor.

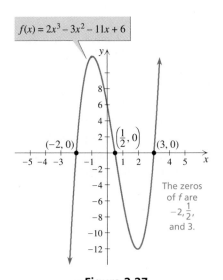

$f(x) = 2x^3 - 3x^2 - 11x + 6$

The zeros of f are $-2, \dfrac{1}{2}$, and 3.

Figure 2.27

$$2x^3 - 3x^2 - 11x + 6 = 0$$

$$\left(x - \frac{1}{2}\right)(2x^2 - 2x - 12) = 0 \qquad \text{Factor.}$$

$$(2x - 1)(x^2 - x - 6) = 0 \qquad \text{Factor 2 out of the second "factor" and distribute it into the first factor.}$$

$$(2x - 1)(x - 3)(x + 2) = 0 \qquad \text{Factor completely.}$$

$$2x - 1 = 0 \quad \text{or} \quad x - 3 = 0 \quad \text{or} \quad x + 2 = 0$$

$$x = \frac{1}{2} \quad \text{or} \qquad x = 3 \quad \text{or} \qquad x = -2$$

The zeros of the function f are $1/2$, 3, and -2. Note that each zero of f corresponds to an x-intercept on the graph of f shown in Fig. 2.27. Because this polynomial had three rational zeros, we could have found them all by using synthetic division or by examining the calculator graph. However, it is good to factor the polynomial to see the correspondence between the three zeros, the three factors, and the three x-intercepts.

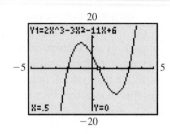

■ **Figure 2.28**

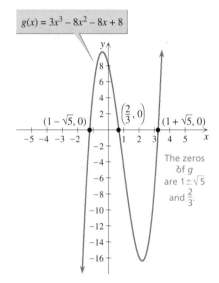

$g(x) = 3x^3 - 8x^2 - 8x + 8$

The zeros
of g
are $1 \pm \sqrt{5}$
and $\frac{2}{3}$.

■ **Figure 2.29**

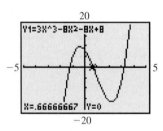

■ **Figure 2.30**

 You could speed up this process of finding a zero with the calculator graph shown in Fig. 2.28. It is not too hard to discover that $1/2$ is a zero by looking at the graph and the list of possible rational zeros that we listed in Example 5. If you use a graph to find that $1/2$ is a zero, you still need to do synthetic division to factor the polynomial.

b. The possible rational zeros of $g(x)$ are listed in Example 5(b). First check $2/3$ to see whether it produces a remainder of 0.

$$\begin{array}{r} \frac{2}{3} \enclose{verticalstroke}{\begin{array}{rrrr} 3 & -8 & -8 & 8 \\ & 2 & -4 & -8 \\ \hline 3 & -6 & -12 & 0 \end{array}} \end{array}$$

Since the remainder in the synthetic division is 0, $2/3$ is a zero of $g(x)$. By the factor theorem, $x - 2/3$ is a factor of the polynomial. The quotient is the other factor.

$$3x^3 - 8x^2 - 8x + 8 = 0$$

$$\left(x - \frac{2}{3}\right)(3x^2 - 6x - 12) = 0$$

$$(3x - 2)(x^2 - 2x - 4) = 0$$

$$3x - 2 = 0 \quad \text{or} \quad x^2 - 2x - 4 = 0$$

$$x = \frac{2}{3} \quad \text{or} \qquad\qquad x = \frac{2 \pm \sqrt{20}}{2}$$

$$x = \frac{2}{3} \quad \text{or} \qquad\qquad x = 1 \pm \sqrt{5}$$

There are one rational and two irrational roots to the equation. So the zeros of the function g are $2/3$, $1 + \sqrt{5}$, and $1 - \sqrt{5}$. Each zero corresponds to an x-intercept on the graph of g shown in Fig. 2.29.

 You could graph the function with a calculator as shown in Fig. 2.30. Keeping in mind the list of possible rational zeros, it is not hard to discover that $2/3$ is a zero. ■

Note that in Example 6(a) all of the zeros were rational. All three could have been found by continuing to check the possible rational zeros using synthetic division. In Example 6(b) we would be wasting time if we continued to check the possible rational zeros, because there is only one. When we get to a quadratic polynomial, it is best to either factor the quadratic polynomial or use the quadratic formula to find the remaining zeros.

For Thought

True or False? Explain.

1. The function $f(x) = 1/x$ has at least one zero.

2. If $P(x) = x^4 - 6x^2 - 8$ is divided by $x^2 - 2$, then the remainder is $P\!\left(\sqrt{2}\right)$.

3. If $1 - 2i$ and $1 + 2i$ are zeros of
$P(x) = x^3 - 5x^2 + 11x - 15$, then $x - 1 - 2i$ and $x - 1 + 2i$ are factors of $P(x)$.

4. If we divide $x^5 - 1$ by $x - 2$, then the remainder is 31.

5. Every polynomial function has at least one zero.

6. If $P(x) = x^3 - x^2 + 4x - 5$ and b is the remainder from division of $P(x)$ by $x - c$, then $b^3 - b^2 + 4b - 5 = c$.

7. If $P(x) = x^3 - 5x^2 + 4x - 15$, then $P(4) = 0$.

8. The equation $\pi^2 x^4 - \dfrac{1}{\sqrt{2}} x^3 + \dfrac{1}{\sqrt{7} + \pi} = 0$ has at least one complex solution.

9. The binomial $x - 1$ is a factor of $x^5 + x^4 - x^3 - x^2 - x + 1$.

10. The binomial $x + 3$ is a factor of $3x^4 - 5x^3 + 7x^2 - 9x - 2$.

2.3 Exercises

Use ordinary division of polynomials to find the quotient and remainder when the first polynomial is divided by the second.

1. $x^2 - 5x + 7, x - 2$ **2.** $x^2 - 3x + 9, x - 4$

3. $-2x^3 + 4x - 9, x + 3$ **4.** $-4w^3 + 5w^2 - 7, w - 3$

5. $s^4 - 3s^2 + 6, s^2 - 5$ **6.** $h^4 + 3h^3 + h - 5, h^2 - 3$

Use synthetic division to find the quotient and remainder when the first polynomial is divided by the second.

7. $x^2 + 4x + 1, x - 2$ **8.** $2x^2 - 3x + 6, x - 5$

9. $-x^3 + x^2 - 4x + 9, x + 3$

10. $-3x^3 + 5x^2 - 6x + 1, x + 1$

11. $4x^3 - 5x + 2, x - \dfrac{1}{2}$ **12.** $-6x^3 + 25x^2 - 9, x - \dfrac{3}{2}$

13. $2a^3 - 3a^2 + 4a + 3, a + \dfrac{1}{2}$

14. $-3b^3 - b^2 - 3b - 1, b + \dfrac{1}{3}$

15. $x^4 - 3, x - 1$ **16.** $x^4 - 16, x - 2$

17. $x^5 - 6x^3 + 4x - 5, x - 2$

18. $2x^5 - 5x^4 - 5x + 7, x - 3$

Let $f(x) = x^5 - 1, g(x) = x^3 - 4x^2 + 8,$ and $h(x) = 2x^4 + x^3 - x^2 + 3x + 3$. Find the following function values by using synthetic division. Check by using substitution.

19. $f(1)$ **20.** $f(-1)$ **21.** $f(-2)$ **22.** $f(3)$

23. $g(1)$ **24.** $g(-1)$ **25.** $g\left(-\dfrac{1}{2}\right)$ **26.** $g\left(\dfrac{1}{2}\right)$

27. $h(-1)$ **28.** $h(2)$ **29.** $h(1)$ **30.** $h(-3)$

Determine whether the given binomial is a factor of the polynomial following it. If it is a factor, then factor the polynomial completely.

31. $x + 3, x^3 + 4x^2 + x - 6$

32. $x + 5, x^3 + 8x^2 + 11x - 20$

33. $x - 4, x^3 + 4x^2 - 17x - 60$

34. $x - 2, x^3 - 12x^2 + 44x - 48$

Determine whether each given number is a zero of the polynomial function following the number.

35. $3, f(x) = 2x^3 - 5x^2 - 4x + 3$

36. $-2, g(x) = 3x^3 - 6x^2 - 3x - 19$

37. $-2, g(d) = d^3 + 2d^2 + 3d + 1$

38. $-1, w(x) = 3x^3 + 2x^2 - 2x - 1$

39. $-1, P(x) = x^4 + 2x^3 + 4x^2 + 6x + 3$

40. $3, G(r) = r^4 + 4r^3 + 5r^2 + 3r + 17$

41. $\dfrac{1}{2}, H(x) = x^3 + 3x^2 - 5x + 7$

42. $-\dfrac{1}{2}, T(x) = 2x^3 + 3x^2 - 3x - 2$

Use the rational zero theorem to find all possible rational zeros for each polynomial function.

43. $f(x) = x^3 - 9x^2 + 26x - 24$

44. $g(x) = x^3 - 2x^2 - 5x + 6$

45. $h(x) = x^3 - x^2 - 7x + 15$

46. $m(x) = x^3 + 4x^2 + 4x + 3$

47. $P(x) = 8x^3 - 36x^2 + 46x - 15$

48. $T(x) = 18x^3 - 9x^2 - 5x + 2$

49. $M(x) = 18x^3 - 21x^2 + 10x - 2$

50. $N(x) = 4x^3 - 10x^2 + 4x + 5$

Find all of the real and imaginary zeros for each polynomial function.

51. $f(x) = x^3 - 9x^2 + 26x - 24$

52. $g(x) = x^3 - 2x^2 - 5x + 6$

53. $h(x) = x^3 - x^2 - 7x + 15$

54. $m(x) = x^3 + 4x^2 + 4x + 3$

55. $P(a) = 8a^3 - 36a^2 + 46a - 15$

56. $T(b) = 18b^3 - 9b^2 - 5b + 2$

57. $M(t) = 18t^3 - 21t^2 + 10t - 2$

58. $N(t) = 4t^3 - 10t^2 + 4t + 5$

59. $S(w) = w^4 + w^3 - w^2 + w - 2$

60. $W(v) = 2v^4 + 5v^3 + 3v^2 + 15v - 9$

61. $V(x) = x^4 + 2x^3 - x^2 - 4x - 2$

62. $U(x) = x^4 - 4x^3 + x^2 + 12x - 12$

63. $f(x) = 24x^3 - 26x^2 + 9x - 1$

64. $f(x) = 30x^3 - 47x^2 - x + 6$

65. $y = 16x^3 - 33x^2 + 82x - 5$

66. $y = 15x^3 - 37x^2 + 44x - 14$

67. $f(x) = 21x^4 - 31x^3 - 21x^2 - 31x - 42$

68. $f(x) = 119x^4 - 5x^3 + 214x^2 - 10x - 48$

69. $f(x) = (x^2 + 9)(x^3 + 6x^2 + 3x - 10)$

70. $f(x) = (x^2 - 5)(x^3 - 5x^2 - 12x + 36)$

71. $f(x) = (x^2 - 4x + 1)(x^3 - 9x^2 + 23x - 15)$

72. $f(x) = (x^2 - 4x + 13)(x^3 - 4x^2 - 17x + 60)$

Use division to write each rational expression in the form quotient + remainder/divisor. Use synthetic division when possible.

73. $\dfrac{2x + 1}{x - 2}$

74. $\dfrac{x - 1}{x + 3}$

75. $\dfrac{a^2 - 3a + 5}{a - 3}$

76. $\dfrac{2b^2 - 3b + 1}{b + 2}$

77. $\dfrac{c^2 - 3c - 4}{c^2 - 4}$

78. $\dfrac{2h^2 + h - 2}{h^2 - 1}$

79. $\dfrac{4t - 5}{2t + 1}$

80. $\dfrac{6y - 1}{3y - 1}$

Solve each problem.

81. *Drug Testing* The concentration of a drug (in parts per million) in a patient's bloodstream t hours after administration of the drug is given by the function

$$P(t) = -t^4 + 12t^3 - 58t^2 + 132t.$$

a. Use the formula to determine when the drug will be totally eliminated from the bloodstream.

b. Use the graph to estimate the maximum concentration of the drug.

c. Use the graph to estimate the time at which the maximum concentration occurs.

d. Use the graph to estimate the amount of time for which the concentration is above 80 ppm.

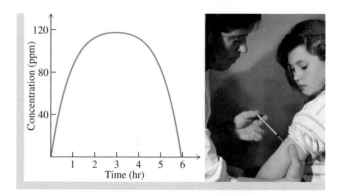

■ **Figure for Exercise 81**

82. *Open-Top Box* Joan intends to make an 18-in.^{3} open-top box out of a 6 in. by 7 in. piece of copper by cutting equal squares

(x in. by x in.) from the corners and folding up the sides. Write the difference between the intended volume and the actual volume as a function of x. For what value of x is there no difference between the intended volume and the actual volume?

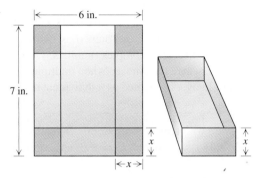

Figure for Exercise 82

83. *Cartridge Box* The height of a box containing an HP Laser Jet III printer cartridge is 4 in. more than the width and the length is 9 in. more than the width. If the volume of the box is 630 in.^{3}, then what are the dimensions of the box?

84. *Computer Case* The width of the case for a 733 megahertz Pentium computer is 4 in. more than twice the height and the depth is 1 in. more than the width. If the volume of the case is 1632 in.^{3}, then what are the dimensions of the case?

Thinking Outside the Box XIII

Moving a Refrigerator A box containing a refrigerator is 3 ft wide, 3 ft deep, and 6 ft high. To move it, Wally lays it on its side, then on its top, then on its other side, and finally stands it upright as shown in the figure. Exactly how far has point A traveled in going from its initial location to its final location?

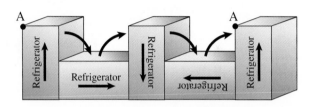

Figure for Thinking Outside the Box XIII

2.3 Pop Quiz

1. Use ordinary division to find the quotient and remainder when $x^3 - 5x + 7$ is divided by $x + 4$.

2. Use synthetic division to find the quotient and remainder when $x^2 - 3x + 9$ is divided by $x - 5$.

3. Use synthetic division to find $f(3)$ if $f(x) = x^3 - 2x^2 + 4x - 1$.

4. List the possible rational zeros for $f(x) = 2x^3 - 3x + 8$.

5. Find all real and imaginary zeros for $f(x) = 2x^3 - x^2 + 18x - 9$.

2.4 The Theory of Equations

One of the main goals in algebra is to keep expanding our knowledge of solving equations. The solutions (roots) of a polynomial equation $P(x) = 0$ are precisely the zeros of a polynomial function $y = P(x)$. Therefore the theorems of Section 2.3 concerning zeros of polynomial functions apply also to the roots of polynomial equations. In this section we study several additional theorems that are useful in solving polynomial equations.

The Number of Roots of a Polynomial Equation

When a polynomial equation is solved by factoring, a factor may occur more than once. For example, $x^2 - 10x + 25 = 0$ is equivalent to $(x - 5)^2 = 0$. Since the factor $x - 5$ occurs twice, we say that 5 is a root of the equation with *multiplicity* 2.

Definition: Multiplicity

If the factor $x - c$ occurs k times in the complete factorization of the polynomial $P(x)$, then c is called a root of $P(x) = 0$ with **multiplicity** k.

If a quadratic equation has a single root, as in $x^2 - 10x + 25 = 0$, then that root has multiplicity 2. If a root with multiplicity 2 is counted as two roots, then every quadratic equation has two roots in the set of complex numbers. This situation is generalized in the following theorem, where the phrase "when multiplicity is considered" means that a root with multiplicity k is counted as k individual roots.

n-Root Theorem

If $P(x) = 0$ is a polynomial equation with real or complex coefficients and positive degree n, then, when multiplicity is considered, $P(x) = 0$ has n roots.

PROOF By the fundamental theorem of algebra, the polynomial equation $P(x) = 0$ with degree n has at least one complex root c_1. By the factor theorem, $P(x) = 0$ is equivalent to

$$(x - c_1)Q_1(x) = 0,$$

where $Q_1(x)$ is a polynomial with degree $n - 1$ (the quotient when $P(x)$ is divided by $x - c_1$). Again, by the fundamental theorem of algebra, there is at least one complex root c_2 of $Q_1(x) = 0$. By the factor theorem, $P(x) = 0$ can be written as

$$(x - c_1)(x - c_2)Q_2(x) = 0,$$

where $Q_2(x)$ is a polynomial with degree $n - 2$. Reasoning in this manner n times, we get a quotient polynomial that has 0 degree, n factors for $P(x)$, and n complex roots, not necessarily all different.   ▪

Example **1** Finding all roots of a polynomial equation

State the degree of each polynomial equation. Find all real and imaginary roots of each equation, stating multiplicity when it is greater than one.

a. $6x^5 + 24x^3 = 0$ **b.** $(x - 3)^2(x + 14)^5 = 0$

Solution

a. This fifth-degree equation can be solved by factoring:

$$6x^3(x^2 + 4) = 0$$

$$6x^3 = 0 \quad \text{or} \quad x^2 + 4 = 0$$

$$x^3 = 0 \quad \text{or} \quad x^2 = -4$$

$$x = 0 \quad \text{or} \quad x = \pm 2i$$

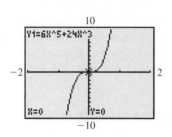

Figure 2.31

The roots are $\pm 2i$ and 0. Since there are two imaginary roots and 0 is a root with multiplicity 3, there are five roots when multiplicity is considered.

Because 0 is the only real root, the graph of $y = 6x^5 + 24x^3$ has only one x-intercept at $(0, 0)$ as shown in Fig. 2.31.   □

b. The highest power of x in $(x - 3)^2$ is 2, and in $(x + 14)^5$ is 5. By the product rule for exponents, the highest power of x in this equation is 7. The only roots of

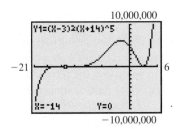

Figure 2.32

this seventh-degree equation are 3 and -14. The root 3 has multiplicity 2, and -14 has multiplicity 5. So there are seven roots when multiplicity is considered. Because the equation has two real solutions, the graph of $y = (x - 3)^2(x + 14)^5$ has two x-intercepts at $(3, 0)$ and $(-14, 0)$ as shown in Fig. 2.32. ■

Note that graphing polynomial functions and solving polynomial equations go hand in hand. The solutions to the equation can help us find an appropriate viewing window for the graph as they did in Example 1, and the graph can help us find solutions to the equation.

The Conjugate Pairs Theorem

For second-degree polynomial equations, the imaginary roots occur in pairs. For example, the roots of $x^2 - 2x + 5 = 0$ are

$$x = \frac{2 \pm \sqrt{(-2)^2 - 4(1)(5)}}{2} = 1 \pm 2i.$$

The roots $1 - 2i$ and $1 + 2i$ are complex conjugates. The $\pm$ symbol in the quadratic formula causes the complex solutions of a quadratic equation with real coefficients to occur in conjugate pairs. The conjugate pairs theorem indicates that this situation occurs also for polynomial equations of higher degree.

Conjugate Pairs Theorem

> If $P(x) = 0$ is a polynomial equation with real coefficients and the complex number $a + bi$ $(b \neq 0)$ is a root, then $a - bi$ is also a root.

Example **2** **Using the conjugate pairs theorem**

Find a polynomial equation with real coefficients that has 2 and $1 - i$ as roots.

Solution

If the polynomial has real coefficients, then its imaginary roots occur in conjugate pairs. So a polynomial with these two roots must actually have at least three roots: 2, $1 - i$, and $1 + i$. Since each root of the equation corresponds to a factor of the polynomial, we can write the following equation.

$$(x - 2)[x - (1 - i)][x - (1 + i)] = 0$$
$$(x - 2)[(x - 1) + i][(x - 1) - i] = 0 \quad \text{Regroup.}$$
$$(x - 2)[(x - 1)^2 - i^2] = 0 \quad {\scriptstyle (a + b)(a - b) = a^2 - b^2}$$
$$(x - 2)[x^2 - 2x + 1 + 1] = 0 \quad {\scriptstyle i^2 = -1}$$
$$(x - 2)(x^2 - 2x + 2) = 0$$
$$x^3 - 4x^2 + 6x - 4 = 0$$

This equation has the required roots and the smallest degree. Any multiple of this equation would also have the required roots but would not be as simple. ■

Descartes's Rule of Signs

Descartes's Rule of Signs

None of the theorems in this chapter tells us how to find all of the n roots to a polynomial equation of degree n. However, the theorems and rules presented here add to our knowledge of polynomial equations and help us to predict the type and number of solutions to expect for a particular equation. Descartes's rule of signs is a method for determining the number of positive, negative, and imaginary solutions. For this rule, a solution with multiplicity k is counted as k solutions.

When a polynomial is written in descending order, a **variation of sign** occurs when the signs of consecutive terms change. For example, if

$$P(x) = 3x^5 - 7x^4 - 8x^3 - x^2 + 3x - 9,$$

there are sign changes in going from the first to the second term, from the fourth to the fifth term, and from the fifth to the sixth term. So there are three variations of sign for $P(x)$. This information determines the number of positive real solutions to $P(x) = 0$. Descartes's rule requires that we look at $P(-x)$ and also count the variations of sign after it is simplified:

$$P(-x) = 3(-x)^5 - 7(-x)^4 - 8(-x)^3 - (-x)^2 + 3(-x) - 9$$
$$= -3x^5 - 7x^4 + 8x^3 - x^2 - 3x - 9$$

In $P(-x)$ the signs of the terms change from the second to the third term and again from the third to the fourth term. So there are two variations of sign for $P(-x)$. This information determines the number of negative real solutions to $P(x) = 0$.

> **Descartes's Rule of Signs**
>
> Suppose $P(x) = 0$ is a polynomial equation with real coefficients and with terms written in descending order.
>
> ▪ The number of positive real roots of the equation is either equal to the number of variations of sign of $P(x)$ or less than that by an even number.
> ▪ The number of negative real roots of the equation is either equal to the number of variations of sign of $P(-x)$ or less than that by an even number.

The proof of Descartes's rule of signs is beyond the scope of this text, but we can apply the rule to polynomial equations. Descartes's rule of signs is especially helpful when the number of variations of sign is 0 or 1.

Example 3 Using Descartes's rule of signs

Discuss the possibilities for the roots to $2x^3 - 5x^2 - 6x + 4 = 0$.

Solution

The number of variations of sign in

$$P(x) = 2x^3 - 5x^2 - 6x + 4$$

is 2. By Descartes's rule, the number of positive real roots is either 2 or 0. Since

$$P(-x) = 2(-x)^3 - 5(-x)^2 - 6(-x) + 4$$
$$= -2x^3 - 5x^2 + 6x + 4,$$

there is one variation of sign in $P(-x)$. So there is exactly one negative real root.

The equation must have three roots, because it is a third-degree polynomial equation. Since there must be three roots and one is negative, the other two roots must be either both imaginary numbers or both positive real numbers. Table 2.1 summarizes these two possibilities.

Table 2.1 Number of roots

Positive	Negative	Imaginary
2	1	0
0	1	2

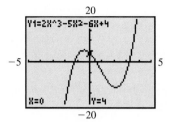

Figure 2.33

The graph of $y = 2x^3 - 5x^2 - 6x + 4$ shown in Fig. 2.33 crosses the positive x-axis twice and the negative x-axis once. So the first case in Table 2.1 is actually correct. ■

Example **4** Using Descartes's rule of signs

Discuss the possibilities for the roots to $3x^4 - 5x^3 - x^2 - 8x + 4 = 0$.

Solution

There are two variations of sign in the polynomial

$$P(x) = 3x^4 - 5x^3 - x^2 - 8x + 4.$$

According to Descartes's rule, there are either two or zero positive real roots to the equation. Since

$$P(-x) = 3(-x)^4 - 5(-x)^3 - (-x)^2 - 8(-x) + 4$$
$$= 3x^4 + 5x^3 - x^2 + 8x + 4,$$

there are two variations of sign in $P(-x)$. So the number of negative real roots is either two or zero. Since the degree of the polynomial is 4, there must be four roots. Each line of Table 2.2 gives a possible distribution of the type of those four roots. Note that the number of imaginary roots is even in each case, as we would expect from the conjugate pairs theorem.

Table 2.2 Number of roots

Positive	Negative	Imaginary
2	2	0
2	0	2
0	2	2
0	0	4

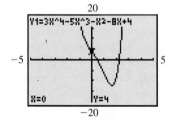

Figure 2.34

The calculator graph of $y = 3x^4 - 5x^3 - x^2 - 8x + 4$ shown in Fig. 2.34 shows two positive intercepts and no negative intercepts. However, we might not have the appropriate viewing window. The negative intercepts might be less than -5. In this case the graph did not allow us to conclude which line in Table 2.2 is correct. ■

Example **5** Using all of the theorems about roots

Find all of the solutions to $2x^3 - 5x^2 - 6x + 4 = 0$.

Solution

In Example 3 we used Descartes's rule of signs on this equation to determine that it has either two positive roots and one negative root or one negative root and two imaginary roots. From the rational zero theorem, the possible rational roots are $\pm 1, \pm 2, \pm 4$, and $\pm 1/2$. Use synthetic division to start checking the possible rational roots. The only possible rational root that is actually a root is $1/2$:

$$\frac{1}{2} \begin{array}{|rrrr} 2 & -5 & -6 & 4 \\ & 1 & -2 & -4 \\ \hline 2 & -4 & -8 & 0 \end{array}$$

Since $1/2$ is a root, $x - 1/2$ is a factor of the polynomial. The last line in the synthetic division indicates that the other factor is $2x^2 - 4x - 8$.

$$\left(x - \frac{1}{2}\right)(2x^2 - 4x - 8) = 0$$

$$(2x - 1)(x^2 - 2x - 4) = 0$$

$$2x - 1 = 0 \quad \text{or} \quad x^2 - 2x - 4 = 0$$

$$x = \frac{1}{2} \quad \text{or} \quad x = \frac{2 \pm \sqrt{4 - 4(1)(-4)}}{2} = 1 \pm \sqrt{5}$$

There are two positive roots, $1/2$ and $1 + \sqrt{5}$. The negative root is $1 - \sqrt{5}$, which is irrational. The roots guaranteed by Descartes's rule of signs are real numbers but not necessarily rational numbers.

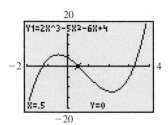

Figure 2.35

The graph of $y = 2x^3 - 5x^2 - 6x + 4$ in Fig. 2.35 supports these conclusions, because its x-intercepts appear to be $\left(1 - \sqrt{5}, 0\right)$, $(1/2, 0)$, and $\left(1 + \sqrt{5}, 0\right)$. ∎

For Thought

True or False? Explain.

1. The number 1 is a root of $x^3 - 1 = 0$ with multiplicity 3.

2. The equation $x^3 = 125$ has three complex number solutions.

3. For $(x + 1)^3(x^2 - 2x + 1) = 0$, -1 is a root with multiplicity 3.

4. For $(x - 5)^3(x^2 - 3x - 10) = 0$, 5 is a root with multiplicity 3.

5. If $4 - 5i$ is a solution to a polynomial equation with real coefficients, then $5i - 4$ is also a solution to the equation.

6. If $P(x) = 0$ is a polynomial equation with real coefficients and i, $2 - 3i$, and $5 + 7i$ are roots, then the degree of $P(x)$ is at least 6.

7. Both $-3 - i\sqrt{5}$ and $3 - i\sqrt{5}$ are solutions to $5x^3 - 9x^2 + 17x - 23 = 0$.

8. Both $3/2$ and 2 are solutions to $2x^5 - 4x^3 - 6x^2 - 3x - 6 = 0$.

9. The equation $x^3 - 5x^2 + 6x - 1 = 0$ has no negative roots.

10. The equation $5x^3 - 171 = 0$ has two imaginary solutions.

2.4 Exercises

State the degree of each polynomial equation. Find all of the real and imaginary roots of each equation, stating multiplicity when it is greater than one.

1. $x^2 - 10x + 25 = 0$
2. $x^2 - 18x + 81 = 0$

3. $x^5 - 9x^3 = 0$
4. $x^6 + x^4 = 0$

5. $x^4 - 2x^3 + x^2 = 0$
6. $x^5 - 6x^4 + 9x^3 = 0$

7. $(2x - 3)^2(3x + 4)^2 = 0$
8. $(2x^2 + x)^2(3x - 1)^4 = 0$

9. $x^3 - 4x^2 - 6x = 0$
10. $-x^3 + 8x^2 - 14x = 0$

Find each product.

11. $(x - 3i)(x + 3i)$
12. $(x + 6i)(x - 6i)$

13. $\left[x - \left(1 + \sqrt{2}\right)\right]\left[x - \left(1 - \sqrt{2}\right)\right]$

14. $\left[x - \left(3 - \sqrt{5}\right)\right]\left[x - \left(3 + \sqrt{5}\right)\right]$

15. $[x - (3 + 2i)][x - (3 - 2i)]$

16. $[x - (3 - i)][x - (3 + i)]$

17. $(x - 2)[x - (3 + 4i)][x - (3 - 4i)]$

18. $(x + 1)(x - (1 - i))(x - (1 + i))$

Find a polynomial equation with real coefficients that has the given roots.

19. $-3, 5$
20. $6, -1$
21. $-4i, 4i$
22. $-9i, 9i$

23. $3 - i$
24. $4 + i$
25. $-2, i$
26. $4, -i$

27. $0, i\sqrt{3}$
28. $-2, i\sqrt{2}$
29. $3, 1 - i$

30. $5, 4 - 3i$
31. $1, 2, 3$
32. $-1, 2, -3$

33. $1, 2 - 3i$
34. $-1, 4 - 2i$
35. $\dfrac{1}{2}, \dfrac{1}{3}, \dfrac{1}{4}$

36. $-\dfrac{1}{2}, -\dfrac{1}{3}, 1$
37. $i, 1 + i$
38. $3i, 3 - i$

Use Descartes's rule of signs to discuss the possibilities for the roots of each equation. Do not solve the equation.

39. $x^3 + 5x^2 + 7x + 1 = 0$
40. $2x^3 - 3x^2 + 5x - 6 = 0$

41. $-x^3 - x^2 + 7x + 6 = 0$
42. $-x^4 - 5x^2 - x + 7 = 0$

43. $y^4 + 5y^2 + 7 = 0$
44. $-3y^4 - 6y^2 + 7 = 0$

45. $t^4 - 3t^3 + 2t^2 - 5t + 7 = 0$

46. $-5r^4 + 4r^3 + 7r - 16 = 0$

47. $x^5 + x^3 + 5x = 0$
48. $x^4 - x^2 + 1 = 0$

Use the rational zero theorem and Descartes's rule of signs to assist you in finding all real and imaginary roots to each equation.

49. $x^3 - 4x^2 - 7x + 10 = 0$

50. $x^3 + 9x^2 + 26x + 24 = 0$

51. $x^3 - 10x - 3 = 0$
52. $2x^3 - 7x^2 - 16 = 0$

53. $x^4 + 2x^3 - 7x^2 + 2x - 8 = 0$

54. $x^4 - 4x^3 + 7x^2 - 16x + 12 = 0$

55. $6x^3 + 25x^2 - 24x + 5 = 0$

56. $6x^3 - 11x^2 - 46x - 24 = 0$

57. $x^4 + 2x^3 - 3x^2 - 4x + 4 = 0$

58. $x^5 + 3x^3 + 2x = 0$

59. $x^4 - 6x^3 + 12x^2 - 8x = 0$

60. $x^4 + 9x^3 + 27x^2 + 27x = 0$

61. $x^6 - x^5 - x^4 + x^3 - 12x^2 + 12x = 0$

62. $2x^7 - 2x^6 + 7x^5 - 7x^4 - 4x^3 + 4x^2 = 0$

63. $8x^5 + 2x^4 - 33x^3 + 4x^2 + 25x - 6 = 0$

64. $6x^5 + x^4 - 28x^3 - 3x^2 + 16x - 4 = 0$

Solve each problem.

65. *Growth Rate for Bacteria*   A car's speedometer indicates velocity at every instant in time. The instantaneous growth rate of a population is the rate at which it is growing at every instant in time. The instantaneous growth rate r of a colony of bacteria t hours after the start of an experiment is given by the function

$$r = 0.01t^3 - 0.08t^2 + 0.11t + 0.20$$

for $0 \le t \le 7$. Find the times for which the instantaneous growth rate is zero.

66. *Retail Store Profit*   The manager of a retail store has figured that her monthly profit P (in thousands of dollars) is determined by her monthly advertising expense x (in tens of thousands of dollars) according to the formula

$$P = x^3 - 20x^2 + 100x \quad \text{for} \quad 0 \le x \le 4.$$

For what value of x does she get $147,000 in profit?

67. *Designing Fireworks*   Marshall is designing a rocket for the Red Rocket Fireworks Company. The rocket will consist of a cardboard circular cylinder with a height that is four times as large as the radius. On top of the cylinder will be a cone with a height of 2 in. and a radius equal to the radius of the base as shown in the figure. If he wants to fill the cone and the cylinder with a total of 114π in.^{3} of powder, then what should be the radius of the cylinder?

Figure for Exercise 67

68. *Heating and Air*   An observatory is built in the shape of a right circular cylinder with a hemispherical roof as shown in the figure. The heating and air contractor has figured the volume of the structure as 3168π ft^3. If the height of the cylindrical walls is 2 ft more than the radius of the building, then what is the radius of the building?

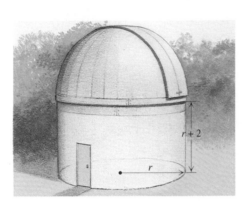

Figure for Exercise 68

Thinking Outside the Box XIV

Packing Billiard Balls   There are several ways to tightly pack nine billiard balls each with radius 1 into a rectangular box. Find the volume of the box in each of the following cases and determine which box has the least volume.

a. Four balls are placed so that they just fit into the bottom of the box, then another layer of four, then one ball in the middle tangent to all four in the second layer, as shown in this side view.

b. Four balls are placed so that they just fit into the bottom of the box as in (a), then one is placed in the middle on top of the first four. Finally, four more are placed so that they just fit at the top of the box.

c. The box is packed with layers of four, one, and four as in (b), but the box is required to be cubic. In this case, the four balls in the bottom layer will not touch each other and the four balls in the top layer will not touch each other. The ball in the middle will be tangent to all of the other eight balls.

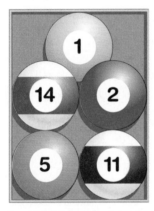

Figure for Thinking Outside the Box XIV

2.4 Pop Quiz

1. Find all real and imaginary roots to $x^5 - x^3 = 0$, including multiplicities.

2. Find a polynomial equation with real coefficients that has the roots $-4i$ and 5.

3. By Descartes's rule of signs, how many positive roots can $x^4 + x^3 - 3x^2 + 5x + 9 = 0$ have?

4. By Descartes's rule of signs, how many negative roots can $3x^3 + 5x^2 - x + 9 = 0$ have?

5. Find all real and imaginary roots to $x^3 - 3x^2 - 6x + 8 = 0$.

Miscellaneous Equations

In Section 2.4 we learned that an nth-degree polynomial equation has n roots. However, it is not always obvious how to find them. In this section we will solve polynomial equations using some new techniques and we will solve several other types of equations. Unfortunately, we cannot generally predict the number of roots to non-polynomial equations.

Factoring Higher-Degree Equations

We usually use factoring to solve quadratic equations. However, since we can also factor many higher-degree polynomials, we can solve many higher-degree equations by factoring. Factoring is often the fastest method for solving an equation.

Example **1** **Solving an equation by factoring**

Solve $x^3 + 3x^2 + x + 3 = 0$.

Solution

Factor the polynomial on the left-hand side by grouping.

$$x^2(x + 3) + 1(x + 3) = 0 \quad \text{Factor by grouping.}$$
$$(x^2 + 1)(x + 3) = 0 \quad \text{Factor out } x + 3.$$
$$x^2 + 1 = 0 \quad \text{or} \quad x + 3 = 0 \quad \text{Zero factor property}$$
$$x^2 = -1 \quad \text{or} \quad x = -3$$
$$x = \pm i \quad \text{or} \quad x = -3$$

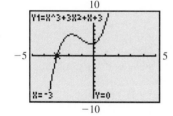

Figure 2.36

The solution set is $\{-3, -i, i\}$.

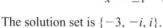 The graph in Fig. 2.36 supports these solutions. Because there is only one real solution, the graph crosses the x-axis only once. ■

Example **2** **Solving an equation by factoring**

Solve $2x^5 = 16x^2$.

Solution

Write the equation with 0 on the right-hand side, then factor completely.

$$2x^5 - 16x^2 = 0$$
$$2x^2(x^3 - 8) = 0 \quad \text{Factor out the greatest common factor.}$$
$$2x^2(x - 2)(x^2 + 2x + 4) = 0 \quad \text{Factor the difference of two cubes.}$$
$$2x^2 = 0 \quad \text{or} \quad x - 2 = 0 \quad \text{or} \quad x^2 + 2x + 4 = 0$$
$$x = 0 \quad \text{or} \quad x = 2 \quad \text{or} \quad x = \frac{-2 \pm \sqrt{-12}}{2} = -1 \pm i\sqrt{3}$$

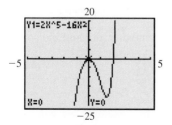

Figure 2.37

The solution set is $\left\{0, 2, -1 \pm i\sqrt{3}\right\}$. Since 0 is a root with multiplicity 2, there are five roots, counting multiplicity, to this fifth-degree equation.

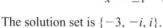 The graph in Fig. 2.37 supports this solution. ■

Note that in Example 2, if we had divided each side by x^2 as our first step, we would have lost the solution $x = 0$. *We do not usually divide each side of an equation by a variable expression.* Instead, bring all expressions to the same side and factor out the common factors.

Equations Involving Square Roots

Recall that $\sqrt{x}$ represents the nonnegative square root of x. To solve $\sqrt{x} = 3$ we can use the definition of square root. Since the nonnegative square root of 9 is 3, the solution to $\sqrt{x} = 3$ is 9. To solve $\sqrt{x} = -3$ we again use the definition of square root. Because $\sqrt{x}$ is nonnegative while -3 is negative, this equation has no solution.

More complicated equations involving square roots are usually solved by squaring both sides. However, squaring both sides does not always lead to an equivalent equation. If we square both sides of $\sqrt{x} = 3$, we get $x = 9$, which is equivalent to $\sqrt{x} = 3$. But if we square both sides of $\sqrt{x} = -3$, we also get $x = 9$, which is not equivalent to $\sqrt{x} = -3$. Because 9 appeared in the attempt to solve $\sqrt{x} = -3$, but does not satisfy the equation, it is called an *extraneous root*. This same situation can occur with an equation involving a fourth root or any other even root. So if you raise each side of an equation to an even power, you must check for extraneous roots.

Example **3** Squaring each side to solve an equation

Solve $\sqrt{x} + 2 = x$.

Solution

Isolate the radical before squaring each side.

$$\sqrt{x} = x - 2$$
$$\left(\sqrt{x}\right)^2 = (x - 2)^2 \qquad \text{Square each side.}$$
$$x = x^2 - 4x + 4 \qquad \text{Use the special product } (a - b)^2 = a^2 - 2ab + b^2.$$
$$0 = x^2 - 5x + 4 \qquad \text{Write in the form } ax^2 + bx + c = 0.$$
$$0 = (x - 4)(x - 1) \qquad \text{Factor the quadratic polynomial.}$$
$$x - 4 = 0 \quad \text{or} \quad x - 1 = 0 \qquad \text{Zero factor property}$$
$$x = 4 \quad \text{or} \quad x = 1$$

Checking $x = 4$, we get $\sqrt{4} + 2 = 4$, which is correct. Checking $x = 1$, we get $\sqrt{1} + 2 = 1$, which is incorrect. So 1 is an extraneous root and the solution set is $\{4\}$.

The graph in Fig. 2.38 supports this solution. ■

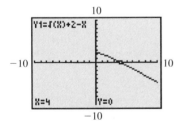

■ **Figure 2.38**

The next example involves two radicals. In this example we will isolate the more complicated radical before squaring each side. But not all radicals are eliminated upon squaring each side. So we isolate the remaining radical and square each side again.

Example **4** Squaring each side twice

Solve $\sqrt{2x + 1} - \sqrt{x} = 1$.

Solution

First we write the equation so that the more complicated radical is isolated. Then we square each side. On the left side, when we square $\sqrt{2x + 1}$, we get $2x + 1$. On the right side, when we square $1 + \sqrt{x}$, we use the special product rule $(a + b)^2 = a^2 + 2ab + b^2$.

$$\sqrt{2x + 1} = 1 + \sqrt{x}$$

$$(\sqrt{2x + 1})^2 = (1 + \sqrt{x})^2 \qquad \text{Square each side.}$$

$$2x + 1 = 1 + 2\sqrt{x} + x$$

$$x = 2\sqrt{x} \qquad \text{All radicals are not eliminated by the first squaring.}$$

$$x^2 = (2\sqrt{x})^2 \qquad \text{Square each side a second time.}$$

$$x^2 = 4x$$

$$x^2 - 4x = 0$$

$$x(x - 4) = 0$$

$$x = 0 \quad \text{or} \quad x - 4 = 0$$

$$x = 0 \quad \text{or} \qquad x = 4$$

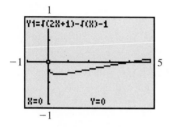

Figure 2.39

Both 0 and 4 satisfy the original equation. So the solution set is $\{0, 4\}$. The graph in Fig. 2.39 supports this solution.   ∎

Use the following strategy when solving equations involving square roots.

STRATEGY   **Solving Equations Involving Square Roots**

1. Isolate the radical if there is only one. Separate the radicals on opposite sides of the equation if there is more than one.
2. Square both sides and simplify.
3. Isolate or separate any remaining radicals and square again.
4. Check all solutions because squaring can produce extraneous solutions.

Equations with Rational Exponents

To solve equations of the form $x^{m/n} = k$ in which m and n are positive integers and m/n is in lowest terms, we adapt the methods of Examples 3 and 4 of raising each side to a power. Cubing each side of $x^{2/3} = 4$, yields $(x^{2/3})^3 = 4^3$ or $x^2 = 64$. By the square root property, $x = \pm 8$. We can shorten this solution by raising each side of the equation to the power $3/2$ (the reciprocal of $2/3$) and inserting the $\pm$ symbol to obtain the two square roots.

$$x^{2/3} = 4$$

$$(x^{2/3})^{3/2} = \pm 4^{3/2} \qquad \text{Raise each side to the power } 3/2 \text{ and insert } \pm.$$

$$x = \pm 8$$

The equation $x^{2/3} = 4$ has two solutions because the numerator of the exponent $2/3$ is an even number. An equation such as $x^{-3/2} = 1/8$ has only one real solution because the numerator of the exponent $-3/2$ is odd. To solve $x^{-3/2} = 1/8$, raise each side to the power $-2/3$ (the reciprocal of $-3/2$).

$$x^{-3/2} = \frac{1}{8}$$

$$(x^{-3/2})^{-2/3} = \left(\frac{1}{8}\right)^{-2/3} \quad \text{\small Raise each side to the power } -2/3.$$

$$x = 4$$

In the next example we solve two more equations of this type by raising each side to a fractional power. Note that we use the $\pm$ symbol only when the numerator of the original exponent is even.

Example ⑤ Equations with rational exponents

Solve each equation.

a. $x^{4/3} = 625$ **b.** $(y - 2)^{-5/2} = 32$

Solution

a. Raise each side of the equation to the power $3/4$. Use the $\pm$ symbol because the numerator of $4/3$ is even.

$$x^{4/3} = 625$$

$$(x^{4/3})^{3/4} = \pm 625^{3/4}$$

$$x = \pm 125$$

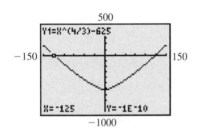

Figure 2.40

Check in the original equation. The solution set is $\{-125, 125\}$.
⊞ The graph in Fig. 2.40 supports this solution.   □

b. Raise each side to the power $-2/5$. Because the numerator in $-5/2$ is an odd number, there is only one real solution.

$$(y - 2)^{-5/2} = 32$$

$$((y - 2)^{-5/2})^{-2/5} = 32^{-2/5} \quad \text{\small Raise each side to the power } -2/5.$$

$$y - 2 = \frac{1}{4}$$

$$y = 2 + \frac{1}{4} = \frac{9}{4}$$

Check $9/4$ in the original equation. The solution set is $\left\{\frac{9}{4}\right\}$.

⊞ The graph in Fig. 2.41 supports this solution.   ■

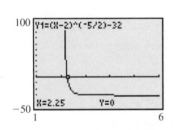

Figure 2.41

Equations of Quadratic Type

In some cases, an equation can be converted to a quadratic equation by substituting a single variable for a more complicated expression. Such equations are called **equations of quadratic type.** An equation of quadratic type has the form $au^2 + bu + c = 0$, where $a \neq 0$ and u is an algebraic expression.

In the next example, the expression x^2 in a fourth-degree equation is replaced by u, yielding a quadratic equation. After the quadratic equation is solved, u is replaced by x^2 so that we find values for x that satisfy the original fourth-degree equation.

Example 6 Solving a fourth-degree polynomial equation

Solve $x^4 - 14x^2 + 45 = 0$.

Solution

We let $u = x^2$ so that $u^2 = (x^2)^2 = x^4$.

$$(x^2)^2 - 14x^2 + 45 = 0$$

$$u^2 - 14u + 45 = 0 \qquad \text{Replace } x^2 \text{ by } u.$$

$$(u - 9)(u - 5) = 0$$

$$u - 9 = 0 \quad \text{or} \quad u - 5 = 0$$

$$u = 9 \quad \text{or} \quad u = 5$$

$$x^2 = 9 \quad \text{or} \quad x^2 = 5 \qquad \text{Replace } u \text{ by } x^2.$$

$$x = \pm 3 \quad \text{or} \quad x = \pm\sqrt{5}$$

Check in the original equation. The solution set is $\{-3, -\sqrt{5}, \sqrt{5}, 3\}$.
 The graph in Fig. 2.42 supports this solution. ■

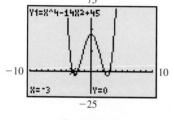

Figure 2.42

Note that the equation of Example 6 could be solved by factoring without doing substitution, because $x^4 - 14x^2 + 45 = (x^2 - 9)(x^2 - 5)$. Since the next example involves a more complicated algebraic expression, we use substitution to simplify it, although it too could be solved by factoring, without substitution.

Example 7 Another equation of quadratic type

Solve $(x^2 - x)^2 - 18(x^2 - x) + 72 = 0$.

Solution

If we let $u = x^2 - x$, then the equation becomes a quadratic equation.

$$(x^2 - x)^2 - 18(x^2 - x) + 72 = 0$$

$$u^2 - 18u + 72 = 0 \qquad \text{Replace } x^2 - x \text{ by } u.$$

$$(u - 6)(u - 12) = 0$$

$$u - 6 = 0 \quad \text{or} \quad u - 12 = 0$$

$$u = 6 \quad \text{or} \quad u = 12$$

$$x^2 - x = 6 \quad \text{or} \quad x^2 - x = 12 \qquad \text{Replace } u \text{ by } x^2 - x.$$

$$x^2 - x - 6 = 0 \quad \text{or} \quad x^2 - x - 12 = 0$$

$$(x - 3)(x + 2) = 0 \quad \text{or} \quad (x - 4)(x + 3) = 0$$

$$x = 3 \quad \text{or} \quad x = -2 \quad \text{or} \quad x = 4 \quad \text{or} \quad x = -3$$

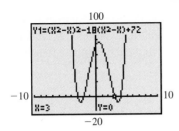

Figure 2.43

Check in the original equation. The solution set is $\{-3, -2, 3, 4\}$.
The graph in Fig. 2.43 supports this solution. ■

The next equations of quadratic type have rational exponents.

Example **8** **Quadratic type and rational exponents**

Find all real solutions to each equation.

a. $x^{2/3} - 9x^{1/3} + 8 = 0$ **b.** $(11x^2 - 18)^{1/4} = x$

Solution

a. If we let $u = x^{1/3}$, then $u^2 = (x^{1/3})^2 = x^{2/3}$.

$$u^2 - 9u + 8 = 0 \qquad \text{Replace } x^{2/3} \text{ by } u^2 \text{ and } x^{1/3} \text{ by } u.$$

$$(u - 8)(u - 1) = 0$$

$$u = 8 \qquad \text{or} \qquad u = 1$$

$$x^{1/3} = 8 \qquad \text{or} \qquad x^{1/3} = 1 \qquad \text{Replace } u \text{ by } x^{1/3}.$$

$$(x^{1/3})^3 = 8^3 \qquad \text{or} \quad (x^{1/3})^3 = 1^3$$

$$x = 512 \quad \text{or} \qquad x = 1$$

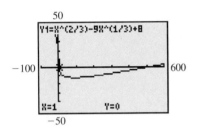

Figure 2.44

Check in the original equation. The solution set is $\{1, 512\}$.
The graph in Fig. 2.44 supports this solution. □

b. $\qquad (11x^2 - 18)^{1/4} = x$

$$((11x^2 - 18)^{1/4})^4 = x^4 \qquad \text{Raise each side to the power 4.}$$

$$11x^2 - 18 = x^4$$

$$x^4 - 11x^2 + 18 = 0$$

$$(x^2 - 9)(x^2 - 2) = 0$$

$$x^2 = 9 \qquad \text{or} \quad x^2 = 2$$

$$x = \pm 3 \quad \text{or} \quad x = \pm\sqrt{2}$$

Since the exponent $1/4$ means principal fourth root, the right-hand side of the equation cannot be negative. So -3 and $-\sqrt{2}$ are extraneous roots. Since 3 and $\sqrt{2}$ satisfy the original equation, the solution set is $\{\sqrt{2}, 3\}$.
The graph in Fig. 2.45 supports this solution. ■

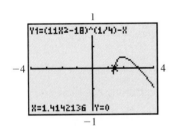

Figure 2.45

Equations Involving Absolute Value

We solved basic absolute value equations in Section 1.1. In the next two examples we solve some more complicated absolute value equations.

Example **9** **An equation involving absolute value**

Solve $|x^2 - 2x - 16| = 8$.

Solution

First write an equivalent statement without using absolute value symbols.

$$x^2 - 2x - 16 = 8 \quad \text{or} \quad x^2 - 2x - 16 = -8$$

$$x^2 - 2x - 24 = 0 \quad \text{or} \quad x^2 - 2x - 8 = 0$$

$$(x - 6)(x + 4) = 0 \quad \text{or} \quad (x - 4)(x + 2) = 0$$

$$x = 6 \quad \text{or} \quad x = -4 \quad \text{or} \quad x = 4 \quad \text{or} \quad x = -2$$

The solution set is $\{-4, -2, 4, 6\}$.

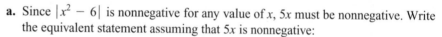 The graph in Fig. 2.46 supports this solution. ■

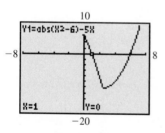

Figure 2.46

In the next example we have an equation in which an absolute value expression is equal to an expression that could be positive or negative and an equation with two absolute value expressions.

Example **10** **More equations involving absolute value**

Solve each equation.

a. $|x^2 - 6| = 5x$ **b.** $|a - 1| = |2a - 3|$

Solution

a. Since $|x^2 - 6|$ is nonnegative for any value of x, $5x$ must be nonnegative. Write the equivalent statement assuming that $5x$ is nonnegative:

$$x^2 - 6 = 5x \quad \text{or} \quad x^2 - 6 = -5x$$

$$x^2 - 5x - 6 = 0 \quad \text{or} \quad x^2 + 5x - 6 = 0$$

$$(x - 6)(x + 1) = 0 \quad \text{or} \quad (x + 6)(x - 1) = 0$$

$$x = 6 \quad \text{or} \quad x = -1 \quad \text{or} \quad x = -6 \quad \text{or} \quad x = 1$$

The expression $|x^2 - 6|$ is nonnegative for any real number x. But $5x$ is negative if $x = -1$ or if $x = -6$. So -1 and -6 are extraneous roots. They do not satisfy the original equation. The solution set is $\{1, 6\}$.

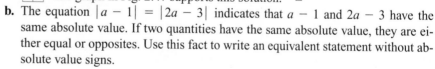 The graph in Fig. 2.47 supports this solution. □

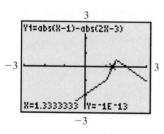

Figure 2.47

b. The equation $|a - 1| = |2a - 3|$ indicates that $a - 1$ and $2a - 3$ have the same absolute value. If two quantities have the same absolute value, they are either equal or opposites. Use this fact to write an equivalent statement without absolute value signs.

$$a - 1 = 2a - 3 \quad \text{or} \quad a - 1 = -(2a - 3)$$

$$a + 2 = 2a \quad \text{or} \quad a - 1 = -2a + 3$$

$$2 = a \quad \text{or} \quad a = \frac{4}{3}$$

Check that both 2 and $\frac{4}{3}$ satisfy the original absolute value equation. The solution set is $\left\{\frac{4}{3}, 2\right\}$.

The graph in Fig. 2.48 supports this solution. ■

Figure 2.48

For Thought

True or False? Explain.

1. Squaring each side of $\sqrt{x-1} + \sqrt{x} = 6$ yields $x - 1 + x = 36$.

2. The equations $(2x - 1)^2 = 9$ and $2x - 1 = 3$ are equivalent.

3. The equations $x^{2/3} = 9$ and $x = 27$ have the same solution set.

4. To solve $2x^{1/4} - x^{1/2} + 3 = 0$, we let $u = x^{1/2}$ and $u^2 = x^{1/4}$.

5. If $(x - 1)^{-2/3} = 4$, then $x = 1 \pm 4^{-3/2}$.

6. No negative number satisfies $x^{-2/5} = 4$.

7. The solution set to $|2x + 10| = 3x$ is $\{-2, 10\}$.

8. No negative number satisfies $|x^2 - 3x + 2| = 7x$.

9. The equation $|2x + 1| = |x|$ is equivalent to $2x + 1 = x$ or $2x + 1 = -x$.

10. The equation $x^9 - 5x^3 + 6 = 0$ is an equation of quadratic type.

2.5   Exercises

Find all real and imaginary solutions to each equation. Check your answers.

1. $x^3 + 3x^2 - 4x - 12 = 0$

2. $x^3 - x^2 - 5x + 5 = 0$

3. $2x^3 + 1000x^2 - x - 500 = 0$

4. $3x^3 - 1200x^2 - 2x + 800 = 0$

5. $a^3 + 5a = 15a^2$

6. $b^3 + 20b = 9b^2$

7. $3y^4 - 12y^2 = 0$

8. $5m^4 - 10m^3 + 5m^2 = 0$

9. $a^4 - 16 = 0$

10. $w^4 + 8w = 0$

Find all real solutions to each equation. Check your answers.

11. $\sqrt{x + 1} = x - 5$

12. $\sqrt{x - 1} = x - 7$

13. $\sqrt{x - 2} = x - 22$

14. $3 + \sqrt{x} = 1 + x$

15. $w = \dfrac{\sqrt{1 - 3w}}{2}$

16. $t = \dfrac{\sqrt{2 - 3t}}{3}$

17. $\dfrac{1}{z} = \dfrac{3}{\sqrt{4z + 1}}$

18. $\dfrac{1}{p} - \dfrac{2}{\sqrt{9p + 1}} = 0$

19. $\sqrt{x^2 - 2x - 15} = 3$

20. $\sqrt{3x^2 + 5x - 3} = x$

21. $\sqrt{x + 40} - \sqrt{x} = 4$

22. $\sqrt{x} + \sqrt{x - 36} = 2$

23. $\sqrt{n + 4} + \sqrt{n - 1} = 5$

24. $\sqrt{y + 10} - \sqrt{y - 2} = 2$

25. $\sqrt{2x + 5} + \sqrt{x + 6} = 9$

26. $\sqrt{3x - 2} - \sqrt{x - 2} = 2$

Find all real solutions to each equation. Check your answers.

27. $x^{2/3} = 2$

28. $x^{2/3} = \dfrac{1}{2}$

29. $w^{-4/3} = 16$

30. $w^{-3/2} = 27$

31. $t^{-1/2} = 7$

32. $t^{-1/2} = \dfrac{1}{2}$

33. $(s - 1)^{-1/2} = 2$

34. $(s - 2)^{-1/2} = \dfrac{1}{3}$

Find all real and imaginary solutions to each equation. Check your answers.

35. $x^4 - 12x^2 + 27 = 0$

36. $x^4 + 10 = 7x^2$

37. $\left(\dfrac{2c - 3}{5}\right)^2 + 2\left(\dfrac{2c - 3}{5}\right) = 8$

38. $\left(\dfrac{b-5}{6}\right)^2 - \left(\dfrac{b-5}{6}\right) - 6 = 0$

39. $\dfrac{1}{(5x-1)^2} + \dfrac{1}{5x-1} - 12 = 0$

40. $\dfrac{1}{(x-3)^2} + \dfrac{2}{x-3} - 24 = 0$

41. $(v^2 - 4v)^2 - 17(v^2 - 4v) + 60 = 0$

42. $(u^2 + 2u)^2 - 2(u^2 + 2u) - 3 = 0$

43. $x - 4\sqrt{x} + 3 = 0$ **44.** $2x + 3\sqrt{x} - 20 = 0$

45. $q - 7q^{1/2} + 12 = 0$ **46.** $h + 1 = 2h^{1/2}$

47. $x^{2/3} + 10 = 7x^{1/3}$ **48.** $x^{1/2} - 3x^{1/4} + 2 = 0$

Solve each absolute value equation.

49. $|w^2 - 4| = 3$ **50.** $|a^2 - 1| = 1$

51. $|v^2 - 3v| = 5v$ **52.** $|z^2 - 12| = z$

53. $|x^2 - x - 6| = 6$ **54.** $|2x^2 - x - 2| = 1$

55. $|x + 5| = |2x + 1|$ **56.** $|3x - 4| = |x|$

Solve each equation. Find imaginary solutions when possible.

57. $\sqrt{16x + 1} - \sqrt{6x + 13} = -1$

58. $\sqrt{16x + 1} - \sqrt{6x + 13} = 1$

59. $v^6 - 64 = 0$ **60.** $t^4 - 1 = 0$

61. $(7x^2 - 12)^{1/4} = x$ **62.** $(10x^2 - 1)^{1/4} = 2x$

63. $\sqrt[3]{2 + x - 2x^2} = x$

64. $\sqrt{48 + \sqrt{x}} - 4 = \sqrt[4]{x}$

65. $\left(\dfrac{x-2}{3}\right)^2 - 2\left(\dfrac{x-2}{3}\right) + 10 = 0$

66. $\dfrac{1}{(x+1)^2} - \dfrac{2}{x+1} + 2 = 0$

67. $(3u - 1)^{2/5} = 2$ **68.** $(2u + 1)^{2/3} = 3$

69. $x^2 - 11\sqrt{x^2 + 1} + 31 = 0$

70. $2x^2 - 3\sqrt{2x^2 - 3} - 1 = 0$

71. $|x^2 - 2x| = |3x - 6|$

72. $|x^2 + 5x| = |3 - x^2|$

73. $(3m + 1)^{-3/5} = -\dfrac{1}{8}$ **74.** $(1 - 2m)^{-5/3} = -\dfrac{1}{32}$

75. $|x^2 - 4| = x - 2$ **76.** $|x^2 + 7x| = x^2 - 4$

Solve each problem.

77. *Maximum Sail Area* According to the International America's Cup Rules, the maximum sail area S for a boat with length L (in meters) and displacement D (in cubic meters) is determined by the equation

$$L + 1.25S^{1/2} - 9.8D^{1/3} = 16.296$$

(America's Cup, www.americascup.org). Find S for a boat with length 21.24 m and displacement 18.34 m³.

$L + 1.25S^{1/2} - 9.8D^{1/3} = 16.296$

■ **Figure for Exercises 77 and 78**

78. *Minimum Displacement for a Yacht* The minimum displacement D for a boat with length 21.52 m and a sail area of 310.64 m² is determined by the equation $L + 1.25S^{1/2} - 9.8D^{1/3} = 16.296$. Find this boat's minimum displacement.

79. *Square Roots* Find two numbers that differ by 6 and whose square roots differ by 1.

80. *Right Triangle* One leg of a right triangle is 1 cm longer than the other leg. What is the length of the short leg if the total length of the hypotenuse and the short leg is 10 cm?

81. *Sail Area-Displacement Ratio* The sail area-displacement ratio S is defined as

$$S = A\left(\dfrac{d}{64}\right)^{-2/3},$$

where A is the sail area in square feet and d is the displacement in pounds. The Oceanis 381 is a 39 ft sailboat with a sail area-displacement ratio of 14.26 and a sail area of 598.9 ft². Find the displacement for the Oceanis 381.

82. *Capsize Screening Value* The capsize screening value C is defined as

$$C = b\left(\frac{d}{64}\right)^{-1/3},$$

where b is the beam (or width) in feet and d is the displacement in pounds. The Bahia 46 is a 46 ft catamaran with a capsize screening value of 3.91 and a beam of 26.1 ft. Find the displacement for the Bahia 46.

83. *Radius of a Pipe* A large pipe is placed next to a wall and a 1-foot high block is placed 5 feet from the wall to keep the pipe in place as shown in the accompanying figure. What is the radius of the pipe?

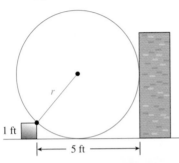

Figure for Exercise 83

84. *Radius of a Pipe* A large pipe is held in place on level ground by using a 1-foot high block on one side and a 2-foot high block on the other side. If the distance between the blocks is 6 feet, then what is the radius of the pipe?

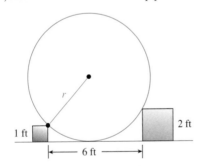

Figure for Exercise 84

■ Foreshadowing Calculus

The situation in the next exercise is studied also in calculus. However, in calculus we find the route that minimizes the total time for the trip.

85. *Time Swimming and Running* Lauren is competing in her town's cross-country competition. Early in the event, she must race from point A on the Greenbriar River to point B, which is 5 mi downstream and on the opposite bank. The Greenbriar is 1 mi wide. In planning her strategy, Lauren knows she can use any combination of running and swimming. She can run 10 mph and swim 8 mph. How long would it take if she ran 5 mi downstream and then swam across? Find the time it would take if she swam diagonally from A to B. Find x so that she could run x miles along the bank, swim diagonally to B, and complete the race in 36 min. (Ignore the current in the river.)

Figure for Exercise 85

86. *Storing Supplies* An army sergeant wants to use a 20-ft by 40-ft piece of canvas to make a two-sided tent for holding supplies as shown in the figure. Write the volume of the tent as a function of b. For what value of b is the volume 1600 ft³? Use a graphing calculator to find the values for b and h that will maximize the volume of the tent.

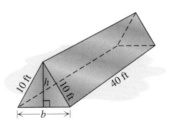

Figure for Exercise 86

Thinking Outside the Box XV

Painting Problem A painter has seven 3-ft by 5-ft rectangular drop cloths. If he lays each drop cloth on the carpet as a 3-ft by 5-ft rectangle, without folding, cutting, or tearing them, then what is the maximum area that he can cover with these drop cloths in an 8-ft by 13-ft room?

2.5 Pop Quiz

Solve each equation. Find imaginary solutions when possible.

1. $x^3 + x^2 + x + 1 = 0$

2. $\sqrt{x + 4} = x - 2$

3. $x^{-2/3} = 4$

4. $x^4 - 3x^2 = 4$

5. $|x + 3| = |2x - 5|$

Graphs of Polynomial Functions

In Chapter 1 we learned that the graph of a polynomial function of degree 0 or 1 is a straight line. In Section 2.1 we learned that the graph of a second-degree polynomial function is a parabola. In this section we will concentrate on graphs of polynomial functions of degree greater than 2.

Drawing Good Graphs

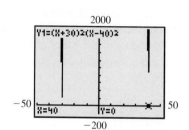

Figure 2.49

A graph of an equation is a picture of all of the ordered pairs that satisfy the equation. However, it is impossible to draw a perfect picture of any set of ordered pairs. We usually find a few important features of the graph and make sure that our picture brings out those features. For example, you can make a good graph of a linear function by drawing a line through the intercepts using a ruler and a sharp pencil. A good parabola should look smooth and symmetric and pass through the vertex and intercepts.

A graphing calculator is a tremendous aid in graphing because it can quickly plot many points. However, the calculator does not know if it has drawn a graph that shows the important features. For example, the graph of $y = (x + 30)^2(x - 40)^2$ has x-intercepts at $(-30, 0)$ and $(40, 0)$, but they do not appear on the graph in Fig. 2.49. ☐

An important theorem for understanding the graphs of polynomial functions is the intermediate value theorem (IVT). The IVT says that *a polynomial function takes on every value between any two of its values.* For example, consider $f(x) = x^3$, for which $f(1) = 1$ and $f(2) = 8$. Select any number between 1 and 8, say, 5. By the IVT there is a number c in the interval $(1, 2)$ such that $f(c) = 5$. In this case, c is easy to find: $c = \sqrt[3]{5}$. The IVT allows us to "connect the dots" when drawing a graph. The curve cannot go from $(1, 1)$ to $(2, 8)$ without hitting every y-coordinate between 1 and 8. If one of the values is positive and the other negative, the IVT guarantees that the curve crosses the x-axis on the interval. For example, $f(-2) = -8$ and $f(2) = 8$. By the IVT there is a c in the interval $(-2, 2)$ for which $f(c) = 0$. Of course in this case $c = 0$ and the x-intercept is $(0, 0)$.

Note that the greatest integer function, which is not a polynomial function, does not obey the IVT. The greatest integer function jumps from one integer to the next without taking on any values between the integers.

We will not prove the IVT here. It is proved in calculus. The theorem is stated symbolically as follows.

The Intermediate Value Theorem

Suppose that f is a polynomial function and $[a, b]$ is an interval for which $f(a) \neq f(b)$. If k is a number between $f(a)$ and $f(b)$ then there is a number c in the interval (a, b) such that $f(c) = k$.

Symmetry

Symmetry is a very special property of graphs of some functions but not others. Recognizing that the graph of a function has some symmetry usually cuts in half the work required to obtain the graph and also helps cut down on errors in graphing. So far we have discussed the following types of symmetry.

SUMMARY   **Types of Symmetry**

1. The graph of a function $f(x)$ is *symmetric about the y-axis* and *f is an even function* if $f(-x) = f(x)$ for any value of x in the domain of the function. (Section 1.7)

2. The graph of a function $f(x)$ is *symmetric about the origin* and *f is an odd function* if $f(-x) = -f(x)$ for any value of x in the domain of the function. (Section 1.7)

3. The graph of a quadratic function $f(x) = ax^2 + bx + c$ is *symmetric about its axis of symmetry, $x = -b/(2a)$.* (Section 2.1)

The graphs of $f(x) = x^2$ and $f(x) = x^3$ shown in Figs. 2.50 and 2.51 are nice examples of symmetry about the y-axis and symmetry about the origin, respectively. The axis of symmetry of $f(x) = x^2$ is the y-axis. The symmetry of the other quadratic functions comes from the fact that the graph of every quadratic function is a transformation of the graph of $f(x) = x^2$.

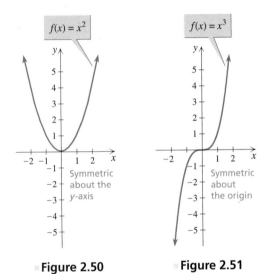

Figure 2.50   **Figure 2.51**

Example **1** **Determining the symmetry of a graph**

Discuss the symmetry of the graph of each polynomial function.

a. $f(x) = 5x^3 - x$ **b.** $g(x) = 2x^4 - 3x^2$ **c.** $h(x) = x^2 - 3x + 6$

d. $j(x) = x^4 - x^3$

Solution

a. Replace x by $-x$ in $f(x) = 5x^3 - x$ and simplify:

$$f(-x) = 5(-x)^3 - (-x)$$
$$= -5x^3 + x$$

Since $f(-x)$ is the opposite of $f(x)$, the graph is symmetric about the origin. Fig. 2.52 supports this conclusion.   □

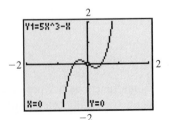

Figure 2.52

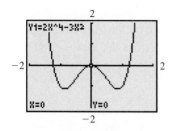

Figure 2.53

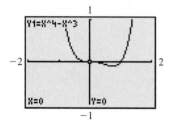

Figure 2.54

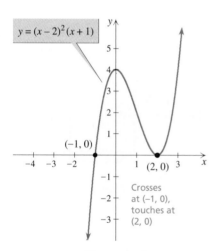

Crosses
at (–1, 0),
touches at
(2, 0)

Figure 2.55

b. Replace x by $-x$ in $g(x) = 2x^4 - 3x^2$ and simplify:

$$g(-x) = 2(-x)^4 - 3(-x)^2$$
$$= 2x^4 - 3x^2$$

Since $g(-x) = g(x)$, the graph is symmetric about the y-axis.
Fig. 2.53 supports this conclusion. □

c. Because h is a quadratic function, its graph is symmetric about the line $x = -b/(2a)$, which in this case is the line $x = 3/2$.

d. In this case

$$j(-x) = (-x)^4 - (-x)^3$$
$$= x^4 + x^3.$$

So $j(-x) \neq j(x)$ and $j(-x) \neq -j(x)$. The graph of j is not symmetric about the y-axis and is not symmetric about the origin.
Fig. 2.54 supports this conclusion. ■

Behavior at the *x*-Intercepts

The x-intercepts are key points for the graph of a polynomial function, as they are for any function. Consider the graph of $y = (x - 2)^2(x + 1)$ in Fig. 2.55. Near 2 the values of y are positive, as shown in Fig. 2.56, and the graph does not cross the x-axis. Near -1 the values of y are negative for $x < -1$ and positive for $x > -1$, as shown in Fig. 2.57, and the graph crosses the x-axis. The reason for this behavior is the exponents in $(x - 2)^2$ and $(x + 1)^1$. Because $(x - 2)^2$ has an even exponent, $(x - 2)^2$ cannot be negative. Because $(x + 1)^1$ has an odd exponent, $(x + 1)^1$ changes sign at -1 but does not change sign at 2. So the product $(x - 2)^2(x + 1)$ does not change sign at 2, but does change sign at -1.

X	Y1	
1.7	.243	
1.8	.112	
1.9	.029	
2	0	
2.1	.031	
2.2	.128	
2.3	.297	
X=2		

Figure 2.56

X	Y1	
-1.3	-3.267	
-1.2	-2.048	
-1.1	-.961	
-1	0	
-.9	.841	
-.8	1.568	
-.7	2.187	
X=-1		

Figure 2.57

Every x-intercept corresponds to a factor of the polynomial. Whether that factor occurs an odd or even number of times determines the behavior of the graph at that intercept.

Theorem: Behavior at the *x*-intercepts

The graph of a polynomial function crosses the x-axis at an x-intercept if the factor corresponding to that intercept is raised to an odd power.
The graph touches but does not cross the x-axis if the factor is raised to an even power.

As another example, consider the graphs of $f(x) = x^2$ and $f(x) = x^3$ shown in Figs. 2.50 and 2.51. Each has only one x-intercept, $(0, 0)$. The factor corresponding to that intercept is x. Note that $f(x) = x^2$ does not cross the x-axis at $(0, 0)$, but $f(x) = x^3$ does cross the x-axis at $(0, 0)$.

Example 2 Crossing at the x-intercepts

Find the x-intercepts and determine whether the graph of the function crosses the x-axis at each x-intercept.

a. $f(x) = (x - 1)^2(x - 3)$ **b.** $f(x) = x^3 + 2x^2 - 3x$

Solution

a. The x-intercepts are found by solving $(x - 1)^2(x - 3) = 0$. The x-intercepts are $(1, 0)$ and $(3, 0)$. The graph does not cross the x-axis at $(1, 0)$ because the factor $x - 1$ occurs to an even power. The graph crosses the x-axis at $(3, 0)$ because $x - 3$ occurs to an odd power.

⎙ The graph in Fig. 2.58 supports these conclusions. ☐

b. The x-intercepts are found by solving $x^3 + 2x^2 - 3x = 0$. By factoring, we get $x(x + 3)(x - 1) = 0$. The x-intercepts are $(0, 0)$, $(-3, 0)$, and $(1, 0)$. Since each factor occurs an odd number of times (once), the graph crosses the x-axis at each of the x-intercepts.

⎙ The graph in Fig. 2.59 supports these conclusions.

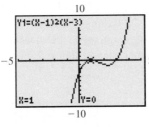
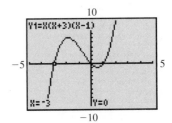

Figure 2.58   **Figure 2.59**   ■

The Leading Coefficient Test

We now consider the behavior of a polynomial function as the x-coordinate goes to or approaches infinity or negative infinity. In symbols, $x \rightarrow \infty$ or $x \rightarrow -\infty$. Since we seek only an intuitive understanding of the ideas presented here, we will not give precise definitions of these terms. Precise definitions are given in a calculus course.

To say that $x \rightarrow \infty$ means that x gets larger and larger without bound. For our purposes we can think of x assuming the values 1, 2, 3, and so on, without end. Similarly, $x \rightarrow -\infty$ means that x gets smaller and smaller without bound. Think of x assuming the values $-1, -2, -3$, and so on, without end.

As x approaches ∞ the y-coordinates of any polynomial function approach ∞ or $-\infty$. Likewise, when $x \rightarrow -\infty$ the y-coordinates approach ∞ or $-\infty$. The direction that y goes is determined by the degree of the polynomial and the sign of the

■ **Foreshadowing Calculus**

This discussion of the properties of the graph of a function is continued in calculus. Using calculus we can determine precisely where the function is increasing and decreasing, and even describe the direction in which the curve is turning.

leading coefficient. The four possible types of behavior are illustrated in the next example.

Example 3 Behavior as $x \to \infty$ or $x \to -\infty$

Determine the behavior of the graph of each function as $x \to \infty$ or $x \to -\infty$.

a. $y = x^3 - x$ **b.** $y = -x^3 + 1$

c. $y = x^4 - 4x^2$ **d.** $y = -x^4 + 4x^2 + x$

Solution

a. Considering the following table. If you have a graphing calculator, make a table like this and scroll through it.

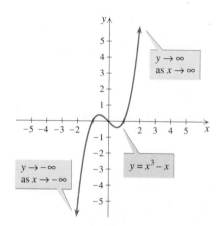

Figure 2.60

x	-30	-20	-10	0	10	20	30
$y = x^3 - x$	$-26{,}970$	-7980	-990	0	990	7980	$26{,}970$

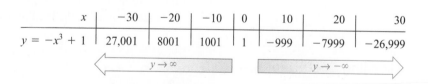

The graph of $y = x^3 - x$ is shown in Fig. 2.60. As x gets larger and larger $(x \to \infty)$, y increases without bound $(y \to \infty)$. As x gets smaller and smaller $(x \to -\infty)$, y decreases without bound $(y \to -\infty)$. Notice that the degree of the polynomial is odd and the sign of the leading coefficient is positive. The behavior of this function is stated with limit notation as

$$\lim_{x \to \infty} x^3 - x = \infty \qquad \text{and} \qquad \lim_{x \to \infty} x^3 - x = -\infty.$$

The notation $\lim_{x \to \infty} x^3 - x = \infty$ is read as "the limit as x approaches ∞ of $x^3 - x$ is ∞." We could also write $\lim_{x \to \infty} y = \infty$ if it is clear that $y = x^3 - x$. For more information on limits see the Concepts of Calculus at the end of Chapter 1.

b. Consider the following table. If you have a graphing calculator, make a table like this and scroll through it.

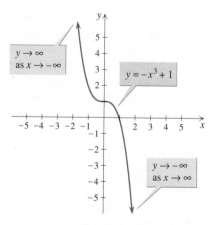

Figure 2.61

x	-30	-20	-10	0	10	20	30
$y = -x^3 + 1$	$27{,}001$	8001	1001	1	-999	-7999	$-26{,}999$

The graph of $y = -x^3 + 1$ is shown in Fig. 2.61. As x gets larger and larger $(x \to \infty)$, y decreases without bound $(y \to -\infty)$. As x gets smaller and smaller $(x \to -\infty)$, y increases without bound $(y \to \infty)$. Notice that the degree of this polynomial is odd and the sign of the leading coefficient is negative. The behavior of this function is stated with limit notation as

$$\lim_{x \to \infty} -x^3 + 1 = -\infty \qquad \text{and} \qquad \lim_{x \to -\infty} -x^3 + 1 = \infty.$$

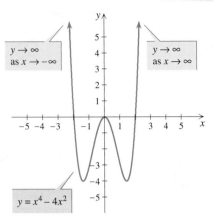

Figure 2.62

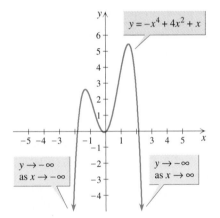

Figure 2.63

c. Consider the following table. If you have a graphing calculator, make a table like this and scroll through it.

x	0	± 10	± 20	± 30	± 40
$y = x^4 - 4x^2$	0	9600	158,400	806,400	2,553,600

$y \to \infty$

The graph of $y = x^4 - 4x^2$ is shown in Fig. 2.62. As $x \to \infty$ or $x \to -\infty$, y increases without bound ($y \to \infty$). Notice that the degree of this polynomial is even and the sign of the leading coefficient is positive. The behavior of this function is stated with limit notation as

$$\lim_{x\to\infty} x^4 - 4x^2 = \infty \qquad \text{and} \qquad \lim_{x\to-\infty} x^4 - 4x^2 = \infty.$$

d. Consider the following table. If you have a graphing calculator, make a table like this and scroll through it.

x	-20	-10	0	10	20
$y = -x^4 + 4x^2 + x$	$-158,420$	-9610	0	-9590	$-158,380$

$y \to -\infty$ $y \to -\infty$

The graph of $y = -x^4 + 4x^2 + x$ is shown in Fig. 2.63. As $x \to \infty$ or $x \to -\infty$, y decreases without bound ($y \to -\infty$). Notice that the degree of this polynomial is even and the sign of the leading coefficient is negative. The behavior of this function is stated with limit notation as

$$\lim_{x\to\infty} -x^4 + 4x^2 + x = -\infty \qquad \text{and} \qquad \lim_{x\to-\infty} -x^4 + 4x^2 + x = -\infty. \qquad ■$$

As $x \to \infty$ or $x \to -\infty$ all first-degree polynomial functions have "end behavior" like the lines $y = x$ or $y = -x$, all second-degree polynomial functions behave like the parabolas $y = x^2$ or $y = -x^2$, and all third-degree polynomial functions behave like $y = x^3$ or $y = -x^3$, and so on. The end behavior is determined by the degree and the sign of the first term. The smaller-degree terms in the polynomial determine the number of "hills" and "valleys" between the ends of the curve. With degree n there are at most $n - 1$ hills and valleys. The end behavior of polynomial functions is summarized in the **leading coefficient test.**

Leading Coefficient Test

If $f(x) = a_n x^n + a_{n-1}x^{n-1} + \cdots + a_1 x + a_0$, the behavior of the graph of f to the left and right is determined as follows:

For n odd and $a_n > 0$, $\lim_{x\to\infty} f(x) = \infty$ and $\lim_{x\to-\infty} f(x) = -\infty.$

For n odd and $a_n < 0$, $\lim_{x\to\infty} f(x) = -\infty$ and $\lim_{x\to-\infty} f(x) = \infty.$

For n even and $a_n > 0$, $\lim_{x\to\infty} f(x) = \infty$ and $\lim_{x\to-\infty} f(x) = \infty.$

For n even and $a_n < 0$, $\lim_{x\to\infty} f(x) = -\infty$ and $\lim_{x\to-\infty} f(x) = -\infty.$

The leading coefficient test is shown visually in Fig. 2.64, which shows only the "ends" of the graphs of the polynomial functions.

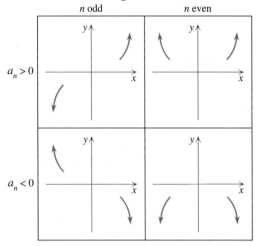

■ **Figure 2.64**

Sketching Graphs of Polynomial Functions

A good graph of a polynomial function should include the features that we have been discussing. The following strategy will help you graph polynomial functions.

STRATEGY   **Graphing a Polynomial Function**

1. Check for symmetry.
2. Find all real zeros of the polynomial function.
3. Determine the behavior at the corresponding x-intercepts.
4. Determine the behavior as $x \to \infty$ and as $x \to -\infty$.
5. Calculate several ordered pairs including the y-intercept to verify your suspicions about the shape of the graph.
6. Draw a smooth curve through the points to make the graph.

Example **4** Graphing polynomial functions

Sketch the graph of each polynomial function.

a. $f(x) = x^3 - 5x^2 + 7x - 3$ **b.** $f(x) = x^4 - 200x^2 + 10{,}000$

Solution

a. First find $f(-x)$ to determine symmetry and the number of negative roots.

$$f(-x) = (-x)^3 - 5(-x)^2 + 7(-x) - 3$$
$$= -x^3 - 5x^2 - 7x - 3$$

From $f(-x)$, we see that the graph has neither type of symmetry. Because $f(-x)$ has no sign changes, $x^3 - 5x^2 + 7x - 3 = 0$ has no negative roots by Descartes's rule of signs. The only possible rational roots are 1 and 3.

$$
\begin{array}{r|rrrr}
1 & 1 & -5 & 7 & -3 \\
  &   & 1 & -4 & 3 \\
\hline
  & 1 & -4 & 3 & 0 \\
\end{array}
$$

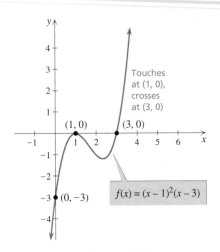

Figure 2.65

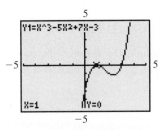

Figure 2.66

From the synthetic division we know that 1 is a root and we can factor $f(x)$:

$$f(x) = (x - 1)(x^2 - 4x + 3)$$
$$= (x - 1)^2(x - 3) \qquad \text{Factor completely.}$$

The x-intercepts are $(1, 0)$ and $(3, 0)$. The graph of f does not cross the x-axis at $(1, 0)$ because $x - 1$ occurs to an even power, while the graph crosses at $(3, 0)$ because $x - 3$ occurs to an odd power. The y-intercept is $(0, -3)$. Since the leading coefficient is positive and the degree is odd, $y \to \infty$ as $x \to \infty$ and $y \to -\infty$ as $x \to -\infty$. Calculate two more ordered pairs for accuracy, say $(2, -1)$ and $(4, 9)$. Draw a smooth curve as in Fig. 2.65.

The calculator graph shown in Fig. 2.66 supports these conclusions. □

b. First find $f(-x)$:

$$f(-x) = (-x)^4 - 200(-x)^2 + 10{,}000$$
$$= x^4 - 200x^2 + 10{,}000$$

Since $f(x) = f(-x)$, the graph is symmetric about the y-axis. We can factor the polynomial as follows.

$$f(x) = x^4 - 200x^2 + 10{,}000$$
$$= (x^2 - 100)(x^2 - 100)$$
$$= (x - 10)(x + 10)(x - 10)(x + 10)$$
$$= (x - 10)^2(x + 10)^2$$

The x-intercepts are $(10, 0)$ and $(-10, 0)$. Since each factor for these intercepts has an even power, the graph does not cross the x-axis at the intercepts. The y-intercept is $(0, 10{,}000)$. Since the leading coefficient is positive and the degree is even, $y \to \infty$ as $x \to \infty$ or as $x \to -\infty$. The graph also goes through $(-20, 90{,}000)$ and $(20, 90{,}000)$. Draw a smooth curve through these points and the intercepts as shown in Fig. 2.67.

The calculator graph shown in Fig. 2.68 supports these conclusions. ■

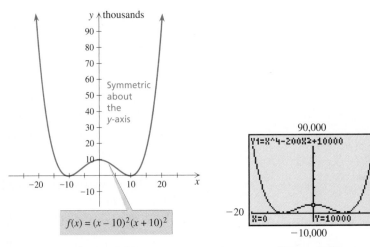

Figure 2.67 **Figure 2.68**

Polynomial Inequalities

We can solve polynomial inequalities by the same methods that we used on quadratic inequalities in Section 2.1. The next example illustrates the test-point method. This method depends on the intermediate value theorem. A polynomial function can change sign only at a zero of the function.

Example **5** **Solving a polynomial inequality with test points**

Solve $x^4 + x^3 - 15x^2 - 3x + 36 < 0$ using the test-point method.

Solution

Use synthetic division to see that 3 and -4 are zeros of the function $f(x) = x^4 + x^3 - 15x^2 - 3x + 36$:

$$
\begin{array}{r|rrrrr}
3 & 1 & 1 & -15 & -3 & 36 \\
  &   & 3 & 12 & -9 & -36 \\
\hline
-4 & 1 & 4 & -3 & -12 & 0 \\
   &   & -4 & 0 & 12 & \\
\hline
   & 1 & 0 & -3 & 0 &
\end{array}
$$

Since $x^4 + x^3 - 15x^2 - 3x + 36 = (x - 3)(x + 4)(x^2 - 3)$, the other two zeros are $\pm\sqrt{3}$. The four zeros determine five intervals on the number line in Fig. 2.69. Select an arbitrary test point in each of these intervals. The selected points -5, -3, 0, 2, and 4 are shown in red in the figure:

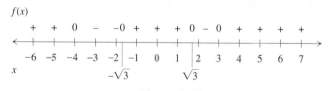

■ **Figure 2.69**

Now evaluate $f(x) = x^4 + x^3 - 15x^2 - 3x + 36$ at each test point:

$$f(-5) = 176, \quad f(-3) = -36, \quad f(0) = 36, \quad f(2) = -6, \quad f(4) = 104$$

These values indicate that the signs of the function are +, −, +, −, and + on the intervals shown in Fig. 2.69. The values of x that satisfy the original inequality are the values of x for which $f(x)$ is negative. So the solution set is $\left(-4, -\sqrt{3}\right) \cup \left(\sqrt{3}, 3\right)$. The calculator graph of $y = x^4 + x^3 - 15x^2 - 3x + 36$ in Fig. 2.70 confirms that y is negative for x in $\left(-4, -\sqrt{3}\right) \cup \left(\sqrt{3}, 3\right)$. Since the multiplicity of each zero of the function is one, the graph crosses the x-axis at each intercept and the y-coordinates change sign at each intercept. ■

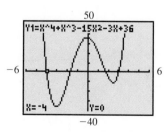

■ **Figure 2.70**

Function Gallery: **Polynomial Functions**

Linear: $f(x) = mx + b$

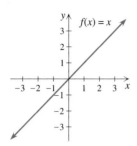

$f(x) = x$

Slope 1, y-intercept $(0, 0)$

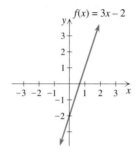

$f(x) = 3x - 2$

Slope 3, y-intercept $(0, -2)$

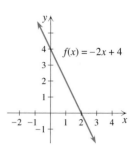

$f(x) = -2x + 4$

Slope -2, y-intercept $(0, 4)$

Quadratic: $f(x) = ax^2 + bx + c$ or $f(x) = a(x - h)^2 + k$

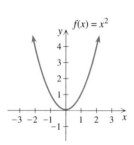

$f(x) = x^2$

Vertex $(0, 0)$
Range $[0, \infty)$

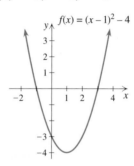

$f(x) = (x - 1)^2 - 4$

Vertex $(1, -4)$
Range $[-4, \infty)$

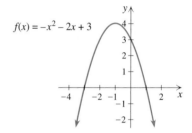

$f(x) = -x^2 - 2x + 3$

Vertex $(-1, 4)$
Range $(-\infty, 4]$

Cubic: $f(x) = ax^3 + bx^2 + cx + d$

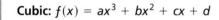

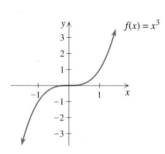

$f(x) = x^3$

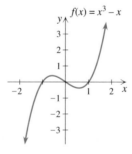

$f(x) = x^3 - x$

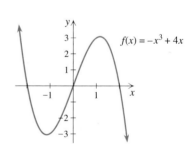

$f(x) = -x^3 + 4x$

Quartic or Fourth-Degree: $f(x) = ax^4 + bx^3 + cx^2 + dx + e$

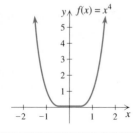

$f(x) = x^4$

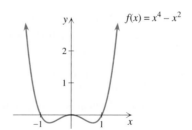

$f(x) = x^4 - x^2$

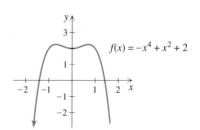

$f(x) = -x^4 + x^2 + 2$

For Thought

True or False? Explain.

1. If P is a function for which $P(2) = 8$ and $P(-2) = -8$, then the graph of P is symmetric about the origin.

2. If $y = -3x^3 + 4x^2 - 6x + 9$, then $y \to -\infty$ as $x \to \infty$.

3. If the graph of $y = P(x)$ is symmetric about the origin and $P(8) = 4$, then $-P(-8) = 4$.

4. If $f(x) = x^3 - 3x$, then $f(x) = f(-x)$ for any value of x.

5. If $f(x) = x^4 - x^3 + x^2 - 6x + 7$, then $f(-x) = x^4 + x^3 + x^2 + 6x + 7$.

6. The graph of $f(x) = x^2 - 6x + 9$ has only one x-intercept.

7. The x-intercepts for $P(x) = (x - 1)^2(x + 1)$ are $(0, 1)$ and $(0, -1)$.

8. The y-intercept for $P(x) = 4(x - 3)^2 + 2$ is $(0, 2)$.

9. The graph of $f(x) = x^2(x + 8)^2$ has no points in quadrants III and IV.

10. The graph of $f(x) = x^3 - 1$ has three x-intercepts.

2.6 Exercises

Discuss the symmetry of the graph of each polynomial function. See the summary of the types of symmetry on page 162.

1. $f(x) = x^6$

2. $f(x) = x^5 - x$

3. $f(x) = x^2 - 3x + 5$

4. $f(x) = 5x^2 + 10x + 1$

5. $f(x) = 3x^6 - 5x^2 + 3x$

6. $f(x) = x^6 - x^4 + x^2 - 8$

7. $f(x) = 4x^3 - x$

8. $f(x) = 7x^3 + x^2$

9. $f(x) = (x - 5)^2$

10. $f(x) = (x^2 - 1)^2$

11. $f(x) = -x$

12. $f(x) = 3x$

Find the x-intercepts and discuss the behavior of the graph of each polynomial function at its x-intercepts.

13. $f(x) = (x - 4)^2$

14. $f(x) = (x - 1)^2(x + 3)^2$

15. $f(x) = (2x - 1)^3$

16. $f(x) = x^6$

17. $f(x) = 4x - 1$

18. $f(x) = x^2 - 5x - 6$

19. $f(x) = x^2 - 3x + 10$

20. $f(x) = x^4 - 16$

21. $f(x) = x^3 - 3x^2$

22. $f(x) = x^3 - x^2 - x + 1$

23. $f(x) = 2x^3 - 5x^2 + 4x - 1$

24. $f(x) = x^3 - 3x^2 + 4$

25. $f(x) = -2x^3 - 8x^2 + 6x + 36$

26. $f(x) = -x^3 + 7x - 6$

For each function use the leading coefficient test to determine whether $y \to \infty$ or $y \to -\infty$ as $x \to \infty$.

27. $y = 2x^3 - x^2 + 9$

28. $y = -3x + 7$

29. $y = -3x^4 + 5$

30. $y = 6x^4 - 5x^2 - 1$

31. $y = x - 3x^3$

32. $y = 5x - 7x^4$

For each function use the leading coefficient test to determine whether $y \to \infty$ or $y \to -\infty$ as $x \to -\infty$.

33. $y = -2x^5 - 3x^2$

34. $y = x^3 + 8x + \sqrt{2}$

35. $y = 3x^6 - 999x^3$

36. $y = -12x^4 - 5x$

For each graph discuss its symmetry, indicate whether the graph crosses the x-axis at each x-intercept, and determine whether $y \to \infty$ or $y \to -\infty$ as $x \to \infty$ and $x \to -\infty$.

37.

38.

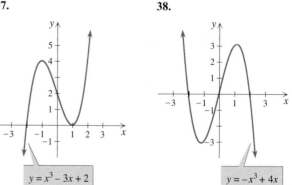

$y = x^3 - 3x + 2$ $y = -x^3 + 4x$

39.

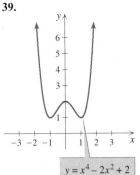

$$y = x^4 - 2x^2 + 2$$

40.

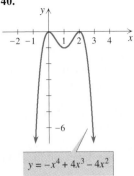

$$y = -x^4 + 4x^3 - 4x^2$$

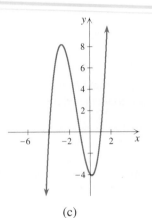

(c)

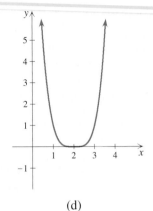

(d)

Determine whether each limit is equal to ∞ or −∞.

41. $\lim\limits_{x \to \infty} x^2 - 4$

42. $\lim\limits_{x \to \infty} x^3 + x$

43. $\lim\limits_{x \to \infty} -x^5 - x^2$

44. $\lim\limits_{x \to \infty} -3x^4 + 9x^2$

45. $\lim\limits_{x \to -\infty} -3x$

46. $\lim\limits_{x \to -\infty} x^3 - 5$

47. $\lim\limits_{x \to -\infty} -2x^2 + 1$

48. $\lim\limits_{x \to -\infty} 6x^4 - x$

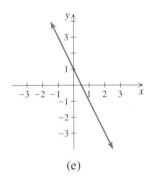

(e)

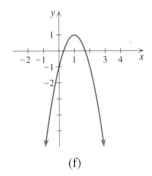

(f)

For each given function make a rough sketch of the graph that shows the behavior at the x-intercepts and the behavior as x approaches ∞ and −∞.

49. $f(x) = (x - 1)^2(x + 3)$

50. $f(x) = (x + 2)^2(x - 5)^2$

51. $f(x) = -2(2x - 1)^2(x + 1)^3$

52. $f(x) = -3(3x - 4)^2(2x + 1)^4$

Match each polynomial function with its graph (a)–(h).

53. $f(x) = -2x + 1$

54. $f(x) = -2x^2 + 1$

55. $f(x) = -2x^3 + 1$

56. $f(x) = -2x^2 + 4x - 1$

57. $f(x) = -2x^4 + 6$

58. $f(x) = -2x^4 + 6x^2$

59. $f(x) = x^3 + 4x^2 - x - 4$

60. $f(x) = (x - 2)^4$

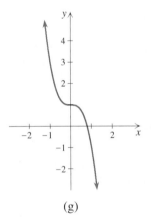

(g)

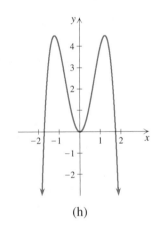

(h)

Sketch the graph of each function. See the strategy for graphing polynomial functions on page 167.

61. $f(x) = x - 30$

62. $f(x) = 40 - x$

63. $f(x) = (x - 30)^2$

64. $f(x) = (40 - x)^2$

65. $f(x) = x^3 - 40x^2$

66. $f(x) = x^3 - 900x$

67. $f(x) = (x - 20)^2(x + 20)^2$

68. $f(x) = (x - 20)^2(x + 12)$

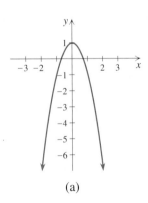

(a)

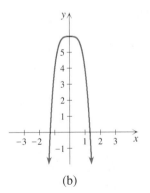

(b)

69. $f(x) = -x^3 - x^2 + 5x - 3$

70. $f(x) = -x^4 + 6x^3 - 9x^2$

71. $f(x) = x^3 - 10x^2 - 600x$

72. $f(x) = -x^4 + 24x^3 - 144x^2$

73. $f(x) = x^3 + 18x^2 - 37x + 60$

74. $f(x) = x^3 - 7x^2 - 25x - 50$

75. $f(x) = -x^4 + 196x^2$

76. $f(x) = -x^4 + x^2 + 12$

77. $f(x) = x^3 + 3x^2 + 3x + 1$

78. $f(x) = -x^3 + 3x + 2$

79. $f(x) = (x - 3)^2(x + 5)^2(x + 7)$

80. $f(x) = x(x + 6)^2(x^2 - x - 12)$

Solve each polynomial inequality using the test-point method.

81. $x^3 - 3x > 0$ **82.** $-x^3 + 3x + 2 < 0$

83. $2x^2 - x^4 \leq 0$ **84.** $-x^4 + x^2 + 12 \geq 0$

85. $x^3 + 4x^2 - x - 4 > 0$ **86.** $x^3 + 2x^2 - 2x - 4 < 0$

87. $x^3 - 4x^2 - 20x + 48 \geq 0$ **88.** $x^3 + 7x^2 - 36 \leq 0$

89. $x^3 - x^2 + x - 1 < 0$ **90.** $x^3 + x^2 + 2x - 4 > 0$

91. $x^4 - 19x^2 + 90 \leq 0$

92. $x^4 - 5x^3 + 3x^2 + 15x - 18 \geq 0$

Determine which of the given functions is shown in the accompanying graph.

93. a. $f(x) = (x - 3)(x + 2)$ **b.** $f(x) = (x + 3)(x - 2)$

  c. $f(x) = \dfrac{1}{3}(x - 3)(x + 2)$ **d.** $f(x) = \dfrac{1}{3}(x + 3)(x - 2)$

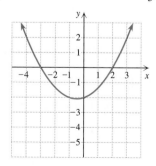

▪ **Figure for Exercise 93**

94. a. $f(x) = (x + 3)(x + 1)(x - 1)$

  b. $f(x) = (x - 3)(x^2 - 1)$

  c. $f(x) = \dfrac{2}{3}(x + 3)(x + 1)(x - 1)$

  d. $f(x) = -\dfrac{2}{3}(x + 3)(x^2 - 1)$

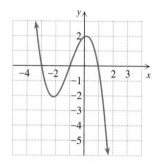

▪ **Figure for Exercise 94**

Solve each problem.

95. *Maximum Volume*   An open-top box is to be made from a 6 in. by 7 in. piece of copper by cutting equal squares (x in. by x in.) from each corner and folding up the sides. Write the volume of the box as a function of x. Use a graphing calculator to find the maximum possible volume to the nearest hundredth of a cubic inch.

▪ Foreshadowing Calculus

Making an open-top box as in Exercise 95 is a classic problem in calculus. However, in calculus we usually find the value of x that maximizes the volume of the box, without using a calculator.

96. *Maximizing Volume*   A manufacturer wants to make a metal can with a top and bottom such that its surface area is 3 square feet. (For this right circular cylinder $V = \pi r^2 h$ and $S = 2\pi r^2 + 2\pi rh$.)
  a. Write the height of the can as a function of the radius.

  b. Write the volume of the can as a function of the radius.

  c. Use a calculator graph of the function in part (b) to find the radius and height (to the nearest tenth of a foot) that maximizes the volume of the can.

97. *Packing Cheese*   Workers at the Green Bay Cheese Factory are trying to cover a block of cheese with an 8 in. by 12 in. piece of foil paper as shown in the figure on the next page.

The ratio of the length and width of the block must be 4 to 3 to accommodate the label. Find a polynomial function that gives the volume of the block of cheese covered in this manner as a function of the thickness *x*. Use a graphing calculator to find the dimensions of the block that will maximize the volume.

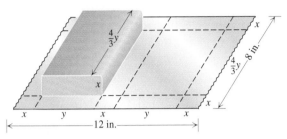

■ **Figure for Exercise 97**

 98. *Giant Teepee*   A casino designer is planning a giant teepee that is 80 ft in diameter and 120 ft high as shown in the figure. Inside the teepee is to be a cylindrical room for slot machines. Write the volume of the cylindrical room as a function of its radius. Use a graphing calculator to find the radius that maximizes the volume of the cylindrical room.

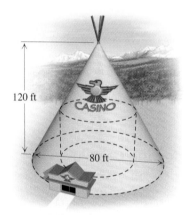

■ **Figure for Exercise 98**

Thinking Outside the Box XVI

Leaning Ladder   A 7-ft ladder is leaning against a vertical wall. There is a point near the bottom of the ladder that is 1 ft from the ground and 1 ft from the wall. Find the exact or approximate distance from the top of the ladder to the ground.

■ **Figure for Thinking Outside the Box XVI**

2.6 Pop Quiz

1. Discuss the symmetry of the graph of $y = x^4 - 3x^2$.

2. Discuss the symmetry of the graph of $y = x^3 - 3x$.

3. Does $f(x) = (x - 4)^3$ cross the *x*-axis at $(4, 0)$?

4. If $y = x^4 - 3x^3$, does *y* go to ∞ or $-\infty$ as $x \to \infty$?

5. If $y = -2x^4 + 5x^2$, does *y* go to ∞ or $-\infty$ as $x \to -\infty$?

6. Solve $(x - 3)^2(x + 1)^3 > 0$.

Rational Functions and Inequalities

In this section we will use our knowledge of polynomial functions to study functions that are ratios of polynomial functions.

Rational Functions and Their Domains

Functions such as

$$y = \frac{1}{x}, \quad f(x) = \frac{x - 3}{x - 1}, \quad \text{and} \quad g(x) = \frac{2x - 3}{x^2 - 4}$$

are rational functions.

**Definition:
Rational Function**

> If $P(x)$ and $Q(x)$ are polynomials, then a function of the form
>
> $$f(x) = \frac{P(x)}{Q(x)}$$
>
> is called a **rational function,** provided that $Q(x)$ is not the zero polynomial.

To simplify discussions of rational functions we will assume that $f(x)$ is in lowest terms ($P(x)$ and $Q(x)$ have no common factors) unless it is stated otherwise.

The domain of a polynomial function is the set of all real numbers, while the domain of a rational function is restricted to real numbers that do not cause the denominator to have a value of 0. The domain of $y = 1/x$ is the set of all real numbers except 0.

Example **1** **The domain of a rational function**

Find the domain of each rational function.

a. $f(x) = \dfrac{x - 3}{x - 1}$ **b.** $g(x) = \dfrac{2x - 3}{x^2 - 4}$

Solution

a. Since $x - 1 = 0$ only for $x = 1$, the domain of f is the set of all real numbers except 1. The domain is written in interval notation as $(-\infty, 1) \cup (1, \infty)$.
b. Since $x^2 - 4 = 0$ for $x = \pm 2$, any real number except 2 and -2 can be used for x. So the domain of g is $(-\infty, -2) \cup (-2, 2) \cup (2, \infty)$. ■

Horizontal and Vertical Asymptotes

The graph of a rational function such as $f(x) = 1/x$ does not look like the graph of a polynomial function. The domain of $f(x) = 1/x$ is the set of all real numbers except 0. However, 0 is an important number for the graph of this function because of the behavior of the graph when x is close to 0. The following table shows ordered pairs in which x is close to 0.

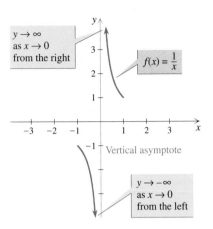

Figure 2.71

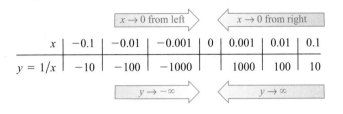

	$x \to 0$ from left				$x \to 0$ from right		
x	-0.1	-0.01	-0.001	0	0.001	0.01	0.1
$y = 1/x$	-10	-100	-1000		1000	100	10
		$y \to -\infty$			$y \to \infty$		

Notice that the closer x is to 0, the farther y is from 0. In symbols $y \to \infty$ as $x \to 0$ from the right. Using limit notation we write $\lim\limits_{x \to 0^+} \frac{1}{x} = \infty$, where the plus symbol indicates that x is approaching 0 from above or from the right. If $x \to 0$ from the left, $y \to -\infty$. Using limit notation we write $\lim\limits_{x \to 0^-} \frac{1}{x} = -\infty$, where the negative symbol indicates that x is approaching 0 from below or from the left. Plotting the ordered pairs from the table suggest a curve that gets closer and closer to the vertical line $x = 0$ (the y-axis) but never touches it, as shown in Fig. 2.71. The y-axis is called a *vertical asymptote* for this curve.

The following table shows ordered pairs in which x is far from 0.

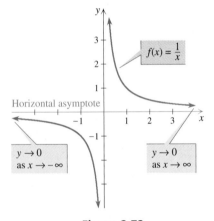

Figure 2.72

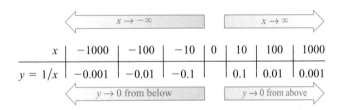

	$x \to -\infty$				$x \to \infty$		
x	-1000	-100	-10	0	10	100	1000
$y = 1/x$	-0.001	-0.01	-0.1		0.1	0.01	0.001
		$y \to 0$ from below			$y \to 0$ from above		

Notice that the farther x is from 0, the closer y is to 0. Using limit notation we write $\lim\limits_{x \to \infty} \frac{1}{x} = 0$ and $\lim\limits_{x \to -\infty} \frac{1}{x} = 0$. Plotting the ordered pairs from the table suggests a curve that lies just above the positive x-axis and just below the negative x-axis. The x-axis is called a *horizontal asymptote* for the graph of f. The complete graph of $f(x) = 1/x$ is shown in Fig. 2.72. Since $f(-x) = 1/(-x) = -f(x)$, the graph is symmetric about the origin.

For any rational function expressed in lowest terms, a horizontal asymptote is determined by the value approached by the rational expression as $|x| \to \infty$ ($x \to \infty$ or $x \to -\infty$). A vertical asymptote occurs for every number that causes the denominator of the function to have a value of 0, provided the rational function is in lowest terms. As we will learn shortly, not every rational function has a vertical and a horizontal asymptote. We can give a formal definition of asymptotes as described below.

Definition: Vertical and Horizontal Asymptotes

Let $f(x) = P(x)/Q(x)$ be a rational function written in lowest terms.

If $|f(x)| \to \infty$ as $x \to a$, then the vertical line $x = a$ is a **vertical asymptote.** Using limit notation, $x = a$ is a vertical asymptote if $\lim\limits_{x \to a} |f(x)| = \infty$.

The line $y = a$ is a **horizontal asymptote** if $f(x) \to a$ as $x \to \infty$ or $x \to -\infty$. Using limit notation, $y = a$ is a horizontal asymptote if $\lim\limits_{x \to \infty} f(x) = a$ or $\lim\limits_{x \to -\infty} f(x) = a$.

To find a horizontal asymptote we need to approximate the value of a rational expression when x is arbitrarily large. If x is large, then expressions such as

$$\frac{500}{x}, \qquad -\frac{14}{x}, \qquad \frac{3}{x^2}, \qquad \frac{6}{x^3}, \qquad \text{and} \qquad \frac{4}{x-5},$$

which consist of a fixed number over a polynomial, are approximately zero. To approximate a ratio of two polynomials that both involve x, such as $\frac{x-2}{2x+3}$, we rewrite the expression by dividing by the highest power of x:

$$\frac{x-2}{2x+3} = \frac{\dfrac{x}{x} - \dfrac{2}{x}}{\dfrac{2x}{x} + \dfrac{3}{x}} = \frac{1 - \dfrac{2}{x}}{2 + \dfrac{3}{x}}$$

Since $2/x$ and $3/x$ are approximately zero when x is large, the approximate value of this expression is $1/2$. We use this idea in the next example.

Example **2** Identifying horizontal and vertical asymptotes

Find the horizontal and vertical asymptotes for each rational function.

a. $f(x) = \dfrac{3}{x^2 - 1}$ **b.** $g(x) = \dfrac{x}{x^2 - 4}$ **c.** $h(x) = \dfrac{2x + 1}{x + 3}$

Solution

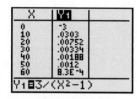

Figure 2.73

a. The denominator $x^2 - 1$ has a value of 0 if $x = \pm 1$. So the lines $x = 1$ and $x = -1$ are vertical asymptotes. As $x \to \infty$ or $x \to -\infty$, $x^2 - 1$ gets larger, making $3/(x^2 - 1)$ the ratio of 3 and a large number. Thus $3/(x^2 - 1) \to 0$ and the x-axis is a horizontal asymptote.

The calculator table shown in Fig. 2.73 supports the conclusion that the x-axis is a horizontal asymptote. ☐

b. The denominator $x^2 - 4$ has a value of 0 if $x = \pm 2$. So the lines $x = 2$ and $x = -2$ are vertical asymptotes. As $x \to \infty$, $x/(x^2 - 4)$ is a ratio of two large numbers. The approximate value of this ratio is not clear. However, if we divide the numerator and denominator by x^2, the highest power of x, we can get a clearer picture of the value of this ratio:

$$g(x) = \frac{x}{x^2 - 4} = \frac{\dfrac{x}{x^2}}{\dfrac{x^2}{x^2} - \dfrac{4}{x^2}} = \frac{\dfrac{1}{x}}{1 - \dfrac{4}{x^2}}$$

As $|x| \to \infty$, the values of $1/x$ and $4/x^2$ go to 0. So

$$g(x) \to \frac{0}{1 - 0} = 0.$$

So the x-axis is a horizontal asymptote.

The calculator table shown in Fig. 2.74 supports the conclusion that the x-axis is a horizontal asymptote. ☐

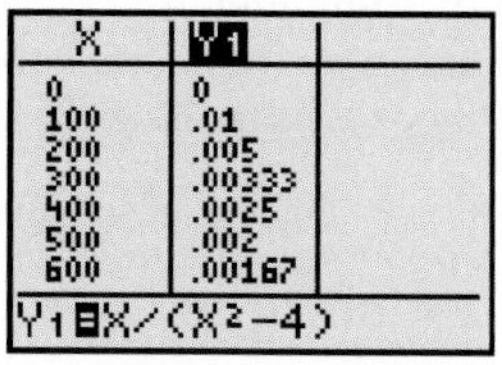

Figure 2.74

c. The denominator $x + 3$ has a value of 0 if $x = -3$. So the line $x = -3$ is a vertical asymptote. To find any horizontal asymptotes, rewrite the rational expression by dividing the numerator and denominator by the highest power of x:

$$h(x) = \frac{2x + 1}{x + 3} = \frac{\dfrac{2x}{x} + \dfrac{1}{x}}{\dfrac{x}{x} + \dfrac{3}{x}} = \frac{2 + \dfrac{1}{x}}{1 + \dfrac{3}{x}}$$

As $x \to \infty$ or $x \to -\infty$ the values of $1/x$ and $3/x$ approach 0. So

$$h(x) \to \frac{2 + 0}{1 + 0} = 2.$$

So the line $y = 2$ is a horizontal asymptote.

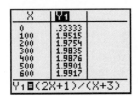

■ **Figure 2.75**

The calculator table shown in Fig. 2.75 supports the conclusion that $y = 2$ is a horizontal asymptote. ■

If the degree of the numerator of a rational function is less than the degree of the denominator, as in Example 2(a) and (b), then the x-axis is a horizontal asymptote for the graph of the function. If the degree of the numerator is equal to the degree of the denominator, as in Example 2(c), then the x-axis is not a horizontal asymptote. We can see this clearly if we use division to rewrite the expression as quotient + remainder/divisor. For the function of Example 2(c), we get

$$h(x) = \frac{2x + 1}{x + 3} = 2 + \frac{-5}{x + 3}.$$

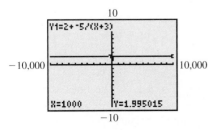

■ **Figure 2.76**

The graph of h is a translation two units upward of the graph of $y = -5/(x + 3)$, which has the x-axis as a horizontal asymptote. So $y = 2$ is a horizontal asymptote for h. Note that 2 is simply the ratio of the leading coefficients.

 Since the graph of h is very close to $y = 2$, the view of h in Fig. 2.76 looks like $y = 2$. Note how this choice of viewing window causes some of the important features of the graph to disappear. □

Oblique Asymptotes

Each rational function of Example 2 had one horizontal asymptote and had a vertical asymptote for each zero of the polynomial in the denominator. The horizontal asymptote $y = 0$ occurs because the y-coordinate gets closer and closer to 0 as $x \to \infty$ or $x \to -\infty$. Some rational functions have a nonhorizontal line for an asymptote. An asymptote that is neither horizontal nor vertical is called an **oblique asymptote** or **slant asymptote.** Oblique asymptotes are determined by using long division or synthetic division of polynomials. A review of long division can be found in Section A.2 of the Appendix.

■ **Foreshadowing Calculus**

The asymptotic behavior of functions is studied more extensively in calculus under the topic of *limits*.

Example **3** **A rational function with an oblique asymptote**

Determine all of the asymptotes for $g(x) = \dfrac{2x^2 + 3x - 5}{x + 2}$.

Solution

If $x + 2 = 0$, then $x = -2$. So the line $x = -2$ is a vertical asymptote. Because the degree of the numerator is larger than the degree of the denominator, we use long division (or in this case, synthetic division) to rewrite the function as quotient +

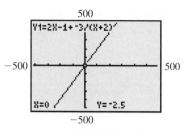

■ **Figure 2.77**

remainder/divisor (dividing the numerator and denominator by x^2 will not work in this case):

$$g(x) = \frac{2x^2 + 3x - 5}{x + 2} = 2x - 1 + \frac{-3}{x + 2}$$

If $|x| \to \infty$, then $-3/(x + 2) \to 0$. So the value of $g(x)$ approaches $2x - 1$ as $|x| \to \infty$. The line $y = 2x - 1$ is an oblique asymptote for the graph of g. You may look ahead to Fig. 2.85 to see the graph of this function with its oblique asymptote. Note that the wide view of g in Fig. 2.77 looks like the line $y = 2x - 1$. ■

If the degree of $P(x)$ is 1 greater than the degree of $Q(x)$ and the degree of $Q(x)$ is at least 1, then the rational function has an oblique asymptote. In this case use division to rewrite the function as quotient + remainder/divisor. The graph of the equation formed by setting y equal to the quotient is an oblique asymptote. We conclude this discussion of asymptotes with a summary.

SUMMARY **Finding Asymptotes for a Rational Function**

Let $f(x) = P(x)/Q(x)$ be a rational function in lowest terms with the degree of $Q(x)$ at least 1.

1. The graph of f has a vertical asymptote corresponding to each root of $Q(x) = 0$.
2. If the degree of $P(x)$ is less than the degree of $Q(x)$, then the x-axis is a horizontal asymptote.
3. If the degree of $P(x)$ equals the degree of $Q(x)$, then the horizontal asymptote is determined by the ratio of the leading coefficients.
4. If the degree of $P(x)$ is greater than the degree of $Q(x)$, then use division to rewrite the function as quotient + remainder/divisor. The graph of the equation formed by setting y equal to the quotient is an asymptote. This asymptote is an oblique or slant asymptote if the degree of $P(x)$ is 1 larger than the degree of $Q(x)$.

Sketching Graphs of Rational Functions

We now use asymptotes and symmetry to help us sketch the graphs of the rational functions discussed in Examples 2 and 3. Use the following steps to graph a rational function.

PROCEDURE **Graphing a Rational Function**

To graph a rational function in lowest terms:

1. Determine the asymptotes and draw them as dashed lines.
2. Check for symmetry.
3. Find any intercepts.
4. Plot several selected points to determine how the graph approaches the asymptotes.
5. Draw curves through the selected points, approaching the asymptotes.

Example **4** **Functions with horizontal and vertical asymptotes**

Sketch the graph of each rational function.

a. $f(x) = \dfrac{3}{x^2 - 1}$ **b.** $g(x) = \dfrac{x}{x^2 - 4}$ **c.** $h(x) = \dfrac{2x + 1}{x + 3}$

Solution

a. From Example 2(a), $x = 1$ and $x = -1$ are vertical asymptotes, and the x-axis is a horizontal asymptote. Draw the vertical asymptotes using dashed lines. Since all of the powers of x are even, $f(-x) = f(x)$ and the graph is symmetric about the y-axis. The y-intercept is $(0, -3)$. There are no x-intercepts because $f(x) = 0$ has no solution. Evaluate the function at $x = 0.9$ and $x = 1.1$ to see how the curve approaches the asymptote at $x = 1$. Evaluate at $x = 2$ and $x = 3$ to see whether the curve approaches the horizontal asymptote from above or below. We get $(0.9, -15.789)$, $(1.1, 14.286)$, $(2, 1)$, and $(3, 3/8)$. From these points you can see that the curve is going downward toward $x = 1$ from the left and upward toward $x = 1$ from the right. It is approaching its horizontal asymptote from above. Now use the symmetry with respect to the y-axis to draw the curve approaching its asymptotes as shown in Fig. 2.78.

The calculator graph in dot mode in Fig. 2.79 supports these conclusions. ☐

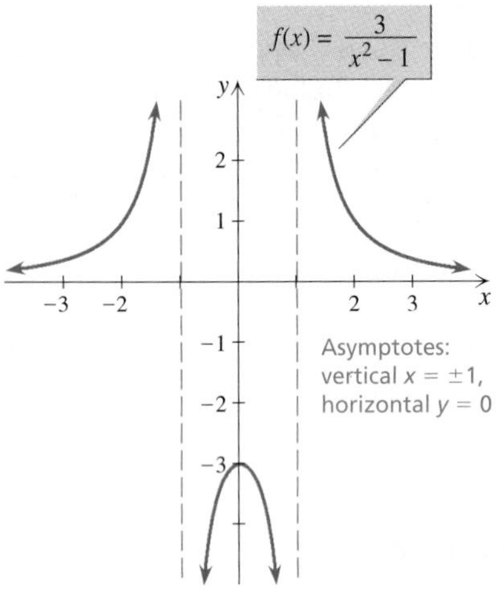

b. Draw the vertical asymptotes $x = 2$ and $x = -2$ from Example 2(b) as dashed lines. The x-axis is a horizontal asymptote. Because $f(-x) = -f(x)$, the graph is symmetric about the origin. The x-intercept is $(0, 0)$. Evaluate the function near the vertical asymptote $x = 2$, and for larger values of x to see how the curve approaches the horizontal asymptote. We get $(1.9, -4.872)$, $(2.1, 5.122)$, $(3, 3/5)$, and $(4, 1/3)$. From these points you can see that the curve is going downward toward $x = 2$ from the left and upward toward $x = 2$ from the right. It is approaching its horizontal asymptote from above. Now use the symmetry with respect to the origin to draw the curve approaching its asymptotes as shown in Fig. 2.80.

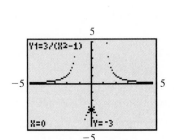

Asymptotes: vertical $x = \pm 1$, horizontal $y = 0$

$f(x) = \dfrac{3}{x^2 - 1}$

■ **Figure 2.78**

■ **Figure 2.79**

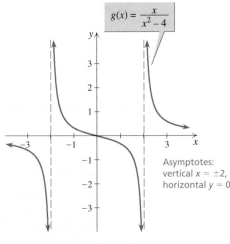

$g(x) = \dfrac{x}{x^2 - 4}$

Asymptotes: vertical $x = \pm 2$, horizontal $y = 0$

■ **Figure 2.80**

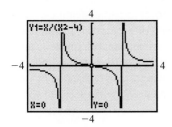

Figure 2.81

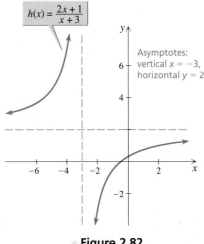

Figure 2.82

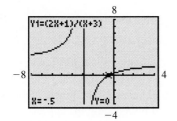

Figure 2.83

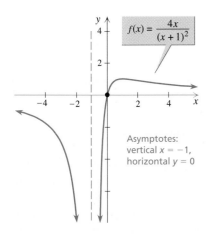

Figure 2.84

The calculator graph in connected mode in Fig. 2.81 confirms these conclusions. Note how the calculator appears to draw the vertical asymptotes as it connects the points that are close to but on opposite sides of the asymptotes. ☐

c. Draw the vertical asymptote $x = -3$ and the horizontal asymptote $y = 2$ from Example 2(c) as dashed lines. The x-intercept is $(-1/2, 0)$ and the y-intercept is $(0, 1/3)$. The points $(-2, -3)$, $(7, 1.5)$, $(-4, 7)$, and $(-13, 2.5)$ are also on the graph. From these points we can conclude that the curve goes upward toward $x = -3$ from the left and downward toward $x = -3$ from the right. It approaches $y = 2$ from above as x goes to $-\infty$ and from below as x goes to ∞. Draw the graph approaching its asymptotes as shown in Fig. 2.82.

The calculator graph in Fig. 2.83 supports these conclusions. ■

The graph of a rational function cannot cross a vertical asymptote, but it can cross a nonvertical asymptote. The graph of a rational function gets closer and closer to a nonvertical asymptote as $|x| \to \infty$, but it can also cross a nonvertical asymptote as illustrated in the next example.

Example **5** **A rational function that crosses an asymptote**

Sketch the graph of $f(x) = \dfrac{4x}{(x + 1)^2}$.

Solution

The graph of f has a vertical asymptote at $x = -1$, and since the degree of the numerator is less than the degree of the denominator, the x-axis is a horizontal asymptote. If $x > 0$ then $f(x) > 0$, and if $x < 0$ then $f(x) < 0$. So the graph approaches the horizontal axis from above for $x > 0$, and from below for $x < 0$. However, the x-intercept is $(0, 0)$. So the graph crosses its horizontal asymptote at $(0, 0)$. The graph shown in Fig. 2.84 goes through $(1, 1)$, $(2, 8/9)$, $(-2, -8)$, and $(-3, -3)$. ■

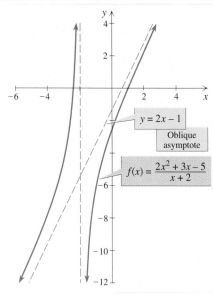

Figure 2.85

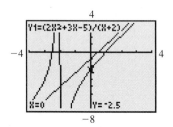

Figure 2.86

The graph of any function should illustrate its most important features. For a rational function that is asymptotic behavior. The graph of a rational function does not cross its vertical asymptotes and the graph approaches but does not touch its nonvertical asymptotes as x approaches ∞ or $-\infty$. So be careful with calculator graphs. On some calculators the graph will cross its vertical asymptote in connected mode and, unless the viewing window is selected carefully, the graph will appear to touch its nonvertical asymptotes.

Example **6** **Graphing a function with an oblique asymptote**

Sketch the graph of $k(x) = \dfrac{2x^2 + 3x - 5}{x + 2}$.

Solution

Draw the vertical asymptote $x = -2$ and the oblique asymptote $y = 2x - 1$ determined in Example 3 as dashed lines. The x-intercepts, $(1, 0)$ and $(-2.5, 0)$, are found by solving $2x^2 + 3x - 5 = 0$. The y-intercept is $(0, -2.5)$. The points $(-1, -6)$, $(4, 6.5)$, and $(-3, -4)$ are also on the graph. From these points we can conclude that the curve goes upward toward $x = -2$ from the left and downward toward $x = -2$ from the right. It approaches $y = 2x - 1$ from above as x goes to $-\infty$ and from below as x goes to ∞. Draw the graph approaching its asymptotes as shown in Fig. 2.85.

The calculator graph in Fig. 2.86 supports these conclusions. ■

All rational functions so far have been given in lowest terms. In the next example the numerator and denominator have a common factor. In this case the graph does not have as many vertical asymptotes as you might expect. The graph is almost identical to the graph of the rational function obtained by reducing the expression to lowest terms.

Example **7** **A graph with a hole in it**

Sketch the graph of $f(x) = \dfrac{x - 2}{x^2 - 4}$.

Solution

Since $x^2 - 4 = (x - 2)(x + 2)$, the domain of f is the set of all real numbers except 2 and -2. Since the rational expression can be reduced, the function f could also be defined as

$$f(x) = \frac{1}{x + 2} \qquad \text{for } x \neq 2 \text{ and } x \neq -2.$$

Note that the domain is determined before the function is simplified. The graph of $y = 1/(x + 2)$ has only one vertical asymptote $x = -2$. In fact, the graph of $y = 1/(x + 2)$ is a translation two units to the left of $y = 1/x$. Since $f(2)$ is undefined, the point $(2, 1/4)$ that would normally be on the graph of $y = 1/(x + 2)$ is omitted. The missing point is indicated on the graph in Fig. 2.87 as a small open circle. Note that the graph made by a graphing calculator will usually not show the missing point. ■

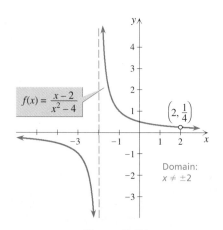

Figure 2.87

Rational Inequalities

An inequality that involves a rational expression, such as $\frac{x+3}{x-2} \le 2$, is called a **rational inequality.** Our first thought for solving this inequality might be to clear the denominator by multiplying each side by $x - 2$. But, when we multiply each side of an inequality by a real number, we must know whether the real number is positive or negative. Whether $x - 2$ is positive or negative depends on x. So multiplying by $x - 2$ is not a good idea. *We usually do not multiply a rational inequality by an expression that involves a variable.* However, we can solve rational inequalities using sign graphs as we did with quadratic inequalities in Section 2.1.

Example **8** **Solving a rational inequality with a sign graph of the factors**

Solve $\frac{x+3}{x-2} \le 2$. State the solution set using interval notation.

Solution

As with quadratic inequalities, we must have 0 on one side. Note that we do not multiply each side by $x - 2$.

$$\frac{x+3}{x-2} \le 2$$

$$\frac{x+3}{x-2} - 2 \le 0$$

$$\frac{x+3}{x-2} - \frac{2(x-2)}{x-2} \le 0 \quad \text{Get a common denominator.}$$

$$\frac{x+3-2(x-2)}{x-2} \le 0 \quad \text{Subtract.}$$

$$\frac{7-x}{x-2} \le 0 \quad \text{Combine like terms.}$$

The sign graph in Fig. 2.88 shows that $7 - x > 0$ if $x < 7$, and $7 - x < 0$ if $x > 7$. If $x > 2$, then $x - 2 > 0$, and if $x < 2$, then $x - 2 < 0$.

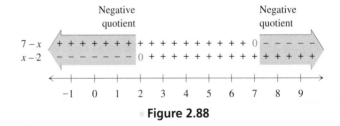

Figure 2.88

The inequality $\frac{7-x}{x-2} \le 0$ is satisfied if the quotient of $7 - x$ and $x - 2$ is negative or 0. Fig. 2.88 indicates that $7 - x$ and $x - 2$ have opposite signs and therefore a negative quotient for $x > 7$ or $x < 2$. If $x = 7$ then $7 - x = 0$ and the quotient is 0. If $x = 2$ then $x - 2 = 0$ and the quotient is undefined. So 7 is in the solution set but 2 is not. The solution set is $(-\infty, 2) \cup [7, \infty)$.

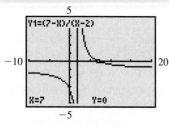

Figure 2.89

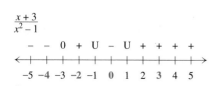

Figure 2.90

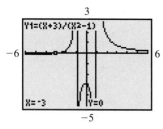

Figure 2.91

The graph of $y = (7 - x)/(x - 2)$ in Fig. 2.89 supports our conclusion that the inequality is satisfied if $x < 2$ or if $x \geq 7$. ■

Making a sign graph of the factors of the numerator and denominator as in Example 8 shows how the signs of the linear factors determine the solution to a rational inequality. Of course that method works only if you can factor the numerator and denominator. The test-point method works on any rational inequality for which we can find all zeros of the numerator and denominator. The test-point method depends on the fact that the graph of a rational function can go from one side of the x-axis to the other only at a vertical asymptote (where the function is undefined) or at an x-intercept (where the value of the function is 0).

Example **9** **Solving a rational inequality with the test-point method**

Solve $\dfrac{x + 3}{x^2 - 1} \geq 0$. State the solution set using interval notation.

Solution

Since the denominator can be factored, the inequality is equivalent to

$$\frac{x + 3}{(x - 1)(x + 1)} \geq 0.$$

The rational expression is undefined if $x = \pm 1$ and has value 0 if $x = -3$. Above each of these numbers on a number line put a 0 or a "U" (for undefined) as shown in Fig. 2.90. Now we select -4, -2, 0, and 3 as test points (shown in red in Fig. 2.90). Let $R(x) = \dfrac{x + 3}{x^2 - 1}$ and evaluate $R(x)$ at each test point to determine the sign of $R(x)$ in the interval of the test point:

$$R(-4) = -\frac{1}{15}, \qquad R(-2) = \frac{1}{3}, \qquad R(0) = -3, \qquad R(3) = \frac{3}{4}$$

So $-$, $+$, $-$, and $+$ are the signs of the rational expression in the four intervals in Fig. 2.90. The inequality is satisfied whenever the rational expression has a positive or 0 value. Since $R(-1)$ and $R(1)$ are undefined, -1 and 1 are not in the solution set. The solution set is $[-3, -1) \cup (1, \infty)$.

The graph of $y = (x + 3)/(x^2 - 1)$ in Fig. 2.91 supports the conclusion that the inequality is satisfied if $-3 \leq x < -1$ or $x > 1$. ■

Applications

Rational functions can occur in many applied situations. A horizontal asymptote might indicate that the average cost of producing a product approaches a fixed value in the long run. A vertical asymptote might show that the cost of a project goes up astronomically as a certain barrier is approached.

Example **10** Average cost of a handbook

Eco Publishing spent \$5000 to produce an environmental handbook and \$8 each for printing. Write a function that gives the average cost to the company per printed handbook. Graph the function for $0 < x \le 500$. What happens to the average cost if the book becomes very, very popular?

Solution

The total cost of producing and printing x handbooks is $8x + 5000$. To find the average cost per book, divide the total cost by the number of books:

$$C = \frac{8x + 5000}{x}$$

This rational function has a vertical asymptote at $x = 0$ and a horizontal asymptote $C = 8$. The graph is shown in Fig. 2.92. As x gets larger and larger, the \$5000 production cost is spread out over more and more books, and the average cost per book approaches \$8, as shown by the horizontal asymptote. Using limit notation we write $\lim_{x \to \infty} C = 8$. ∎

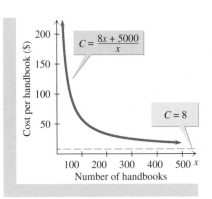

■ **Figure 2.92**

Function Gallery: **Some Basic Rational Functions**

Horizontal Asymptote *x*-axis and Vertical Asymptote *y*-axis

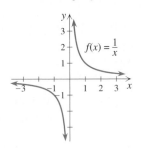

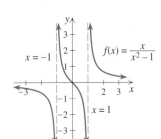

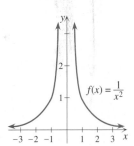

Various Asymptotes

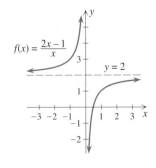

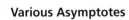

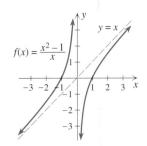

For Thought

True or False? Explain.

1. The function $f(x) = \dfrac{1}{\sqrt{x} - 3}$ is a rational function.

2. The domain of $f(x) = \dfrac{x + 2}{x - 2}$ is
$(-\infty, -2) \cup (-2, 2) \cup (2, \infty)$.

3. The number of vertical asymptotes for a rational function equals the degree of the denominator.

4. The graph of $f(x) = \dfrac{1}{x^4 - 4x^2}$ has four vertical asymptotes.

5. The x-axis is a horizontal asymptote for
$$f(x) = \frac{x^2 - 2x + 7}{x^3 - 4x}.$$

6. The x-axis is a horizontal asymptote for the graph of
$$f(x) = \frac{5x - 1}{x + 2}.$$

7. The line $y = x - 3$ is an asymptote for the graph of
$$f(x) = x - 3 + \frac{1}{x - 1}.$$

8. The graph of a rational function cannot intersect its asymptote.

9. The graph of $f(x) = \dfrac{4}{x^2 - 16}$ is symmetric about the y-axis.

10. The graph of $f(x) = \dfrac{2x + 6}{x^2 - 9}$ has only one vertical asymptote.

2.7 Exercises

Find the domain of each rational function.

1. $f(x) = \dfrac{4}{x + 2}$

2. $f(x) = \dfrac{-1}{x - 2}$

3. $f(x) = \dfrac{-x}{x^2 - 4}$

4. $f(x) = \dfrac{2}{x^2 - x - 2}$

5. $f(x) = \dfrac{2x + 3}{x - 3}$

6. $f(x) = \dfrac{4 - x}{x + 2}$

7. $f(x) = \dfrac{x^2 - 2x + 4}{x}$

8. $f(x) = \dfrac{x^3 + 2}{x^2}$

9. $f(x) = \dfrac{3x^2 - 1}{x^3 - x}$

10. $f(x) = \dfrac{x^2 + 1}{8x^3 - 2x}$

11. $f(x) = \dfrac{-x^2 + x}{x^2 + 5x + 6}$

12. $f(x) = \dfrac{-x^2 + 2x - 3}{x^2 + x - 12}$

Determine the domain and the equations of the asymptotes for the graph of each rational function.

13.

14.

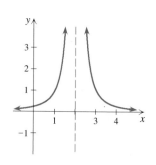

15.

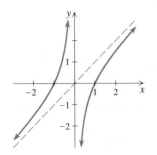

16.

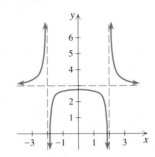

Determine the equations of all asymptotes for the graph of each function. See the summary for finding asymptotes for a rational function on page 179.

17. $f(x) = \dfrac{5}{x - 2}$

18. $f(x) = \dfrac{-1}{x + 12}$

19. $f(x) = \dfrac{-x}{x^2 - 9}$

20. $f(x) = \dfrac{-2}{x^2 - 5x + 6}$

21. $f(x) = \dfrac{2x + 4}{x - 1}$

22. $f(x) = \dfrac{5 - x}{x + 5}$

23. $f(x) = \dfrac{x^2 - 2x + 1}{x}$

24. $f(x) = \dfrac{x^3 - 8}{x^2}$

25. $f(x) = \dfrac{3x^2 + 4}{x + 1}$

26. $f(x) = \dfrac{x^2}{x - 9}$

27. $f(x) = \dfrac{-x^2 + 4x}{x + 2}$

28. $f(x) = \dfrac{-x^2 + 3x - 7}{x - 3}$

Find all asymptotes, x-intercepts, and y-intercepts for the graph of each rational function and sketch the graph of the function. See the procedure for graphing a rational function on page 179.

29. $f(x) = \dfrac{-1}{x}$

30. $f(x) = \dfrac{1}{x^2}$

31. $f(x) = \dfrac{1}{x - 2}$

32. $f(x) = \dfrac{-1}{x + 1}$

33. $f(x) = \dfrac{1}{x^2 - 4}$

34. $f(x) = \dfrac{1}{x^2 - 2x + 1}$

35. $f(x) = \dfrac{-1}{(x + 1)^2}$

36. $f(x) = \dfrac{-2}{x^2 - 9}$

37. $f(x) = \dfrac{2x + 1}{x - 1}$

38. $f(x) = \dfrac{3x - 1}{x + 1}$

39. $f(x) = \dfrac{x - 3}{x + 2}$

40. $f(x) = \dfrac{2 - x}{x + 2}$

41. $f(x) = \dfrac{x}{x^2 - 1}$

42. $f(x) = \dfrac{-x}{x^2 - 9}$

43. $f(x) = \dfrac{4x}{x^2 - 2x + 1}$

44. $f(x) = \dfrac{-2x}{x^2 + 6x + 9}$

45. $f(x) = \dfrac{8 - x^2}{x^2 - 9}$

46. $f(x) = \dfrac{2x^2 + x - 8}{x^2 - 4}$

47. $f(x) = \dfrac{2x^2 + 8x + 2}{x^2 + 2x + 1}$

48. $f(x) = \dfrac{-x^2 + 7x - 9}{x^2 - 6x + 9}$

Use a graph or a table to find each limit.

49. $\lim\limits_{x \to \infty} \dfrac{1}{x^2}$

50. $\lim\limits_{x \to -\infty} \dfrac{1}{x^2}$

51. $\lim\limits_{x \to \infty} \dfrac{2x - 3}{x - 1}$

52. $\lim\limits_{x \to \infty} \dfrac{3x^2 - 1}{x^2 - x}$

53. $\lim\limits_{x \to 0^+} \dfrac{1}{x^2}$

54. $\lim\limits_{x \to 0^-} \dfrac{1}{x^2}$

55. $\lim\limits_{x \to 1^+} \dfrac{2}{x - 1}$

56. $\lim\limits_{x \to 1^-} \dfrac{2}{x - 1}$

Find the oblique asymptote and sketch the graph of each rational function.

57. $f(x) = \dfrac{x^2 + 1}{x}$

58. $f(x) = \dfrac{x^2 - 1}{x}$

59. $f(x) = \dfrac{x^3 - 1}{x^2}$

60. $f(x) = \dfrac{x^3 + 1}{x^2}$

61. $f(x) = \dfrac{x^2}{x + 1}$

62. $f(x) = \dfrac{x^2}{x - 1}$

63. $f(x) = \dfrac{2x^2 - x}{x - 1}$

64. $f(x) = \dfrac{-x^2 + x + 1}{x + 1}$

65. $f(x) = \dfrac{x^3 - x^2 - 4x + 5}{x^2 - 4}$

66. $f(x) = \dfrac{x^3 + 2x^2 + x - 2}{x^2 - 1}$

67. $f(x) = \dfrac{-x^3 + x^2 + 5x - 4}{x^2 + x - 2}$

68. $f(x) = \dfrac{x^3 + x^2 - 16x - 24}{x^2 - 2x - 8}$

Match each rational function with its graph (a)–(h) on the next page, without using a graphing calculator.

69. $f(x) = -\dfrac{3}{x}$

70. $f(x) = \dfrac{1}{3 - x}$

71. $f(x) = \dfrac{x}{x - 3}$

72. $f(x) = \dfrac{x - 3}{x}$

73. $f(x) = \dfrac{1}{x^2 - 3x}$

74. $f(x) = \dfrac{x^2}{x^2 - 9}$

75. $f(x) = \dfrac{x^2 - 3}{x}$

76. $f(x) = \dfrac{-x^3 + 1}{x^2}$

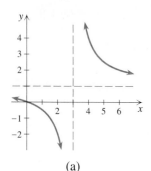

(a)

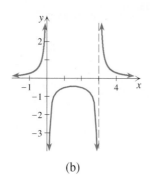

(b)

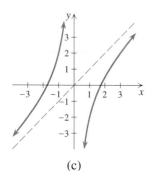

(c)

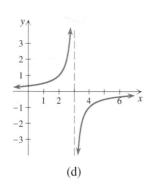

(d)

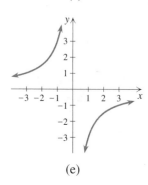

(e)

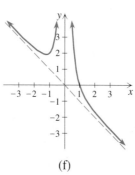

(f)

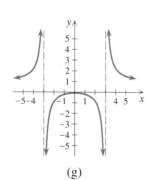

(g)

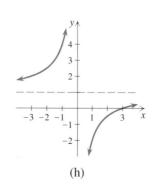

(h)

Sketch the graph of each rational function. Note that the functions are not in lowest terms. Find the domain first.

77. $f(x) = \dfrac{x + 1}{x^2 - 1}$

78. $f(x) = \dfrac{x}{x^2 + 2x}$

79. $f(x) = \dfrac{x^2 - 1}{x - 1}$

80. $f(x) = \dfrac{x^2 - 5x + 6}{x - 2}$

Solve with a sign graph of the factors. State the solution set using interval notation.

81. $\dfrac{x - 4}{x + 2} \le 0$

82. $\dfrac{x + 3}{x + 5} \ge 0$

83. $\dfrac{q - 2}{q + 3} < 2$

84. $\dfrac{p + 1}{2p - 1} \ge 1$ –

85. $\dfrac{w^2 - w - 6}{w - 6} \ge 0$

86. $\dfrac{z - 5}{z^2 + 2z - 8} \le 0$

87. $\dfrac{1}{x + 2} > \dfrac{1}{x - 3}$

88. $\dfrac{1}{x + 1} > \dfrac{2}{x - 1}$

89. $x < \dfrac{3x - 8}{5 - x}$

90. $\dfrac{2}{x + 3} \ge \dfrac{1}{x - 1}$

Solve with the test-point method. State the solution set using interval notation.

91. $\dfrac{(x - 3)(x + 1)}{x - 5} \ge 0$

92. $\dfrac{(x + 2)(x - 1)}{(x + 4)^2} \le 0$

93. $\dfrac{x^2 - 7}{2 - x^2} \le 0$

94. $\dfrac{x^2 + 1}{5 - x^2} \ge 0$

95. $\dfrac{x^2 + 2x + 1}{x^2 - 2x - 15} \ge 0$

96. $\dfrac{x^2 - 2x - 8}{x^2 + 10x + 25} \le 0$

97. $\dfrac{1}{w} > \dfrac{1}{w^2}$

98. $\dfrac{1}{w} > w^2$

99. $w > \dfrac{w - 5}{w - 3}$

100. $w < \dfrac{w - 2}{w + 1}$

Solve each problem.

101. *Admission to the Zoo*   Winona paid $100 for a lifetime membership to Friends of the Zoo, so that she could gain admittance to the zoo for only $1 per visit. Write Winona's average cost per visit C as a function of the number of visits when she has visited x times. What is her average cost per visit when she has visited the zoo 100 times? Graph the function for $x > 0$. What happens to her average cost per visit if she starts when she is young and visits the zoo every day?

102. *Renting a Car*   The cost of renting a car for one day is $19 plus 30 cents per mile. Write the average cost per mile C as a function of the number of miles driven in one day x. Graph the function for $x > 0$. What happens to C as the number of miles gets very large?

103. *Average Speed of an Auto Trip*   A 200-mi trip by an electric car must be completed in 4 hr. Let x be the number of hours it takes to travel the first half of the distance and write the average speed for the second half of the trip as a function of x. Graph this function for $0 < x < 4$. What is the significance of the vertical asymptote?

104. *Billboard Advertising*   An Atlanta marketing agency figures that the monthly cost of a billboard advertising campaign depends on the fraction of the market p that the client wishes to reach. For $0 \le p < 1$ the cost in dollars is determined by the formula $C = (4p - 1200)/(p - 1)$. What is the monthly cost for a campaign intended to reach 95% of the market? Graph this function for $0 \le p < 1$. What happens to the cost for a client who wants to reach 100% of the market?

105. *Balancing the Costs*   A furniture maker buys foam rubber x times per year. The delivery charge is $400 per purchase regardless of the amount purchased. The annual cost of storage is figured as $10,000/x$, because the more frequent the purchase, the less it costs for storage. So the annual cost of delivery and storage is given by

$$C = 400x + \frac{10,000}{x}.$$

   a. Graph the function with a graphing calculator.

   b. Find the number of purchases per year that minimizes the annual cost of delivery and storage.

106. *Making a Gas Tank*   An engineer is designing a cylindrical metal tank that is to hold 500 ft³ of gasoline.
   a. The volume of the tank is given by $V = \pi r^2 h$, where r is the radius and h is the height. Write h as a function of r.

   b. The surface area of a tank is given by $S = 2\pi r^2 + 2\pi rh$. Use the result of part (a) to write S as a function of r and graph it.

   c. Use the minimum feature of a graphing calculator to find the radius to the nearest tenth of a foot that minimizes the surface area. Ignore the thickness of the metal.

   d. If the tank costs $8 per square foot to construct, then what is the minimum cost for making the tank?

Thinking Outside the Box XVII

Filling a Triangle   Fiber-optic cables just fit inside a triangular pipe as shown in the figure. The cables have circular cross sections and the cross section of the pipe is an equilateral triangle with sides of length 1. Suppose that there are n cables of the same size in the bottom row, $n - 1$ of that size in the next row, and so on.
   a. Write the total cross-sectional area of the cables as a function of n.

   b. As n approaches infinity, will the triangular pipe get totally filled with cables? Explain.

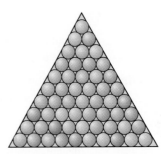

▪ **Figure for Thinking Outside the Box XVII**

2.7 Pop Quiz

1. What is the domain of $f(x) = \frac{x - 1}{x + 4}$?

2. Find the equations of all asymptotes for the graph of $y = \frac{x - 5}{x + 2}$.

3. Find the x-intercepts and y-intercept for $f(x) = \frac{x^2 - 9}{x^2 - 1}$.

4. What is the horizontal asymptote for $y = \frac{x - 8}{x + 3}$?

5. What is the oblique asymptote for $f(x) = \frac{x^2 - 2x}{x - 3}$?

6. Solve $\frac{x - 1}{x + 4} \ge 0$.

■■■ Highlights

2.1 Quadratic Functions and Inequalities

Quadratic Function	$y = ax^2 + bx + c$ where $a \neq 0$ Graph is a parabola opening upward for $a > 0$ and downward for $a < 0$.	$y = 2x^2 + 8x - 1$ Opens upward
Vertex	$x = -b/(2a)$ is the x-coordinate of the vertex. Use $y = ax^2 + bx + c$ to find the y-coordinate.	$x = -8/(2 \cdot 2) = -2$ Vertex: $(-2, -9)$
Maximum Minimum	The y-coordinate of the vertex is the max value of the function if $a < 0$ and the min if $a > 0$.	$y = 2x^2 + 8x - 1$ Min y-value is -9.
Two Forms	By completing the square, $y = ax^2 + bx + c$ can be written as $y = a(x - h)^2 + k$, which has vertex (h, k).	$y = 2x^2 + 8x - 1$ $y = 2(x + 2)^2 - 9$
Inequalities	To solve $ax^2 + bx + c > 0$ find all roots to $ax^2 + bx + c = 0$, then test a point in each interval determined by the roots.	$(x - 2)(x + 3) > 0$ Roots are -3 and 2. Solution set: $(-\infty, -3) \cup (2, \infty)$

2.2 Complex Numbers

Standard Form	Numbers of the form $a + bi$ where a and b are real numbers, $i = \sqrt{-1}$, and $i^2 = -1$	$2 + 3i, -\pi + i\sqrt{2}, 6, 0, \frac{1}{2}i$
Add, Subtract, Multiply	Add, subtract, and multiply like binomials with variable i, using $i^2 = -1$ to simplify.	$(3 - 2i)(4 + 5i)$ $= 12 + 7i - 10i^2$ $= 22 + 7i$
Divide	Divide by multiplying the numerator and denominator by the complex conjugate of the denominator.	$6/(1 + i)$ $= \dfrac{6(1 - i)}{(1 + i)(1 - i)} = 3 - 3i$
Square Roots of Negative Numbers	Square roots of negative numbers must be converted to standard form using $\sqrt{-b} = i\sqrt{b}$, for $b > 0$, before doing computations.	$\sqrt{-4} \cdot \sqrt{-9} = 2i \cdot 3i = -6$

2.3 Zeros of Polynomial Functions

Remainder Theorem	The remainder when $P(x)$ is divided by $x - c$ is $P(c)$.	$P(x) = x^2 + 3x - 4$ $P(x)$ divided by $x + 1$ has remainder $P(-1)$ or -6.
Synthetic Division	An abbreviated version of long division, used only for dividing a polynomial by $x - c$.	$\begin{array}{r|rrr} -1 & 1 & 3 & -4 \\ & & -1 & -2 \\ \hline & 1 & 2 & -6 \end{array}$ Dividend: $P(x)$, divisor $x + 1$ quotient $x + 2$, remainder -6
Factor Theorem	c is a zero of $y = P(x)$ if and only if $x - c$ is a factor of $P(x)$.	Since $x - 2$ is a factor of $P(x) = x^2 + x - 6$, $P(2) = 0$.

| **Fundamental Theorem of Algebra** | If $y = P(x)$ is a polynomial function of positive degree, then $y = P(x)$ has at least one zero in the set of complex numbers. | $f(x) = x^7 - x^5 + 3x^2 - 9$ has at least one complex zero and in fact has seven of them. |
| **Rational Zero Theorem** | If p/q is a rational zero in lowest terms for $y = P(x)$ with integral coefficients, then p is a factor of the constant term and q is a factor of the leading coefficient. | $P(x) = 6x^2 + x - 15$ $P(3/2) = 0$ 3 is a factor of -15, 2 is a factor of 6 |

2.4 The Theory of Equations

n-Root Theorem	If $P(x)$ has positive degree and complex coefficients, then $P(x) = 0$ has n roots counting multiplicity.	$(x - 2)^3(x^4 - 9) = 0$ has seven complex solutions counting multiplicity.
Conjugate Pairs Theorem	If $P(x)$ has real coefficients, then the imaginary roots of $P(x) = 0$ occur in conjugate pairs.	$x^2 - 4x + 5 = 0$ $x = 2 \pm i$
Descartes's Rule of Signs	The changes in sign of $P(x)$ determine the number of positive roots and the changes in sign of $P(-x)$ determine the number of negative roots to $P(x) = 0$.	$P(x) = x^5 - x^2$ has 1 positive root, no negative roots

2.5 Miscellaneous Equations

Higher Degree	Solve by factoring.	$x^4 - x^2 - 6 = 0$ $(x^2 - 3)(x^2 + 2) = 0$
Squaring Each Side	Possibly extraneous roots	$\sqrt{x - 3} = -2, x - 3 = 4$ $x = 7$ extraneous root
Rational Exponents	Raise each side to a whole number power and then apply the even or odd root property.	$x^{2/3} = 4, x^2 = 64, x = \pm 8$
Quadratic Type	Make a substitution to get a quadratic equation.	$x^{1/3} + x^{1/6} - 12 = 0$ $a^2 + a - 12 = 0$ if $a = x^{1/6}$
Absolute Value	Write equivalent equations without absolute value.	$|x^2 - 4| = 2$ $x^2 - 4 = 2$ or $x^2 - 4 = -2$

2.6 Graphs of Polynomial Functions

Axis of Symmetry	The parabola $y = ax^2 + bx + c$ is symmetric about the line $x = -b/(2a)$.	Axis of symmetry for $y = x^2 - 4x$ is $x = 2$.
Behavior at the x-Intercepts	A polynomial function crosses the x-axis at $(c, 0)$ if $x - c$ occurs with an odd power or touches the x-axis if $x - c$ occurs with an even power.	$f(x) = (x - 3)^2(x + 1)^3$ crosses at $(-1, 0)$, does not cross at $(3, 0)$
End Behavior	The leading coefficient and the degree of the polynomial determine the behavior as $x \to -\infty$ or $x \to \infty$.	$y = x^3$ $y \to \infty$ as $x \to \infty$ $y \to -\infty$ as $x \to -\infty$
Polynomial Inequality	Locate all zeros and then test a point in each interval determined by the zeros.	$x^3 - 4x > 0$ when x is in $(-2, 0) \cup (2, \infty)$

2.7 Rational Functions and Inequalities

Vertical Asymptote	A rational function in lowest terms has a vertical asymptote wherever the denominator is zero.	$y = x/(x^2 - 4)$ has vertical asymptotes $x = -2$ and $x = 2$.
Horizontal Asymptote	If degree of denominator exceeds degree of numerator, then the x-axis is the horizontal asymptote. If degree of numerator equals degree of denominator, then the ratio of leading coefficients is the horizontal asymptote.	$y = x/(x^2 - 4)$ Horizontal asymptote: $y = 0$ $y = (2x - 1)/(3x - 2)$ Horizontal asymptote: $y = 2/3$
Oblique or Slant Asymptote	If degree of numerator exceeds degree of denominator by 1, then use division to determine the slant asymptote.	$y = \dfrac{x^2}{x - 1} = x + 1 + \dfrac{1}{x - 1}$ Slant asymptote: $y = x + 1$
Rational Inequality	Test a point in each interval determined by the zeros and the horizontal asymptotes.	$\dfrac{x - 2}{x + 3} \le 0$ when x is in the interval $(-3, 2]$.

■ ■ ■ Chapter 2 Review Exercises

Solve each problem.

1. Write the function $f(x) = 3x^2 - 2x + 1$ in the form $f(x) = a(x - h)^2 + k$.

2. Write the function $f(x) = -4\left(x - \frac{1}{3}\right)^2 - \frac{1}{2}$ in the form $f(x) = ax^2 + bx + c$.

3. Find the vertex, axis of symmetry, x-intercepts, and y-intercept for the parabola $y = 2x^2 - 4x - 1$.

4. Find the maximum value of the function $y = -x^2 + 3x - 5$.

5. Write the equation of a parabola that has x-intercepts $(-1, 0)$ and $(3, 0)$ and y-intercept $(0, 6)$.

6. Write the equation of the parabola that has vertex $(1, 2)$ and y-intercept $(0, 5)$.

Write each expression in the form $a + bi$, where a and b are real numbers.

7. $(3 - 7i) + (-4 + 6i)$

8. $(-6 - 3i) - (3 - 2i)$

9. $(4 - 5i)^2$

10. $7 - i(2 - 3i)^2$

11. $(1 - 3i)(2 + 6i)$

12. $(0.3 + 2i)(0.3 - 2i)$

13. $(2 - 3i) \div i$

14. $(-2 + 4i) \div (-i)$

15. $(1 - i) \div (2 + i)$

16. $(3 + 6i) \div (4 - i)$

17. $\dfrac{6 + \sqrt{-8}}{2}$

18. $\dfrac{-2 - \sqrt{-18}}{2}$

19. $i^{34} + i^{19}$

20. $\sqrt{6} + \sqrt{-3}\sqrt{-2}$

Find all the real and imaginary zeros for each polynomial function.

21. $f(x) = 3x - 1$

22. $g(x) = 7$

23. $h(x) = x^2 - 8$

24. $m(x) = x^3 - 8$

25. $n(x) = 8x^3 - 1$

26. $C(x) = 3x^2 - 2$

27. $P(t) = t^4 - 100$

28. $S(t) = 25t^4 - 1$

29. $R(s) = 8s^3 - 4s^2 - 2s + 1$

30. $W(s) = s^3 + s^2 + s + 1$

31. $f(x) = x^3 + 2x^2 - 6x$

32. $f(x) = 2x^3 - 4x^2 + 3x$

For each polynomial, find the indicated value in two different ways.

33. $P(x) = 4x^3 - 3x^2 + x - 1, P(3)$

34. $P(x) = 2x^3 + 5x^2 - 3x - 2, P(-2)$

35. $P(x) = -8x^5 + 2x^3 - 6x + 2, P\left(-\dfrac{1}{2}\right)$

36. $P(x) = -4x^4 + 3x^2 + 1, P\left(\dfrac{1}{2}\right)$

Use the rational zero theorem to list all possible rational zeros for each polynomial function.

37. $f(x) = -3x^3 + 6x^2 + 5x - 2$

38. $f(x) = 2x^4 + 9x^2 - 8x - 3$

39. $f(x) = 6x^4 - x^2 - 9x + 3$

40. $f(x) = 4x^3 - 5x^2 - 13x - 8$

Find a polynomial equation with integral coefficients (and lowest degree) that has the given roots.

41. $-\dfrac{1}{2}, 3$

42. $\dfrac{1}{2}, -5$

43. $3 - 2i$

44. $4 + 2i$

45. $2, 1 - 2i$

46. $-3, 3 - 4i$

47. $2 - \sqrt{3}$

48. $1 + \sqrt{2}$

Use Descartes's rule of signs to discuss the possibilities for the roots to each equation. Do not solve the equation.

49. $x^8 + x^6 + 2x^2 = 0$

50. $-x^3 - x - 3 = 0$

51. $4x^3 - 3x^2 + 2x - 9 = 0$

52. $5x^5 + x^3 + 5x = 0$

53. $x^3 + 2x^2 + 2x + 1 = 0$

54. $-x^4 - x^3 + 3x^2 + 5x - 8 = 0$

Find all real and imaginary solutions to each equation, stating multiplicity when it is greater than one.

55. $x^3 - 6x^2 + 11x - 6 = 0$

56. $x^3 + 7x^2 + 16x + 12 = 0$

57. $6x^4 - 5x^3 + 7x^2 - 5x + 1 = 0$

58. $6x^4 + 5x^3 + 25x^2 + 20x + 4 = 0$

59. $x^3 - 9x^2 + 28x - 30 = 0$

60. $x^3 - 4x^2 + 6x - 4 = 0$

61. $x^3 - 4x^2 + 7x - 6 = 0$

62. $2x^3 - 5x^2 + 10x - 4 = 0$

63. $2x^4 - 5x^3 - 2x^2 + 2x = 0$

64. $2x^5 - 15x^4 + 26x^3 - 12x^2 = 0$

Find all real solutions to each equation.

65. $|2v - 1| = 3v$

66. $|2h - 3| = |h|$

67. $x^4 + 7x^2 = 18$

68. $2x^{-2} + 5x^{-1} = 12$

69. $\sqrt{x + 6} - \sqrt{x - 5} = 1$

70. $\sqrt{2x - 1} = \sqrt{x - 1} + 1$

71. $\sqrt{y} + \sqrt[4]{y} = 6$

72. $\sqrt[3]{x^2} + \sqrt[3]{x} = 2$

73. $x^4 - 3x^2 - 4 = 0$

74. $(y - 1)^2 - (y - 1) = 2$

75. $(x - 1)^{2/3} = 4$

76. $(2x - 3)^{-1/2} = \dfrac{1}{2}$

77. $(x + 3)^{-3/4} = -8$

78. $\left(\dfrac{1}{x - 3}\right)^{-1/4} = \dfrac{1}{2}$

79. $\sqrt[3]{3x - 7} = \sqrt[3]{4 - x}$

80. $\sqrt[3]{x + 1} = \sqrt[6]{4x + 9}$

Discuss the symmetry of the graph of each function.

81. $f(x) = 2x^2 - 3x + 9$

82. $f(x) = -3x^2 + 12x - 1$

83. $f(x) = -3x^4 - 2$

84. $f(x) = \dfrac{-x^3}{x^2 - 1}$

85. $f(x) = \dfrac{x}{x^2 + 1}$

86. $f(x) = 2x^4 + 3x^2 + 1$

Find the domain of each rational function.

87. $f(x) = \dfrac{x^2 - 4}{2x + 5}$

88. $f(x) = \dfrac{4x + 1}{x^2 - x - 6}$

89. $f(x) = \dfrac{1}{x^2 + 1}$

90. $f(x) = \dfrac{x - 9}{x^2 - 1}$

Find the x-intercepts, y-intercept, and asymptotes for the graph of each function and sketch the graph.

91. $f(x) = x^2 - x - 2$

92. $f(x) = -2(x - 1)^2 + 6$

93. $f(x) = x^3 - 3x - 2$

94. $f(x) = x^3 - 3x^2 + 4$

95. $f(x) = \dfrac{1}{2}x^3 - \dfrac{1}{2}x^2 - 2x + 2$

96. $f(x) = \dfrac{1}{2}x^3 - 3x^2 + 4x$

97. $f(x) = \dfrac{1}{4}x^4 - 2x^2 + 4$

98. $f(x) = \dfrac{1}{2}x^4 + 2x^3 + 2x^2$

99. $f(x) = \dfrac{2}{x + 3}$

100. $f(x) = \dfrac{1}{2 - x}$

101. $f(x) = \dfrac{2x}{x^2 - 4}$

102. $f(x) = \dfrac{2x^2}{x^2 - 4}$

103. $f(x) = \dfrac{x^2 - 2x + 1}{x - 2}$

104. $f(x) = \dfrac{-x^2 + x + 2}{x - 1}$

105. $f(x) = \dfrac{2x - 1}{2 - x}$ **106.** $f(x) = \dfrac{1 - x}{x + 1}$

107. $f(x) = \dfrac{x^2 - 4}{x - 2}$ **108.** $f(x) = \dfrac{x^3 + x}{x}$

Solve each inequality. State the solution set using interval notation.

109. $8x^2 + 1 < 6x$ **110.** $x^2 + 2x < 63$

111. $(3 - x)(x + 5) \geq 0$ **112.** $-x^2 - 2x + 15 < 0$

113. $4x^3 - 400x^2 - x + 100 \geq 0$

114. $x^3 - 49x^2 - 52x + 100 < 0$

115. $\dfrac{x + 10}{x + 2} < 5$ **116.** $\dfrac{x - 6}{2x + 1} \geq 1$

117. $\dfrac{12 - 7x}{x^2} > -1$ **118.** $x - \dfrac{2}{x} \leq -1$

119. $\dfrac{x^2 - 3x + 2}{x^2 - 7x + 12} \geq 0$ **120.** $\dfrac{x^2 + 4x + 3}{x^2 - 2x - 15} \leq 0$

Solve each problem.

121. Find the quotient and remainder when $x^3 - 6x^2 + 9x - 15$ is divided by $x - 3$.

122. Find the quotient and remainder when $3x^3 + 4x^2 + 2x - 4$ is divided by $3x - 2$.

123. *Altitude of a Rocket* If the altitude in feet of a model rocket is given by the equation $S = -16t^2 + 156t$, where t is the time in seconds after ignition, then what is the maximum height attained by the rocket?

124. *Bonus Room* A homeowner wants to put a room in the attic of her house. The house is 48 ft wide and the roof has a 7–12 pitch. (The roof rises 7 ft in a horizontal distance of 12 ft.) Find the dimensions of the room that will maximize the area of the cross section shown in the figure.

48 ft

■ Figure for Exercise 124

125. *Maximizing Area* An isosceles triangle has one vertex at the origin and a horizontal base below the x-axis with its end-points on the curve $y = x^2 - 16$. See the figure. Let (a, b) be the vertex in the fourth quadrant and write the area of the triangle as a function of a. Use a graphing calculator to find the point (a, b) for which the triangle has the maximum possible area.

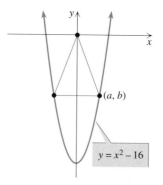

(a, b)

$y = x^2 - 16$

■ Figure for Exercise 125

126. *Limiting Velocity* As a skydiver falls, his velocity keeps increasing. However, because of air resistance, the rate at which the velocity increases keeps decreasing and there is a limit velocity that the skydiver cannot exceed. To see this behavior, graph

$$V = \frac{1000t}{5t + 8},$$

where V is the velocity in feet per second and t is the time in seconds.

a. What is the velocity at time $t = 10$ sec?

b. What is the horizontal asymptote for this graph?

c. What is the limiting velocity that cannot be exceeded?

■ Figure for Exercise 126

Thinking Outside the Box XVIII

Polynomial Equation Find a polynomial equation with integral coefficients for which $\sqrt{3} + \sqrt{5}$ is a root.

Concepts of Calculus

Instantaneous rate of change

Suppose that you start at Amarillo and drive west on I-40 for 500 miles. If this trip takes you 10 hours, then the average rate at which your location is changing is 50 miles per hour. At every instant of your trip your speedometer shows the instantaneous rate at which your location is changing. Your instantaneous rate of change might range from 0 miles per hour to 70 miles per hour or more.

We defined the average rate of change of a function $f(x)$ on the interval $[x, x + h]$ as $\dfrac{f(x + h) - f(x)}{h}$. We cannot use a speedometer to find the instantaneous rate of change of a function. We use the idea of limits, which was discussed in the Concepts of Calculus at the end of Chapter 1. For the instantaneous rate of change we shrink the interval to nothing. So we define the instantaneous rate of change of $f(x)$ as

$$\lim_{h \to 0} \frac{f(x + h) - f(x)}{h}.$$

Exercises

1. Let $f(x) = x^2$.

    a. Find $\dfrac{f(2 + h) - f(2)}{h}$.

    b. Find $\lim\limits_{h \to 0} \dfrac{f(2 + h) - f(2)}{h}$.

2. Let $f(x) = x^2 - 2x$.

    a. Find $\dfrac{f(x + h) - f(x)}{h}$.

    b. Find $\lim\limits_{h \to 0} \dfrac{f(x + h) - f(x)}{h}$.

    c. Find the instantaneous rate of change of $f(x) = x^2 - 2x$ when $x = 5$.

3. Let $f(x) = \sqrt{x}$.

    a. Find $\dfrac{f(x + h) - f(x)}{h}$ and write your answer with a rational numerator.

    b. Find $\lim\limits_{h \to 0} \dfrac{f(x + h) - f(x)}{h}$.

    c. Find the instantaneous rate of change of $f(x) = \sqrt{x}$ when $x = 9$.

4. A ball is tossed into the air from ground level at 128 feet per second. The function $f(t) = -16t^2 + 128t$ gives the height above ground (in feet) as a function of time (in seconds).

    a. Find $f(0)$ and $f(3)$.

    b. How far did the ball travel in the first three seconds of its flight?

    c. Find the average rate of change of the height for the time interval $[0, 3]$.

    d. Find $\dfrac{f(t + h) - f(t)}{h}$.

    e. Find $\lim\limits_{h \to 0} \dfrac{f(t + h) - f(t)}{h}$.

    f. Find the instantaneous rate of change of the height (or the *instantaneous velocity*) of the ball at times $t = 0, 2, 4, 6,$ and 8 seconds.

The average rate of change of the function f on the interval $[c, x]$ is $\dfrac{f(x) - f(c)}{x - c}$.

5. Let $f(x) = \dfrac{1}{x}$.

    a. Find and simplify $\dfrac{f(x) - f(2)}{x - 2}$.

    b. Find $\lim\limits_{x \to 2} \dfrac{f(x) - f(2)}{x - 2}$.

    c. Find the instantaneous rate of change of $f(x)$ when $x = 2$.

6. Let $f(x) = x^3$.

    a. Find and simplify $\dfrac{f(x) - f(c)}{x - c}$.

    b. Find $\lim\limits_{x \to c} \dfrac{f(x) - f(c)}{x - c}$.

    c. Find the instantaneous rate of change of $f(x)$ when $x = c$.

3 Trigonometric Functions

The Verrazano-Narrows Bridge

Location   Brooklyn and Staten Island, New York

Completion Date   1964

Cost $325 million

Length   13,700 feet

Longest Single Span   4260 feet

Engineer   Othmar H. Ammann

With an overall length of more than 2 miles, the Verrazano-Narrows bridge was the longest suspension bridge in the world in 1964. Today it is only in the top 10. Engineer Othmar Ammann, perhaps the greatest bridge engineer of all time, was also responsible for four other New York City bridges, the Triborough, Bronx-Whitestone, Throgs Neck, and George Washington. Because the bridge's 690 foot towers are so high and far apart, Ammann had to take into account the curvature of the earth when designing the bridge. Like all steel bridges, the Verrazano-Narrows bridge expands and contracts with changes in temperature. The bridge roadway is 12 feet lower in the summer than during the winter.

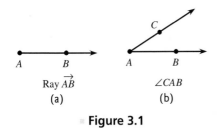

3.1 Angles and Their Measurements

Figure 3.1

Trigonometry was first studied by the Greeks, Egyptians, and Babylonians and used in surveying, navigation, and astronomy. Using trigonometry, they had a powerful tool for finding areas of triangular plots of land, as well as lengths of sides and measures of angles, without physically measuring them. We begin our study of trigonometry by studying angles and their measurements.

Degree Measure of Angles

In geometry a **ray** is defined as a point on a line together with all points of the line on one side of that point. Figure 3.1(a) shows ray $\overrightarrow{AB}$. An **angle** is defined as the union of two rays with a common endpoint, the **vertex.** The angle shown in Fig. 3.1(b) is named $\angle A$, $\angle BAC$, or $\angle CAB$. (Read the symbol $\angle$ as "angle.") Angles are also named using Greek letters such as α (alpha), β (beta), γ (gamma), or θ (theta).

An angle is often thought of as being formed by rotating one ray away from a fixed ray as indicated by angle α and the arrow in Fig. 3.2(a). The fixed ray is the **initial side** and the rotated ray is the **terminal side.** An angle whose vertex is the center of a circle, as shown in Fig. 3.2(b), is a **central angle,** and the arc of the circle through which the terminal side moves is the **intercepted arc.** An angle in **standard position** is located in a rectangular coordinate system with the vertex at the origin and the initial side on the positive x-axis as shown in Fig. 3.2(c).

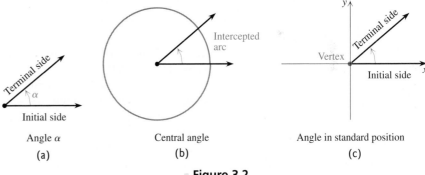

Angle α
(a)

Central angle
(b)

Angle in standard position
(c)

Figure 3.2

The measure $m(\alpha)$, of an angle α indicates the amount of rotation of the terminal side from the initial position. It is found using any circle centered at the vertex. The circle is divided into 360 equal arcs and each arc is one **degree** ($1°$).

Definition:
Degree Measure

> The **degree measure of an angle** is the number of degrees in the intercepted arc of a circle centered at the vertex. The degree measure is positive if the rotation is counterclockwise and negative if the rotation is clockwise.

Figure 3.3 on the next page shows the positions of the terminal sides of some angles in standard position with common positive measures between $0°$ and $360°$. An angle with measure between $0°$ and $90°$ is an **acute angle.** An angle with measure

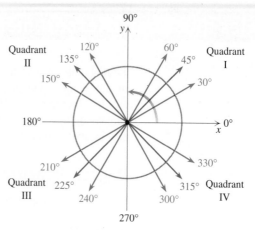

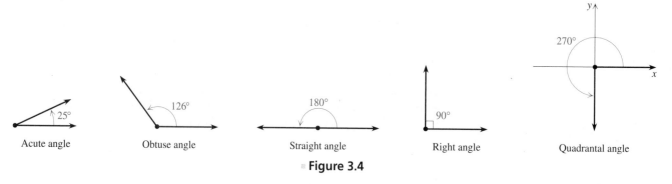

Figure 3.4

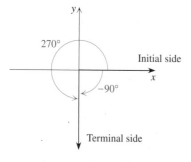

Coterminal angles

Figure 3.5

between 90° and 180° is an **obtuse angle.** An angle of exactly 180° is a **straight angle.** See Fig. 3.4. A 90° angle is a **right angle.** An angle in standard position is said to lie in the quadrant where its terminal side lies. If the terminal side is on an axis, the angle is a **quadrantal angle.** We often think of the degree measure of an angle as the angle itself. For example, we write $m(\alpha) = 60°$ or $\alpha = 60°$, and we say that 60° is an acute angle.

The initial side of an angle may be rotated in a positive or negative direction to get to the position of the terminal side. For example, if the initial side shown in Fig. 3.5 rotates clockwise for one-quarter of a revolution to get to the terminal position, then the measure of the angle is −90°. If the initial side had rotated counterclockwise to get to the position of the terminal side, then the measure of the angle would be 270°. If the initial side had rotated clockwise for one and a quarter revolutions to get to the terminal position, then the angle would be −450°. **Coterminal angles** are angles in standard position that have the same initial side and the same terminal side. The angles −90°, −450°, and 270° are examples of coterminal angles. Any two coterminal angles have degree measures that differ by a multiple of 360° (one complete revolution).

Coterminal Angles

Angles α and β in standard position are coterminal if and only if there is an integer k such that $m(\beta) = m(\alpha) + k360°$.

Note that any angle formed by two rays can be thought of as one angle with infinitely many different measures, or infinitely many different coterminal angles.

Example **1** **Finding coterminal angles**

Find two positive angles and two negative angles that are coterminal with $-50°$.

Solution

Since any angle of the form $-50° + k360°$ is coterminal with $-50°$, there are infinitely many possible answers. For simplicity, we choose the positive integers 1 and 2 and the negative integers -1 and -2 for k to get the following angles:

$$-50° + 1 \cdot 360° = 310°$$

$$-50° + 2 \cdot 360° = 670°$$

$$-50° + (-1)360° = -410°$$

$$-50° + (-2)360° = -770°$$

The angles $310°$, $670°$, $-410°$, and $-770°$ are coterminal with $-50°$. ■

To determine whether two angles are coterminal we must determine whether they differ by a multiple of $360°$.

Example **2** **Determining whether angles are coterminal**

Determine whether angles in standard position with the given measures are coterminal.

a. $m(\alpha) = 190°$, $m(\beta) = -170°$

b. $m(\alpha) = 150°$, $m(\beta) = 880°$

Solution

a. If there is an integer k such that $190 + 360k = -170$, then α and β are coterminal.

$$190 + 360k = -170$$

$$360k = -360$$

$$k = -1$$

Since the equation has an integral solution, α and β are coterminal.

b. If there is an integer k such that $150 + 360k = 880$, then α and β are coterminal.

$$150 + 360k = 880$$

$$360k = 730$$

$$k = \frac{73}{36}$$

Since there is no integral solution to the equation, α and β are not coterminal. ■

The quadrantal angles, such as $90°$, $180°$, $270°$, and $360°$, have terminal sides on an axis and do not lie in any quadrant. Any angle that is not coterminal with a quadrantal angle lies in one of the four quadrants. To determine the quadrant in which an angle lies, add or subtract multiples of $360°$ (one revolution) to obtain a coterminal angle with a measure between $0°$ and $360°$. A nonquadrantal angle is in the quadrant in which its terminal side lies.

Example **3** Determining in which quadrant
an angle lies

Name the quadrant in which each angle lies.

a. 230° **b.** −580° **c.** 1380°

Solution

a. Since 180° < 230° < 270°, a 230° angle lies in quadrant III.
b. We must add 2(360°) to −580° to get an angle between 0° and 360°:

$$-580° + 2(360°) = 140°$$

So 140° and −580° are coterminal. Since 90° < 140° < 180°, 140° lies in quadrant II and so does a −580° angle.

c. From 1380° we must subtract 3(360°) to obtain an angle between 0° and 360°:

$$1380° - 3(360°) = 300°$$

So 1380° and 300° are coterminal. Since 270° < 300° < 360°, 300° lies in quadrant IV and so does 1380°. ■

Each degree is divided into 60 equal parts called **minutes,** and each minute is divided into 60 equal parts called **seconds.** A minute (min) is 1/60 of a degree (deg), and a second (sec) is 1/60 of a minute or 1/3600 of a degree. An angle with measure 44°12′30″ is an angle with a measure of 44 degrees, 12 minutes, and 30 seconds. Historically, angles were measured by using degrees-minutes-seconds, but with calculators it is convenient to have the fractional parts of a degree written as a decimal number such as 7.218°. Some calculators can handle angles in degrees-minutes-seconds and even convert them to decimal degrees.

Example **4** Converting degrees-minutes-seconds
to decimal degrees

Convert the measure 44°12′30″ to decimal degrees.

Solution

Since 1 degree = 60 minutes and 1 degree = 3600 seconds, we get

$$12 \text{ min} = 12 \text{ min} \cdot \frac{1 \text{ deg}}{60 \text{ min}} = \frac{12}{60} \text{ deg} \quad \text{and}$$

$$30 \text{ sec} = 30 \text{ sec} \cdot \frac{1 \text{ deg}}{3600 \text{ sec}} = \frac{30}{3600} \text{ deg}.$$

So

$$44°12′30″ = \left(44 + \frac{12}{60} + \frac{30}{3600}\right)° \approx 44.2083°.$$

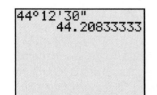

■ **Figure 3.6**

A graphing calculator can convert degrees-minutes-seconds to decimal degrees as shown in Fig. 3.6. ■

Note that the conversion of Example 4 was done by "cancellation of units." Minutes in the numerator canceled with minutes in the denominator to give the result in degrees, and seconds in the numerator canceled with seconds in the denominator to give the result in degrees.

Example **5** **Converting decimal degrees to degrees-minutes-seconds**

Convert the measure 44.235° to degrees-minutes-seconds.

Solution

First convert 0.235° to minutes. Since 1 degree = 60 minutes,

$$0.235 \text{ deg} = 0.235 \text{ deg} \cdot \frac{60 \text{ min}}{1 \text{ deg}} = 14.1 \text{ min.}$$

So 44.235° = 44°14.1′. Now convert 0.1′ to seconds. Since 1 minute = 60 seconds,

$$0.1 \text{ min} = 0.1 \text{ min} \cdot \frac{60 \text{ sec}}{1 \text{ min}} = 6 \text{ sec.}$$

So 44.235° = 44°14′6″.

A graphing calculator can convert to degrees-minutes-seconds as shown in Fig. 3.7. ■

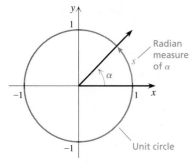

■ **Figure 3.7**

Note again how the original units canceled in Example 5, giving the result in the desired units of measurement.

Radian Measure of Angles

Degree measure of angles is used mostly in applied areas such as surveying, navigation, and engineering. Radian measure of angles is used more in scientific fields and results in simpler formulas in trigonometry and calculus.

For radian measures of angles we use a **unit circle** (a circle with radius 1) centered at the origin. The radian measure of an angle in standard position is simply the length of the intercepted arc on the unit circle. See Fig. 3.8. Since the radius of the unit circle is the real number 1 without any dimension (such as feet or inches), the length of an intercepted arc is a real number without any dimension and so the radian measure of an angle is also a real number without any dimension. One **radian** (abbreviated 1 rad) is the real number 1.

■ **Figure 3.8**

Definition: Radian Measure

> To find the **radian measure** of the angle α in standard position, find the length of the intercepted arc on the unit circle. If the rotation is counterclockwise, the radian measure is the length of the arc. If the rotation is clockwise, the radian measure is the opposite of the length of the arc.

Radian measure is called a **directed length** because it is positive or negative depending on the direction of rotation of the initial side. If s is the length of the intercepted arc on the unit circle for an angle α, as shown in Fig. 3.8, we write $m(\alpha) = s$. To emphasize that s is the length of an arc on a unit circle, we may write $m(\alpha) = s$ radians.

■ **Figure 3.9**

Because the circumference of a circle with radius r is $2\pi r$, the circumference of the unit circle is 2π. If the initial side rotates 360° (one complete revolution), then the length of the intercepted arc is 2π. So an angle of 360° has a radian measure of 2π radians. We express this relationship as 360° = 2π rad or simply 360° = 2π. Dividing each side by 2 yields 180° = π, which is the basic relationship to remember for conversion of one unit of measurement to the other.

⊿⊽ Use the MODE key on a graphing calculator to set the calculator to radian or degree mode as shown in Fig. 3.9. □

Degree-Radian Conversion

> Conversion from degrees to radians or radians to degrees is based on
>
> $$180 \text{ degrees} = \pi \text{ radians.}$$

To convert degrees to radians or radians to degrees, we use 180 deg = π rad and cancellation of units. For example,

$$1 \text{ deg} = 1 \text{ deg} \cdot \frac{\pi \text{ rad}}{180 \text{ deg}} = \frac{\pi}{180} \text{ rad} \approx 0.01745 \text{ rad}$$

and

$$1 \text{ rad} = 1 \text{ rad} \cdot \frac{180 \text{ deg}}{\pi \text{ rad}} = \frac{180}{\pi} \text{ deg} \approx 57.3 \text{ deg.}$$

⊿⊽ If your calculator is in radian mode, as in Fig. 3.10(a), pressing ENTER converts degrees to radians. When the calculator is in degree mode, as in Fig. 3.10(b), pressing ENTER converts radians to degrees. □

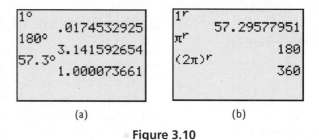

(a)   (b)

■ **Figure 3.10**

Figure 3.11(a) shows an angle of 1°, and Fig. 3.11(b) shows an angle of 1 radian. An angle of 1 radian intercepts an arc on the unit circle equal in length to the radius of the unit circle.

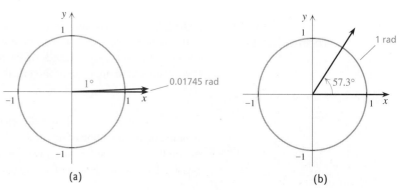

(a)   (b)

■ **Figure 3.11**

Example **6** **Converting from degrees to radians**

Convert the degree measures to radians.

a. 270°   **b.** −23.6°

Solution

a. To convert degrees to radians, multiply the degree measure by π rad/180 deg:

$$270° = 270 \text{ deg} \cdot \frac{\pi \text{ rad}}{180 \text{ deg}} = \frac{3\pi}{2} \text{ rad}$$

The exact value, $3\pi/2$ rad, is approximately 4.71 rad; but when a measure in radians is a simple multiple of π, we usually write the exact value.

    Check this result with a calculator in radian mode as shown in Fig. 3.12.   □

b. $-23.6° = -23.6 \text{ deg} \cdot \dfrac{\pi \text{ rad}}{180 \text{ deg}} \approx -0.412$ rad.   ■

Example **7** **Converting from radians to degrees**

Convert the radian measures to degrees.

a. $\dfrac{7\pi}{6}$ **b.** 12.3

Solution

Multiply the radian measure by 180 deg/π rad:

a. $\dfrac{7\pi}{6} = \dfrac{7\pi}{6} \text{ rad} \cdot \dfrac{180 \text{ deg}}{\pi \text{ rad}} = 210°$ **b.** $12.3 = 12.3 \text{ rad} \cdot \dfrac{180 \text{ deg}}{\pi \text{ rad}} \approx 704.7°$

    Check these answers in degree mode as shown in Fig. 3.13.   ■

     Figure 3.14 shows angles with common measures in standard position. Coterminal angles in standard position have radian measures that differ by an integral multiple of 2π (their degree measures differ by an integral multiple of 360°).

Figure 3.12 (calculator screen)

270°
    4.71238898
3π/2
    4.71238898

■ **Figure 3.12**

(7π/6)ʳ
       210
12.3ʳ
   704.738088

■ **Figure 3.13**

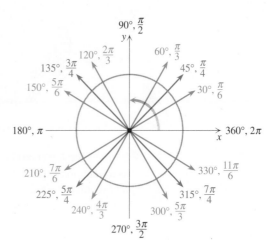

■ **Figure 3.14**

Example **8** **Finding coterminal angles using radian measure**

Find two positive and two negative angles that are coterminal with $\pi/6$.

Solution

All angles coterminal with $\pi/6$ have a radian measure of the form $\pi/6 + k(2\pi)$, where k is an integer.

$$\frac{\pi}{6} + 1(2\pi) = \frac{\pi}{6} + \frac{12\pi}{6} = \frac{13\pi}{6}$$

$$\frac{\pi}{6} + 2(2\pi) = \frac{\pi}{6} + \frac{24\pi}{6} = \frac{25\pi}{6}$$

$$\frac{\pi}{6} + (-1)(2\pi) = \frac{\pi}{6} - \frac{12\pi}{6} = -\frac{11\pi}{6}$$

$$\frac{\pi}{6} + (-2)(2\pi) = \frac{\pi}{6} - \frac{24\pi}{6} = -\frac{23\pi}{6}$$

The angles $13\pi/6$, $25\pi/6$, $-11\pi/6$, and $-23\pi/6$ are coterminal with $\pi/6$. ■

Arc Length

Radian measure of a central angle of a circle can be used to easily find the length of the intercepted arc of the circle. For example, an angle of $\pi/2$ radians positioned at the center of the earth, as shown in Fig. 3.15, intercepts an arc on the surface of the earth that runs from the equator to the North Pole. Using 3950 miles as the approximate radius of the earth yields a circumference of $2\pi(3950) = 7900\pi$ miles. Since $\pi/2$ radians is 1/4 of a complete circle, the length of the intercepted arc is 1/4 of the circumference. So the distance from the equator to the North Pole is $7900\pi/4$ miles, or about 6205 miles.

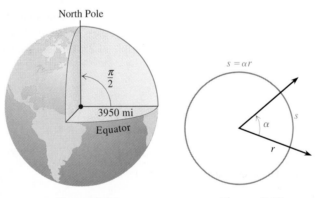

■ **Figure 3.15**   ■ **Figure 3.16**

In general, a central angle α in a circle of radius r intercepts an arc whose length s is a fraction of the circumference of the circle, as shown in Fig. 3.16. Since a complete revolution is 2π radians, that fraction is $\alpha/(2\pi)$. Since the circumference is $2\pi r$, we get

$$s = \frac{\alpha}{2\pi} \cdot 2\pi r = \alpha r.$$

Theorem: Length of an Arc

The length s of an arc intercepted by a central angle of α *radians* on a circle of radius r is given by

$$s = \alpha r.$$

If α is negative, the formula $s = \alpha r$ gives a negative number for the length of the arc. So s is a directed length. Note that the formula $s = \alpha r$ applies only if α is in radians.

Example **9** **Finding the length of an arc**

The wagon wheel shown in Fig. 3.17 has a diameter of 28 inches and an angle of 30° between the spokes. What is the length of the arc s between two adjacent spokes?

Solution

First convert 30° to radians:

$$30° = 30 \text{ deg} \cdot \frac{\pi \text{ rad}}{180 \text{ deg}} = \frac{\pi}{6} \text{ rad}$$

Now use $r = 14$ inches and $\alpha = \pi/6$ in the formula for arc length $s = \alpha r$:

$$s = \frac{\pi}{6} \cdot 14 \text{ in.} = \frac{7\pi}{3} \text{ in.} \approx 7.33 \text{ in.}$$

Since the radian measure of an angle is a dimensionless real number, the product of radians and inches is given in inches.

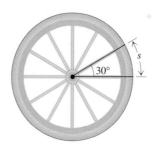

■ **Figure 3.17**   ■

For Thought

True or False? Explain.

1. An angle is a union of two rays with a common endpoint.

2. The lengths of the rays of $\angle A$ determine the degree measure of $\angle A$.

3. Angles of 5° and −365° are coterminal.

4. The radian measure of an angle cannot be negative.

5. An angle of $38\pi/4$ radians is a quadrantal angle.

6. If $m(\angle A) = 210°$, then $m(\angle A) = 5\pi/6$ radians.

7. $25°20'40'' = 25.34°$

8. $2\pi + \pi/3 = 7\pi/3$

9. A central angle of $\pi/3$ radians in a circle of radius 3 ft intercepts an arc of length π feet.

10. A central angle of 1 rad in a circle of radius r intercepts an arc of length r.

3.1 Exercises

Find two positive angles and two negative angles that are coterminal with each given angle.

1. 60°   **2.** 45°   **3.** −16°   **4.** −90°

Determine whether the angles in each given pair are coterminal.

5. 123.4°, −236.6°   **6.** 744°, −336°

7. 1055°, 155°   **8.** 0°, 359.9°

Name the quadrant in which each angle lies.

9. 85°   **10.** 110°   **11.** −125°   **12.** −200°

13. 300°   **14.** 205°   **15.** 750°   **16.** −980°

Match each of the following angles with one of the degree measures: 30°, 45°, 60°, 120°, 135°, 150°.

17.   **18.**

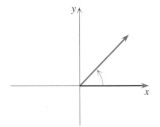

19.   **20.**

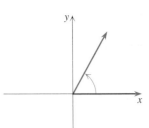

21.   **22.**

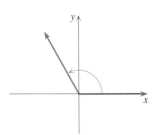

Find the measure in degrees of the least positive angle that is coterminal with each given angle.

23. 400°   **24.** 540°   **25.** −340°

26. −180°   **27.** −1100°   **28.** −840°

Convert each angle to decimal degrees. When necessary, round to four decimal places.

29. 13°12′   **30.** 45°6′   **31.** −8°30′18″

32. −5°45′30″   **33.** 28°5′9″   **34.** 44°19′32″

Convert each angle to degrees-minutes-seconds. Round to the nearest whole number of seconds.

35. 75.5°   **36.** 39.4°   **37.** −17.33°

38. −9.12°   **39.** 18.123°   **40.** 122.786°

Convert each degree measure to radian measure. Give exact answers.

41. 30°   **42.** 45°

43. 18°   **44.** 48°

45. −67.5°   **46.** −105°

47. 630°   **48.** 495°

Convert each degree measure to radian measure. Use the value of π found on a calculator and round answers to three decimal places.

49. 37.4°   **50.** 125.3°   **51.** −13°47′

52. −99°15′   **53.** −53°37′6″   **54.** 187°49′36″

Convert each radian measure to degree measure. Use the value of π found on a calculator and round approximate answers to three decimal places.

55. $\dfrac{5\pi}{12}$ **56.** $\dfrac{17\pi}{12}$ **57.** $\dfrac{7\pi}{4}$ **58.** $\dfrac{13\pi}{6}$

59. −6π   **60.** −9π   **61.** 2.39   **62.** 0.452

Using radian measure, find two positive angles and two negative angles that are coterminal with each given angle.

63. $\dfrac{\pi}{3}$ **64.** $\dfrac{\pi}{4}$ **65.** $-\dfrac{\pi}{6}$ **66.** $-\dfrac{2\pi}{3}$

Find the measure in radians of the least positive angle that is coterminal with each given angle.

67. 3π **68.** 6π **69.** $\dfrac{9\pi}{2}$ **70.** $\dfrac{19\pi}{2}$

71. $-\dfrac{5\pi}{3}$ **72.** $-\dfrac{7\pi}{6}$ **73.** $-\dfrac{13\pi}{3}$ **74.** $-\dfrac{19\pi}{4}$

75. 8.32 **76.** -23.55

Determine whether the angles in each given pair are coterminal.

77. $\dfrac{3\pi}{4}, \dfrac{29\pi}{4}$ **78.** $-\dfrac{\pi}{3}, \dfrac{5\pi}{3}$

79. $\dfrac{7\pi}{6}, -\dfrac{5\pi}{6}$ **80.** $\dfrac{3\pi}{2}, -\dfrac{9\pi}{2}$

Name the quadrant in which each angle lies.

81. $\dfrac{5\pi}{12}$ **82.** $\dfrac{13\pi}{12}$ **83.** $-\dfrac{6\pi}{7}$ **84.** $-\dfrac{39\pi}{20}$

85. $\dfrac{13\pi}{8}$ **86.** $-\dfrac{11\pi}{8}$ **87.** -7.3 **88.** 23.1

Find the missing degree or radian measure for each position of the terminal side shown. For Exercise 89, use degrees between 0° and 360° and radians between 0 and 2π. For Exercise 90, use degrees between −360° and 0° and radians between −2π and 0. Practice these two exercises until you have memorized the degree and radian measures corresponding to these common angles.

89.

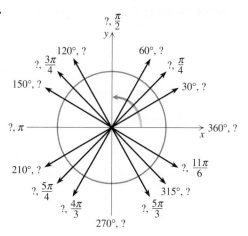

90.

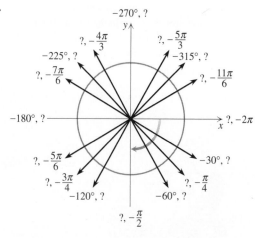

Find the length of the arc intercepted by the given central angle α in a circle of radius r.

91. $\alpha = \dfrac{\pi}{4}, r = 12$ ft **92.** $\alpha = 1, r = 4$ cm

93. $\alpha = 3°, r = 4000$ mi **94.** $\alpha = 60°, r = 2$ m

Find the radius of the circle in which the given central angle α intercepts an arc of the given length s.

95. $\alpha = 1, s = 1$ mi **96.** $\alpha = 0.004, s = 99$ km

97. $\alpha = 180°, s = 10$ km **98.** $\alpha = 360°, s = 8$ m

Solve each problem.

99. *Distance between Towers* The towers of the Verrazano-Narrows bridge are 693 ft tall and the length of the arc joining the bases of the towers is 4260 ft as shown in the accompanying figure. The radius of the earth is 4000 mi. How much greater in length is the arc joining the tops of the towers than the arc joining the bases?

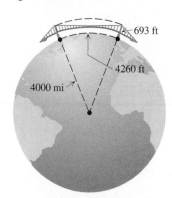

▪ **Figure for Exercise 99**

100. *Distance to the Helper* A surveyor sights her 6-ft 2-in. helper on a nearby hill as shown in the figure. If the angle of sight between the helper's feet and head is 0°37′, then approximately how far away is the helper?

▪ **Figure for Exercise 100**

101. *Eratosthenes Measures the Earth* Over 2200 years ago Eratosthenes read in the Alexandria library that at noon on June 21 a vertical stick in Syene cast no shadow. So on June 21 at noon Eratosthenes set out a vertical stick in Alexandria and found an angle of 7° in the position shown in the accompanying drawing. Eratosthenes reasoned that since the sun is so far away, sunlight must be arriving at the earth in parallel rays. With this assumption he concluded that the earth is round and the central angle in the drawing must also be 7°. He then paid a man to pace off the distance between Syene and Alexandria and found it to be 800 km. From these facts, calculate the circumference of the earth as Eratosthenes did and compare his answer with the circumference calculated by using the currently accepted radius of 6378 km.

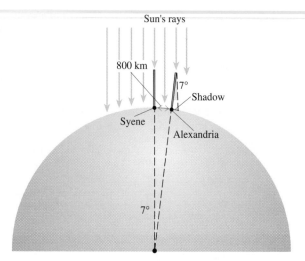

▪ **Figure for Exercise 101**

Thinking Outside the Box XIX

Triangles and Circles Each of the five circles in the accompanying diagram has radius *r*. The four right triangles are congruent and each hypotenuse has length 1. Each line segment that appears to intersect a circle intersects it at exactly one point (a point of tangency). Find *r*.

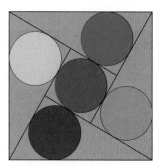

▪ **Figure for Thinking Outside the Box XIX**

3.1 Pop Quiz

1. Find the degree measure of the least positive angle that is coterminal with 1267°.

2. In which quadrant does −120° lie?

3. Convert 70°30′36″ to decimal degrees.

4. Convert 270° to radian measure.

5. Convert $7\pi/4$ to degree measure.

6. Are $-3\pi/4$ and $5\pi/4$ coterminal?

7. Find the exact length of the arc intercepted by a central angle of 60° in a circle with radius 30 feet.

3.2 The Sine and Cosine Functions

In Section 3.1 we learned that angles can be measured in degrees or radians. Now we define two trigonometric functions whose domain is the set of all angles (measured in degrees or radians). These two functions are unlike any functions defined in algebra, but they form the foundation of trigonometry.

Definition

If α is an angle in standard position whose terminal side intersects the unit circle at point (x, y), as shown in Fig. 3.18, then **sine of α**—abbreviated $\sin(\alpha)$ or $\sin \alpha$—is the y-coordinate of that point, and **cosine of α**—abbreviated $\cos(\alpha)$ or $\cos \alpha$—is the x-coordinate. Sine and cosine are called **trigonometric functions.**

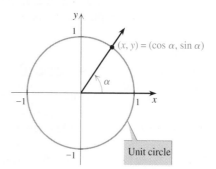

■ **Figure 3.18**

Definition: Sine and Cosine

If α is an angle in standard position and (x, y) is the point of intersection of the terminal side and the unit circle, then

$$\sin \alpha = y \quad \text{and} \quad \cos \alpha = x.$$

The domain of the sine function and the cosine function is the set of angles in standard position, but since each angle has a measure in degrees or radians, we generally use the set of degree measures or the set of radian measures as the domain. If (x, y) is on the unit circle, then $-1 \le x \le 1$ and $-1 \le y \le 1$, so the range of each of these functions is the interval $[-1, 1]$.

Example **1** Evaluating sine and cosine at a multiple of 90°

Find the exact values of the sine and cosine functions for each angle.

a. 90°   **b.** $\pi/2$ **c.** 180°   **d.** $-5\pi/2$ **e.** $-720°$

Solution

a. Consider the unit circle shown in Fig. 3.19. Since the terminal side of 90° intersects the unit circle at $(0, 1)$, $\sin(90°) = 1$ and $\cos(90°) = 0$.

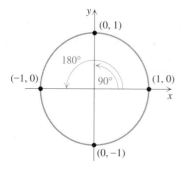

■ **Figure 3.19**

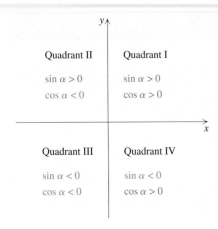

Figure 3.20

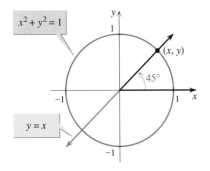

Figure 3.21

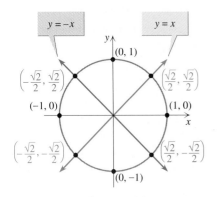

Figure 3.22

b. Since $\pi/2$ is coterminal with $90°$, we have $\sin(\pi/2) = 1$ and $\cos(\pi/2) = 0$.

c. The terminal side for $180°$ intersects the unit circle at $(-1, 0)$. So $\sin(180°) = 0$ and $\cos(180°) = -1$.

d. Since $-5\pi/2$ is coterminal with $3\pi/2$, the terminal side of $-5\pi/2$ lies on the negative y-axis and intersects the unit circle at $(0, -1)$. So $\sin(-5\pi/2) = -1$ and $\cos(-5\pi/2) = 0$.

e. Since $-720°$ is coterminal with $0°$, the terminal side of $-720°$ lies on the positive x-axis and intersects the unit circle at $(1, 0)$. So $\sin(-720°) = 0$ and $\cos(-720°) = 1$. ■

The *signs* of the sine and cosine functions depend on the quadrant in which the angle lies. For any point (x, y) on the unit circle in quadrant I, the x- and y-coordinates are positive. So if α is an angle in quadrant I, $\sin\alpha > 0$ and $\cos\alpha > 0$. Since the x-coordinate of any point in quadrant II is negative, the cosine of any angle in quadrant II is negative. Figure 3.20 gives the signs of sine and cosine for each of the four quadrants.

Sine and Cosine of a Multiple of 45°

In Example 1, exact values of the sine and cosine functions were found for some angles that were multiples of $90°$ (quadrantal angles). We can also find the exact values of $\sin(45°)$ and $\cos(45°)$ and then use that information to find the exact values of these functions for any multiple of $45°$.

Since the terminal side for $45°$ lies on the line $y = x$, as shown in Fig. 3.21, the x- and y-coordinates at the point of intersection with the unit circle are equal. From Section 1.3, the equation of the unit circle is $x^2 + y^2 = 1$. Because $y = x$, we can write $x^2 + x^2 = 1$ and solve for x:

$$2x^2 = 1$$

$$x^2 = \frac{1}{2}$$

$$x = \pm\sqrt{\frac{1}{2}} = \pm\frac{\sqrt{2}}{2}$$

For a $45°$ angle in quadrant I, x and y are both positive numbers. So $\sin(45°) = \cos(45°) = \sqrt{2}/2$.

There are four points where the lines $y = x$ and $y = -x$ intersect the unit circle. The coordinates of these points, shown in Fig. 3.22, can all be found as above or by the symmetry of the unit circle. Figure 3.22 shows the coordinates of the key points for determining the sine and cosine of any angle that is an integral multiple of $45°$.

Example **2** **Evaluating sine and cosine at a multiple of 45°**

Find the exact value of each expression.

a. $\sin(135°)$ **b.** $\sin\left(\dfrac{5\pi}{4}\right)$ **c.** $2\sin\left(-\dfrac{9\pi}{4}\right)\cos\left(-\dfrac{9\pi}{4}\right)$

Solution

a. The terminal side for 135° lies in quadrant II, halfway between 90° and 180°. As shown in Fig. 3.22, the point on the unit circle halfway between 90° and 180° is $\left(-\sqrt{2}/2,\ \sqrt{2}/2\right)$. So $\sin(135°) = \sqrt{2}/2$.

b. The terminal side for $5\pi/4$ is in quadrant III, halfway between $4\pi/4$ and $6\pi/4$. From Fig. 3.22, $\sin(5\pi/4)$ is the y-coordinate of $\left(-\sqrt{2}/2,\ -\sqrt{2}/2\right)$. So $\sin(5\pi/4) = -\sqrt{2}/2$.

c. Since $-8\pi/4$ is one clockwise revolution, $-9\pi/4$ is coterminal with $-\pi/4$. From Fig. 3.22, we have $\cos(-9\pi/4) = \cos(-\pi/4) = \sqrt{2}/2$ and $\sin(-9\pi/4) = -\sqrt{2}/2$. So

$$2 \sin\left(-\frac{9\pi}{4}\right)\cos\left(-\frac{9\pi}{4}\right) = 2 \cdot \left(-\frac{\sqrt{2}}{2}\right) \cdot \frac{\sqrt{2}}{2} = -1.$$ ■

Sine and Cosine of a Multiple of 30°

Exact values for the sine and cosine of 60° and the sine and cosine of 30° are found with a little help from geometry. The terminal side of a 60° angle intersects the unit circle at a point (x, y), as shown in Fig. 3.23(a), where x and y are the lengths of the legs of a 30-60-90 triangle whose hypotenuse is length 1.

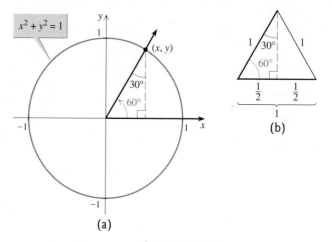

(a)

(b)

■ **Figure 3.23**

Since two congruent 30-60-90 triangles are formed by the altitude of an equilateral triangle, as shown in Fig. 3.23(b), the side opposite the 30° angle is half the length of the hypotenuse. Since the length of the hypotenuse is 1, the side opposite 30° is 1/2, that is, $x = 1/2$. Use the fact that $x^2 + y^2 = 1$ for the unit circle to find y:

$$\left(\frac{1}{2}\right)^2 + y^2 = 1 \qquad \text{Replace } x \text{ with } \tfrac{1}{2} \text{ in } x^2 + y^2 = 1.$$

$$y^2 = \frac{3}{4}$$

$$y = \pm\sqrt{\frac{3}{4}} = \pm\frac{\sqrt{3}}{2}$$

Since $y > 0$ (in quadrant I), $\cos(60°) = 1/2$ and $\sin(60°) = \sqrt{3}/2$.

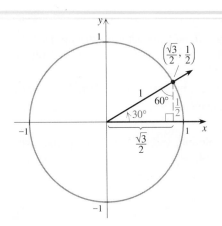

■ **Figure 3.24**

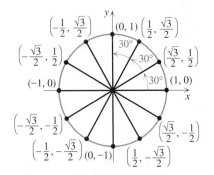

■ **Figure 3.25**

The point of intersection of the terminal side for a 30° angle with the unit circle determines a 30-60-90 triangle exactly the same size as the one described above. In this case the longer leg is on the x-axis, as shown in Fig. 3.24, and the point on the unit circle is $\left(\sqrt{3}/2, 1/2\right)$. So $\cos(30°) = \sqrt{3}/2$, and $\sin(30°) = 1/2$.

Using the symmetry of the unit circle and the values just found for the sine and cosine of 30° and 60°, we can label points on the unit circle, as shown in Fig. 3.25. This figure shows the coordinates of every point on the unit circle where the terminal side of a multiple of 30° intersects the unit circle. For example, the terminal side of a 120° angle in standard position intersects the unit circle at $\left(-1/2, \sqrt{3}/2\right)$, because $120° = 4 \cdot 30°$.

In the next example, we use Fig. 3.25 to evaluate expressions involving multiples of 30° $(\pi/6)$. Note that the square of $\sin\alpha$ could be written as $(\sin\alpha)^2$, but for simplicity it is written as $\sin^2\alpha$. Likewise, $\cos^2\alpha$ is used for $(\cos\alpha)^2$.

Example **3** **Evaluating sine and cosine at a multiple of 30°**

Find the exact value of each expression.

a. $\sin\left(\dfrac{7\pi}{6}\right)$ **b.** $\cos^2(-240°) - \sin^2(-240°)$

Solution

a. Since $\pi/6 = 30°$, we have $7\pi/6 = 210°$. To determine the location of the terminal side of $7\pi/6$, notice that $7\pi/6$ is 30° larger than the straight angle 180° and so it lies in quadrant III. From Fig. 3.25, $7\pi/6$ intersects the unit circle at $\left(-\sqrt{3}/2, -1/2\right)$ and $\sin(7\pi/6) = -1/2$.

b. Since $-240°$ is coterminal with 120°, the terminal side for $-240°$ lies in quadrant II and intersects the unit circle at $\left(-1/2, \sqrt{3}/2\right)$, as indicated in Fig. 3.25. So $\cos(-240°) = -1/2$ and $\sin(-240°) = \sqrt{3}/2$. Now use these values in the original expression:

$$\cos^2(-240°) - \sin^2(-240°) = \left(-\frac{1}{2}\right)^2 - \left(\frac{\sqrt{3}}{2}\right)^2 = \frac{1}{4} - \frac{3}{4} = -\frac{1}{2} \quad ■$$

A good way to remember the sines and cosines for the common angles is to note the following patterns:

$$\sin(30°) = \frac{\sqrt{1}}{2}, \quad \sin(45°) = \frac{\sqrt{2}}{2}, \quad \sin(60°) = \frac{\sqrt{3}}{2}$$

$$\cos(30°) = \frac{\sqrt{3}}{2}, \quad \cos(45°) = \frac{\sqrt{2}}{2}, \quad \cos(60°) = \frac{\sqrt{1}}{2}$$

Sine and Cosine of an Arc

In calculus we study functions such as $f(x) = x^2 + \sin(x)$. For this function to make sense, x must be a real number and not degrees. So in calculus we generally use the set of real numbers or radians as the domain of the trigonometric functions. Since the radian measure of a central angle in the unit circle is the same as the length

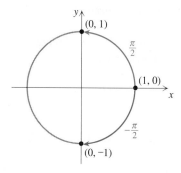

Figure 3.26

of the intercepted arc, we also use arcs on the unit circle with initial point (1, 0) and terminal point (x, y) as the domain of the trigonometric functions.

Example 4 Sine and cosine of an arc

Find $\sin(\pi/2)$ and $\cos(-\pi/2)$.

Solution

Since the circumference of the unit circle is 2π, an arc of length $\pi/2$ is one-quarter of the circumference. Arcs of length $\pi/2$ and $-\pi/2$ are shown in Fig. 3.26. Since sine is the y-coordinate and cosine is the x-coordinate at the terminal point of an arc, $\sin(\pi/2) = 1$ and $\cos(-\pi/2) = 0$. ■

Approximate Values for Sine and Cosine

The sine and cosine for any angle that is a multiple of 30° or 45° can be found exactly. These angles are so common that it is important to know these exact values. However, for most other angles or real numbers we use approximate values for the sine and cosine, found with the help of a scientific calculator or a graphing calculator. Calculators can evaluate $\sin \alpha$ or $\cos \alpha$ if α is a real number (radian) or α is the degree measure of an angle. Generally, there is a mode key that sets the calculator to degree mode or radian mode. Consult your calculator manual.

Example 5 Evaluating sine and cosine with a calculator

Find each function value rounded to four decimal places.

a. $\sin(4.27)$ **b.** $\cos(-39.46°)$

Solution

a. With the calculator in radian mode, we get $\sin(4.27) \approx -0.9037$.
b. With the calculator in degree mode, we get $\cos(-39.46°) \approx 0.7721$.

    If you use the symbol for radians or degrees, as shown in Fig. 3.27, then it is not necessary to change the mode with a graphing calculator. ■

```
sin(4.27ʳ)
        -.9037315214
cos(-39.46°)
         .7720684617
```

Figure 3.27

The Fundamental Identity

An identity is an equation that is satisfied for all values of the variable for which both sides are defined. The most fundamental identity in trigonometry involves the squares of the sine and cosine functions. If α is an angle in standard position, then $\sin \alpha = y$ and $\cos \alpha = x$, where (x, y) is the point of intersection of the terminal side of α and the unit circle. Since every point on the unit circle satisfies the equation $x^2 + y^2 = 1$, we get the following identity.

The Fundamental Identity of Trigonometry

If α is any angle or real number,

$$\sin^2\alpha + \cos^2\alpha = 1.$$

(sin(5.9))²+(cos
(5.9))²
                    1
(sin(48))²+(cos(
48))²
                    1

Figure 3.28

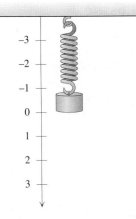

Figure 3.29

The fundamental identity can be illustrated with a graphing calculator as shown in Fig. 3.28. □

If we know the value of the sine or cosine of an angle, then we can use the fundamental identity to find the value of the other function of the angle.

Example 6 Using the fundamental identity

Find $\cos\alpha$, given that $\sin\alpha = 3/5$ and α is an angle in quadrant II.

Solution

Use the fundamental identity $\sin^2\alpha + \cos^2\alpha = 1$ to find $\cos\alpha$:

$$\left(\frac{3}{5}\right)^2 + \cos^2\alpha = 1 \qquad \text{Replace } \sin\alpha \text{ with } 3/5.$$

$$\cos^2\alpha = \frac{16}{25} \qquad 1 - \frac{9}{25} = \frac{16}{25}$$

$$\cos\alpha = \pm\sqrt{\frac{16}{25}} = \pm\frac{4}{5}$$

Since $\cos\alpha < 0$ for any α in quadrant II, we choose the negative sign and get $\cos\alpha = -4/5$. ■

Modeling the Motion of a Spring

The sine and cosine functions are used in modeling the motion of a spring. If a weight is at rest while hanging from a spring, as shown in Fig. 3.29, then it is at the **equilibrium** position, or 0 on a vertical number line. If the weight is set in motion with an initial velocity v_0 from location x_0, then the location at time t is given by

$$x = \frac{v_0}{\omega}\sin(\omega t) + x_0\cos(\omega t).$$

The letter ω (omega) is a constant that depends on the stiffness of the spring and the amount of weight on the spring. For positive values of x the weight is below equilibrium, and for negative values it is above equilibrium. The initial velocity is considered to be positive if it is in the downward direction and negative if it is upward. Note that the domain of the sine and cosine functions in this formula is the nonnegative real numbers, and this application has nothing to do with angles.

Example 7 Motion of a spring

A weight on a certain spring is set in motion with an upward velocity of 3 centimeters per second from a position 2 centimeters below equilibrium. Assume that for this spring and weight combination the constant ω has a value of 1. Write a formula that gives the location of the weight in centimeters as a function of the time t in seconds, and find the location of the weight 2 seconds after the weight is set in motion.

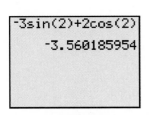

■ **Figure 3.30**

Solution

Since upward velocity is negative and locations below equilibrium are positive, use $v_0 = -3$, $x_0 = 2$, and $\omega = 1$ in the formula for the motion of a spring:

$$x = -3 \sin t + 2 \cos t$$

If $t = 2$ seconds, then $x = -3 \sin(2) + 2 \cos(2) \approx -3.6$ centimeters, which is 3.6 centimeters above the equilibrium position.

Figure 3.30 shows the calculation for x with a graphing calculator in radian mode. ■

For Thought

True or False? Explain.

1. $\cos(90°) = 1$

2. $\cos(90) = 0$

3. $\sin(45°) = 1/\sqrt{2}$

4. $\sin(-\pi/3) = \sin(\pi/3)$

5. $\cos(-\pi/3) = -\cos(\pi/3)$ **6.** $\sin(390°) = \sin(30°)$

7. If $\sin(\alpha) < 0$ and $\cos(\alpha) > 0$, then α is an angle in quadrant III.

8. If $\sin^2(\alpha) = 1/4$ and α is in quadrant III, then $\sin(\alpha) = 1/2$.

9. If $\cos^2(\alpha) = 3/4$ and α is an angle in quadrant I, then $\alpha = \pi/6$.

10. For any angle α, $\cos^2(\alpha) = (1 - \sin(\alpha))(1 + \sin(\alpha))$.

3.2 Exercises

Redraw each diagram and label the indicated points with the proper coordinates. Exercise 1 shows the points where the terminal side of every multiple of 45° intersects the unit circle. Exercise 2 shows the points where the terminal side of every multiple of 30° intersects the unit circle. Repeat Exercises 1 and 2 until you can do them from memory.

1.

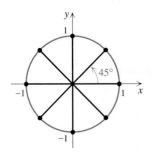

2.

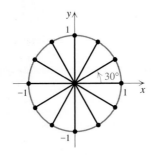

Use the diagrams drawn in Exercises 1 and 2 to help you find the exact value of each function. Do not use a calculator.

3. $\sin(0)$ **4.** $\cos(\pi)$ **5.** $\cos(-90°)$

6. $\sin(0°)$ **7.** $\sin(2\pi)$ **8.** $\cos(-\pi)$

9. $\cos(3\pi/2)$ **10.** $\sin(-5\pi/2)$ **11.** $\sin(135°)$

12. $\cos(-135°)$ **13.** $\sin(-\pi/4)$ **14.** $\sin(3\pi/4)$

15. $\sin(30°)$ **16.** $\sin(120°)$ **17.** $\cos(-60°)$

18. $\cos(-120°)$ **19.** $\cos(7\pi/6)$ **20.** $\cos(-2\pi/3)$

21. $\sin(-4\pi/3)$ **22.** $\sin(5\pi/6)$ **23.** $\sin(390°)$

24. $\sin(765°)$ **25.** $\cos(-420°)$ **26.** $\cos(-450°)$

27. $\cos(13\pi/6)$ **28.** $\cos(-7\pi/3)$

Evaluate the trigonometric functions for the given arc lengths on the unit circle.

29. $\sin(\pi/4)$ **30.** $\sin(-\pi/4)$ **31.** $\cos(-\pi)$

32. $\cos(\pi)$ **33.** $\sin(\pi/3)$ **34.** $\sin(\pi/6)$

35. $\cos(-\pi/3)$ **36.** $\cos(-\pi/6)$

Find the exact value of each expression without using a calculator. Check your answer with a calculator.

37. $\dfrac{\cos(\pi/3)}{\sin(\pi/3)}$ **38.** $\dfrac{\sin(-5\pi/6)}{\cos(-5\pi/6)}$

39. $\dfrac{\sin(7\pi/4)}{\cos(7\pi/4)}$ **40.** $\dfrac{\sin(-3\pi/4)}{\cos(-3\pi/4)}$

41. $\sin\left(\dfrac{\pi}{3} + \dfrac{\pi}{6}\right)$ **42.** $\cos\left(\dfrac{\pi}{3} - \dfrac{\pi}{6}\right)$

43. $\dfrac{1 - \cos(5\pi/6)}{\sin(5\pi/6)}$ **44.** $\dfrac{\sin(5\pi/6)}{1 + \cos(5\pi/6)}$

45. $\sin(\pi/4) + \cos(\pi/4)$ **46.** $\sin^2(\pi/6) + \cos^2(\pi/6)$

Determine whether each of the following expressions is positive (+) or negative (−) without using a calculator.

47. $\sin(121°)$ **48.** $\cos(157°)$ **49.** $\cos(359°)$

50. $\sin(213°)$ **51.** $\sin(7\pi/6)$ **52.** $\cos(7\pi/3)$

53. $\cos(-3\pi/4)$ **54.** $\sin(-7\pi/4)$

Use a calculator to find the value of each function. Round answers to four decimal places.

55. $\cos(-359.4°)$ **56.** $\sin(344.1°)$

57. $\sin(23°48')$ **58.** $\cos(49°13')$

59. $\sin(-48°3'12'')$ **60.** $\cos(-9°4'7'')$

61. $\sin(1.57)$ **62.** $\cos(3.14)$

63. $\cos(7\pi/12)$ **64.** $\sin(-13\pi/8)$

Find the exact value of each expression for the given value of θ. Do not use a calculator.

65. $\sin(2\theta)$ if $\theta = \pi/4$

66. $\sin(2\theta)$ if $\theta = \pi/6$

67. $\cos(2\theta)$ if $\theta = \pi/6$

68. $\cos(2\theta)$ if $\theta = \pi/3$

69. $\sin(\theta/2)$ if $\theta = 3\pi/2$

70. $\sin(\theta/2)$ if $\theta = 2\pi/3$

71. $\cos(\theta/2)$ if $\theta = \pi/3$

72. $\cos(\theta/2)$ if $\theta = \pi/2$

Solve each problem.

73. Find $\cos(\alpha)$, given that $\sin(\alpha) = 5/13$ and α is in quadrant II.

74. Find $\sin(\alpha)$, given that $\cos(\alpha) = -4/5$ and α is in quadrant III.

75. Find $\sin(\alpha)$, given that $\cos(\alpha) = 3/5$ and α is in quadrant IV.

76. Find $\cos(\alpha)$, given that $\sin(\alpha) = -12/13$ and α is in quadrant IV.

77. Find $\cos(\alpha)$, given that $\sin(\alpha) = 1/3$ and $\cos(\alpha) > 0$.

78. Find $\sin(\alpha)$, given that $\cos(\alpha) = 2/5$ and $\sin(\alpha) < 0$.

Solve each problem.

79. *Motion of a Spring*   A weight on a vertical spring is given an initial downward velocity of 4 cm/sec from a point 3 cm above equilibrium. Assuming that the constant ω has a value of 1, write the formula for the location of the weight at time t, and find its location 3 sec after it is set in motion.

80. *Motion of a Spring*   A weight on a vertical spring is given an initial upward velocity of 3 in./sec from a point 1 in. below equilibrium. Assuming that the constant ω has a value of $\sqrt{3}$, write the formula for the location of the weight at time t, and find its location 2 sec after it is set in motion.

81. *Spacing Between Teeth*   The length of an arc intercepted by a central angle of θ radians in a circle of radius r is $r\theta$. The length of the chord, c, joining the endpoints of that arc is given by $c = r\sqrt{2 - 2\cos\theta}$. Find the actual distance between the tips of two adjacent teeth on a 12-in.-diameter carbide-tipped circular saw blade with 22 equally spaced teeth. Compare your answer with the length of a circular arc joining two adjacent teeth on a circle 12 in. in diameter. Round to three decimal places.

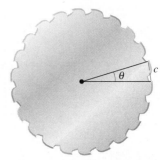

■ **Figure for Exercise 81**

82. *Throwing a Javelin* The formula

$$d = \frac{1}{32} v_0^2 \sin(2\theta)$$

gives the distance d in feet that a projectile will travel when its launch angle is θ and its initial velocity is v_0 feet per second. Approximately what initial velocity in miles per hour does it take to throw a javelin 367 feet with launch angle 43°, which is a typical launch angle (www.canthrow.com)?

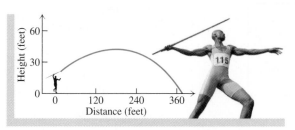

■ **Figure for Exercise 82**

Thinking Outside the Box XX

Telling Time At 12 noon the hour hand, minute hand, and second hand of a clock are all pointing straight upward.

a. Find the first time after 12 noon to the nearest tenth of a second at which the angle between the hour hand and minute hand is 120°.

b. Is there a time between 12 noon and 1 P.M. at which the three hands divide the face of the clock into thirds (that is, the angles between the hour hand, minute hand, and second hand are equal and each is 120°)?

c. Does the alignment of the three hands described in part (b) ever occur?

3.2 Pop Quiz

Find the exact value.

1. $\sin(0°)$

2. $\sin(-60°)$

3. $\sin(3\pi/4)$

4. $\cos(90°)$

5. $\cos(-2\pi/3)$

6. $\cos(11\pi/6)$

7. $\sin(\alpha)$, if $\cos(\alpha) = -3/5$ and α is in quadrant III

3.3 The Graphs of the Sine and Cosine Functions

In Section 3.2 we studied the sine and cosine functions. In this section we will study their graphs. The graphs of trigonometric functions are important for understanding their use in modeling physical phenomena such as radio, sound, and light waves, and the motion of a spring or a pendulum.

The Graph of $y = \sin(x)$

X	Y₁
0	0
.7854	.70711
1.5708	1
2.3562	.70711
3.1416	0
3.927	-.7071
4.7124	-1

X=0

■ **Figure 3.31**

Until now, s or α has been used as the independent variable in writing $\sin(s)$ or $\sin(\alpha)$. When graphing functions in an xy-coordinate system, it is customary to use x as the independent variable and y as the dependent variable, as in $y = \sin(x)$. We assume that x is a real number or radian measure unless it is stated that x is the degree measure of an angle. We often omit the parentheses and write $\sin x$ for $\sin(x)$.

Consider some ordered pairs that satisfy $y = \sin x$ for x in $[0, 2\pi]$, as shown in Fig. 3.31. These ordered pairs are best understood by recalling that $\sin x$ is the second coordinate of the terminal point on the unit circle for an arc of length x, as

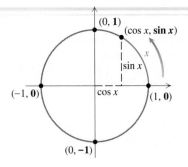

■ **Figure 3.32**

shown in Fig. 3.32. As the arc length x increases from length 0 to $\pi/2$, the second coordinate of the endpoint increases to 1. As the length of x increases from $\pi/2$ to π, the second coordinate of the endpoint decreases to 0. As the arc length x increases from π to 2π, the second coordinate decreases to -1 and then increases to 0. The calculator graph in Fig. 3.33 shows 95 accurately plotted points on the graph of $y = \sin(x)$ between 0 and 2π. The actual graph of $y = \sin(x)$ is a smooth curve through those 95 points, as shown in Fig. 3.34.

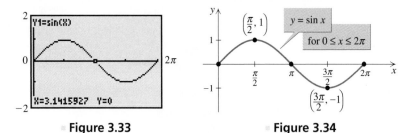

■ **Figure 3.33**   ■ **Figure 3.34**

Since the x-intercepts and the maximum and minimum values of the function occur at multiples of π, we usually label the x-axis with multiples of π, as in Fig. 3.34. There are five key points on the graph of $y = \sin x$ between 0 and 2π. Their exact coordinates are given in the following table.

x	0	$\pi/2$	π	$3\pi/2$	2π
$y = \sin x$	0	1	0	-1	0

Note that these five points divide the interval $[0, 2\pi]$ into four equal parts.

Since the domain of $y = \sin x$ is the set of all real numbers (or radian measures of angles), we must consider values of x outside the interval $[0, 2\pi]$. As x increases from 2π to 4π, the values of $\sin x$ again increase from 0 to 1, decrease to -1, then increase to 0. Because $\sin(x + 2\pi) = \sin x$, the exact shape that we saw in Fig. 3.34 is repeated for x in intervals such as $[2\pi, 4\pi]$, $[-2\pi, 0]$, $[-4\pi, -2\pi]$, and so on. So the curve shown in Fig. 3.35 continues indefinitely to the left and right. The range of $y = \sin x$ is $[-1, 1]$. The graph $y = \sin x$ or any transformation of $y = \sin x$ is called a **sine wave**, a **sinusoidal wave**, or a **sinusoid.**

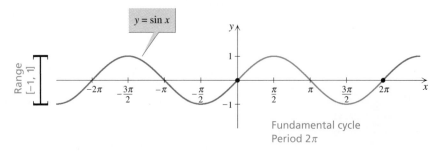

■ **Figure 3.35**

Since the shape of $y = \sin x$ for x in $[0, 2\pi]$ is repeated infinitely often, $y = \sin x$ is called a *periodic function.*

Definition:	If $y = f(x)$ is a function and a is a nonzero constant such that $f(x) = f(x + a)$
Periodic Function	for every x in the domain of f, then f is called a **periodic function**. The smallest such positive constant a is the **period** of the function.

For the sine function, the smallest value of a such that $\sin x = \sin(x + a)$ is $a = 2\pi$, and so the period of $y = \sin x$ is 2π. To prove that 2π is actually the *smallest* value of a for which $\sin x = \sin(x + a)$ is a bit complicated, and we will omit the proof. The graph of $y = \sin x$ over any interval of length 2π is called a **cycle** of the sine wave. The graph of $y = \sin x$ over $[0, 2\pi]$ is called the **fundamental cycle** of $y = \sin x$.

Example 1 Graphing a periodic function

Sketch the graph of $y = 2\sin x$ for x in the interval $[-2\pi, 2\pi]$.

Solution

We can obtain the graph of $y = 2\sin x$ from the graph of $y = \sin x$ by stretching the graph of $y = \sin x$ by a factor of 2. In other words, double the y-coordinates of the five key points on the graph of $y = \sin x$ to obtain five key points on the graph of $y = 2\sin x$ for the interval $[0, 2\pi]$, as shown in the following table.

x	0	$\pi/2$	π	$3\pi/2$	2π
$y = 2\sin x$	0	2	0	-2	0

The x-intercepts for $y = 2\sin x$ on $[0, 2\pi]$ are $(0, 0)$, $(\pi, 0)$ and $(2\pi, 0)$. Midway between the intercepts are found the highest point $(\pi/2, 2)$ and the lowest point $(3\pi/2, -2)$. Draw one cycle of $y = 2\sin x$ through these five points, as shown in Fig. 3.36. Draw another cycle of the sine wave for x in $[-2\pi, 0]$ to complete the graph of $y = 2\sin x$ for x in $[-2\pi, 2\pi]$.

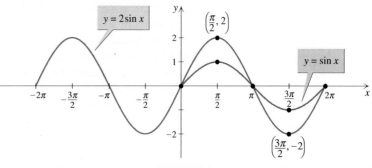

■ **Figure 3.36**

■ **Figure 3.37**

◸◹ Use $\pi/2$ as the x-scale on your calculator graph as we did in Fig. 3.36. The calculator graph in Fig. 3.37 supports our conclusions.   ■

The amplitude of a sine wave is a measure of the "height" of the wave. When an oscilloscope is used to get a picture of the sine wave corresponding to a sound, the amplitude of the sine wave corresponds to the intensity or loudness of the sound.

Definition: Amplitude

> The **amplitude** of a sine wave, or the amplitude of the function, is the absolute value of half the difference between the maximum and minimum y-coordinates on the wave.

Example **2** **Finding amplitude**

Find the amplitude of the functions $y = \sin x$ and $y = 2 \sin x$.

Solution

For $y = \sin x$, the maximum y-coordinate is 1 and the minimum is -1. So the amplitude is

$$\left| \frac{1}{2} [1 - (-1)] \right| = 1.$$

For $y = 2 \sin x$, the maximum y-coordinate is 2 and the minimum is -2. So the amplitude is

$$\left| \frac{1}{2} [2 - (-2)] \right| = 2. \qquad \blacksquare$$

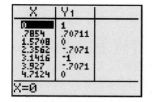

■ **Figure 3.38**

The Graph of $y = \cos(x)$

The graph of $y = \cos x$ is best understood by recalling that $\cos x$ is the first coordinate of the terminal point on the unit circle for an arc of length x, as shown in Fig. 3.32. As the arc length x increases from length 0 to $\pi/2$, the first coordinate of the endpoint decreases from 1 to 0. Some ordered pairs that satisfy $y = \cos x$ are shown in Fig. 3.38. As the length of x increases from $\pi/2$ to π, the first coordinate decreases from 0 to -1. As the length increases from π to 2π, the first coordinate increases from -1 to 1. The graph of $y = \cos x$ has exactly the same shape as the graph of $y = \sin x$. If the graph of $y = \sin x$ is shifted a distance of $\pi/2$ to the left, the graphs would coincide. For this reason the graph of $y = \cos x$ is also called a sine wave with amplitude 1 and period 2π. The graph of $y = \cos x$ over $[0, 2\pi]$ is called the **fundamental cycle** of $y = \cos x$. Since $\cos x = \cos(x + 2\pi)$, the fundamental cycle of $y = \cos x$ is repeated on $[2\pi, 4\pi]$, $[-2\pi, 0]$, and so on. The graph of $y = \cos x$ is shown in Fig. 3.39.

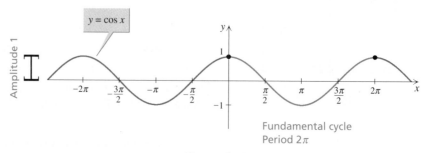

■ **Figure 3.39**

Note that there are five key points on the graph of $y = \cos x$ between 0 and 2π. These points give us the highest and lowest points in the cycle as well as the x-intercepts. The exact coordinates are given in the following table.

x	0	$\pi/2$	π	$3\pi/2$	2π
$y = \cos x$	1	0	-1	0	1

These five points divide the fundamental cycle into four equal parts.

Example 3 Graphing another periodic function

Sketch the graph of $y = -3 \cos x$ for x in the interval $[-2\pi, 2\pi]$ and find its amplitude.

Solution

Make a table of ordered pairs for x in $[0, 2\pi]$ to get one cycle of the graph. Note that the five x-coordinates in the table divide the interval $[0, 2\pi]$ into four equal parts. Multiply the y-coordinates of $y = \cos x$ by -3 to obtain the y-coordinates for $y = -3 \cos x$.

x	0	$\pi/2$	π	$3\pi/2$	2π
$y = -3 \cos x$	-3	0	3	0	-3

Draw one cycle of $y = -3 \cos x$ through these five points, as shown in Fig. 3.40. Repeat the same shape for x in the interval $[-2\pi, 0]$ to get the graph of $y = -3 \cos x$ for x in $[-2\pi, 2\pi]$. The amplitude is $|0.5(3 - (-3))|$ or 3. ■

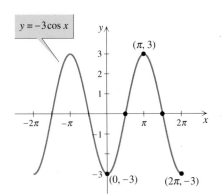

$y = -3\cos x$

Figure 3.40

Transformations of Sine and Cosine

In Section 1.7 we discussed how various changes in a formula affect the graph of the function. We know the changes that cause horizontal or vertical translations, reflections, and stretching or shrinking. The graph of $y = 2 \sin x$ from Example 1 can be obtained by stretching the graph of $y = \sin x$. The graph of $y = -3 \cos x$ in Example 3 can be obtained by stretching and reflecting the graph of $y = \cos x$. The amplitude of $y = 2 \sin x$ is 2 and the amplitude of $y = -3 \sin x$ is 3. In general, the amplitude is determined by the coefficient of the sine or cosine function.

Theorem: Amplitude

> The amplitude of $y = A \sin x$ or $y = A \cos x$ is $|A|$.

In Section 1.7 we saw that the graph of $y = f(x - C)$ is a horizontal translation of the graph of $y = f(x)$, to the right if $C > 0$ and to the left if $C < 0$. So the graphs of $y = \sin(x - C)$ and $y = \cos(x - C)$ are horizontal translations of $y = \sin x$ and $y = \cos x$, respectively, to the right if $C > 0$ and to the left if $C < 0$.

Definition: Phase Shift

> The **phase shift** of the graph of $y = \sin(x - C)$ or $y = \cos(x - C)$ is C.

Example 4 Horizontal translation

Graph two cycles of $y = \sin(x + \pi/6)$, and determine the phase shift of the graph.

Solution

Since $x + \pi/6 = x - (-\pi/6)$ and $C < 0$, the graph of $y = \sin(x + \pi/6)$ is obtained by moving $y = \sin x$ a distance of $\pi/6$ to the left. Since the phase shift is $-\pi/6$, label the x-axis with multiples of $\pi/6$, as shown in Fig. 3.41. Concentrate on moving the fundamental cycle of $y = \sin x$. The three x-intercepts $(0, 0)$, $(\pi, 0)$, and $(2\pi, 0)$ move to $(-\pi/6, 0)$, $(5\pi/6, 0)$ and $(11\pi/6, 0)$. The high and low points, $(\pi/2, 1)$ and $(3\pi/2, -1)$, move to $(\pi/3, 1)$ and $(4\pi/3, -1)$. Draw one cycle through these five points and continue the pattern for another cycle, as shown in Fig. 3.41. The second cycle could be drawn to the right or left of the first cycle.

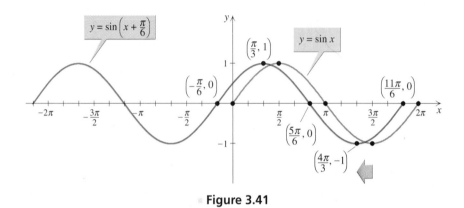

■ **Figure 3.41**

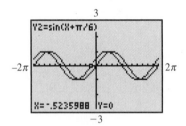

■ **Figure 3.42**

▱ Use $\pi/6$ as the x-scale on your calculator graph as we did in Fig. 3.41. The calculator graph in Fig. 3.42 supports the conclusion that $y_2 = \sin(x + \pi/6)$ has shifted $\pi/6$ to the left of $y_1 = \sin(x)$. ∎

The graphs of $y = \sin(x) + D$ and $y = \cos(x) + D$ are vertical translations of $y = \sin x$ and $y = \cos x$, respectively. The translation is upward for $D > 0$ and downward for $D < 0$. The next example combines a phase shift and a vertical translation. Note how we follow the five basic points of the fundamental cycle to see where they go in the transformation.

Example 5 Horizontal and vertical translation

Graph two cycles of $y = \cos(x - \pi/4) + 2$, and determine the phase shift of the graph.

Solution

The graph of $y = \cos(x - \pi/4) + 2$ is obtained by moving $y = \cos x$ a distance of $\pi/4$ to the right and two units upward. Since the phase shift is $\pi/4$, label the x-axis with multiples of $\pi/4$, as shown in Fig. 3.43. Concentrate on moving the fundamental cycle of $y = \cos x$. The points $(0, 1)$, $(\pi, -1)$, and $(2\pi, 1)$ move to $(\pi/4, 3)$, $(5\pi/4, 1)$, and $(9\pi/4, 3)$. The x-intercepts $(\pi/2, 0)$ and $(3\pi/2, 0)$ move to $(3\pi/4, 2)$ and $(7\pi/4, 2)$. Draw one cycle through these five points and continue the pattern for another cycle, as shown in Fig. 3.43.

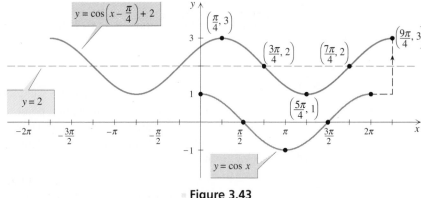

Figure 3.43

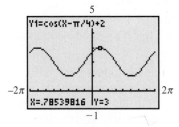

Figure 3.44

Use $\pi/4$ as the x-scale on your calculator graph as we did in Fig. 3.43. The calculator graph in Fig. 3.44 supports the conclusion that the shift is $\pi/4$ to the right and two units upward. ■

Changing the Period

We can alter the sine and cosine functions in a way that we did not alter the functions of algebra in Chapter 1. The period of a periodic function can be changed by replacing x by a multiple of x.

Example **6** Changing the period

Graph two cycles of $y = \sin(2x)$ and determine the period of the function.

Solution

The graph of $y = \sin x$ completes its fundamental cycle for $0 \le x \le 2\pi$. So $y = \sin(2x)$ completes one cycle for $0 \le 2x \le 2\pi$, or $0 \le x \le \pi$. So the period is π. For $0 \le x \le \pi$, the x-intercepts are $(0, 0)$, $(\pi/2, 0)$, and $(\pi, 0)$. Midway between 0 and $\pi/2$, at $x = \pi/4$, the function reaches a maximum value of 1, and the function attains its minimum value of -1 at $x = 3\pi/4$. Draw one cycle of the graph through the x-intercepts and through $(\pi/4, 1)$ and $(3\pi/4, -1)$. Then graph another cycle of $y = \sin(2x)$ as shown in Fig. 3.45. ■

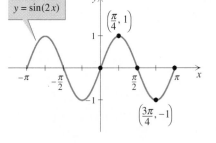

Figure 3.45

Note that in Example 6 the period of $y = \sin(2x)$ is the period of $y = \sin x$ divided by 2. In general, one complete cycle of $y = \sin(Bx)$ or $y = \cos(Bx)$ for $B > 0$ occurs for $0 \le Bx \le 2\pi$, or $0 \le x \le 2\pi/B$.

Theorem: Period of
$y = \sin(Bx)$ **and** $y = \cos(Bx)$

The period P of $y = \sin(Bx)$ and $y = \cos(Bx)$ is given by

$$P = \frac{2\pi}{B}.$$

Note that the period is a natural number (that is, not a multiple of π) when B is a multiple of π.

Example **7** A period that is not a multiple of π

Determine the period of $y = \cos\left(\frac{\pi}{2}x\right)$ and graph two cycles of the function.

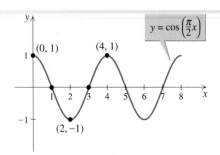

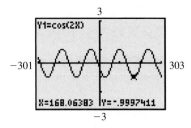

■ **Figure 3.46**

■ **Figure 3.47**

Solution

For this function, $B = \pi/2$. To find the period, use $P = 2\pi/B$:

$$P = \frac{2\pi}{\pi/2} = 4$$

So one cycle of $y = \cos\left(\frac{\pi}{2}x\right)$ is completed for $0 \le x \le 4$. The cycle starts at $(0, 1)$ and ends at $(4, 1)$. A minimum point occurs halfway in between, at $(2, -1)$. The x-intercepts are $(1, 0)$ and $(3, 0)$. Draw a curve through these five points to get one cycle of the graph. Continue this pattern from 4 to 8 to get a second cycle, as shown in Fig. 3.46.   ■

A calculator graph for a periodic function can be very misleading. For example, consider $y = \cos(2x)$, shown in Fig. 3.47. From the figure it appears that $y = \cos(2x)$ has a period of about 150 and is a left shift of $y = \cos(x)$. However, we know that the period is π and that there is no shift. What we see in Fig. 3.47 is the pattern formed by choosing 95 equally spaced points on the graph of $y = \cos(2x)$. Equally spaced points on the graph of a periodic function will usually have some kind of pattern, but the pattern may not be a good graph of the function. Because x ranges from -301 to 303 in Fig. 3.47, the spaces between these points are approximately 6 units each, which is enough for about two cycles of $y = \cos(2x)$. So the viewing window is much too large to show the relatively small features of $y = \cos(2x)$.   □

The General Sine Wave

We can use any combination of translating, reflecting, phase shifting, stretching, shrinking, or period changing in a single trigonometric function.

The General Sine Wave

> The graph of
>
> $$y = A \sin[B(x - C)] + D \qquad \text{or} \qquad y = A \cos[B(x - C)] + D$$
>
> is a sine wave with an amplitude $|A|$, period $2\pi/B$ $(B > 0)$, phase shift C, and vertical translation D.

We assume that $B > 0$, because any general sine or cosine function can be rewritten with $B > 0$ using identities from Section 3.7. Notice that A and B affect the shape of the curve, while C and D determine its location.

PROCEDURE   **Graphing a Sine Wave**

To graph $y = A \sin[B(x - C)] + D$ or $y = A \cos[B(x - C)] + D$:

1. Sketch one cycle of $y = \sin Bx$ or $y = \cos Bx$ on $[0, 2\pi/B]$.

2. Change the amplitude of the cycle according to the value of A.

3. If $A < 0$, reflect the curve in the x-axis.

4. Translate the cycle $|C|$ units to the right if $C > 0$ or to the left if $C < 0$.

5. Translate the cycle $|D|$ units upward if $D > 0$ or downward if $D < 0$.

Example **8** A transformation of $y = \sin(x)$

Determine amplitude, period, and phase shift, and sketch two cycles of $y = 2\sin(3x + \pi) + 1$.

Solution

First we rewrite the function in the form $y = A\sin[B(x - C)] + D$ by factoring 3 out of $3x + \pi$:

$$y = 2\sin\left[3\left(x + \frac{\pi}{3}\right)\right] + 1$$

From this equation we get $A = 2$, $B = 3$, and $C = -\pi/3$. So the amplitude is 2, the period is $2\pi/3$, and the phase shift is $-\pi/3$. The period change causes the fundamental cycle of $y = \sin x$ on $[0, 2\pi]$ to shrink to the interval $[0, 2\pi/3]$. Now draw one cycle of $y = \sin 3x$ on $[0, 2\pi/3]$, as shown in Fig. 3.48. Stretch the cycle vertically so that it has an amplitude of 2. The numbers $\pi/3$ and 1 shift the cycle a distance of $\pi/3$ to the left and up one unit. So one cycle of the function occurs on $[-\pi/3, \pi/3]$. Check by evaluating the function at the endpoints and midpoint of the interval $[-\pi/3, \pi/3]$ to get $(-\pi/3, 1)$, $(0, 1)$, and $(\pi/3, 1)$. Since the graph is shifted one unit upward, these points are the points where the curve intersects the line $y = 1$. Evaluate the function midway between these points to get $(-\pi/6, 3)$ and $(\pi/6, -1)$, the highest and lowest points of this cycle. One cycle is drawn through these five points and continued for another cycle, as shown in Fig. 3.48.

The calculator graph in Fig. 3.49 supports our conclusions about amplitude, period, and phase shift. Note that it is easier to obtain the amplitude, period, and phase shift from the equation than from the calculator graph.

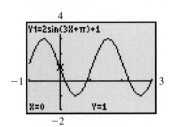

■ **Figure 3.49**

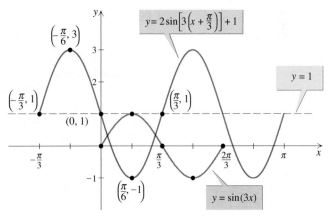

■ **Figure 3.48**   ■

Example **9** A transformation of $y = \cos(x)$

Determine amplitude, period, and phase shift, and sketch one cycle of $y = -3\cos(2x - \pi) - 1$.

Solution

Rewrite the function in the general form as

$$y = -3\cos[2(x - \pi/2)] - 1.$$

The amplitude is 3. Since the period is $2\pi/B$, the period is $2\pi/2$ or π. The fundamental cycle is shrunk to the interval $[0, \pi]$. Sketch one cycle of $y = \cos 2x$ on $[0, \pi]$, as shown in Fig. 3.50. This cycle is reflected in the x-axis and stretched by a factor of 3. A shift of $\pi/2$ to the right means that one cycle of the original function occurs for $\pi/2 \le x \le 3\pi/2$. Evaluate the function at the endpoints and midpoint of the interval $[\pi/2, 3\pi/2]$ to get $(\pi/2, -4)$, $(\pi, 2)$, and $(3\pi/2, -4)$ for the endpoints and midpoint of this cycle. Since the graph is translated downward one unit, midway between these maximum and minimum points we get points where the graph intersects the line $y = -1$. These points are $(3\pi/4, -1)$ and $(5\pi/4, -1)$. Draw one cycle of the graph through these five points, as shown in Fig. 3.50.

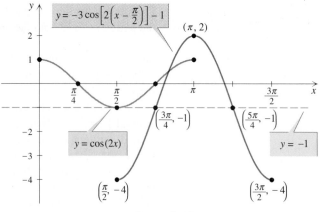

■ **Figure 3.50**

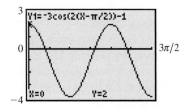

■ **Figure 3.51**

△△ The calculator graph in Fig. 3.51 supports our conclusions.   ■

Frequency

Sine waves are used to model physical phenomena such as radio, sound, or light waves. A high-frequency radio wave is a wave that has a large number of cycles per second. If we think of the x-axis as a time axis, then the period of a sine wave is the amount of time required for the wave to complete one cycle, and the reciprocal of the period is the number of cycles per unit of time. For example, the sound wave for middle C on a piano completes 262 cycles per second. The period of the wave is $1/262$ second, which means that one cycle is completed in $1/262$ second.

Definition: Frequency   The **frequency** F of a sine wave with period P is defined by $F = 1/P$.

Example **10** **Frequency of a sine wave**

Find the frequency of the sine wave given by $y = \sin(524\pi x)$.

Solution

First find the period:

$$P = \frac{2\pi}{524\pi} = \frac{1}{262} \approx 0.004$$

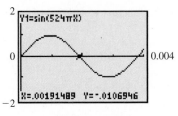

Figure 3.52

Since $F = 1/P$, the frequency is 262. A sine wave with this frequency completes 262 cycles for x in $[0, 1]$ or approximately one cycle in $[0, 0.004]$.

We cannot draw a graph showing 262 cycles in an interval of length one, but we can see the cycle that occurs in the interval $[0, 0.004]$ by looking at Fig. 3.52. ■

Note that if B is a large positive number in $y = \sin(Bx)$ or $y = \cos(Bx)$, then the period is short and the frequency is high.

Function Gallery: **The Sine and Cosine Functions**

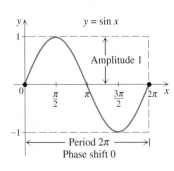

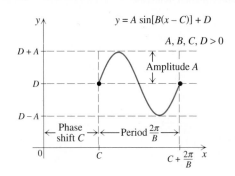

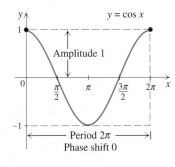

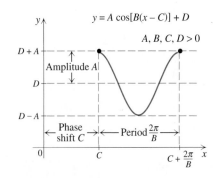

For Thought

True or False? Explain.

1. The period of $y = \cos(2\pi x)$ is π.

2. The range of $y = 4\sin(x) + 3$ is $[-4, 4]$.

3. The graph of $y = \sin(2x + \pi/6)$ has a period of π and a phase shift $\pi/6$.

4. The points $(5\pi/6, 0)$ and $(11\pi/6, 0)$ are on the graph of $y = \cos(x - \pi/3)$.

5. The frequency of the sine wave $y = \sin x$ is $1/(2\pi)$.

6. The period for $y = \sin(0.1\pi x)$ is 20.

7. The graphs of $y = \sin x$ and $y = \cos(x + \pi/2)$ are identical.

8. The period of $y = \cos(4x)$ is $\pi/2$.

9. The maximum value of the function $y = -2\cos(3x) + 4$ is 6.

10. The range of the function $y = 3\sin(5x - \pi) + 2$ is $[-1, 5]$.

3.3 Exercises

Match each graph with one of the functions $y = 3\sin(x)$, $y = 3\cos(x)$, $y = -2\sin(x)$, and $y = -2\cos(x)$. Determine the amplitude of each function.

1.

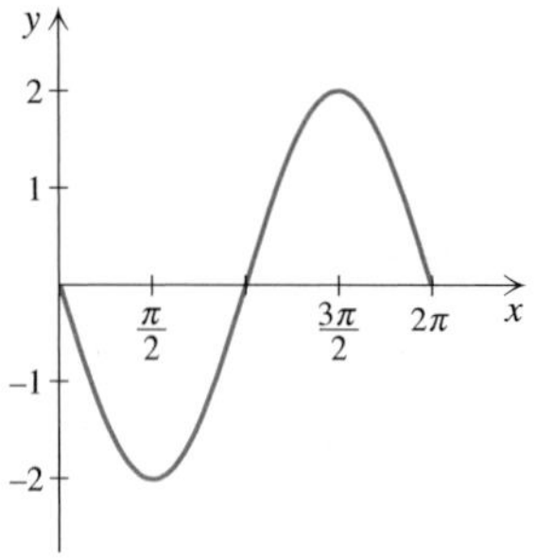

2.

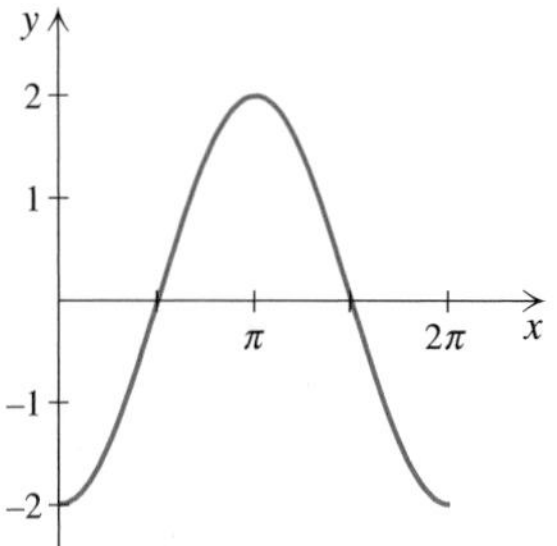

3.

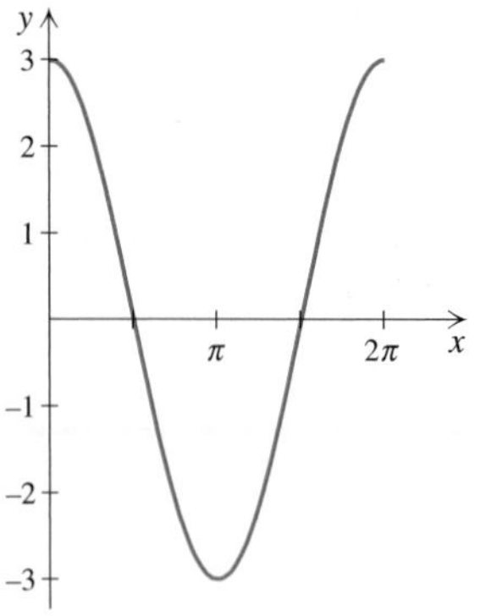

4.

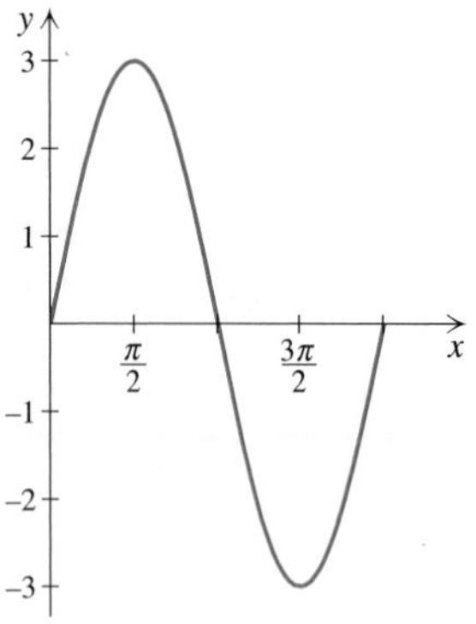

Determine the amplitude, period, and phase shift for each function.

5. $y = -2\cos x$

6. $y = -4\cos x$

7. $f(x) = \cos(x - \pi/2)$

8. $f(x) = \sin(x + \pi/2)$

9. $y = -2\sin(x + \pi/3)$

10. $y = -3\sin(x - \pi/6)$

Determine the amplitude and phase shift for each function, and sketch at least one cycle of the graph. Label five points as done in the examples.

11. $y = -\sin x$

12. $y = -\cos x$

13. $y = -3\sin x$

14. $y = 4\sin x$

15. $y = \frac{1}{2}\cos x$

16. $y = \frac{1}{3}\cos x$

17. $y = \sin(x + \pi)$

18. $y = \cos(x - \pi)$

19. $y = \cos(x - \pi/3)$

20. $y = \cos(x + \pi/4)$

21. $f(x) = \cos(x) + 2$

22. $f(x) = \cos(x) - 3$

23. $y = -\sin(x) - 1$

24. $y = -\sin(x) + 2$

25. $y = \sin(x + \pi/4) + 2$

26. $y = \sin(x - \pi/2) - 2$

27. $y = 2\cos(x + \pi/6) + 1$

28. $y = 3\cos(x + 2\pi/3) - 2$

29. $f(x) = -2\sin(x - \pi/3) + 1$

30. $f(x) = -3\cos(x + \pi/3) - 1$

Determine the amplitude, period, and phase shift for each function.

31. $y = 3\sin(4x)$

32. $y = -2\cos(3x)$

33. $y = -\cos\left(\frac{x}{2}\right) + 3$

34. $y = \sin\left(\frac{x}{3}\right) - 5$

35. $y = 2\sin(x - \pi) + 3$

36. $y = -5\cos(x + 4) + \pi$

37. $y = -2\cos\left(2x + \frac{\pi}{2}\right) - 1$

38. $y = 4\cos\left(3x - \frac{\pi}{4}\right)$

39. $y = -2 \cos\left(\dfrac{\pi}{2} x + \pi\right)$

40. $y = 8 \sin\left(\dfrac{\pi}{3} x - \dfrac{\pi}{2}\right)$

Find a function of the form $y = A \sin[B(x - C)] + D$ with the given period, phase shift, and range. Answers may vary.

41. $\pi, -\pi/2, [3, 7]$ **42.** $\pi/2, \pi, [-2, 4]$

43. $2, 2, [-1, 9]$ **44.** $4, 7, [5, 25]$

45. $\dfrac{1}{2}, -\pi, [-9, 3]$ **46.** $\dfrac{1}{3}, 2\pi, [-6, 2]$

Find the equation for each curve in its final position.

47. The graph of $y = \sin(x)$ is shifted a distance of $\pi/4$ to the right, reflected in the x-axis, then translated one unit upward.

48. The graph of $y = \cos(x)$ is shifted a distance of $\pi/6$ to the left, reflected in the x-axis, then translated two units downward.

49. The graph of $y = \cos(x)$ is stretched by a factor of 3, shifted a distance of π to the right, translated two units downward, then reflected in the x-axis.

50. The graph of $y = \sin(x)$ is shifted a distance of $\pi/2$ to the left, translated one unit upward, stretched by a factor of 4, then reflected in the x-axis.

Sketch at least one cycle of the graph of each function. Determine the period, phase shift, and range of the function. Label five points on the graph as done in the examples. See the procedure for graphing a sine wave on page 224.

51. $y = \sin(3x)$ **52.** $y = \cos(x/3)$

53. $y = -\sin(2x)$ **54.** $y = -\cos(3x)$

55. $y = \cos(4x) + 2$ **56.** $y = \sin(3x) - 1$

57. $y = 2 - \sin(x/4)$ **58.** $y = 3 - \cos(x/5)$

59. $y = \sin\left(\dfrac{\pi}{3} x\right)$ **60.** $y = \sin\left(\dfrac{\pi}{4} x\right)$

61. $f(x) = \sin\left[2\left(x - \dfrac{\pi}{2}\right)\right]$

62. $f(x) = \sin\left[3\left(x + \dfrac{\pi}{3}\right)\right]$

63. $f(x) = \sin\left(\dfrac{\pi}{2} x + \dfrac{3\pi}{2}\right)$

64. $f(x) = \cos\left(\dfrac{\pi}{3} x - \dfrac{\pi}{3}\right)$

65. $y = 2 \cos\left[2\left(x + \dfrac{\pi}{6}\right)\right] + 1$

66. $y = 3 \cos\left[4\left(x - \dfrac{\pi}{2}\right)\right] - 1$

67. $y = -\dfrac{1}{2} \sin\left[3\left(x - \dfrac{\pi}{6}\right)\right] - 1$

68. $y = -\dfrac{1}{2} \sin\left[4\left(x + \dfrac{\pi}{4}\right)\right] + 1$

Write an equation of the form $y = A \sin[B(x - C)] + D$ whose graph is the given sine wave.

69.

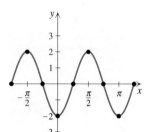

70.

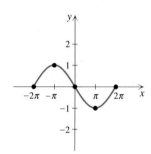

71.

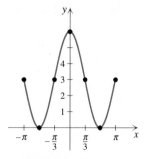

72.

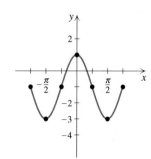

Solve each problem.

73. What is the frequency of the sine wave determined by $y = \sin(200\pi x)$, where x is time in seconds?

74. What is the frequency of the sine wave determined by $y = \cos(0.001\pi x)$, where x is time in seconds?

75. If the period of a sine wave is 0.025 hr, then what is the frequency?

76. If the frequency of a sine wave is 40,000 cycles per second, then what is the period?

77. *Motion of a Spring*   A weight hanging on a vertical spring is set in motion with a downward velocity of 6 cm/sec from its equilibrium position. Assume that the constant ω for this particular spring and weight combination is 2. Write the formula that gives the location of the weight in centimeters as a function of the time t in seconds. Find the amplitude and period of the function and sketch its graph for t in the interval $[0, 2\pi]$. (See Section 3.2 for the general formula that describes the motion of a spring.)

78. *Motion of a Spring*   A weight hanging on a vertical spring is set in motion with an upward velocity of 4 cm/sec from its equilibrium position. Assume that the constant ω for this particular spring and weight combination is π. Write the formula that gives the location of the weight in centimeters as a function of the time t in seconds. Find the period of the function and sketch its graph for t in the interval $[0, 4]$.

79. *Sun Spots*   Astronomers have been recording sunspot activity for over 130 years. The number of sunspots per year varies like a periodic function over time, as shown in the graph. What is the approximate period of this function?

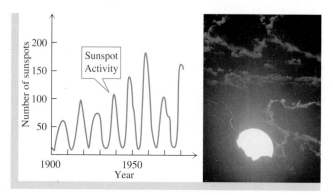

Figure for Exercise 79

80. *First Pulsar*   In 1967, Jocelyn Bell, a graduate student at Cambridge University, England, found the peculiar pattern shown in the graph on a paper chart from a radio telescope. She had made the first discovery of a pulsar, a very small neutron star that emits beams of radiation as it rotates as fast as 1000

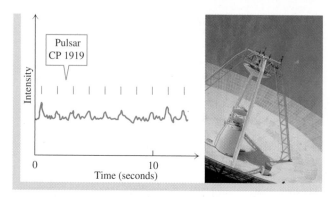

Figure for Exercise 80

times per second. From the graph shown here, estimate the period of the first discovered pulsar, now known as CP 1919.

81. *Lung Capacity*   The volume of air v in cubic centimeters in the lungs of a certain distance runner is modeled by the equation $v = 400 \sin(60\pi t) + 900$, where t is time in minutes.
   a. What are the maximum and minimum volumes of air in the runner's lungs at any time?

   b. How many breaths does the runner take per minute?

82. *Blood Velocity*   The velocity v of blood at a valve in the heart of a certain rodent is modeled by the equation $v = -4\cos(6\pi t) + 4$, where v is in centimeters per second and t is time in seconds.
   a. What are the maximum and minimum velocities of the blood at this valve?

   b. What is the rodent's heart rate in beats per minute?

83. *Periodic Revenue*   For the past three years, the manager of The Toggery Shop has observed that revenue reaches a high of about \$40,000 in December and a low of about \$10,000 in June, and that a graph of the revenue looks like a sinusoid. If the months are numbered 1 through 36 with 1 corresponding to January, then what are the period, amplitude, and phase shift for this sinusoid? What is the vertical translation? Write a formula for the curve and find the approximate revenue for April.

84. *Periodic Cost*   For the past three years, the manager of The Toggery Shop has observed that the utility bill reaches a high of about \$500 in January and a low of about \$200 in July, and the graph of the utility bill looks like a sinusoid. If the months are numbered 1 through 36 with 1 corresponding to January, then what are the period, amplitude, and phase shift for this sinusoid? What is the vertical translation? Write a formula for the curve and find the approximate utility bill for November.

85. *Ocean Waves*   Scientists use the same types of terms to describe ocean waves that we use to describe sine waves. The *wave period* is the time between crests and the *wavelength* is the distance between crests. The wave *height* is the vertical distance from the trough to the crest. The accompanying figure shows a *swell* in a coordinate system. Write an equation for the swell, assuming that its shape is that of a sinusoid.

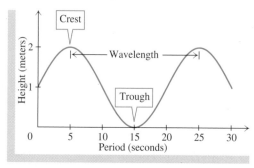

Figure for Exercise 85

86. *Large Ocean Waves*   A *tsunami* is a series of large waves caused by an earthquake. The wavelength for a tsunami can be as long as several hundred kilometers. The accompanying figure shows a tsunami in a coordinate system. Write an equation for the tsunami, assuming that its shape is that of a sinusoid.

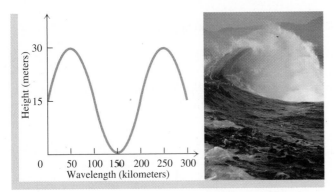

Figure for Exercise 86

Thinking Outside the Box XXI

The Survivor   There are 13 contestants on a reality television show. They are instructed to each take a seat at a circular table containing 13 chairs that are numbered consecutively with the numbers 1 through 13. The producer then starts at number 1 and tells that contestant that he is a survivor. That contestant then leaves the table. The producer skips a contestant and tells the contestant in chair number 3 that he is a survivor. That contestant then leaves the table. The producer continues around the table skipping a contestant and telling the next contestant that he is a survivor. Each survivor leaves the table. The last person left at the table is *not* a survivor and must leave the show.

a. For $n = 13$ find the unlucky number k, for which the person sitting in chair k must leave the show.

b. Find k for $n = 8, 16$, and 41.

c. Find a formula for k.

3.3   Pop Quiz

1. Determine the amplitude, period, and phase shift for $y = -5\sin(2x + 2\pi/3)$.

2. List the coordinates for the five key points for one cycle of $y = 3\sin(2x)$.

3. If $y = \cos(x)$ is shifted $\pi/2$ to the right, reflected in the x-axis, and shifted 3 units upward, then what is the equation of the curve in its final position?

4. What is the range of $f(x) = -4\sin(x - 3) + 2$?

5. What is the frequency of the sine wave determined by $y = \sin(500\pi x)$, where x is time in minutes?

3.4   The Other Trigonometric Functions and Their Graphs

So far we have studied two of the six trigonometric functions. In this section we will define the remaining four functions.

Definitions

The tangent, cotangent, secant, and cosecant functions are all defined in terms of the unit circle. We use the abbreviation tan for tangent, cot for cotangent, sec for secant, and csc for cosecant. As usual, we may think of α as an angle, the measure of an angle in degrees, the measure of an angle in radians, or a real number.

Definition: Tangent, Cotangent, Secant, and Cosecant Functions

If α is an angle in standard position and (x, y) is the point of intersection of the terminal side and the unit circle, we define the **tangent, cotangent, secant,** and **cosecant** functions as

$$\tan \alpha = \frac{y}{x}, \quad \cot \alpha = \frac{x}{y}, \quad \sec \alpha = \frac{1}{x}, \quad \text{and} \quad \csc \alpha = \frac{1}{y}.$$

We exclude from the domain of each function any value of α for which the denominator is 0.

Since $\sin \alpha = y$ and $\cos \alpha = x$, we can rewrite the definitions of tangent, cotangent, secant, and cosecant to get the following identities for these four functions.

Identities from the Definitions

If α is any angle or real number

$$\tan \alpha = \frac{\sin \alpha}{\cos \alpha}, \quad \cot \alpha = \frac{\cos \alpha}{\sin \alpha}, \quad \sec \alpha = \frac{1}{\cos \alpha}, \quad \text{and} \quad \csc \alpha = \frac{1}{\sin \alpha},$$

provided no denominator is zero.

■ **Foreshadowing Calculus**

In trigonometry we see that all of the trigonometric functions are related to each other. These relationships are important in calculus when we study rates of change of the trigonometric functions.

The domain of the tangent and secant functions is the set of angles except those for which $\cos \alpha = 0$. The only points on the unit circle where the first coordinate is 0 are $(0, 1)$ and $(0, -1)$. Angles such as $\pi/2, 3\pi/2, 5\pi/2$, and so on, have terminal sides through either $(0, 1)$ or $(0, -1)$. These angles are of the form $\pi/2 + k\pi$, where k is any integer. So the domain of tangent and secant is

$$\left\{ \alpha \,\middle|\, \alpha \neq \frac{\pi}{2} + k\pi, \text{ where } k \text{ is an integer} \right\}.$$

The only points on the unit circle where the second coordinate is 0 are $(1, 0)$ and $(-1, 0)$. Angles that are multiples of π, such as $0, \pi, 2\pi$, and so on, have terminal sides that go through one of these points. Any angle of the form $k\pi$, where k is any integer, has a terminal side that goes through either $(1, 0)$ or $(-1, 0)$. So $\sin(k\pi) = 0$ for any integer k. By definition, the zeros of the sine function are excluded from the domain of cotangent and cosecant; thus the domain of cotangent and cosecant is

$$\{ \alpha \,|\, \alpha \neq k\pi, \text{ where } k \text{ is an integer} \}.$$

To find the values of the six trigonometric functions for an angle α, first find $\sin \alpha$ and $\cos \alpha$. Then use the identities from the definitions to find values of the other four functions. Of course, $\cot \alpha$ can be found also by using $1/\tan \alpha$, but not if $\tan \alpha$ is undefined like it is for $\alpha = \pm\pi/2$. If $\tan \alpha$ is undefined, then $\cot \alpha = 0$, and if $\tan \alpha = 0$, then $\cot \alpha$ is undefined.

The signs of the six trigonometric functions for angles in each quadrant are shown in Fig. 3.53. It is not necessary to memorize these signs, because they can be easily obtained by knowing the signs of the sine and cosine functions in each quadrant.

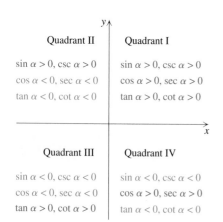

Quadrant II

$\sin \alpha > 0, \csc \alpha > 0$
$\cos \alpha < 0, \sec \alpha < 0$
$\tan \alpha < 0, \cot \alpha < 0$

Quadrant I

$\sin \alpha > 0, \csc \alpha > 0$
$\cos \alpha > 0, \sec \alpha > 0$
$\tan \alpha > 0, \cot \alpha > 0$

Quadrant III

$\sin \alpha < 0, \csc \alpha < 0$
$\cos \alpha < 0, \sec \alpha < 0$
$\tan \alpha > 0, \cot \alpha > 0$

Quadrant IV

$\sin \alpha < 0, \csc \alpha < 0$
$\cos \alpha > 0, \sec \alpha > 0$
$\tan \alpha < 0, \cot \alpha < 0$

■ **Figure 3.53**

Example **1** **Evaluating the trigonometric functions**

Find the values of all six trigonometric functions for each angle.

a. $\pi/4$ **b.** $150°$

Solution

a. Use $\sin(\pi/4) = \sqrt{2}/2$ and $\cos(\pi/4) = \sqrt{2}/2$ to find the other values:

$$\tan(\pi/4) = \frac{\sin(\pi/4)}{\cos(\pi/4)} = \frac{\sqrt{2}/2}{\sqrt{2}/2} = 1 \quad \cot(\pi/4) = \frac{\cos(\pi/4)}{\sin(\pi/4)} = 1$$

$$\sec(\pi/4) = \frac{1}{\cos(\pi/4)} = \frac{2}{\sqrt{2}} = \sqrt{2} \quad \csc(\pi/4) = \frac{1}{\sin(\pi/4)} = \frac{2}{\sqrt{2}} = \sqrt{2}$$

b. The reference angle for $150°$ is $30°$. So $\sin(150°) = \sin(30°) = 1/2$ and $\cos(150°) = -\cos(30°) = -\sqrt{3}/2$. Use these values to find the other four values:

$$\tan(150°) = \frac{1/2}{-\sqrt{3}/2} = -\frac{1}{\sqrt{3}} = -\frac{\sqrt{3}}{3} \quad \cot(150°) = -\frac{3}{\sqrt{3}} = -\sqrt{3}$$

$$\sec(150°) = \frac{1}{\cos(150°)} = -\frac{2}{\sqrt{3}} = -\frac{2\sqrt{3}}{3} \quad \csc(150°) = \frac{1}{\sin(150°)} = 2 \quad ∎$$

Most scientific calculators have keys for the sine, cosine, and tangent functions only. To find values for the other three functions, we use the definitions. Keys labeled $\sin^{-1}$, $\cos^{-1}$, and $\tan^{-1}$ on a calculator are for the inverse trigonometric functions, which we will study in the next section. These keys do *not* give the values of $1/\sin x$, $1/\cos x$, or $1/\tan x$.

Example **2** **Evaluating with a calculator**

Use a calculator to find approximate values rounded to four decimal places.

a. $\sec(\pi/12)$ **b.** $\csc(123°)$ **c.** $\cot(-12.4)$

Solution

a. $\sec(\pi/12) = \dfrac{1}{\cos(\pi/12)} \approx 1.0353$

b. $\csc(123°) = \dfrac{1}{\sin(123°)} \approx 1.1924$

c. $\cot(-12.4) = \dfrac{1}{\tan(-12.4)} \approx 5.9551$

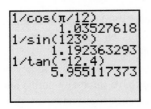

▪ **Figure 3.54**

These expressions are evaluated with a graphing calculator in radian mode as shown in Fig. 3.54. Note that in radian mode the degree symbol is used for part (b). ∎

Graph of $y = \tan(x)$

Consider some ordered pairs that satisfy $y = \tan x$ for x in $(-\pi/2, \pi/2)$. Note that $-\pi/2$ and $\pi/2$ are not in the domain of $y = \tan x$, but x can be chosen close to $\pm\pi/2$.

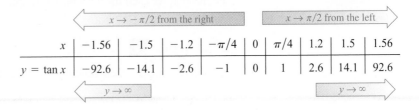

	$x \to -\pi/2$ from the right						$x \to \pi/2$ from the left		
x	-1.56	-1.5	-1.2	$-\pi/4$	0	$\pi/4$	1.2	1.5	1.56
$y = \tan x$	-92.6	-14.1	-2.6	-1	0	1	2.6	14.1	92.6
	$y \to \infty$						$y \to \infty$		

The graph of $y = \tan x$ includes the points $(-\pi/4, -1)$, $(0, 0)$, and $(\pi/4, 1)$. Since $\tan x = \sin x/\cos x$, the graph of $y = \tan x$ has a vertical asymptote for every zero of the cosine function. So the vertical lines $x = \pi/2 + k\pi$ for any integer k are the **vertical asymptotes.** The behavior of $y = \tan x$ near the asymptotes $x = \pm\pi/2$ can be seen from the table of ordered pairs. As x approaches $\pi/2$ (approximately 1.57) from the left, $\tan x \to \infty$. Using limit notation, $\lim\limits_{x \to \pi/2^-} \tan(x) = \infty$. As x approaches $-\pi/2$ from the right, $\tan x \to -\infty$ Using limit notation, $\lim\limits_{x \to -\pi/2^+} \tan(x) = -\infty$. In the interval $(-\pi/2, \pi/2)$, the function is increasing. The graph of $y = \tan x$ is shown in Fig. 3.55. The shape of the tangent curve betwen $-\pi/2$ and $\pi/2$ is repeated between each pair of consecutive asymptotes, as shown in the figure. The period of $y = \tan x$ is π, and the fundamental cycle is the portion of the graph between $-\pi/2$ and $\pi/2$. Since the range of $y = \tan x$ is $(-\infty, \infty)$, the concept of amplitude is not defined for this function.

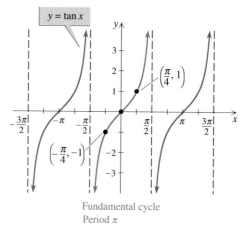

Fundamental cycle
Period π

■ **Figure 3.55**

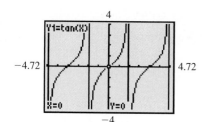

■ **Figure 3.56**

📉 The calculator graph of $y = \tan(x)$ in connected mode is shown in Fig. 3.56. In connected mode the calculator connects two points on opposite sides of each asymptote and appears to draw the vertical asymptotes. □

To help us understand the sine and cosine curve, we identified five key points on the fundamental cycle. For the tangent curve, we have three key points and the asymptotes (where the function is undefined). The exact coordinates are given in the following table.

x	$-\pi/2$	$-\pi/4$	0	$\pi/4$	$\pi/2$
$y = \tan x$	undefined	-1	0	1	undefined

We can transform the graph of the tangent function by using the same techniques that we used for the sine and cosine functions.

Example **3** **A tangent function with a transformation**

Sketch two cycles of the function $y = \tan(2x)$ and determine the period.

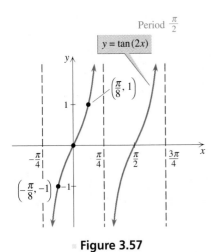

Figure 3.57

Solution

The graph of $y = \tan x$ completes one cycle for $-\pi/2 < x < \pi/2$. So the graph of $y = \tan(2x)$ completes one cycle for $-\pi/2 < 2x < \pi/2$, or $-\pi/4 < x < \pi/4$. The graph includes the points $(-\pi/8, -1)$, $(0, 0)$, and $(\pi/8, 1)$. The graph is similar to the graph of $y = \tan x$, but the asymptotes are $x = \pi/4 + k\pi/2$ for any integer k, and the period of $y = \tan(2x)$ is $\pi/2$. Two cycles of the graph are shown in Fig. 3.57. ■

By understanding what happens to the fundamental cycles of the sine, cosine, and tangent functions when the period changes, we can easily determine the location of the new function and sketch its graph quickly. We start with the fundamental cycles of $y = \sin x$, $y = \cos x$, and $y = \tan x$, which occur over the intervals $[0, 2\pi]$, $[0, 2\pi]$, and $[-\pi/2, \pi/2]$, respectively. For $y = \sin Bx$, $y = \cos Bx$, and $y = \tan Bx$ (for $B > 0$) these fundamental cycles move to the intervals $[0, 2\pi/B]$, $[0, 2\pi/B]$, and $[-\pi/(2B), \pi/(2B)]$, respectively. In every case, divide the old period by B to get the new period. Note that only the nonzero endpoints of the intervals are changed. The following Function Gallery summarizes the period change for the sine, cosine, and tangent functions with $B > 1$.

Function Gallery: Periods of Sine, Cosine, and Tangent ($B > 1$)

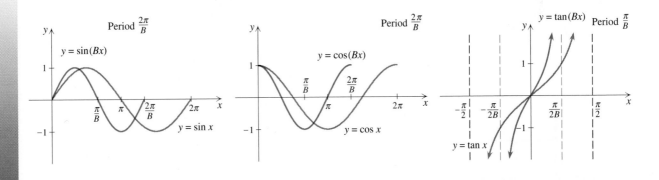

Fundamental cycles

Graph of $y = \cot(x)$

We know that $\cot x = 1/\tan x$, provided $\tan x$ is defined and not 0. So we can use $y = \tan x$ to graph $y = \cot x$. Since $\tan x = 0$ for $x = k\pi$, where k is any integer, the vertical lines $x = k\pi$ for an integer k are the vertical asymptotes of $y = \cot x$.

Consider some ordered pairs that satisfy $y = \cot x$ between the asymptotes $x = 0$ and $x = \pi$:

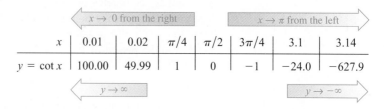

x	0.01	0.02	$\pi/4$	$\pi/2$	$3\pi/4$	3.1	3.14
$y = \cot x$	100.00	49.99	1	0	-1	-24.0	-627.9

As $x \to 0$ from the right, $\cot x \to \infty$. Using limit notation, $\lim\limits_{x \to 0^+} \cot(x) = \infty$. As $x \to \pi$ from the left, $\cot x \to -\infty$. Using limit notation, $\lim\limits_{x \to \pi^-} \cot(x) = -\infty$. In the interval $(0, \pi)$, the function is decreasing. The graph of $y = \cot x$ is shown in Fig. 3.58. The shape of the curve between 0 and π is repeated between each pair of consecutive asymptotes, as shown in the figure. The period of $y = \cot x$ is π, and the fundamental cycle is the portion of the graph between 0 and π. The range of $y = \cot x$ is $(-\infty, \infty)$.

Because $\cot x = 1/\tan x$, $\cot x$ is large when $\tan x$ is small, and vice versa. The graph of $y = \cot x$ has an x-intercept wherever $y = \tan x$ has a vertical asymptote, and a vertical asymptote wherever $y = \tan x$ has an x-intercept. So for every integer k, $(\pi/2 + k\pi, 0)$ is an x-intercept of $y = \cot x$, and the vertical line $x = k\pi$ is an asymptote, as shown in Fig. 3.58.

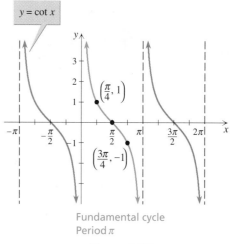

Fundamental cycle
Period π

▪ **Figure 3.58**

▪ **Figure 3.59**

◁▷ To see the graph of $y = \cot(x)$ on a graphing calculator, you can graph $y = 1/\tan(x)$, as shown in Fig. 3.59. ☐

For the cotangent curve, note the three key points and the asymptotes (where the function is undefined). The exact coordinates are given in the following table.

x	0	$\pi/4$	$\pi/2$	$3\pi/4$	π
$y = \cot x$	undefined	1	0	-1	undefined

When we graph a transformation of the cotangent function, as in the next example, we must determine what happens to these five features.

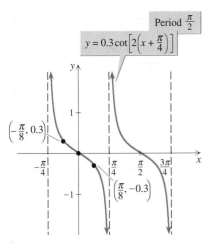

Period $\frac{\pi}{2}$

$y = 0.3 \cot\left[2\left(x + \frac{\pi}{4}\right)\right]$

$\left(-\frac{\pi}{8}, 0.3\right)$

$\left(\frac{\pi}{8}, -0.3\right)$

▪ Figure 3.60

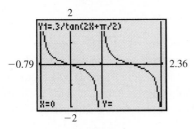

▪ Figure 3.61

Example **4** **A cotangent function with a transformation**

Sketch two cycles of the function $y = 0.3 \cot(2x + \pi/2)$, and determine the period.

Solution

Factor out 2 to write the function as

$$y = 0.3 \cot\left[2\left(x + \frac{\pi}{4}\right)\right].$$

Since $y = \cot x$ completes one cycle for $0 < x < \pi$, the graph of $y = 0.3 \cot[2(x + \pi/4)]$ completes one cycle for

$$0 < 2\left(x + \frac{\pi}{4}\right) < \pi, \quad \text{or} \quad 0 < x + \frac{\pi}{4} < \frac{\pi}{2}, \quad \text{or} \quad -\frac{\pi}{4} < x < \frac{\pi}{4}.$$

The interval $(-\pi/4, \pi/4)$ for the fundamental cycle can also be obtained by dividing the period π of $y = \cot x$ by 2 to get $\pi/2$ as the period, and then shifting the interval $(0, \pi/2)$ a distance of $\pi/4$ to the left. The factor 0.3 shrinks the y-coordinates. The graph goes through $(-\pi/8, 0.3)$, $(0, 0)$, and $(\pi/8, -0.3)$ as it approaches its vertical asymptotes $x = \pm\pi/4$. The graph for two cycles is shown in Fig. 3.60.
The calculator graph in Fig. 3.61 supports these conclusions. ▪

Since the period for $y = \cot x$ is π, the period for $y = \cot(Bx)$ with $B > 0$ is π/B and $y = \cot(Bx)$ completes one cycle on the interval $(0, \pi/B)$. The left-hand asymptote for the fundamental cycle of $y = \cot x$ remains fixed at $x = 0$, while the right-hand asymptote changes to $x = \pi/B$.

Graph of $y = \sec(x)$

Since $\sec x = 1/\cos x$, the values of $\sec x$ are large when the values of $\cos x$ are small. For any x such that $\cos x$ is 0, $\sec x$ is undefined and the graph of $y = \sec x$ has a vertical asymptote. Because of the reciprocal relationship between $\sec x$ and $\cos x$, we first draw the graph of $y = \cos x$ for reference when graphing $y = \sec x$. At every x-intercept of $y = \cos x$, we draw a vertical asymptote, as shown in Fig. 3.62. If $\cos x = \pm 1$, then $\sec x = \pm 1$. So every maximum or minimum point on the

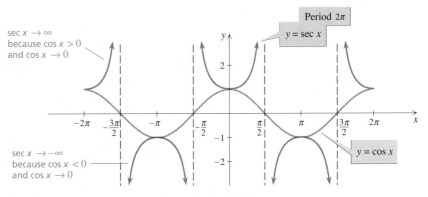

▪ Figure 3.62

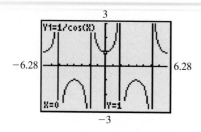

Figure 3.63

graph of $y = \cos x$ is also on the graph of $y = \sec x$. If $\cos x > 0$ and $\cos x \to 0$, then $\sec x \to \infty$. If $\cos x < 0$ and $\cos x \to 0$, then $\sec x \to -\infty$. These two facts cause the graph of $y = \sec x$ to approach its asymptotes in the manner shown in Fig. 3.62. The period of $y = \sec x$ is 2π, the same as the period for $y = \cos x$. The range of $y = \sec x$ is $(-\infty, -1] \cup [1, \infty)$.

The calculator graph in Fig. 3.63 supports these conclusions.   □

Example **5** **A secant function with a transformation**

Sketch two cycles of the function $y = 2 \sec(x - \pi/2)$, and determine the period and the range of the function.

Solution

Since

$$y = 2 \sec\left(x - \frac{\pi}{2}\right) = \frac{2}{\cos(x - \pi/2)},$$

we first graph $y = \cos(x - \pi/2)$, as shown in Fig. 3.64. The function $y = \sec(x - \pi/2)$ goes through the maximum and minimum points on the graph of $y = \cos(x - \pi/2)$, but $y = 2 \sec(x - \pi/2)$ stretches $y = \sec(x - \pi/2)$ by a factor of 2. So the portions of the curve that open up do not go lower than 2, and the portions that open down do not go higher than -2, as shown in the figure. The period is 2π, and the range is $(-\infty, -2] \cup [2, \infty)$.

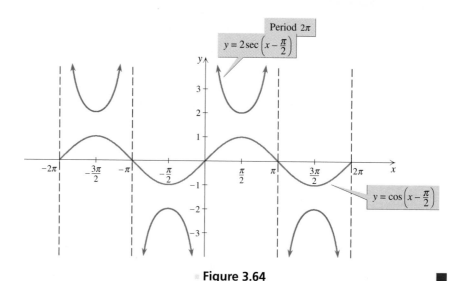

Figure 3.64   ■

Graph of $y = \csc(x)$

Since $\csc x = 1/\sin x$, the graph of $y = \csc x$ is related to the graph of $y = \sin x$ in the same way that the graphs of $y = \sec x$ and $y = \cos x$ are related. To graph $y = \csc x$, first draw the graph of $y = \sin x$ and a vertical asymptote at each x-intercept. Since the graph of $y = \sin x$ can be obtained by shifting $y = \cos x$ a distance of $\pi/2$ to the right, the graph of $y = \csc x$ is obtained from $y = \sec x$ by shifting a distance of $\pi/2$ to the right, as shown in Fig. 3.65. The period of $y = \csc x$ is 2π, the same as the period for $y = \sin x$.

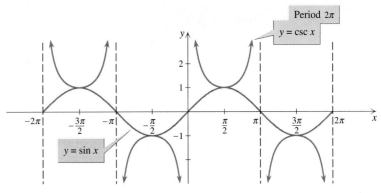

■ **Figure 3.65**

Example **6** A cosecant function with a transformation

Sketch two cycles of the graph of $y = \csc(2x - 2\pi/3)$ and determine the period and the range of the function.

Solution

Since

$$y = \csc(2x - 2\pi/3)$$

$$= \frac{1}{\sin[2(x - \pi/3)]}$$

we first graph $y = \sin[2(x - \pi/3)]$. The period for $y = \sin[2(x - \pi/3)]$ is π with phase shift of $\pi/3$. So the fundamental cycle of $y = \sin x$ is transformed to occur on the interval $[\pi/3, 4\pi/3]$. Draw at least two cycles of $y = \sin[2(x - \pi/3)]$ with a vertical asymptote (for the cosecant function) at every x-intercept, as shown in Fig. 3.66.

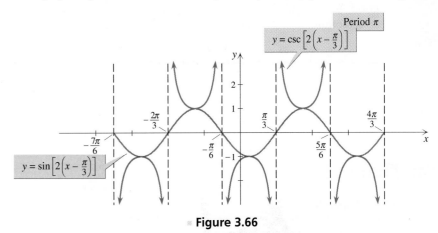

■ **Figure 3.66**

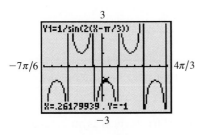

■ **Figure 3.67**

Each portion of $y = \csc[2(x - \pi/3)]$ that opens up has a minimum value of 1, and each portion that opens down has a maximum value of -1. The period of the function is π, and the range is $(-\infty, -1] \cup [1, \infty)$.

The calculator graph in Fig. 3.67 supports these conclusions. ■

The Function Gallery summarizes some of the facts that we have learned about the six trigonometric functions. Also, one cycle of the graph of each trigonometric function is shown.

Function Gallery: **Trigonometric Functions**

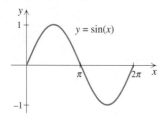

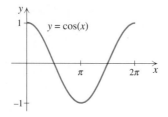

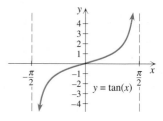

Domain (*k* any integer)	$(-\infty, \infty)$	$(-\infty, \infty)$	$x \neq \dfrac{\pi}{2} + k\pi$
Range	$[-1, 1]$	$[-1, 1]$	$(-\infty, \infty)$
Period	2π	2π	π
Fundamental cycle	$[0, 2\pi]$	$[0, 2\pi]$	$\left[-\dfrac{\pi}{2}, \dfrac{\pi}{2}\right]$

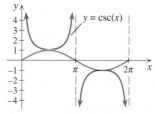

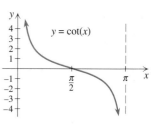

Domain (*k* any integer)	$x \neq k\pi$	$x \neq \dfrac{\pi}{2} + k\pi$	$x \neq k\pi$
Range	$(-\infty, -1] \cup [1, \infty)$	$(-\infty, -1] \cup [1, \infty)$	$(-\infty, \infty)$
Period	2π	2π	π
Fundamental cycle	$[0, 2\pi]$	$[0, 2\pi]$	$[0, \pi]$

For Thought

True or False? Explain.

1. $\sec(\pi/4) = 1/\sin(\pi/4)$ **2.** $\cot(\pi/2) = 1/\tan(\pi/2)$

3. $\csc(60°) = 2\sqrt{3}/3$ **4.** $\tan(5\pi/2) = 0$

5. $\sec(95°) = \sqrt{5}$ **6.** $\csc(120°) = 2/\sqrt{3}$

7. The graphs of $y = 2\csc x$ and $y = 1/(2\sin x)$ are identical.

8. The range of $y = 0.5\csc(13x - 5\pi)$ is $(-\infty, -0.5] \cup [0.5, \infty)$.

9. The graph of $y = \tan(3x)$ has vertical asymptotes at $x = \pm\pi/6$.

10. The graph of $y = \cot(4x)$ has vertical asymptotes at $x = \pm\pi/4$.

3.4 Exercises

Redraw each unit circle and label each indicated point with the proper value of the tangent function. The points in Exercise 1 are the terminal points for arcs with lengths that are multiples of π/4. The points in Exercise 2 are the terminal points for arcs with lengths that are multiples of π/6. Repeat Exercises 1 and 2 until you can do them from memory.

1.

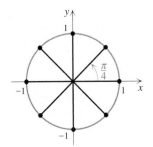

2.

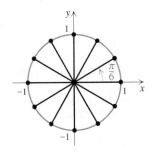

Find the exact value of each of the following expressions without using a calculator.

3. $\tan(\pi/3)$

4. $\tan(\pi/4)$

5. $\tan(-\pi/4)$

6. $\tan(\pi/6)$

7. $\cot(\pi/2)$

8. $\cot(2\pi/3)$

9. $\cot(-\pi/3)$

10. $\cot(0)$

11. $\sec(\pi/6)$

12. $\sec(\pi/3)$

13. $\sec(\pi/2)$

14. $\sec(\pi)$

15. $\csc(-\pi)$

16. $\csc(\pi/6)$

17. $\csc(3\pi/4)$

18. $\csc(-\pi/3)$

19. $\tan(135°)$

20. $\tan(270°)$

21. $\cot(210°)$

22. $\cot(120°)$

23. $\sec(-120°)$

24. $\sec(-90°)$

25. $\csc(315°)$

26. $\csc(240°)$

27. $\cot(-90°)$

28. $\cot(270°)$

Find the approximate value of each expression. Round to four decimal places.

29. $\tan(1.55)$

30. $\tan(1.6)$

31. $\cot(-3.48)$

32. $\cot(22.4)$

33. $\csc(0.002)$

34. $\csc(1.54)$

35. $\sec(\pi/12)$

36. $\sec(-\pi/8)$

37. $\cot(0.09°)$

38. $\cot(179.4°)$

39. $\csc(-44.3°)$

40. $\csc(-124.5°)$

41. $\sec(89.2°)$

42. $\sec(-0.024°)$

43. $\tan(-44.6°)$

44. $\tan(138°)$

Find the exact value of each expression for the given value of θ. Do not use a calculator.

45. $\sec^2(2\theta)$ if $\theta = \pi/6$

46. $\csc^2(2\theta)$ if $\theta = \pi/8$

47. $\tan(\theta/2)$ if $\theta = \pi/3$

48. $\csc(\theta/2)$ if $\theta = \pi/2$

49. $\sec(\theta/2)$ if $\theta = 3\pi/2$

50. $\cot(\theta/2)$ if $\theta = 2\pi/3$

Determine the period and sketch at least one cycle of the graph of each function.

51. $y = \tan(3x)$

52. $y = \tan(4x)$

53. $y = \cot(x + \pi/4)$

54. $y = \cot(x - \pi/6)$

55. $y = \cot(x/2)$

56. $y = \cot(x/3)$

57. $y = \tan(\pi x)$

58. $y = \tan(\pi x/2)$

59. $y = -2\tan x$

60. $y = -\tan(x - \pi/2)$

61. $y = -\cot(x + \pi/2)$

62. $y = 2 + \cot x$

63. $y = \cot(2x - \pi/2)$

64. $y = \cot(3x + \pi)$

65. $y = \tan\left(\dfrac{\pi}{2}x - \dfrac{\pi}{2}\right)$

66. $y = \tan\left(\dfrac{\pi}{4}x + \dfrac{3\pi}{4}\right)$

Determine the period and sketch at least one cycle of the graph of each function. State the range of each function.

67. $y = \sec(2x)$

68. $y = \sec(3x)$

69. $y = \csc(x - \pi/2)$

70. $y = \csc(x + \pi/4)$

71. $y = \csc(x/2)$

72. $y = \csc(x/4)$

73. $y = \sec(\pi x/2)$

74. $y = \sec(\pi x)$

75. $y = 2\sec x$

76. $y = \dfrac{1}{2}\sec x$

77. $y = \csc(2x - \pi/2)$ **78.** $y = \csc(3x + \pi)$

79. $y = -\csc\left(\dfrac{\pi}{2}x + \dfrac{\pi}{2}\right)$ **80.** $y = -2\csc(\pi x - \pi)$

81. $y = 2 + 2\sec(2x)$ **82.** $y = 2 - 2\sec\left(\dfrac{x}{2}\right)$

Determine the period and range of each function.

83. $y = \tan(2x - \pi) + 3$ **84.** $y = 2\cot(3x + \pi) - 8$

85. $y = 2\sec(x/2 - 1) - 1$ **86.** $y = -2\sec(x/3 - 6) + 3$

87. $y = -3\csc(2x - \pi) - 4$ **88.** $y = 4\csc(3x - \pi) + 5$

Write the equation of each curve in its final position.

89. The graph of $y = \tan(x)$ is shifted $\pi/4$ units to the right, stretched by a factor of 3, then translated 2 units upward.

90. The graph of $y = \cot(x)$ is shifted $\pi/2$ units to the left, reflected in the x-axis, then translated 1 unit upward.

91. The graph of $y = \sec(x)$ is shifted π units to the left, reflected in the x-axis, then shifted 2 units upward.

92. The graph of $y = \csc(x)$ is shifted 2 units to the right, translated 3 units downward, then reflected in the x-axis.

Thinking Outside the Box XXII

Counting Votes   Fifteen experts are voting to determine the best convertible of the year. The choices are a Porsche Carrera, a Chrysler Crossfire, and a Nissan Roadster. The experts will rank the three cars 1st, 2nd, and 3rd. There are three common ways to determine the winner.

1. *Plurality:* The car with the most first place votes (preferences) is the winner.

2. *Instant runoff:* The car with the least number of preferences is eliminated. Then the ballots where the eliminated car is first are revised so that the second place car is moved to first. Finally, the car with the most preferences is the winner.

3. *The point system:* Two points are given for each time a car occurs in first place on a ballot, one point for each time the car appears in second place on a ballot, and no points for third place.

When the ballots were cast, the Porsche won when plurality was used, the Chrysler won when instant runoff was used, and the Nissan won when the point system was used. Determine 15 actual votes for which this result would occur.

3.4 Pop Quiz

1. What is the period for $y = \tan(3x)$?

2. Find the equations of all asymptotes for $y = \cot(2x)$.

3. Find the equations of all asymptotes for $y = \sec(2x)$.

4. What is the range of $y = 3\csc(2x)$?

Find the exact value.

5. $\tan(\pi/4)$ **6.** $\tan(120°)$

7. $\cot(-\pi/3)$ **8.** $\sec(60°)$

9. $\csc(-3\pi/4)$

3.5 The Inverse Trigonometric Functions

We have learned how to find the values of the trigonometric functions for angles or real numbers, but to make the trigonometric functions really useful we must be able to reverse this process. In this section we define the inverses of the trigonometric functions.

The Inverse of the Sine Function

In Chapter 1 we learned that only one-to-one functions are invertible. Since $y = \sin x$ with domain $(-\infty, \infty)$ is a periodic function, it is certainly not one-to-one.

However, if we restrict the domain to the interval $[-\pi/2, \pi/2]$, then the restricted function is one-to-one and invertible. Other intervals could be used, but this interval is chosen to keep the inverse function as simple as possible.

The graph of the sine function with domain $[-\pi/2, \pi/2]$ is shown in Fig. 3.68(a). Its range is $[-1, 1]$. We now define the inverse sine function and denote it as $f^{-1}(x) = \sin^{-1}x$ (read "inverse sine of x") or $f^{-1}(x) = \arcsin x$ (read "arc sine of x").

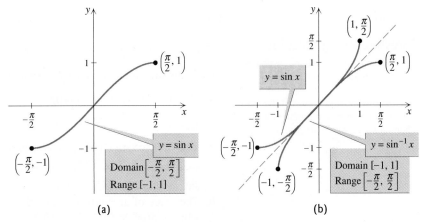

■ **Figure 3.68**

Definition: The Inverse Sine Function

> The function $y = \sin^{-1}x$ or $y = \arcsin x$ is the inverse of the function $y = \sin x$ restricted to $[-\pi/2, \pi/2]$. The domain of $y = \sin^{-1}x$ is $[-1, 1]$ and its range is $[-\pi/2, \pi/2]$.

The graph of $y = \sin^{-1}x$ is a reflection about the line $y = x$ of the graph of $y = \sin x$ on $[-\pi/2, \pi/2]$, as shown in Fig. 3.68(b).

If $y = \sin^{-1}x$, then y is the real number such that $-\pi/2 \le y \le \pi/2$ and $\sin y = x$. Depending on the context, $\sin^{-1}x$ might also be an angle, a measure of an angle in degrees or radians, or the length of an arc of the unit circle. The expression $\sin^{-1}x$ can be read as "the angle whose sine is x" or "the arc length whose sine is x." The notation $y = \arcsin x$ reminds us that y is the arc length whose sine is x. For example, $\arcsin(1)$ is the arc length in $[-\pi/2, \pi/2]$ whose sine is 1. Since we know that $\sin(\pi/2) = 1$, we have $\arcsin(1) = \pi/2$. We will assume that $\sin^{-1}x$ is a real number unless indicated otherwise.

Note that the nth power of the sine function is usually written as $\sin^n(x)$ as a shorthand notation for $(\sin x)^n$, provided $n \ne -1$. The -1 used in $\sin^{-1}x$ indicates the inverse function and does *not* mean reciprocal. To write $1/\sin x$ using exponents, we must write $(\sin x)^{-1}$.

Example **1** **Evaluating the inverse sine function**

Find the exact value of each expression without using a table or a calculator.

a. $\sin^{-1}(1/2)$ **b.** $\arcsin\left(-\sqrt{3}/2\right)$

Solution

a. The value of $\sin^{-1}(1/2)$ is the number α in the interval $[-\pi/2, \pi/2]$ such that $\sin(\alpha) = 1/2$. We recall that $\sin(\pi/6) = 1/2$, and so $\sin^{-1}(1/2) = \pi/6$. Note that $\pi/6$ is the only value of α in $[-\pi/2, \pi/2]$ for which $\sin(\alpha) = 1/2$.

b. The value of $\arcsin(-\sqrt{3}/2)$ is the number α in $[-\pi/2, \pi/2]$ such that $\sin(\alpha) = -\sqrt{3}/2$. Since $\sin(-\pi/3) = -\sqrt{3}/2$, we have $\arcsin(-\sqrt{3}/2) = -\pi/3$. Note that $-\pi/3$ is the only value of α in $[-\pi/2, \pi/2]$ for which $\sin(\alpha) = -\sqrt{3}/2$. ■

■ **Foreshadowing Calculus**

In calculus we discover some interesting relationships between functions. The rates of change of the trigonometric functions are other trigonometric functions, but the rates of change of the inverse trigonometric functions are algebraic functions.

Example **2** **Evaluating the inverse sine function**

Find the exact value of each expression in degrees without using a table or a calculator.

a. $\sin^{-1}\left(\sqrt{2}/2\right)$ **b.** $\arcsin(0)$

Solution

a. The value of $\sin^{-1}\left(\sqrt{2}/2\right)$ in degrees is the angle α in the interval $[-90°, 90°]$ such that $\sin(\alpha) = \sqrt{2}/2$. We recall that $\sin(45°) = \sqrt{2}/2$, and so $\sin^{-1}\left(\sqrt{2}/2\right) = 45°$.

b. The value of $\arcsin(0)$ in degrees is the angle α in the interval $[-90°, 90°]$ for which $\sin(\alpha) = 0$. Since $\sin(0°) = 0$, we have $\arcsin(0) = 0°$. ■

In the next example, we use a calculator to find the degree measure of an angle whose sine is given. To obtain degree measure, make sure the calculator is in degree mode. Scientific calculators usually have a key labeled $\sin^{-1}$ that gives values for the inverse sine function.

Example **3** **Finding an angle given its sine**

Let α be an angle such that $-90° < \alpha < 90°$. In each case, find α to the nearest tenth of a degree.

a. $\sin\alpha = 0.88$ **b.** $\sin\alpha = -0.27$

Solution

a. The value of $\sin^{-1}(0.88)$ is the only angle in $[-90°, 90°]$ with a sine of 0.88. Use a calculator in degree mode to get $\alpha = \sin^{-1}(0.88) \approx 61.6°$.

b. Use a calculator to get $\alpha = \sin^{-1}(-0.27) \approx -15.7°$.

Figure 3.69 shows how to find the angle in part (a) on a graphing calculator and how to check. Make sure that the mode is degrees. ■

```
sin⁻¹(.88)
        61.64236342
sin(Ans)
              .88
```

■ **Figure 3.69**

The Inverse Cosine Function

Since the cosine function is not one-to-one on $(-\infty, \infty)$, we restrict the domain to $[0, \pi]$, where the cosine function is one-to-one and invertible. The graph of the cosine function with this restricted domain is shown in Fig. 3.70(a). Note that the range of the restricted function is $[-1, 1]$. We now define the inverse of $f(x) = \cos x$ for x in $[0, \pi]$ and denote it as $f^{-1}(x) = \cos^{-1} x$ or $f^{-1}(x) = \arccos x$.

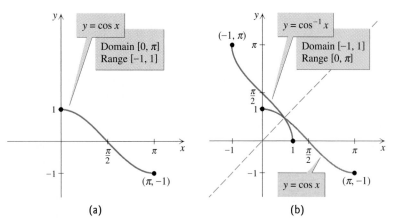

(a)   (b)

■ **Figure 3.70**

Definition: The Inverse Cosine Function

The function $y = \cos^{-1} x$ or $y = \arccos x$ is the inverse of the function $y = \cos x$ restricted to $[0, \pi]$. The domain of $y = \cos^{-1} x$ is $[-1, 1]$ and its range is $[0, \pi]$.

If $y = \cos^{-1} x$, then y is the real number in $[0, \pi]$ such that $\cos y = x$. The expression $\cos^{-1} x$ can be read as "the angle whose cosine is x" or "the arc length whose cosine is x." The graph of $y = \cos^{-1} x$, shown in Fig. 3.70(b), is obtained by reflecting the graph of $y = \cos x$ (restricted to $[0, \pi]$) about the line $y = x$. We will assume that $\cos^{-1} x$ is a real number unless indicated otherwise.

Example **4** **Evaluating the inverse cosine function**

Find the exact value of each expression without using a table or a calculator.

a. $\cos^{-1}(-1)$ **b.** $\arccos(-1/2)$ **c.** $\cos^{-1}\left(\sqrt{2}/2\right)$

Solution

a. The value of $\cos^{-1}(-1)$ is the number α in $[0, \pi]$ such that $\cos(\alpha) = -1$. We recall that $\cos(\pi) = -1$, and so $\cos^{-1}(-1) = \pi$.
b. The value of $\arccos(-1/2)$ is the number α in $[0, \pi]$ such that $\cos(\alpha) = -1/2$. We recall that $\cos(2\pi/3) = -1/2$, and so $\arccos(-1/2) = 2\pi/3$.
c. Since $\cos(\pi/4) = \sqrt{2}/2$, we have $\cos^{-1}\left(\sqrt{2}/2\right) = \pi/4$. ■

In the next example, we use a calculator to find the degree measure of an angle whose cosine is given. Most scientific calculators have a key labeled $\cos^{-1}$ that gives values for the inverse cosine function. To get the degree measure, make sure the calculator is in degree mode.

Example **5** **Finding an angle given its cosine**

In each case find the angle α to the nearest tenth of a degree, given that $0° < \alpha < 180°$.

a. $\cos \alpha = 0.23$ **b.** $\cos \alpha = -0.82$

Solution

a. Since $\cos^{-1}(0.23)$ is the unique angle in $[0°, 180°]$ with a cosine of 0.23, $\alpha = \cos^{-1}(0.23) \approx 76.7°$.

b. Use a calculator to get $\alpha = \cos^{-1}(-0.82) \approx 145.1°$.

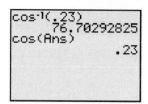

■ **Figure 3.71**

Figure 3.71 shows how to find $\cos^{-1}(0.23)$ on a graphing calculator and how to check. Make sure that the mode is degrees. ■

Inverses of Tangent, Cotangent, Secant, and Cosecant

Since all of the trigonometric functions are periodic, they must all be restricted to a domain where they are one-to-one before the inverse functions can be defined. There is more than one way to choose a domain to get a one-to-one function, but we will use the most common restrictions. The restricted domain for $y = \tan x$ is $(-\pi/2, \pi/2)$, for $y = \csc x$ it is $[-\pi/2, 0) \cup (0, \pi/2]$, for $y = \sec x$ it is $[0, \pi/2) \cup (\pi/2, \pi]$, and for $y = \cot x$ it is $(0, \pi)$. The functions $\tan^{-1}$, $\cot^{-1}$, $\sec^{-1}$, and $\csc^{-1}$ are defined to be the inverses of these restricted functions. The notations arctan, arccot, arcsec, and arccsc are also used for these inverse functions. The graphs of all six inverse functions, with their domains and ranges, are shown in the accompanying Function Gallery.

When studying inverse trigonometric functions, you should first learn to evaluate $\sin^{-1}$, $\cos^{-1}$, and $\tan^{-1}$. Then use the identities

$$\csc \alpha = \frac{1}{\sin \alpha}, \qquad \sec \alpha = \frac{1}{\cos \alpha}, \qquad \text{and} \qquad \cot \alpha = \frac{1}{\tan \alpha}$$

to evaluate $\csc^{-1}$, $\sec^{-1}$, and $\cot^{-1}$. For example, $\sin(\pi/6) = 1/2$ and $\csc(\pi/6) = 2$. So the angle whose cosecant is 2 is the same as the angle whose sine is $1/2$. In symbols,

$$\csc^{-1}(2) = \sin^{-1}\left(\frac{1}{2}\right) = \frac{\pi}{6}.$$

In general, $\csc^{-1} x = \sin^{-1}(1/x)$. Likewise, $\sec^{-1} x = \cos^{-1}(1/x)$. For the inverse cotangent, $\cot^{-1} x = \tan^{-1}(1/x)$ only for $x > 0$, because of the choice of $(0, \pi)$ as the range of the inverse cotangent. We have $\cot^{-1}(0) = \pi/2$ and, if $x < 0$, $\cot^{-1}(x) = \tan^{-1}(1/x) + \pi$.

Function Gallery: **Inverse Trigonometric Functions**

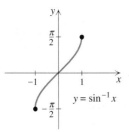

Domain [−1, 1]
Range $\left[-\frac{\pi}{2}, \frac{\pi}{2}\right]$

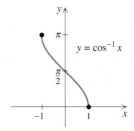

Domain [−1, 1]
Range [0, π]

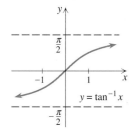

Domain (−∞, ∞)
Range $\left(-\frac{\pi}{2}, \frac{\pi}{2}\right)$

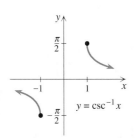

Domain (−∞, −1] ∪ [1, ∞)
Range $\left[-\frac{\pi}{2}, 0\right) \cup \left(0, \frac{\pi}{2}\right]$

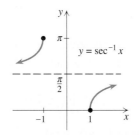

Domain (−∞, −1] ∪ [1, ∞)
Range $\left[0, \frac{\pi}{2}\right) \cup \left(\frac{\pi}{2}, \pi\right]$

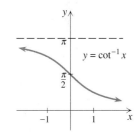

Domain (−∞, ∞)
Range (0, π]

We can see another relationship between $\cot^{-1}$ and $\tan^{-1}$ from their graphs in the Function Gallery. Notice that the graph of $y = \cot^{-1}(x)$ can be obtained by reflecting the graph of $y = \tan^{-1}(x)$ about the x-axis and then translating $\pi/2$ units upward. So $\cot^{-1}(x) = -\tan^{-1}(x) + \pi/2$ or $\cot^{-1}(x) = \pi/2 - \tan^{-1}(x)$.

Identities for the Inverse Functions

$$\csc^{-1}(x) = \sin^{-1}(1/x) \text{ for } |x| \geq 1$$

$$\sec^{-1}(x) = \cos^{-1}(1/x) \text{ for } |x| \geq 1$$

$$\cot^{-1}(x) = \begin{cases} \tan^{-1}(1/x) & \text{for } x > 0 \\ \tan^{-1}(1/x) + \pi & \text{for } x < 0 \\ \pi/2 & \text{for } x = 0 \end{cases}$$

$$\cot^{-1}(x) = \pi/2 - \tan^{-1}(x)$$

Example **6** **Evaluating the inverse functions**

Find the exact value of each expression without using a table or a calculator.

a. $\tan^{-1}(1)$ **b.** $\text{arcsec}(-2)$ **c.** $\csc^{-1}(\sqrt{2})$ **d.** $\text{arccot}(-1/\sqrt{3})$

Solution

a. Since $\tan(\pi/4) = 1$ and since $\pi/4$ is in the range of $\tan^{-1}$, we have

$$\tan^{-1}(1) = \frac{\pi}{4}.$$

b. To evaluate the inverse secant, we use the identity $\sec^{-1}(x) = \cos^{-1}(1/x)$. In this case, the arc whose secant is -2 is the same as the arc whose cosine is $-1/2$. So we must find $\cos^{-1}(-1/2)$. Since $\cos(2\pi/3) = -1/2$ and since $2\pi/3$ is in the range of arccos, we have

$$\text{arcsec}(-2) = \arccos\left(-\frac{1}{2}\right) = \frac{2\pi}{3}.$$

c. To evaluate $\csc^{-1}(\sqrt{2})$, we use the identity $\csc^{-1}(x) = \sin^{-1}(1/x)$ with $x = \sqrt{2}$. So we must find $\sin^{-1}(1/\sqrt{2})$. Since $\sin(\pi/4) = 1/\sqrt{2}$ and since $\pi/4$ is in the range of $\csc^{-1}$,

$$\csc^{-1}(\sqrt{2}) = \sin^{-1}\left(\frac{1}{\sqrt{2}}\right) = \frac{\pi}{4}.$$

d. If x is negative, we use the identity $\cot^{-1}(x) = \tan^{-1}(1/x) + \pi$. Since $x = -1/\sqrt{3}$, we must find $\tan^{-1}(-\sqrt{3})$. Since $\tan^{-1}(-\sqrt{3}) = -\pi/3$, we get

$$\text{arccot}\left(\frac{-1}{\sqrt{3}}\right) = \tan^{-1}(-\sqrt{3}) + \pi = -\frac{\pi}{3} + \pi = \frac{2\pi}{3}.$$

Note that $\cot(-\pi/3) = \cot(2\pi/3) = -1/\sqrt{3}$, but $\text{arccot}(-1/\sqrt{3}) = 2\pi/3$ because $2\pi/3$ is in the range of the function arccot. ■

The functions $\sin^{-1}$, $\cos^{-1}$, and $\tan^{-1}$ are available on scientific and graphing calculators. The calculator values of the inverse functions are given in degrees or radians, depending on the mode setting. We will assume that the values of the inverse functions are to be in radians unless indicated otherwise. Calculators use the domains and ranges of these functions defined here. The functions $\sec^{-1}$, $\csc^{-1}$, and $\cot^{-1}$ are generally not available on a calculator, and expressions involving these functions must be written in terms of $\sin^{-1}$, $\cos^{-1}$, and $\tan^{-1}$ by using the identities.

Example **7** **Evaluating the inverse functions with a calculator**

Find the approximate value of each expression rounded to four decimal places.

a. $\sin^{-1}(0.88)$ **b.** $\text{arccot}(2.4)$ **c.** $\csc^{-1}(4)$ **d.** $\cot^{-1}(0)$

Solution

a. Use the inverse sine function in radian mode to get $\sin^{-1}(0.88) \approx 1.0759$.
b. Use the identity $\cot^{-1}(x) = \tan^{-1}(1/x)$ because $2.4 > 0$:

$$\text{arccot}(2.4) = \tan^{-1}\left(\frac{1}{2.4}\right) \approx 0.3948$$

c. $\csc^{-1}(4) = \sin^{-1}(0.25) \approx 0.2527$

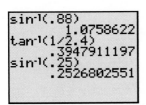

■ Figure 3.72

d. The value of $\cot^{-1}(0)$ cannot be found on a calculator. But we know that $\cot(\pi/2) = 0$ and $\pi/2$ is in the range $(0, \pi)$ for $\cot^{-1}$, so we have $\cot^{-1}(0) = \pi/2$.

The expressions in parts (a), (b), and (c) are shown on a graphing calculator in Fig. 3.72. ■

Compositions of Functions

One trigonometric function can be followed by another to form a composition of functions. We can evaluate an expression such as $\sin(\sin(\alpha))$ because the sine of the real number $\sin(\alpha)$ is defined. However, it is more common to have a composition of a trigonometric function and an inverse trigonometric function. For example, $\tan^{-1}(\alpha)$ is the angle whose tangent is α, and so $\sin(\tan^{-1}(\alpha))$ is the sine of the angle whose tangent is α.

Example **8** **Evaluating compositions of functions**

Find the exact value of each composition without using a table or a calculator.

a. $\sin(\tan^{-1}(0))$ **b.** $\arcsin(\cos(\pi/6))$ **c.** $\tan\!\left(\sec^{-1}\!\left(\sqrt{2}\right)\right)$

Solution

a. Since $\tan(0) = 0$, $\tan^{-1}(0) = 0$. Therefore,

$$\sin(\tan^{-1}(0)) = \sin(0) = 0.$$

b. Since $\cos(\pi/6) = \sqrt{3}/2$, we have

$$\arcsin\!\left(\cos\!\left(\frac{\pi}{6}\right)\right) = \arcsin\!\left(\frac{\sqrt{3}}{2}\right) = \frac{\pi}{3}.$$

c. To find $\sec^{-1}\!\left(\sqrt{2}\right)$, we use the identity $\sec^{-1}(x) = \cos^{-1}(1/x)$. Since $\cos(\pi/4) = 1/\sqrt{2}$, we have $\cos^{-1}\!\left(1/\sqrt{2}\right) = \pi/4$ and $\sec^{-1}\!\left(\sqrt{2}\right) = \pi/4$. Therefore,

$$\tan\!\left(\sec^{-1}\!\left(\sqrt{2}\right)\right) = \tan\!\left(\frac{\pi}{4}\right) = 1.$$

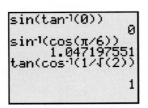

■ Figure 3.73

You can check parts (a), (b), and (c) with a graphing calculator, as shown in Fig. 3.73. Note that $\pi/3 \approx 1.047$. ■

The composition of $y = \sin x$ restricted to $[-\pi/2, \pi/2]$ and $y = \sin^{-1}x$ is the identity function, because they are inverse functions of each other. So

$$\sin^{-1}(\sin x) = x \quad \text{for } x \text{ in } [-\pi/2, \pi/2]$$

and

$$\sin(\sin^{-1} x) = x \quad \text{for } x \text{ in } [-1, 1].$$

Note that if x is not in $[-\pi/2, \pi/2]$, then the sine function followed by the inverse sine function is not the identity function. For example, $\sin^{-1}(\sin(2\pi/3)) = \pi/3$.

The general sine function $f(x) = A\sin[B(x - C)] + D$ is a composition of a trigonometric function and several algebraic functions. Since the sine function has an inverse and the algebraic functions have inverses, we can find the inverse for a general sine function provided it is restricted to a suitable domain.

Example **9** **The inverse of a general sine function**

Find the inverse of $f(x) = 3 \sin(2x) + 5$, where $-\pi/4 \le x \le \pi/4$, and determine the domain of f^{-1}.

Solution

Interchange x and y in $y = 3 \sin(2x) + 5$, and then solve for y:

$$x = 3 \sin(2y) + 5 \qquad \text{Switch } x \text{ and } y.$$

$$\frac{x - 5}{3} = \sin(2y)$$

$$2y = \sin^{-1}\left(\frac{x - 5}{3}\right) \qquad \text{Definition of } \sin^{-1}$$

$$y = \frac{1}{2} \sin^{-1}\left(\frac{x - 5}{3}\right)$$

$$f^{-1}(x) = \frac{1}{2} \sin^{-1}\left(\frac{x - 5}{3}\right)$$

Since $\sin(2x)$ is between -1 and 1, the range of f is $[2, 8]$ and the domain of f^{-1} is $[2, 8]$. ■

For Thought

True or False? Explain.

1. $\sin^{-1}(0) = \sin(0)$

2. $\sin(3\pi/4) = 1/\sqrt{2}$

3. $\cos^{-1}(0) = 1$

4. $\sin^{-1}\left(\sqrt{2}/2\right) = 135°$

5. $\cot^{-1}(5) = \dfrac{1}{\tan^{-1}(5)}$

6. $\sec^{-1}(5) = \cos^{-1}(0.2)$

7. $\sin\left(\cos^{-1}\left(\sqrt{2}/2\right)\right) = 1/\sqrt{2}$

8. $\sec(\sec^{-1}(2)) = 2$

9. The functions $f(x) = \sin^{-1} x$ and $f^{-1}(x) = \sin x$ are inverse functions.

10. The secant and cosecant functions are inverses of each other.

3.5 Exercises

Find the exact value of each expression without using a calculator or table.

1. $\sin^{-1}(-1/2)$

2. $\sin^{-1}(0)$

3. $\arcsin(1/2)$

4. $\arcsin\left(\sqrt{3}/2\right)$

5. $\arcsin\left(\sqrt{2}/2\right)$

6. $\arcsin(1)$

Find the exact value of each expression in degrees without using a calculator or table.

7. $\sin^{-1}\left(-1/\sqrt{2}\right)$

8. $\sin^{-1}\left(\sqrt{3}/2\right)$

9. $\arcsin(1/2)$

10. $\arcsin(-1)$

11. $\sin^{-1}(0)$

12. $\sin^{-1}\left(\sqrt{2}/2\right)$

In each case find α to the nearest tenth of a degree, where $-90° \leq \alpha \leq 90°$.

13. $\sin \alpha = -1/3$ **14.** $\sin \alpha = 0.4138$

15. $\sin \alpha = 0.5682$ **16.** $\sin \alpha = -0.34$

Find the exact value of each expression without using a calculator or table.

17. $\cos^{-1}\left(-\sqrt{2}/2\right)$ **18.** $\cos^{-1}(1)$

19. $\arccos(1/2)$ **20.** $\arccos\left(-\sqrt{3}/2\right)$

21. $\arccos(-1)$ **22.** $\arccos(0)$

Find the exact value of each expression in degrees without using a calculator or table.

23. $\cos^{-1}\left(-\sqrt{2}/2\right)$ **24.** $\cos^{-1}\left(\sqrt{3}/2\right)$

25. $\arccos(-1)$ **26.** $\arccos(0)$

27. $\cos^{-1}(-1/2)$ **28.** $\cos^{-1}(1)$

In each case find α to the nearest tenth of a degree, where $0° \leq \alpha \leq 180°$.

29. $\cos \alpha = -0.993$ **30.** $\cos \alpha = 0.7392$

31. $\cos \alpha = 0.001$ **32.** $\cos \alpha = -0.499$

Find the exact value of each expression without using a calculator or table.

33. $\tan^{-1}(-1)$ **34.** $\cot^{-1}\left(1/\sqrt{3}\right)$

35. $\sec^{-1}(2)$ **36.** $\csc^{-1}\left(2/\sqrt{3}\right)$

37. $\operatorname{arcsec}\left(\sqrt{2}\right)$ **38.** $\arctan\left(-1/\sqrt{3}\right)$

39. $\operatorname{arccsc}(-2)$ **40.** $\operatorname{arccot}\left(-\sqrt{3}\right)$

41. $\tan^{-1}(0)$ **42.** $\sec^{-1}(1)$

43. $\csc^{-1}(1)$ **44.** $\csc^{-1}(-1)$

45. $\cot^{-1}(-1)$ **46.** $\cot^{-1}(0)$

47. $\cot^{-1}\left(-\sqrt{3}/3\right)$ **48.** $\cot^{-1}(1)$

Find the approximate value of each expression with a calculator. Round answers to two decimal places.

49. $\arcsin(0.5682)$ **50.** $\sin^{-1}(-0.4138)$

51. $\cos^{-1}(-0.993)$ **52.** $\cos^{-1}(0.7392)$

53. $\tan^{-1}(-0.1396)$ **54.** $\cot^{-1}(4.32)$

55. $\sec^{-1}(-3.44)$ **56.** $\csc^{-1}(6.8212)$

57. $\operatorname{arcsec}\left(\sqrt{6}\right)$ **58.** $\arctan\left(-2\sqrt{7}\right)$

59. $\operatorname{arccsc}\left(-2\sqrt{2}\right)$ **60.** $\operatorname{arccot}\left(-\sqrt{5}\right)$

61. $\operatorname{arccot}(-12)$ **62.** $\operatorname{arccot}(0.001)$

63. $\cot^{-1}(15.6)$ **64.** $\cot^{-1}(-1.01)$

Find the exact value of each composition without using a calculator or table.

65. $\tan(\arccos(1/2))$ **66.** $\sec\left(\arcsin\left(1/\sqrt{2}\right)\right)$

67. $\sin^{-1}(\cos(2\pi/3))$ **68.** $\tan^{-1}(\sin(\pi/2))$

69. $\cot^{-1}(\cot(\pi/6))$ **70.** $\sec^{-1}(\sec(\pi/3))$

71. $\arcsin(\sin(3\pi/4))$ **72.** $\arccos(\cos(-\pi/3))$

73. $\tan(\arctan(1))$ **74.** $\cot(\operatorname{arccot}(0))$

75. $\cos^{-1}(\cos(3\pi/2))$ **76.** $\sin(\csc^{-1}(-2))$

77. $\cos\left(2\sin^{-1}\left(\sqrt{2}/2\right)\right)$ **78.** $\tan(2\cos^{-1}(1/2))$

79. $\sin^{-1}(2\sin(\pi/6))$ **80.** $\cos^{-1}(0.5\tan(\pi/4))$

Find the inverse of each function and state its domain.

81. $f(x) = \sin(2x)$ for $-\dfrac{\pi}{4} \leq x \leq \dfrac{\pi}{4}$

82. $f(x) = \cos(3x)$ for $0 \leq x \leq \dfrac{\pi}{3}$

83. $f(x) = 3 + \tan(\pi x)$ for $-\dfrac{1}{2} < x < \dfrac{1}{2}$

84. $f(x) = 2 - \sin(\pi x - \pi)$ for $\dfrac{1}{2} \leq x \leq \dfrac{3}{2}$

85. $f(x) = \sin^{-1}(x/2) + 3$ for $-2 \leq x \leq 2$

86. $f(x) = 2\cos^{-1}(5x) + 3$ for $-\dfrac{1}{5} \leq x \leq \dfrac{1}{5}$

In a circle with radius r, a central angle θ intercepts a chord of length c, where $\theta = \cos^{-1}\left(1 - \dfrac{c^2}{2r^2}\right)$. *Use this formula for Exercises 87 and 88.*

87. An airplane at 2000 feet flies directly over a gun that has a range of 2400 feet, as shown in the figure on the next page. What is the measure in degrees of the angle for which the airplane is within range of the gun?

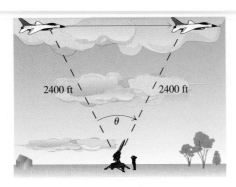

■ **Figure for Exercise 87**

88. A triangle has two sides that are both 5.2 meters long and one side with a length of 1.3 meters. Find the measure in degrees for the smallest angle of the triangle.

Thinking Outside the Box XXIII

Clear Sailing   A sailor plans to install a windshield wiper on a porthole that has radius 1 foot. The wiper blade of length x feet is to be attached to the edge of the porthole as shown in the figure. The area cleaned by the blade is a sector of a circle centered at the point of attachment.

a. Write the area cleaned by the blade as a function of x.

b. If the blade cleans half of the window, then what is the exact length of the blade?

c. Use a graphing calculator to find the length for the blade that would maximize the area cleaned?

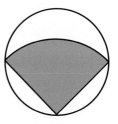

■ **Figure for Thinking Outside the Box XXIII**

3.5 Pop Quiz

1. Find $f^{-1}(x)$ if $f(x) = \cos(2x)$ for $0 \le x \le \pi/2$.

Find the exact value.

2. $\sin^{-1}(-1)$

3. $\sin^{-1}(1/2)$

4. $\arccos(-1)$

5. $\arctan(-1)$

6. $\tan(\arcsin(1/2))$

7. $\sin^{-1}(\sin(3\pi/4))$

3.6 Right Triangle Trigonometry

One reason trigonometry was invented was to determine the measures of sides and angles of geometric figures without actually measuring them. In this section we study right triangles (triangles that have a 90° angle) and see what information is needed to determine the measures of all unknown sides and angles of a right triangle.

Trigonometric Ratios

We defined the sine and cosine functions for an angle in standard position in terms of the point at which the terminal side intersects the unit circle. However, it is often the case that we do not know that point, but we do know a different point on the terminal side. When this situation occurs, we can find the values for the sine and cosine of the angle using trigonometric ratios.

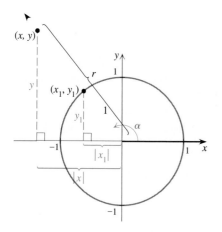

Figure 3.74

Figure 3.74 shows an angle α in quadrant II with the terminal side passing through the point (x, y) and intersecting the unit circle at (x_1, y_1). If we draw vertical line segments down to the x-axis, we form two similar right triangles, as shown in Fig. 3.74. If r is the distance from (x, y) to the origin, then $r = \sqrt{x^2 + y^2}$. The lengths of the legs in the larger triangle are y and $|x|$, and its hypotenuse is r. The lengths of the legs in the smaller triangle are y_1 and $|x_1|$, and its hypotenuse is 1. Since ratios of the lengths of corresponding sides of similar triangles are equal, we have

$$\frac{y_1}{y} = \frac{1}{r} \quad \text{and} \quad \frac{|x_1|}{|x|} = \frac{1}{r}.$$

The first equation can be written as $y_1 = y/r$ and the second as $x_1 = x/r$. (Since x_1 and x are both negative in this case, the absolute value symbols can be omitted.) Since $y_1 = \sin \alpha$ and $x_1 = \cos \alpha$, we get

$$\sin \alpha = \frac{y}{r} \quad \text{and} \quad \cos \alpha = \frac{x}{r}.$$

This argument can be repeated in each quadrant with (x, y) chosen inside, outside, or on the unit circle. These ratios also give the correct sine or cosine if α is a quadrantal angle. Since all other trigonometric functions are related to the sine and cosine, their values can also be obtained from x, y, and r.

Theorem:
Trigonometric Ratios

If (x, y) is any point other than the origin on the terminal side of an angle α in standard position and $r = \sqrt{x^2 + y^2}$, then

$$\sin \alpha = \frac{y}{r}, \qquad \cos \alpha = \frac{x}{r}, \qquad \text{and} \qquad \tan \alpha = \frac{y}{x} \, (x \neq 0).$$

Example **1** Trigonometric ratios

Find the values of the six trigonometric functions of the angle α in standard position whose terminal side passes through $(4, -2)$.

Solution

Use $x = 4$, $y = -2$, and $r = \sqrt{4^2 + (-2)^2} = \sqrt{20} = 2\sqrt{5}$ to get

$$\sin \alpha = \frac{-2}{2\sqrt{5}} = -\frac{\sqrt{5}}{5}, \qquad \cos \alpha = \frac{4}{2\sqrt{5}} = \frac{2\sqrt{5}}{5}, \qquad \text{and}$$

$$\tan \alpha = \frac{-2}{4} = -\frac{1}{2}.$$

Since cosecant, secant, and cotangent are the reciprocals of sine, cosine, and tangent,

$$\csc \alpha = \frac{1}{\sin \alpha} = -\frac{5}{\sqrt{5}} = -\sqrt{5}, \qquad \sec \alpha = \frac{1}{\cos \alpha} = \frac{5}{2\sqrt{5}} = \frac{\sqrt{5}}{2}, \qquad \text{and}$$

$$\cot \alpha = \frac{1}{\tan \alpha} = -2. \qquad ■$$

Right Triangles

So far the trigonometric functions have been tied to a coordinate system and an angle in standard position. Trigonometric ratios can also be used to evaluate the trigonometric functions for an acute angle of a right triangle without having the angle or the triangle located in a coordinate system.

Consider a right triangle with acute angle α, legs of length x and y, and hypotenuse r, as shown in Fig. 3.75(a). If this triangle is positioned in a coordinate system, as in Fig. 3.75(b), then (x, y) is a point on the terminal side of α and

$$\sin(\alpha) = \frac{y}{r}, \qquad \cos(\alpha) = \frac{x}{r}, \qquad \text{and} \qquad \tan(\alpha) = \frac{y}{x}.$$

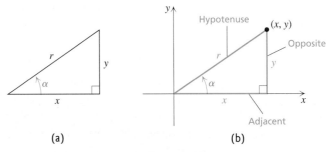

(a)   (b)

Figure 3.75

However, the values of the trigonometric functions are simply ratios of the lengths of the sides of the right triangle and it is not necessary to move the triangle to a coordinate system to find them. Notice that y is the length of the side **opposite** the angle α, x is the length of the side **adjacent** to α, and r is the length of the **hypotenuse.** We use the abbreviations opp, adj, and hyp to represent the lengths of these sides in the following theorem.

Theorem: Trigonometric Functions of an Acute Angle of a Right Triangle

If α is an acute angle of a right triangle, then

$$\sin \alpha = \frac{\text{opp}}{\text{hyp}}, \qquad \cos \alpha = \frac{\text{adj}}{\text{hyp}}, \qquad \text{and} \qquad \tan \alpha = \frac{\text{opp}}{\text{adj}}.$$

Since the cosecant, secant, and cotangent are the reciprocals of the sine, cosine, and tangent, respectively, the values of all six trigonometric functions can be found for an acute angle of a right triangle.

Example **2** **Trigonometric functions in a right triangle**

Find the values of all six trigonometric functions for the angle α of the right triangle with legs of length 1 and 4, as shown in Fig. 3.76.

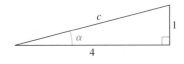

Figure 3.76

Solution

The length of the hypotenuse is $c = \sqrt{4^2 + 1^2} = \sqrt{17}$. Since the length of the side opposite α is 1 and the length of the adjacent side is 4, we have

$$\sin \alpha = \frac{\text{opp}}{\text{hyp}} = \frac{1}{\sqrt{17}} = \frac{\sqrt{17}}{17}, \qquad \cos \alpha = \frac{\text{adj}}{\text{hyp}} = \frac{4}{\sqrt{17}} = \frac{4\sqrt{17}}{17},$$

$$\tan \alpha = \frac{\text{opp}}{\text{adj}} = \frac{1}{4}, \qquad \csc \alpha = \frac{1}{\sin \alpha} = \sqrt{17},$$

$$\sec \alpha = \frac{1}{\cos \alpha} = \frac{\sqrt{17}}{4}, \qquad \cot \alpha = \frac{1}{\tan \alpha} = 4. \qquad ■$$

Solving a Right Triangle

The values of the trigonometric functions for an acute angle of a right triangle are determined by ratios of lengths of sides of the triangle. We can use those ratios along with the inverse trigonometric functions to find missing parts of a right triangle in which some of the measures of angles or lengths of sides are known. Finding all of the unknown lengths of sides or measures of angles is called **solving the triangle.** A triangle can be solved only if enough information is given to determine a unique triangle. For example, the lengths of the sides in a 30-60-90 triangle cannot be found because there are infinitely many such triangles of different sizes. However, the lengths of the missing sides in a 30-60-90 triangle with a hypotenuse of 6 will be found in Example 3.

In solving right triangles, we usually name the acute angles α and β and the lengths of the sides opposite those angles a and b. The 90° angle is γ, and the length of the side opposite γ is c.

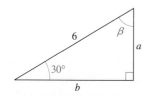

■ **Figure 3.77**

Example **3** Solving a right triangle

Solve the right triangle in which $\alpha = 30°$ and $c = 6$.

Solution

The triangle is shown in Fig. 3.77. Since $\alpha = 30°$, $\gamma = 90°$, and the sum of the measures of the angles of any triangle is 180°, we have $\beta = 60°$. Since $\sin \alpha = \text{opp}/\text{hyp}$, we get $\sin 30° = a/6$ and

$$a = 6 \cdot \sin 30° = 6 \cdot \frac{1}{2} = 3.$$

Since $\cos \alpha = \text{adj}/\text{hyp}$, we get $\cos 30° = b/6$ and

$$b = 6 \cdot \cos 30° = 6 \cdot \frac{\sqrt{3}}{2} = 3\sqrt{3}.$$

The angles of the right triangle are 30°, 60°, and 90°, and the sides opposite those angles are 3, $3\sqrt{3}$, and 6, respectively. ■

Example **4** Solving a right triangle

Solve the right triangle in which $a = 4$ and $b = 6$. Find the acute angles to the nearest tenth of a degree.

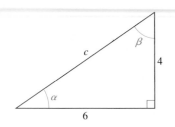

■ **Figure 3.78**

Solution

The triangle is shown in Fig. 3.78. By the Pythagorean theorem, $c^2 = 4^2 + 6^2$, or $c = \sqrt{52} = 2\sqrt{13}$. To find α, first find $\sin\alpha$:

$$\sin\alpha = \frac{\text{opp}}{\text{hyp}} = \frac{4}{2\sqrt{13}} = \frac{2}{\sqrt{13}}$$

Now, α is the angle whose sine is $2/\sqrt{13}$:

$$\alpha = \sin^{-1}\left(\frac{2}{\sqrt{13}}\right) \approx 33.7°$$

Since $\alpha + \beta = 90°$, $\beta = 90° - 33.7° = 56.3°$. The angles of the triangle are 33.7°, 56.3°, and 90°, and the sides opposite those angles are 4, 6, and $2\sqrt{13}$, respectively.   ■

In Example 4 we could have found α by using $\alpha = \tan^{-1}(4/6) \approx 33.7°$. We could then have found c by using $\cos(33.7°) = 6/c$, or $c = 6/\cos(33.7°) \approx 7.2$. There are many ways to solve a right triangle, but the basic strategy is always the same.

STRATEGY   **Solving a Right Triangle**

1. Use the Pythagorean theorem to find the length of a third side when the lengths of two sides are known.
2. Use the trigonometric ratios to find missing sides or angles.
3. Use the fact that the sum of the measures of the angles of a triangle is 180° to determine a third angle when two are known.

Applications

Using trigonometry, we can find the size of an object without actually measuring the object but by measuring an angle. Two common terms used in this regard are **angle of elevation** and **angle of depression.** The angle of elevation α for a point above a horizontal line is the angle formed by the horizontal line and the observer's line of sight through the point, as shown in Fig. 3.79. The angle of depression β for a point below a horizontal line is the angle formed by the horizontal line and the observer's line of sight through the point, as shown in Fig. 3.79. We use these angles and our skills in solving triangles to find the sizes of objects that would be inconvenient to measure.

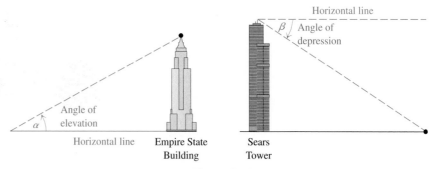

■ **Figure 3.79**

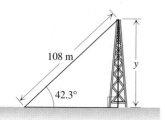

■ **Figure 3.80**

Example **5** **Finding the height of an object**

A guy wire of length 108 meters runs from the top of an antenna to the ground. If the angle of elevation of the top of the antenna, sighting along the guy wire, is 42.3°, then what is the height of the antenna?

Solution

Let y represent the height of the antenna, as shown in Fig. 3.80. Since $\sin(42.3°) = y/108$,

$$y = 108 \cdot \sin(42.3°) \approx 72.7 \text{ meters.} \qquad ■$$

In Example 5 we knew the distance to the top of the antenna, and we found the height of the antenna. If we knew the distance on the ground to the base of the antenna and the angle of elevation of the guy wire, we could still have found the height of the antenna. Both cases involve knowing the distance to the antenna either on the ground or through the air. However, one of the biggest triumphs of trigonometry is being able to find the size of an object or the distance to an object (such as the moon) without going to the object. The next example shows one way to find the height of an object without actually going to it. In Section 3.9, Example 8, we will show another (slightly simpler) solution to the same problem using the law of sines.

Example **6** **Finding the height of an object from a distance**

The angle of elevation of the top of a water tower from point A on the ground is 19.9°. From point B, 50.0 feet closer to the tower, the angle of elevation is 21.8°. What is the height of the tower?

Solution

Let y represent the height of the tower and x represent the distance from point B to the base of the tower, as shown in Fig. 3.81.

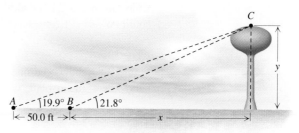

■ **Figure 3.81**

At point B, $\tan 21.8° = y/x$ or

$$x = \frac{y}{\tan 21.8°}.$$

Since the distance to the base of the tower from point A is $x + 50$,

$$\tan 19.9° = \frac{y}{x + 50}$$

or

$$y = (x + 50) \tan 19.9°.$$

To find the value of y we must write an equation that involves only y. Since $x = y/\tan 21.8°$, we can substitute $y/\tan 21.8°$ for x in the last equation:

$$y = \left(\frac{y}{\tan 21.8°} + 50 \right) \tan 19.9°$$

$$y = \frac{y \cdot \tan 19.9°}{\tan 21.8°} + 50 \tan 19.9° \quad \text{Distributive property}$$

$$y - \frac{y \cdot \tan 19.9°}{\tan 21.8°} = 50 \tan 19.9°$$

$$y \left(1 - \frac{\tan 19.9°}{\tan 21.8°} \right) = 50 \tan 19.9° \quad \text{Factor out } y.$$

$$y = \frac{50 \tan 19.9°}{1 - \dfrac{\tan 19.9°}{\tan 21.8°}} \approx 191 \text{ feet}$$

```
50tan(19.9)/(1-t
an(19.9)/tan(21.
8))
        190.6278641
```

■ **Figure 3.82**

This computation is shown on a graphing calculator in Fig. 3.82.   ■

In the next example we combine the solution of a right triangle with the arc length of a circle from Section 3.1 to solve a problem of aerial photography.

Example **7** Photography from a spy plane

In the late 1950s, the Soviets labored to develop a missile that could stop the U-2 spy plane. On May 1, 1960, Nikita S. Khrushchev announced to the world that the Soviets had shot down Francis Gary Powers while Powers was photographing the Soviet Union from a U-2 at an altitude of 14 miles. How wide a path on the earth's surface could Powers see from that altitude? (Use 3950 miles as the earth's radius.)

Solution

Figure 3.83 shows the line of sight to the horizon on the left-hand side and right-hand side of the airplane while flying at the altitude of 14 miles.

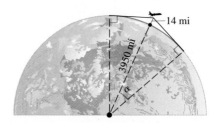

■ **Figure 3.83**

Since a line tangent to a circle (the line of sight) is perpendicular to the radius at the point of tangency, the angle α at the center of the earth in Fig. 3.83 is an acute angle of a right triangle with hypotenuse $3950 + 14$ or 3964. So we have

$$\cos \alpha = \frac{3950}{3964}$$

$$\alpha = \cos^{-1}\left(\frac{3950}{3964}\right) \approx 4.8°.$$

The width of the path seen by Powers is the length of the arc intercepted by the central angle 2α or 9.6°. Using the formula $s = \alpha r$ from Section 3.1, where α is in radians, we get

$$s = 9.6 \text{ deg} \cdot \frac{\pi \text{ rad}}{180 \text{ deg}} \cdot 3950 \text{ miles} \approx 661.8 \text{ miles}.$$

From an altitude of 14 miles, Powers could see a path that was 661.8 miles wide. Actually, he photographed a path that was somewhat narrower, because parts of the photographs near the horizon were not usable. ▪

For Thought

True or False? Explain. For Exercises 1–4, α is an angle in standard position.

1. If the terminal side of α goes through $(5, -10)$, then $\sin \alpha = 10/\sqrt{125}$.

2. If the terminal side of α goes through $(-1, 2)$, then $\sec \alpha = -\sqrt{5}$.

3. If the terminal side of α goes through $(-2, 3)$, then $\alpha = \sin^{-1}\left(3/\sqrt{13}\right)$.

4. If the terminal side of α goes through $(3, 1)$, then $\alpha = \cos^{-1}\left(3/\sqrt{10}\right)$.

5. In a right triangle, $\sin \alpha = \cos \beta$, $\sec \alpha = \csc \beta$, and $\tan \alpha = \cot \beta$.

6. If $a = 4$ and $b = 2$ in a right triangle, then $c = \sqrt{6}$.

7. If $a = 6$ and $b = 2$ in a right triangle, then $\beta = \tan^{-1}(3)$.

8. If $a = 8$ and $\alpha = 55°$ in a right triangle, then $b = 8/\tan(55°)$.

9. In a right triangle with sides of length 3, 4, and 5, the smallest angle is $\cos^{-1}(0.8)$.

10. In a right triangle, $\sin(90°) = \text{hyp/adj}$.

3.6 Exercises

Assume that α is an angle in standard position whose terminal side contains the given point. Find the exact values of $\sin \alpha$, $\cos \alpha$, $\tan \alpha$, $\csc \alpha$, $\sec \alpha$, *and* $\cot \alpha$.

1. $(3, 4)$

2. $(4, 4)$

3. $(-2, 6)$

4. $(-3, 6)$

5. $\left(-2, -\sqrt{2}\right)$

6. $\left(-1, -\sqrt{3}\right)$

7. $\left(\sqrt{6}, -\sqrt{2}\right)$

8. $\left(2\sqrt{3}, -2\right)$

For Exercises 9–14 find exact values of $\sin \alpha$, $\cos \alpha$, $\tan \alpha$, $\sin \beta$, *cos* β, *and* $\tan \beta$ *for the given right triangle.*

9.
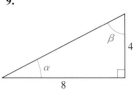

10.
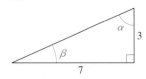

11.
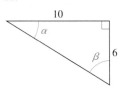

12.
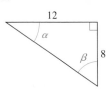

13.

14.
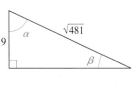

Assume that α *is an angle in standard position whose terminal side contains the given point and that* $0° < \alpha < 90°$. *Find the degree measure of* α *to the nearest tenth of a degree.*

15. (1.5, 9)

16. (4, 5)

17. $\left(\sqrt{2}, \sqrt{6}\right)$

18. (4.3, 6.9)

Assume that α *is an angle in standard position whose terminal side contains the given point and that* $0 < \alpha < \pi/2$. *Find the radian measure of* α *to the nearest tenth of a radian.*

19. (4, 6.3)

20. (1/3, 1/2)

21. $\left(\sqrt{5}, 1\right)$

22. $\left(\sqrt{7}, \sqrt{3}\right)$

Solve each right triangle with the given sides and angles. In each case, make a sketch. Note that, α *is the acute angle opposite leg a and* β *is the acute angle opposite leg b. The hypotenuse is c. See the strategy for solving a right triangle on page 256.*

23. $\alpha = 60°$, $c = 20$

24. $\beta = 45°$, $c = 10$

25. $a = 6$, $b = 8$

26. $a = 10$, $c = 12$

27. $b = 6$, $c = 8.3$

28. $\alpha = 32.4°$, $b = 10$

29. $\alpha = 16°$, $c = 20$

30. $\beta = 47°$, $a = 3$

31. $\alpha = 39°9'$, $a = 9$

32. $\beta = 19°12'$, $b = 60$

33. *Aerial Photography* An aerial photograph from a U-2 spy plane is taken of a building suspected of housing nuclear warheads. The photograph is made when the angle of elevation of the sun is 32°. By comparing the shadow cast by the building to objects of known size in the illustration, analysts determine that the shadow is 80 ft long. How tall is the building?

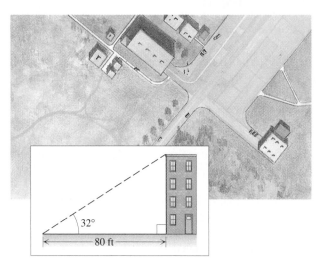

■ **Figure for Exercise 33**

34. *Giant Redwood* A hiker stands 80 feet from a giant redwood tree and sights the top with an angle of elevation of 75°. How tall is the tree to the nearest foot?

35. *Avoiding a Swamp* Muriel was hiking directly toward a long, straight road when she encountered a swamp. She turned 65° to the right and hiked 4 mi in that direction to reach the road. How far was she from the road when she encountered the swamp?

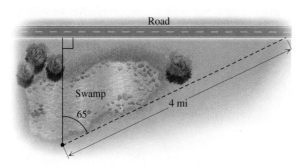

Figure for Exercise 35

36. *Tall Antenna*   A 100 foot guy wire is attached to the top of an antenna. The angle between the guy wire and the ground is 62°. How tall is the antenna to the nearest foot?

37. *Angle of Depression*   From a highway overpass, 14.3 m above the road, the angle of depression of an oncoming car is measured at 18.3°. How far is the car from a point on the highway directly below the observer?

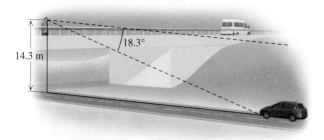

Figure for Exercise 37

38. *Length of a Tunnel*   A tunnel under a river is 196.8 ft below the surface at its lowest point, as shown in the drawing. If the angle of depression of the tunnel is 4.962°, then how far apart on the surface are the entrances to the tunnel? How long is the tunnel?

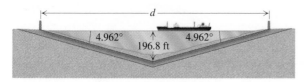

Figure for Exercise 38

39. *Height of a Crosswalk*   The angle of elevation of a pedestrian crosswalk over a busy highway is 8.34°, as shown in the drawing.

If the distance between the ends of the crosswalk measured on the ground is 342 ft, then what is the height *h* of the crosswalk at the center?

Figure for Exercise 39

40. *Shortcut to Snyder*   To get from Muleshoe to Snyder, Harry drives 50 mph for 178 mi south on route 214 to Seminole, then goes east on route 180 to Snyder. Harriet leaves Muleshoe one hour later at 55 mph, but takes US 84, which goes straight from Muleshoe to Snyder through Lubbock. If US 84 intersects route 180 at a 50° angle, then how many more miles does Harry drive?

41. *Installing a Guy Wire*   A 41-m guy wire is attached to the top of a 34.6-m antenna and to a point on the ground. How far is the point on the ground from the base of the antenna, and what angle does the guy wire make with the ground?

42. *Robin and Marian*   Robin Hood plans to use a 30-ft ladder to reach the castle window of Maid Marian. Little John, who made the ladder, advised Robin that the angle of elevation of the ladder must be between 55° and 70° for safety. What are the minimum and maximum heights that can safely be reached by the top of the ladder when it is placed against the 50-ft castle wall?

43. *Height of a Rock*   Inscription Rock rises almost straight upward from the valley floor. From one point the angle of elevation of the top of the rock is 16.7°. From a point 168 m closer to the rock, the angle of elevation of the top of the rock is 24.1°. How high is Inscription Rock?

44. *Height of a Balloon*   A hot air balloon is between two spotters who are 1.2 mi apart. One spotter reports that the angle of elevation of the balloon is 76°, and the other reports that it is 68°. What is the altitude of the balloon in miles?

45. *Passing in the Night*   A boat sailing north sights a lighthouse to the east at an angle of 32° from the north, as shown in the drawing on the next page. After the boat travels one more kilometer, the angle of the lighthouse from the north is 36°. If the boat continues to sail north, then how close will the boat come to the lighthouse?

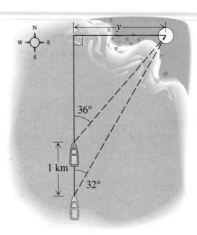

Figure for Exercise 45

46. *Height of a Skyscraper*   For years the Woolworth sky-scraper in New York held the record for the world's tallest office building. If the length of the shadow of the Wool-worth building increases by 17.4 m as the angle of elevation of the sun changes from 44° to 42°, then how tall is the building?

47. *View from Landsat*   The satellite Landsat orbits the earth at an altitude of 700 mi, as shown in the figure. What is the width of the path on the surface of the earth that can be seen by the cameras of Landsat? Use 3950 mi for the radius of the earth.

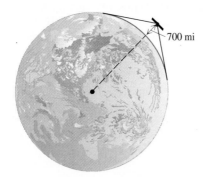

Figure for Exercise 47

48. *Communicating Via Satellite*   A communication satellite is usually put into a synchronous orbit with the earth, which means that it stays above a fixed point on the surface of the earth at all times. The radius of the orbit of such a satellite is 6.5 times the radius of the earth (3950 mi). The satellite is used to relay a signal from one point on the earth to another point on the earth. The sender and receiver of a signal must be in a line of sight with the satellite, as shown in the figure. What is the maximum distance on the surface

of the earth between the sender and receiver for this type of satellite?

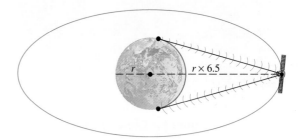

Figure for Exercise 48

49. *Hanging a Pipe*   A contractor wants to pick up a 10-ft diame-ter pipe with a chain of length 40 ft. The chain encircles the pipe and is attached to a hook on a crane. What is the distance between the hook and the pipe? (This is an actual problem that an engineer was asked to solve on the job.)

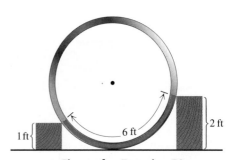

Figure for Exercise 49

50. *Blocking a Pipe*   A large pipe is held in place by using a 1-ft high block on one side and a 2-ft high block on the other side. If the length of the arc between the points where the pipe touches the blocks is 6 ft, then what is the radius of the pipe? Ignore the thickness of the pipe.

Figure for Exercise 50

Thinking Outside the Box XXIV

Kicking a Field Goal   In professional football the ball must be placed between the left and right hash mark when a field goal is to be kicked. The hash marks are 9.25 ft from the center line of the field. The goal post is 30 ft past the goal line. Suppose that θ is the angle between the lines of sight from the ball to the left and right uprights as shown in the figure. The larger the value of θ the easier it is to kick the ball between the uprights. What is the difference (to the nearest thousandth of a degree) between the values of θ at the right hash mark and the value of θ at the center of the field when the ball is 60 ft from the goal line? The two vertical bars on the goal are 18.5 ft apart and the horizontal bar is 10 ft above the ground.

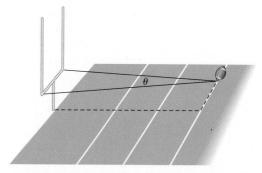

Figure for Thinking Outside the Box XXIV

3.6 Pop Quiz

1. Find $\sin \alpha$, $\cos \alpha$, and $\tan \alpha$ if the terminal side of α in standard position goes through $(-3, 4)$.

2. A right triangle has legs with lengths 3 and 6, and α is the acute angle opposite the smallest leg. Find exact values for $\sin \alpha$, $\cos \alpha$, and $\tan \alpha$.

3. At a distance of 1000 feet from a building the angle of elevation to the top of the building is $36°$. Find the height of the building to the nearest foot.

3.7 Identities

An identity is an equation that is satisfied by *every* number for which both sides are defined. In calculus identities are used to simplify expressions and determine whether expressions are equivalent. In this section we review the identities used in calculus, omitting most of the proofs. The proofs can be found in any standard trigonometry text.

Pythagorean Identities

In Section 3.2 we proved the fundamental identity $\sin^2 \alpha + \cos^2 \alpha = 1$. Dividing each side by $\sin^2 \alpha$ yields a new identity:

$$\frac{\sin^2 \alpha}{\sin^2 \alpha} + \frac{\cos^2 \alpha}{\sin^2 \alpha} = \frac{1}{\sin^2 \alpha}$$

$$1 + \left(\frac{\cos \alpha}{\sin \alpha}\right)^2 = \left(\frac{1}{\sin \alpha}\right)^2$$

$$1 + \cot^2 \alpha = \csc^2 \alpha$$

■ **Foreshadowing Calculus**

Trigonometric identities are used extensively in calculus. We often use identities to change the form or simplify a function when solving calculus problems involving trigonometric functions.

Now divide each side of the fundamental identity by $\cos^2 \alpha$:

$$\frac{\sin^2 \alpha}{\cos^2 \alpha} + \frac{\cos^2 \alpha}{\cos^2 \alpha} = \frac{1}{\cos^2 \alpha}$$

$$\tan^2 \alpha + 1 = \sec^2 \alpha$$

These identities are based on the equation of the unit circle $x^2 + y^2 = 1$, which comes from the Pythagorean theorem. They are called the **Pythagorean identities.**

Pythagorean Identities

$$\sin^2 \alpha + \cos^2 \alpha = 1 \qquad 1 + \cot^2 \alpha = \csc^2 \alpha \qquad \tan^2 \alpha + 1 = \sec^2 \alpha$$

Example **1** Using identities

Simplify $\sin x \cot x \cos x$.

Solution

$$\sin x + \cot x \cos x = \sin x + \frac{\cos x}{\sin x} \cdot \cos x \qquad \text{Rewrite using sines and cosines.}$$

$$= \sin x + \frac{\cos^2 x}{\sin x}$$

$$= \frac{\sin^2 x}{\sin x} + \frac{\cos^2 x}{\sin x} \qquad \text{Multiply } \sin x \text{ by } \frac{\sin x}{\sin x}.$$

$$= \frac{\sin^2 x + \cos^2 x}{\sin x} \qquad \text{Add the fractions.}$$

$$= \frac{1}{\sin x} \qquad \text{Since } \sin^2 x + \cos^2 x = 1$$

$$= \csc x \qquad \text{Definition of cosecant} \qquad ■$$

Odd and Even Identities

An *odd function* is one for which $f(-x) = -f(x)$. An *even function* is one for which $f(-x) = f(x)$. Each of the six trigonometric functions is either odd or even.

Odd and Even Identities

Odd: $\sin(-x) = -\sin(x)$ $\csc(-x) = -\csc(x)$
$\tan(-x) = -\tan(x)$ $\cot(-x) = -\cot(x)$

Even: $\cos(-x) = \cos(x)$ $\sec(-x) = \sec(x)$

Example **2** Using odd and even identities

Simplify $\dfrac{1}{1 + \cos(-x)} + \dfrac{1}{1 - \cos x}$.

Solution

First note that $\cos(-x) = \cos(x)$. Then find a common denominator and add the expressions.

$$\frac{1}{1 + \cos(-x)} + \frac{1}{1 - \cos x} = \frac{1}{1 + \cos x} + \frac{1}{1 - \cos x}$$

$$= \frac{1(1 - \cos x)}{(1 + \cos x)(1 - \cos x)} + \frac{1(1 + \cos x)}{(1 - \cos x)(1 + \cos x)}$$

$$= \frac{1 - \cos x}{1 - \cos^2 x} + \frac{1 + \cos x}{1 - \cos^2 x} \quad \text{Product of a sum and difference}$$

$$= \frac{1 - \cos x}{\sin^2 x} + \frac{1 + \cos x}{\sin^2 x} \quad \text{Pythagorean identity}$$

$$= \frac{1 - \cos x + 1 + \cos x}{\sin^2 x} \quad \text{Add the fractions.}$$

$$= \frac{2}{\sin^2 x}$$

$$= 2 \cdot \frac{1}{\sin^2 x}$$

$$= 2 \csc^2 x \quad \csc x = 1/\sin \quad ■$$

Sum and Difference Identities

The sum and difference identities are used to write a trigonometric function of a sum or difference of angles in terms of trigonometric functions of the angles. Note that each equation in the following box represents two identities. For the first identity we use the top operation in the symbol $\pm$ or $\mp$ and for the second identity we use the bottom operation.

Sum and Difference Identities

$$\cos(\alpha \pm \beta) = \cos \alpha \cos \beta \mp \sin \alpha \sin \beta$$

$$\sin(\alpha \pm \beta) = \sin \alpha \cos \beta \pm \cos \alpha \sin \beta$$

$$\tan(\alpha \pm \beta) = \frac{\tan \alpha \pm \tan \beta}{1 \mp \tan \alpha \tan \beta}$$

Example **3** **The cosine of a sum**

Find the exact value of $\cos(75°)$.

Solution

Use $75° = 30° + 45°$ and the identity for the cosine of a sum.

$$\cos 75° = \cos(30° + 45°)$$

$$= \cos(30°)\cos(45°) - \sin(30°)\sin(45°)$$

$$= \frac{\sqrt{3}}{2} \cdot \frac{\sqrt{2}}{2} - \frac{1}{2} \cdot \frac{\sqrt{2}}{2}$$

$$= \frac{\sqrt{6} - \sqrt{2}}{4}$$

To check, evaluate $\cos 75°$ and $\left(\sqrt{6} - \sqrt{2}\right)/4$ using a calculator. ■

Cofunction Identities

Using $\pi/2$ for one of the angles in the difference identities yields the following cofunction identities.

Cofunction Identities

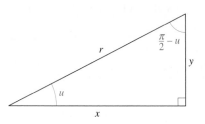

Figure 3.84

$$\sin\left(\frac{\pi}{2} - u\right) = \cos u \qquad \cos\left(\frac{\pi}{2} - u\right) = \sin u$$

$$\tan\left(\frac{\pi}{2} - u\right) = \cot u \qquad \cot\left(\frac{\pi}{2} - u\right) = \tan u$$

$$\sec\left(\frac{\pi}{2} - u\right) = \csc u \qquad \csc\left(\frac{\pi}{2} - u\right) = \sec u$$

If $0 < u < \pi/2$, then u and $\pi/2 - u$ are the measures of the acute angles of a right triangle as shown in Fig. 3.84. The term *cofunction* comes from the fact that these angles are complementary. The cofunction identities indicate that the value of a trigonometric function of one acute angle of a right triangle is equal to the value of the cofunction of the other acute angle. For example, $\sin(20°) = \cos(70°)$ and $\cot(89°) = \tan(1°)$.

Example **4** **Using the odd/even and cofunction identities**

Verify that $\cos(x - \pi/2) = \sin(x)$ is an identity.

Solution

$$\cos\left(x - \frac{\pi}{2}\right) = \cos\left(-\left(\frac{\pi}{2} - x\right)\right) \qquad \text{Because } x - \frac{\pi}{2} = -\left(\frac{\pi}{2} - x\right)$$

$$= \cos\left(\frac{\pi}{2} - x\right) \qquad \text{Cosine is an even function } (\cos(-\alpha) = \cos(\alpha)).$$

$$= \sin(x) \qquad \text{Cofunction identity} \qquad ■$$

Double-Angle and Half-Angle Identities

To get an identity for $\sin 2x$, replace both α and β by x in the identity for $\sin(\alpha + \beta)$:

$$\sin 2x = \sin(x + x) = \sin x \cos x + \cos x \sin x = 2 \sin x \cos x$$

We can repeat this procedure using the identities for the cosine of a sum and the tangent of a sum to get the following **double-angle identities.**

Double-Angle Identities

$$\sin 2x = 2 \sin x \cos x \qquad \tan 2x = \frac{2 \tan x}{1 - \tan^2 x}$$

$$\cos 2x = \cos^2 x - \sin^2 x$$

$$\cos 2x = 2 \cos^2 x - 1$$

$$\cos 2x = 1 - 2 \sin^2 x$$

Be careful to learn the double-angle identities exactly as they are written. A "nice-looking" equation such as $\cos 2x = 2 \cos x$ could be mistaken for an identity if you are not careful. Since $\cos(\pi/2) \neq 2\cos(\pi/4)$, the "nice-looking" equation is not an identity. Remember that an equation is not an identity if at least one permissible value of the variable fails to satisfy the equation.

Example **5** Using the double-angle identities

Find $\sin(120°)$, $\cos(120°)$, and $\tan(120°)$ using double-angle identities.

Solution

Note that $120° = 2 \cdot 60°$ and use the values $\sin(60°) = \sqrt{3}/2$, $\cos(60°) = 1/2$, and $\tan(60°) = \sqrt{3}$ in the appropriate identities:

$$\sin(120°) = 2\sin(60°)\cos(60°) = 2 \cdot \frac{\sqrt{3}}{2} \cdot \frac{1}{2} = \frac{\sqrt{3}}{2}$$

$$\cos(120°) = \cos^2(60°) - \sin^2(60°) = \left(\frac{1}{2}\right)^2 - \left(\frac{\sqrt{3}}{2}\right)^2 = -\frac{1}{2}$$

$$\tan(120°) = \frac{2\tan(60°)}{1 - \tan^2(60°)} = \frac{2 \cdot \sqrt{3}}{1 - (\sqrt{3})^2} = \frac{2\sqrt{3}}{-2} = -\sqrt{3}$$

These results are the well-known values of $\sin(120°)$, $\cos(120°)$, and $\tan(120°)$. ■

The **half-angle identities** are derived from the double-angle identities.

Half-Angle Identities

$$\sin\frac{x}{2} = \pm\sqrt{\frac{1 - \cos x}{2}} \qquad \cos\frac{x}{2} = \pm\sqrt{\frac{1 + \cos x}{2}}$$

$$\tan\frac{x}{2} = \pm\sqrt{\frac{1 - \cos x}{1 + \cos x}} \qquad \tan\frac{x}{2} = \frac{\sin x}{1 + \cos x} \qquad \tan\frac{x}{2} = \frac{1 - \cos x}{\sin x}$$

Example **6** Using half-angle identities

Use the half-angle identity to find the exact value of $\tan(-15°)$.

Solution

Use $x = -30°$ in the half-angle identity $\tan\dfrac{x}{2} = \dfrac{\sin x}{1 + \cos x}$.

$$\tan(-15°) = \tan\left(\frac{-30°}{2}\right)$$

$$= \frac{\sin(-30°)}{1 + \cos(-30°)}$$

$$= \frac{-\dfrac{1}{2}}{1 + \dfrac{\sqrt{3}}{2}}$$

$$= \frac{-1}{2 + \sqrt{3}}$$

$$= \frac{-1(2 - \sqrt{3})}{(2 + \sqrt{3})(2 - \sqrt{3})} = -2 + \sqrt{3} \qquad ■$$

Example **7** **Using the identities**

Find $\sin(\alpha/2)$ if $\sin \alpha = 3/5$ and $\pi/2 < \alpha < \pi$.

Solution

Use the identity $\sin^2\alpha + \cos^2\alpha = 1$ to find $\cos \alpha$:

$$\left(\frac{3}{5}\right)^2 + \cos^2 \alpha = 1$$

$$\cos^2 \alpha = \frac{16}{25}$$

$$\cos \alpha = \pm\frac{4}{5}$$

For $\pi/2 < \alpha < \pi$, we have $\cos \alpha < 0$. So $\cos \alpha = -4/5$. Now use the half-angle identity for sine:

$$\sin \frac{\alpha}{2} = \pm\sqrt{\frac{1 - \cos \alpha}{2}} = \pm\sqrt{\frac{1 + \frac{4}{5}}{2}}$$

$$= \pm\sqrt{\frac{9}{10}} = \pm\frac{\sqrt{90}}{10} = \pm\frac{3\sqrt{10}}{10}$$

If $\pi/2 < \alpha < \pi$, then $\pi/4 < \alpha/2 < \pi/2$. So $\sin(\alpha/2) > 0$ and $\sin(\alpha/2) = 3\sqrt{10}/10$. ■

Product and Sum Identities

The sum and difference identities for sines and cosines can be used to derive the **product-to-sum identities.** These identities are not used as often as other identities. It is not necessary to memorize these identities. Just remember them by name and look them up as necessary.

Product-to-Sum Identities

$$\sin A \cos B = \frac{1}{2}[\sin(A + B) + \sin(A - B)]$$

$$\sin A \sin B = \frac{1}{2}[\cos(A - B) - \cos(A + B)]$$

$$\cos A \sin B = \frac{1}{2}[\sin(A + B) - \sin(A - B)]$$

$$\cos A \cos B = \frac{1}{2}[\cos(A - B) + \cos(A + B)]$$

Example **8** **Expressing a product as a sum**

Use the product-to-sum identities to rewrite each expression.

a. $\sin 12° \cos 9°$ **b.** $\sin(\pi/12) \sin(\pi/8)$

Solution

a. Use the product-to-sum identity for $\sin A \cos B$:

$$\sin 12° \cos 9° = \frac{1}{2}[\sin(12° + 9°) + \sin(12° - 9°)]$$

$$= \frac{1}{2}[\sin 21° + \sin 3°]$$

b. Use the product-to-sum identity for $\sin A \sin B$:

$$\sin\left(\frac{\pi}{12}\right)\sin\left(\frac{\pi}{8}\right) = \frac{1}{2}\left[\cos\left(\frac{\pi}{12} - \frac{\pi}{8}\right) - \cos\left(\frac{\pi}{12} + \frac{\pi}{8}\right)\right]$$

$$= \frac{1}{2}\left[\cos\left(-\frac{\pi}{24}\right) - \cos\left(\frac{5\pi}{24}\right)\right]$$

$$= \frac{1}{2}\left[\cos\frac{\pi}{24} - \cos\frac{5\pi}{24}\right]$$

```
sin(π/12)sin(π/8
)
        .0990457605
.5(cos(π/24)-cos
(5π/24))
        .0990457605
```

■ **Figure 3.85**

Use a calculator to check as in Fig. 3.85.   ■

The **sum-to-product identities** are used to write a sum of two trigonometric functions as a product.

Sum-to-Product Identities

$$\sin x + \sin y = 2\sin\left(\frac{x + y}{2}\right)\cos\left(\frac{x - y}{2}\right)$$

$$\sin x - \sin y = 2\cos\left(\frac{x + y}{2}\right)\sin\left(\frac{x - y}{2}\right)$$

$$\cos x + \cos y = 2\cos\left(\frac{x + y}{2}\right)\cos\left(\frac{x - y}{2}\right)$$

$$\cos x - \cos y = -2\sin\left(\frac{x + y}{2}\right)\sin\left(\frac{x - y}{2}\right)$$

Example **9** **Expressing a sum or difference as a product**

Use the sum-to-product identities to rewrite each expression.

a. $\cos(\pi/5) - \cos(\pi/8)$ **b.** $\sin(6t) - \sin(4t)$

Solution

a. Use the sum-to-product identity for $\cos x - \cos y$:

$$\cos\left(\frac{\pi}{5}\right) - \cos\left(\frac{\pi}{8}\right) = -2\sin\left(\frac{\pi/5 + \pi/8}{2}\right)\sin\left(\frac{\pi/5 - \pi/8}{2}\right)$$

$$= -2\sin\left(\frac{13\pi}{80}\right)\sin\left(\frac{3\pi}{80}\right)$$

b. Use the sum-to-product identity for $\sin x - \sin y$:

$$\sin 6t - \sin 4t = 2 \cos\left(\frac{6t + 4t}{2}\right) \sin\left(\frac{6t - 4t}{2}\right)$$

$$= 2 \cos 5t \sin t \qquad ■$$

The Function $y = a \sin x + b \cos x$

Functions of the form $y = a \sin x + b \cos x$ occur in applications such as the position of a weight in motion due to the force of a spring, the position of a swinging pendulum, and the current in an electrical circuit. The **reduction formula** is used to express this function in terms of a single trigonometric function from which amplitude, period, and phase shift can be determined.

Theorem:
Reduction Formula

> If α is an angle in standard position whose terminal side contains (a, b), then
> $$a \sin x + b \cos x = \sqrt{a^2 + b^2} \sin(x + \alpha)$$
> for any real number x.

To rewrite an expression of the form $a \sin x + b \cos x$ using the reduction formula, we need to find α so that the terminal side of α goes through (a, b). By using trigonometric ratios, we have

$$\sin \alpha = \frac{b}{\sqrt{a^2 + b^2}}, \qquad \cos \alpha = \frac{a}{\sqrt{a^2 + b^2}}, \qquad \tan \alpha = \frac{b}{a}.$$

Since we know a and b, we can find $\sin \alpha$, $\cos \alpha$, or $\tan \alpha$, and then use an inverse trigonometric function to find α. However, because of the ranges of the inverse functions, the angle obtained from an inverse function might not have its terminal side through (a, b) as required. We will address this problem in the next example.

Example 🔟 Using the reduction formula

Use the reduction formula to rewrite $-3 \sin x - 3 \cos x$ in the form $A \sin(x + C)$.

Solution

Because $a = -3$ and $b = -3$, we have

$$\sqrt{a^2 + b^2} = \sqrt{18}$$
$$= 3\sqrt{2}.$$

Since the terminal side of α must go through $(-3, -3)$, we have

$$\cos \alpha = \frac{-3}{3\sqrt{2}} = -\frac{\sqrt{2}}{2}.$$

Now $\cos^{-1}\left(-\sqrt{2}/2\right) = 3\pi/4$, but the terminal side of $3\pi/4$ is in quadrant II, as shown in Fig. 3.86. However, we also have $\cos(5\pi/4) = -\sqrt{2}/2$ and the terminal

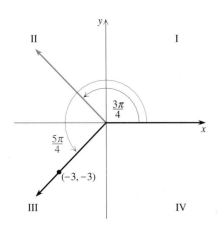

■ **Figure 3.86**

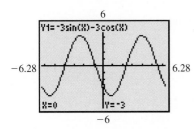

■ **Figure 3.87**

side for $5\pi/4$ does pass through $(-3, -3)$ in quadrant III. So $\alpha = 5\pi/4$. By the reduction formula

$$-3 \sin x - 3 \cos x = 3\sqrt{2} \sin\left(x + \frac{5\pi}{4}\right).$$

The reduction formula explains why the calculator graph of $y = -3 \sin x - 3 \cos x$ in Fig. 3.87 is a sine wave with amplitude $3\sqrt{2}$.   ■

For Thought

True or False? Explain.

1. The equation $(\tan x)(\cot x) = 1$ is an identity.

2. The equation $(\sin x + \cos x)^2 = \sin^2 x + \cos^2 x$ is an identity.

3. $\tan 1 = \sqrt{1 - \sec^2 1}$

4. $\sin^2(-6) = -\sin^2(6)$

5. $\dfrac{\pi}{4} - \dfrac{\pi}{3} = \dfrac{\pi}{12}$

6. $\cos(4) \cos(5) + \sin(4) \sin(5) = \cos(-1)$

7. $\cos(\pi/2 - 5) = \sin(5)$

8. $\sin(7\pi/12) = \sin(\pi/3)\cos(\pi/4) + \cos(\pi/3)\sin(\pi/4)$

9. $\sin 150° = \sqrt{\dfrac{1 - \cos 75°}{2}}$

10. $\cos 4 + \cos 12 = 2 \cos 8 \cos 4$

3.7 Exercises

Use identities to simplify each expression. Do not use a calculator.

1. $\dfrac{1}{\sin^2 x} - \dfrac{1}{\tan^2 x}$

2. $\dfrac{1 + \cos \alpha \tan \alpha \csc \alpha}{\csc \alpha}$

3. $\dfrac{\sin^4 x - \sin^2 x}{\sec x}$

4. $\dfrac{-\tan^2 t - 1}{\sec^2 t}$

5. $\dfrac{\cos w \sin^2 w + \cos^3 w}{\sec w}$

6. $\dfrac{\tan^3 x - \sec^2 x \tan x}{\cot(-x)}$

7. $\sin x + \dfrac{\cos^2 x}{\sin x}$

8. $\dfrac{1}{\sin^3 x} - \dfrac{\cot^2 x}{\sin x}$

9. $\sin(-x) \cot(-x)$

10. $\sec(-x) - \sec(x)$

11. $\dfrac{\sin(x)}{\cos(-x)} + \dfrac{\sin(-x)}{\cos(x)}$

12. $\dfrac{\cos(-x)}{\sin(-x)} - \dfrac{\cos(-x)}{\sin(x)}$

13. $\cos(5) \cos(6) - \sin(5) \sin(6)$

14. $\cos(7.1) \cos(1.4) - \sin(7.1) \sin(1.4)$

15. $\cos 2k \cos k + \sin 2k \sin k$

16. $\cos 3y \cos y - \sin 3y \sin y$

17. $\sin(23°) \cos(67°) + \cos(23°) \sin(67°)$

18. $\sin(2°) \cos(7°) - \cos(2°) \sin(7°)$

19. $\sin(-\pi/2) \cos(\pi/5) + \cos(\pi/2) \sin(-\pi/5)$

20. $\sin(-\pi/6) \cos(-\pi/3) + \cos(-\pi/6) \sin(-\pi/3)$

21. $\dfrac{\tan(\pi/9) + \tan(\pi/6)}{1 - \tan(\pi/9) \tan(\pi/6)}$

22. $\dfrac{\tan(\pi/3) - \tan(\pi/5)}{1 + \tan(\pi/3) \tan(\pi/5)}$

23. $\cos(-3k) \cos(-k) - \cos(\pi/2 - 3k) \sin(-k)$

24. $\cos(y - \pi/2) \cos(y) + \sin(\pi/2 - y) \sin(-y)$

25. $2 \sin 13° \cos 13°$

26. $\sin^2\left(\dfrac{\pi}{5}\right) - \cos^2\left(\dfrac{\pi}{5}\right)$

27. $\dfrac{\tan 15°}{1 - \tan^2(15°)}$

28. $\dfrac{2}{\cot 5(1 - \tan^2 5)}$

29. $2\sin\left(\dfrac{\pi}{9} - \dfrac{\pi}{2}\right)\cos\left(\dfrac{\pi}{2} - \dfrac{\pi}{9}\right)$

30. $2\cos^2\left(\dfrac{\pi}{5} - \dfrac{\pi}{2}\right) - 1$

In each exercise, use identities to find the exact values at α for the remaining five trigonometric functions.

31. $\tan \alpha = 1/2$ and $0 < \alpha < \pi/2$

32. $\sin \alpha = 3/4$ and $\pi/2 < \alpha < \pi$

33. $\cos \alpha = -\sqrt{3}/5$ and α is in quadrant III

34. $\sec \alpha = -4\sqrt{5}/5$ and α is in quadrant II

In each case, find sin α, cos α, tan α, csc α, sec α, and cot α.

35. $\cos 2\alpha = 3/5$ and $0° < 2\alpha < 90°$

36. $\cos 2\alpha = 1/3$ and $360° < 2\alpha < 450°$

37. $\cos(\alpha/2) = -1/4$ and $\pi/2 < \alpha/2 < 3\pi/4$

38. $\sin(\alpha/2) = -1/3$ and $7\pi/4 < \alpha/2 < 2\pi$

Use appropriate identities to find the exact value of each expression.

39. $\sin(15°)$

40. $\sin(75°)$

41. $\cos(5\pi/12)$

42. $\cos(7\pi/12)$

43. $\cos(-\pi/12)$

44. $\cos(-5\pi/12)$

45. $\tan(75°)$

46. $\tan(-15°)$

Prove that each of the following equations is an identity.

47. $\tan x \cos x + \csc x \sin^2 x = 2\sin x$

48. $\cot x \sin x - \cos^2 x \sec x = 0$

49. $2 - \csc \beta \sin \beta = \sin^2 \beta + \cos^2 \beta$

50. $(1 - \sin^2 \beta)(1 + \sin^2 \beta) = 2\cos^2 \beta - \cos^4 \beta$

51. $\dfrac{\sec x}{\tan x} - \dfrac{\tan x}{\sec x} = \cos x \cot x$

52. $\dfrac{1 - \sin^2 x}{1 - \sin x} = \dfrac{\csc x + 1}{\csc x}$

53. $1 + \csc x \sec x = \dfrac{\cos(-x) - \csc(-x)}{\cos(x)}$

54. $\tan^2(-x) - \dfrac{\sin(-x)}{\sin x} = \sec^2 x$

55. $\dfrac{\csc y + 1}{\csc y - 1} = \dfrac{1 + \sin y}{1 - \sin y}$

56. $\dfrac{1 - 2\cos^2 y}{1 - 2\cos y \sin y} = \dfrac{\sin y + \cos y}{\sin y - \cos y}$

57. $\dfrac{1 - \sin^2(-x)}{1 - \sin(-x)} = 1 - \sin x$

58. $\dfrac{1 - \sin^2 x \csc^2 x + \sin^2 x}{\cos^2 x} = \tan^2 x$

59. $\cos(x - \pi/2) = \cos x \tan x$

60. $\dfrac{\cos(x + y)}{\cos x \cos y} = 1 - \tan x \tan y$

61. $\dfrac{\cos(\alpha + \beta)}{\cos \alpha + \sin \beta} = \dfrac{\cos \alpha - \sin \beta}{\cos(\beta - \alpha)}$

62. $\sec(v + t) = \dfrac{\cos v \cos t + \sin v \sin t}{\cos^2 v - \sin^2 t}$

63. $\sin(180° - \alpha) = \dfrac{1 - \cos^2 \alpha}{\sin \alpha}$

64. $\dfrac{\sin(x + y)}{\sin x \cos y} = 1 + \cot x \tan y$

65. $\dfrac{\cos(\alpha + \beta)}{\sin(\alpha - \beta)} = \dfrac{1 - \tan \alpha \tan \beta}{\tan \alpha - \tan \beta}$

66. $\dfrac{\cos(\alpha - \beta)}{\sin(\alpha + \beta)} = \dfrac{1 + \tan \alpha \tan \beta}{\tan \alpha + \tan \beta}$

67. $\cos^4 s - \sin^4 s = \cos 2s$

68. $\sin 2s = -2\sin s \sin(s - \pi/2)$

69. $\dfrac{\cos 2x + \cos 2y}{\sin x + \cos y} = 2\cos y - 2\sin x$

70. $(\sin \alpha - \cos \alpha)^2 = 1 - \sin 2\alpha$

71. $2 \sin^2 \left(\dfrac{u}{2} \right) = \dfrac{\sin^2 u}{1 + \cos u}$

72. $\cos 2y = \dfrac{1 - \tan^2 y}{1 + \tan^2 y}$

73. $\dfrac{1 - \sin^2 \left(\dfrac{x}{2} \right)}{1 + \sin^2 \left(\dfrac{x}{2} \right)} = \dfrac{1 + \cos x}{3 - \cos x}$

74. $\dfrac{1 - \cos^2 \left(\dfrac{x}{2} \right)}{1 - \sin^2 \left(\dfrac{x}{2} \right)} = \dfrac{1 - \cos x}{1 + \cos x}$

75. $\dfrac{\cos x - \cos 3x}{\cos x + \cos 3x} = \tan 2x \tan x$

76. $\dfrac{\cos 5y + \cos 3y}{\cos 5y - \cos 3y} = -\cot 4y \cot y$

Use the product-to-sum identities to rewrite each expression.

77. $\sin 13° \sin 9°$

78. $\cos 34° \cos 39°$

79. $\cos \left(\dfrac{\pi}{6} \right) \cos \left(\dfrac{\pi}{5} \right)$

80. $\sin \left(\dfrac{2\pi}{9} \right) \sin \left(\dfrac{3\pi}{4} \right)$

Find the exact value of each product.

81. $\sin(52.5°) \sin(7.5°)$

82. $\cos(105°) \cos(75°)$

83. $\sin \left(\dfrac{13\pi}{24} \right) \cos \left(\dfrac{5\pi}{24} \right)$

84. $\cos \left(\dfrac{5\pi}{24} \right) \sin \left(-\dfrac{\pi}{24} \right)$

Use the sum-to-product identities to rewrite each expression.

85. $\sin 12° - \sin 8°$

86. $\sin 7° + \sin 11°$

87. $\cos \left(\dfrac{\pi}{3} \right) - \cos \left(\dfrac{\pi}{5} \right)$

88. $\cos \left(\dfrac{1}{2} \right) + \cos \left(\dfrac{2}{3} \right)$

Find the exact value of each sum.

89. $\sin(75°) + \sin(15°)$

90. $\sin(285°) - \sin(15°)$

91. $\cos \left(-\dfrac{\pi}{24} \right) - \cos \left(\dfrac{7\pi}{24} \right)$

92. $\cos \left(\dfrac{5\pi}{24} \right) + \cos \left(\dfrac{\pi}{24} \right)$

Rewrite each expression in the form $A \sin(x + C)$.

93. $\sin x - \cos x$

94. $2 \sin x + 2 \cos x$

95. $-\dfrac{1}{2} \sin x + \dfrac{\sqrt{3}}{2} \cos x$

96. $\dfrac{\sqrt{2}}{2} \sin x - \dfrac{\sqrt{2}}{2} \cos x$

97. $\dfrac{\sqrt{3}}{2} \sin x - \dfrac{1}{2} \cos x$

98. $-\dfrac{\sqrt{3}}{2} \sin x - \dfrac{1}{2} \cos x$

Solve each problem.

99. *Viewing Area*   Find a formula for the viewing area of a television screen in terms of its diagonal and the angle α shown in the figure. Rewrite the formula using a single trigonometric function.

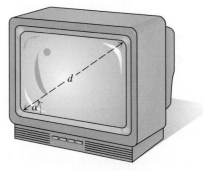

∎ **Figure for Exercises 99 and 100**

100. Use the formula from the previous exercise to find the viewing area for a 32 in. diagonal television for which $\alpha = 37.2°$.

101. *Motion of a Spring*   A block is attached to a spring, as shown in the figure, so that its resting position is the origin of the number line. If the block is set in motion, then it will oscillate about the origin. Suppose that the location of the block at time t in seconds is given in meters by $x = \sqrt{3} \sin t + \cos t$. Use the reduction formula to rewrite this function and find the maximum distance reached by the block from the resting position.

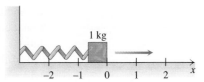

∎ **Figure for Exercise 101**

102. *Motion of a Spring*   A block hanging from a spring, as shown in the figure, oscillates about the origin on a vertical number line. If the block is given an upward velocity of

0.3 m/sec from a point 0.5 m below its resting position, then its position at time t in seconds is given in meters by $x = -0.3 \sin t + 0.5 \cos t$. Use the reduction formula to find the maximum distance that the block travels from the resting position.

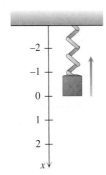

Figure for Exercise 102

Thinking Outside the Box XXV

Tangent Circles   The three large circles in the accompanying diagram are tangent to each other and each has radius 1. The small circle in the middle is tangent to each of the three large circles. Find its radius.

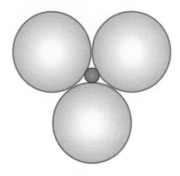

■ **Figure for Thinking Outside the Box XXV**

3.7 Pop Quiz

1. Simplify the expression $\cot x \sec x$.

2. Find exact values for $\cos \alpha$ and $\cot \alpha$ if $\sin \alpha = 1/3$ and $0 < \alpha < \pi/2$.

3. Is $f(x) = \cos(3x)$ even or odd?

4. Simplify $\sin(2x)\cos(x) + \cos(2x)\sin(x)$.

5. Simplify $\dfrac{2}{\sec^2 \alpha} + \dfrac{2}{\csc^2 \alpha}$.

6. Prove that $\sin^4 x - \cos^4 x = -\cos(2x)$.

3.8 Conditional Trigonometric Equations

An identity is satisfied by *all* values of the variable for which both sides are defined. In this section we investigate **conditional equations,** those equations that are not identities but that have *at least one* solution. For example, the equation $\sin x = 0$ is a conditional equation. Because the trigonometric functions are periodic, conditional equations involving trigonometric functions usually have infinitely many solutions. The equation $\sin x = 0$ is satisfied by $x = 0, \pm\pi, \pm 2\pi, \pm 3\pi$, and so on. All of these solutions are of the form $k\pi$, where k is an integer. In this section we will solve conditional equations and see that identities play a fundamental role in their solution.

Cosine Equations

The most basic conditional equation involving cosine is of the form $\cos x = a$, where a is a number in the interval $[-1, 1]$. If a is not in $[-1, 1]$, $\cos x = a$ has no solution. For a in $[-1, 1]$, the equation $x = \cos^{-1} a$ provides one solution in the interval $[0, \pi]$. From this single solution we can determine all of the solutions because of the periodic nature of the cosine function. We must remember also that the cosine of an arc is the x-coordinate of its terminal point on the unit circle. So arcs that terminate at opposite ends of a vertical chord in the unit circle have the same cosine.

Example **1** Solving a cosine equation

Find all real numbers that satisfy each equation.

a. $\cos x = 1$ **b.** $\cos x = 0$ **c.** $\cos x = -1/2$

Solution

a. One solution to $\cos x = 1$ is

$$x = \cos^{-1}(1) = 0.$$

Since the period of cosine is 2π, any integral multiple of 2π can be added to this solution to get additional solutions. So the equation is satisfied by 0, $\pm 2\pi$, $\pm 4\pi$, and so on. Now $\cos x = 1$ is satisifed only if the arc of length x on the unit circle has terminal point $(1, 0)$, as shown in Fig. 3.88. So there are no more solutions. The solution set is written as

$$\{x \mid x = 2k\pi\},$$

where k is any integer.

b. One solution to $\cos x = 0$ is

$$x = \cos^{-1}(0) = \frac{\pi}{2}.$$

The terminal point for the arc of length $\pi/2$ is $(0, 1)$. So any arc length of the form $\pi/2 + 2k\pi$ is a solution to $\cos x = 0$. However, an arc of length x that terminates at $(0, -1)$ also satisfies $\cos x = 0$, as shown in Fig. 3.89, and these arcs are not included in the form $\pi/2 + 2k\pi$. Since the distance between $(0, 1)$ and $(0, -1)$ on the unit circle is π, all arcs that terminate at these points are of the form $\pi/2 + k\pi$. So the solution set is

$$\left\{x \mid x = \frac{\pi}{2} + k\pi\right\},$$

where k is any integer.

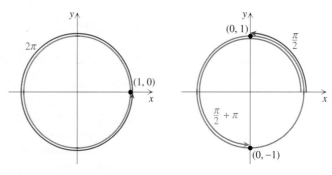

▪ **Figure 3.88**   ▪ **Figure 3.89**

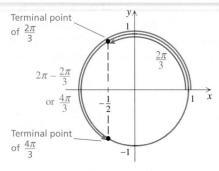

Terminal point of $\frac{2\pi}{3}$

$2\pi - \frac{2\pi}{3}$ or $\frac{4\pi}{3}$

Terminal point of $\frac{4\pi}{3}$

Figure 3.90

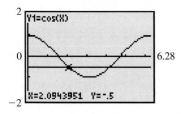

Figure 3.91

c. One solution to the equation $\cos x = -1/2$ is

$$x = \cos^{-1}\left(-\frac{1}{2}\right) = \frac{2\pi}{3}.$$

Since the period of cosine is 2π, all arcs of length $2\pi/3 + 2k\pi$ (where k is any integer) satisfy the equation. The arc of length $4\pi/3$ also terminates at a point with x-coordinate $-1/2$, as shown in Fig. 3.90. So $4\pi/3$ is also a solution to $\cos x = -1/2$, but it is not included in the form $2\pi/3 + 2k\pi$. So the solution set to $\cos x = -1/2$ is

$$\left\{x\,\middle|\,x = \frac{2\pi}{3} + 2k\pi \quad \text{or} \quad x = \frac{4\pi}{3} + 2k\pi\right\},$$

where k is any integer. Note that the arcs of length $2\pi/3$ and $4\pi/3$ terminate at opposite ends of a vertical chord in the unit circle.

Note also that the calculator graph of $y_1 = \cos x$ in Fig. 3.91 crosses the horizontal line $y_2 = -1/2$ twice in the interval $[0, 2\pi]$, at $2\pi/3$ and $4\pi/3$. ■

The procedure used in Example 1(c) can be used to solve $\cos x = a$ for any nonzero a between -1 and 1. First find two values of x between 0 and 2π that satisfy the equation. (In general, $s = \cos^{-1}(a)$ and $2\pi - s$ are the values.) Next write all solutions by adding $2k\pi$ to the first two solutions. The solution sets for $a = -1, 0$, and 1 are easier to find and remember. All of the cases for solving $\cos x = a$ are summarized next. The letter k is used to represent any integer. You should not simply memorize this summary, but you should be able to solve $\cos x = a$ for any real number a with the help of a unit circle.

SUMMARY   **Solving cos x = a**

1. If $-1 < a < 1$ and $a \neq 0$, then the solution set to $\cos x = a$ is $\{x\,|\,x = s + 2k\pi$ or $x = 2\pi - s + 2k\pi\}$, where $s = \cos^{-1} a$.
2. The solution set to $\cos x = 1$ is $\{x\,|\,x = 2k\pi\}$.
3. The solution set to $\cos x = 0$ is $\{x\,|\,x = \pi/2 + k\pi\}$.
4. The solution set to $\cos x = -1$ is $\{x\,|\,x = \pi + 2k\pi\}$.
5. If $|a| > 1$, then $\cos x = a$ has no solution.

Sine Equations

To solve $\sin x = a$, we use the same strategy that we used to solve $\cos x = a$. We look for the smallest nonnegative solution and then build on it to obtain all solutions. However, the "first" solution obtained from $s = \sin^{-1} a$ might be negative because the range for the function $\sin^{-1}$ is $[-\pi/2, \pi/2]$. In this case $s + 2\pi$ is positive and we build on it to find all solutions. Remember that the sine of an arc on the unit circle is the y-coordinate of the terminal point of the arc. So arcs that terminate at opposite ends of a horizontal chord in the unit circle have the same y-coordinate for the terminal point and the same sine. This fact is the key to finding all solutions of a sine equation.

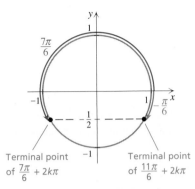

$\frac{7\pi}{6}$

$-\frac{1}{2}$

$\frac{\pi}{6}$

Terminal point of $\frac{7\pi}{6} + 2k\pi$ Terminal point of $\frac{11\pi}{6} + 2k\pi$

■ **Figure 3.92**

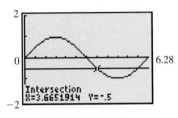

Intersection
X=3.6651914 Y=-.5

■ **Figure 3.93**

Example **2** **Solving a sine equation**

Find all real numbers that satisfy $\sin x = -1/2$.

Solution

One solution to $\sin x = -1/2$ is

$$x = \sin^{-1}\left(-\frac{1}{2}\right) = -\frac{\pi}{6}.$$

See Fig. 3.92. The smallest positive arc with the same terminal point as $-\pi/6$ is $-\pi/6 + 2\pi = 11\pi/6$. Since the period of the sine function is 2π, all arcs of the form $11\pi/6 + 2k\pi$ have the same terminal point and satisfy $\sin x = -1/2$. The smallest positive arc that terminates at the other end of the horizontal chord shown in Fig. 3.92 is $\pi - (-\pi/6)$ or $7\pi/6$. So $7\pi/6$ also satisfies the equation, but it is not included in the form $11\pi/6 + 2k\pi$. So the solution set is

$$\left\{ x \,\middle|\, x = \frac{7\pi}{6} + 2k\pi \quad \text{or} \quad x = \frac{11\pi}{6} + 2k\pi \right\},$$

where k is any integer.

Note that the calculator graph of $y_1 = \sin x$ in Fig. 3.93 crosses the horizontal line $y_2 = -1/2$ twice in the interval $[0, 2\pi]$, at $7\pi/6$ and $11\pi/6$. ■

The procedure used in Example 2 can be used to solve $\sin x = a$ for any nonzero a between -1 and 1. First find two values between 0 and 2π that satisfy the equation. (In general, $s = \sin^{-1} a$ and $\pi - s$ work when s is positive; $s + 2\pi$ and $\pi - s$ work when s is negative.) Next write all solutions by adding $2k\pi$ to the first two solutions. The equations $\sin x = 0$, $\sin x = 1$, and $\sin x = -1$ have solutions that are similar to the corresponding cosine equations. We can summarize the different solution sets to $\sin x = a$ as follows. You should not simply memorize this summary, but you should be able to solve $\sin x = a$ for any real number a with the help of a unit circle.

SUMMARY **Solving sin x = a**

1. If $-1 < a < 1, a \neq 0$, and $s = \sin^{-1} a$, then the solution set to $\sin x = a$ is
   $$\{x \,|\, x = s + 2k\pi \text{ or } x = \pi - s + 2k\pi\} \text{ for } s > 0,$$
   $$\{x \,|\, x = s + 2\pi + 2k\pi \text{ or } x = \pi - s + 2k\pi\} \text{ for } s < 0.$$
2. The solution set to $\sin x = 1$ is $\{x \,|\, x = \pi/2 + 2k\pi\}$.
3. The solution set to $\sin x = 0$ is $\{x \,|\, x = k\pi\}$.
4. The solution set to $\sin x = -1$ is $\{x \,|\, x = 3\pi/2 + 2k\pi\}$.
5. If $|a| > 1$, then $\sin x = a$ has no solution.

Tangent Equations

Tangent equations are a little simpler than sine and cosine equations, because the tangent function is one-to-one in its fundamental cycle, while sine and cosine are not one-to-one in their fundamental cycles. In the next example, we see that the solution set to $\tan x = a$ consists of any single solution plus multiples of π.

Example **3** **Solving a tangent equation**

Find all solutions, in degrees.

a. $\tan \alpha = 1$ **b.** $\tan \alpha = -1.34$

Solution

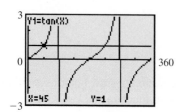

■ **Figure 3.94**

a. Since $\tan^{-1}(1) = 45°$ and the period of tangent is $180°$, all angles of the form $45° + k180°$ satisfy the equation. There are no additional angles that satisfy the equation. The solution set to $\tan \alpha = 1$ is

$$\{\alpha \mid \alpha = 45° + k180°\}.$$

The calculator graphs of $y_1 = \tan x$ and $y_2 = 1$ in Fig. 3.94 support the conclusion that the solutions to $\tan x = 1$ are $180°$ apart.   □

b. Since $\tan^{-1}(-1.34) = -53.3°$, one solution is $\alpha = -53.3°$. Since all solutions to $\tan \alpha = -1.34$ differ by a multiple of $180°$, $-53.3° + 180°$, or $126.7°$ is the smallest positive solution. So the solution set is

$$\{\alpha \mid \alpha = 126.7° + k180°\}.$$ ■

To solve $\tan x = a$ for any real number a, we first find the smallest nonnegative solution. (In general, $s = \tan^{-1} a$ works if $s > 0$, and $s + \pi$ works if $s < 0$.) Next, add on all integral multiples of π. The solution to the equation $\tan x = a$ is summarized as follows.

SUMMARY   **Solving tan x = a**

If a is any real number and $s = \tan^{-1} a$, then the solution set to $\tan x = a$ is $\{x \mid x = s + k\pi\}$ for $s \geq 0$, and $\{x \mid x = s + \pi + k\pi\}$ for $s < 0$.

In the summaries for solving sine, cosine, and tangent equations, the domain of x is the set of real numbers. Similar summaries can be made if the domain of x is the set of degree measures. Start by finding $\sin^{-1} a$, $\cos^{-1} a$, or $\tan^{-1} a$ in degrees, then write the solution set in terms of multiples of $180°$ or $360°$.

Equations Involving Multiple Angles

Equations can involve expressions such as $\sin 2x$, $\cos 3\alpha$, or $\tan(s/2)$. These expressions involve a multiple of the variable rather than a single variable such as x, α, or s. In this case we solve for the multiple just as we would solve for a single variable and then find the value of the single variable in the last step of the solution.

Example **4** **A sine equation involving a double angle**

Find all solutions in degrees to $\sin 2\alpha = 1/\sqrt{2}$.

Solution

The only values for 2α between $0°$ and $360°$ that satisfy the equation are $45°$ and $135°$. (Note that $\sin^{-1}\left(1/\sqrt{2}\right) = 45°$ and $135° = 180° - 45°$.) See Fig. 3.95. So proceed as follows:

$$\sin 2\alpha = \frac{1}{\sqrt{2}}$$

$2\alpha = 45° + k360°$ or $2\alpha = 135° + k360°$

$\alpha = 22.5° + k180°$ or $\alpha = 67.5° + k180°$ Divide each side by 2.

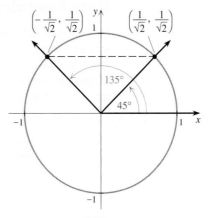

▪ **Figure 3.95**

The solution set is $\{\alpha \,|\, \alpha = 22.5° + k180° \text{ or } \alpha = 67.5° + k180°\}$, where k is any integer. ∎

Note that in Example 4, all possible values for 2α are found and *then* divided by 2 to get all possible values for α. Finding $\alpha = 22.5°$ and then adding on multiples of $360°$ will not produce the same solutions. Observe the same procedure in the next example, where we wait until the final step to divide each side by 3.

Example **5** **A tangent equation involving angle multiples**

Find all solutions to $\tan 3x = -\sqrt{3}$ in the interval $(0, 2\pi)$.

Solution

First find the smallest positive value for $3x$ that satisfies the equation. Then form all of the solutions by adding on multiples of π. Since $\tan^{-1}\left(-\sqrt{3}\right) = -\pi/3$, the smallest positive value for $3x$ that satisfies the equation is $-\pi/3 + \pi$, or $2\pi/3$. So proceed as follows:

$$\tan 3x = -\sqrt{3}$$

$$3x = \frac{2\pi}{3} + k\pi$$

$$x = \frac{2\pi}{9} + \frac{k\pi}{3}$$ Divide each side by 3.

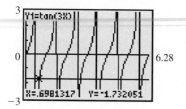

Figure 3.96

The solutions between 0 and 2π occur if $k = 0, 1, 2, 3, 4,$ and 5. If $k = 6$, then x is greater than 2π. So the solution set is

$$\left\{ \frac{2\pi}{9}, \frac{5\pi}{9}, \frac{8\pi}{9}, \frac{11\pi}{9}, \frac{14\pi}{9}, \frac{17\pi}{9} \right\}.$$

The graphs of $y_1 = \tan(3x)$ and $y_2 = -\sqrt{3}$ in Fig. 3.96 show the six solutions in the interval $(0, 2\pi)$.   ■

More Complicated Equations

More complicated equations involving trigonometric functions are solved by first solving for $\sin x$, $\cos x$, or $\tan x$, and then solving for x. In solving for the trigonometric functions, we may use trigonometric identities or properties of algebra, such as factoring or the quadratic formula. In stating formulas for solutions to equations, we will continue to use the letter k to represent any integer.

Example **6** **An equation solved by factoring**

Find all solutions in the interval $[0, 2\pi)$ to the equation

$$\sin 2x = \sin x.$$

Solution

First we use the identity $\sin 2x = 2 \sin x \cos x$ to get all of the trigonometric functions written in terms of the variable x alone. Then we rearrange and factor:

$$\sin 2x = \sin x$$
$$2 \sin x \cos x = \sin x \quad \text{By the double-angle identity}$$
$$2 \sin x \cos x - \sin x = 0 \quad \text{Subtract } \sin x \text{ from each side.}$$
$$\sin x (2 \cos x - 1) = 0 \quad \text{Factor.}$$
$$\sin x = 0 \quad \text{or} \quad 2 \cos x - 1 = 0 \quad \text{Set each factor equal to 0.}$$
$$x = k\pi \quad \text{or} \quad 2 \cos x = 1$$
$$\cos x = \frac{1}{2}$$
$$x = \frac{\pi}{3} + 2k\pi \quad \text{or} \quad x = \frac{5\pi}{3} + 2k\pi$$

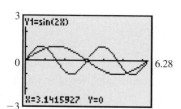

Figure 3.97

The only solutions in the interval $[0, 2\pi)$ are $0, \pi/3, \pi,$ and $5\pi/3$.

The graphs of $y_1 = \sin(2x)$ and $y_2 = \sin(x)$ in Fig. 3.97 show the four solutions to $\sin 2x = \sin x$ in the interval $[0, 2\pi)$.   ■

In algebraic equations, we generally do not divide each side by any expression that involves a variable, and the same rule holds for trigonometric equations. In Example 6, we did not divide by $\sin x$ when it appeared on opposite sides of the equation. If we had, the solutions 0 and π would have been lost.

In the next example, an equation of quadratic type is solved by factoring.

Example **7** **An equation of quadratic type**

Find all solutions to the equation

$$6 \cos^2\left(\frac{x}{2}\right) - 7 \cos\left(\frac{x}{2}\right) + 2 = 0$$

in the interval $[0, 2\pi)$. Round approximate answers to four decimal places.

Solution

Let $y = \cos(x/2)$ to get a quadratic equation:

$$6y^2 - 7y + 2 = 0$$

$$(2y - 1)(3y - 2) = 0 \qquad \text{Factor.}$$

$$2y - 1 = 0 \quad \text{or} \quad 3y - 2 = 0$$

$$y = \frac{1}{2} \quad \text{or} \quad y = \frac{2}{3}$$

$$\cos\frac{x}{2} = \frac{1}{2} \quad \text{or} \quad \cos\frac{x}{2} = \frac{2}{3} \qquad \text{Replace } y \text{ by } \cos(x/2).$$

$$\frac{x}{2} = \frac{\pi}{3} + 2k\pi \text{ or } \frac{5\pi}{3} + 2k\pi \quad \text{or} \quad \frac{x}{2} \approx 0.8411 + 2k\pi \text{ or } 5.4421 + 2k\pi$$

Now multiply by 2 to solve for x:

$$x = \frac{2\pi}{3} + 4k\pi \text{ or } \frac{10\pi}{3} + 4k\pi \quad \text{or} \quad x \approx 1.6821 + 4k\pi \text{ or } 10.8842 + 4k\pi$$

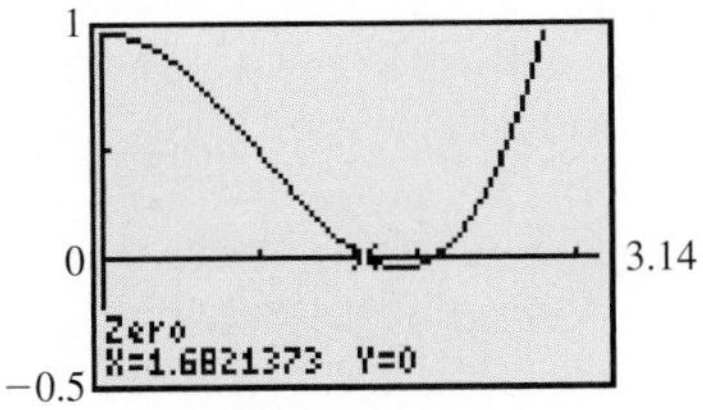

For all k, $10\pi/3 + 4k\pi$ and $10.8842 + 4k\pi$ are outside the interval $[0, 2\pi)$. The solutions in the interval $[0, 2\pi)$ are $2\pi/3$ and 1.6821.

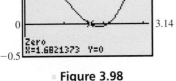 The graph of $y = 6(\cos(x/2))^2 - 7\cos(x/2) + 2$ in Fig. 3.98 appears to cross the x-axis at $2\pi/3$ and 1.6821. ■

For a trigonometric equation to be of quadratic type, it must be written in terms of a trigonometric function and the square of that function. In the next example, an identity is used to convert an equation involving sine and cosine to one with only sine. This equation is of quadratic type like Example 7, but it does not factor.

Example **8** **An equation solved by the quadratic formula**

Find all solutions to the equation

$$\cos^2 \alpha - 0.2 \sin \alpha = 0.9$$

in the interval $[0°, 360°)$. Round answers to the nearest tenth of a degree.

Solution

$$\cos^2 \alpha - 0.2 \sin \alpha = 0.9$$

$$1 - \sin^2 \alpha - 0.2 \sin \alpha = 0.9 \qquad \text{Replace } \cos^2 \alpha \text{ with } 1 - \sin^2 \alpha.$$

$$\sin^2 \alpha + 0.2 \sin \alpha - 0.1 = 0 \qquad \text{An equation of quadratic type}$$

$$\sin \alpha = \frac{-0.2 \pm \sqrt{(0.2)^2 - 4(1)(-0.1)}}{2} \qquad \begin{array}{l}\text{Use } a = 1, b = 0.2, \text{ and } c = -0.1 \\ \text{in the quadratic formula.}\end{array}$$

$$\sin \alpha \approx 0.2317 \quad \text{or} \quad \sin \alpha \approx -0.4317$$

Figure 3.98

Zero
X=1.6821373 Y=0

▪ **Figure 3.99**

Find two positive solutions to $\sin \alpha = 0.2317$ in $[0, 360°)$ by using a calculator to get $\alpha = \sin^{-1}(0.2317) \approx 13.4°$ and $180° - \alpha \approx 166.6°$. Now find two positive solutions to $\sin \alpha = -0.4317$ in $[0, 360°)$. Using a calculator, we get $\alpha = \sin^{-1}(-0.4317) \approx -25.6°$ and $180° - \alpha \approx 205.6°$. Since $-25.6°$ is negative, we use $-25.6° + 360°$ or $334.4°$ along with $205.6°$ as the two solutions. We list all possible solutions as

$$\alpha \approx 13.4°, \ 166.6°, \ 205.6°, \quad \text{or} \quad 334.4° \quad (+k360° \text{ in each case}).$$

The solutions in $[0°, 360°)$ are $13.4°$, $166.6°$, $205.6°$, and $334.4°$.

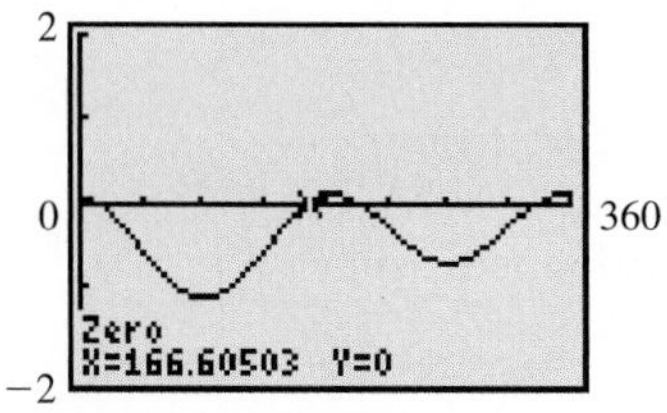

The graph of $y = \cos^2 x - 0.2 \sin x - 0.9$ in Fig. 3.99 appears to cross the x-axis four times in the interval $[0°, 360°)$. ■

The Pythagorean identities $\sin^2 x = 1 - \cos^2 x$, $\csc^2 x = 1 + \cot^2 x$, and $\sec^2 x = 1 + \tan^2 x$ are frequently used to replace one function by the other. If an equation involves sine and cosine, cosecant and cotangent, or secant and tangent, we might be able to square each side and then use these identities.

Example **9** **Square each side of the equation**

Find all values of y in the interval $[0, 360°)$ that satisfy the equation

$$\tan 3y + 1 = \sqrt{2} \sec 3y.$$

Solution

$$(\tan 3y + 1)^2 = \left(\sqrt{2} \sec 3y\right)^2 \qquad \text{Square each side.}$$
$$\tan^2 3y + 2 \tan 3y + 1 = 2 \sec^2 3y$$
$$\tan^2 3y + 2 \tan 3y + 1 = 2(\tan^2 3y + 1) \qquad \text{Since } \sec^2 x = \tan^2 x + 1$$
$$-\tan^2 3y + 2 \tan 3y - 1 = 0 \qquad \text{Subtract } 2\tan^2 3y + 2 \text{ from both sides.}$$
$$\tan^2 3y - 2 \tan 3y + 1 = 0$$
$$(\tan 3y - 1)^2 = 0$$
$$\tan 3y - 1 = 0$$
$$\tan 3y = 1$$
$$3y = 45° + k180°$$

Because we squared each side, we must check for extraneous roots. First check $3y = 45°$. If $3y = 45°$, then $\tan 3y = 1$ and $\sec 3y = \sqrt{2}$. Substituting these values into $\tan 3y + 1 = \sqrt{2} \sec 3y$ gives us

$$1 + 1 = \sqrt{2} \cdot \sqrt{2}. \qquad \text{Correct.}$$

If we add any *even* multiple of $180°$, or any multiple of $360°$, to $45°$, we get the same values for $\tan 3y$ and $\sec 3y$. So for any k, $3y = 45° + k360°$ satisfies the original equation. Now check $45°$ plus *odd* multiples of $180°$. If $3y = 225°$ ($k = 1$), then

$$\tan 3y = 1 \qquad \text{and} \qquad \sec 3y = -\sqrt{2}.$$

These values do not satisfy the original equation. Since $\tan 3y$ and $\sec 3y$ have these same values for $3y = 45° + k180°$ for any odd k, the only solutions are of the form

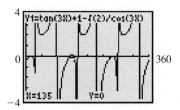

■ **Figure 3.100**

$3y = 45° + k360°$, or $y = 15° + k120°$. The solutions in the interval $[0°, 360°)$ are $15°$, $135°$, and $255°$.

The graph of $y = \tan(3x) + 1 - \sqrt{2}/\cos(3x)$ in Fig. 3.100 appears to touch the x-axis at these three locations. ■

There is no single method that applies to all trigonometric equations, but the following strategy will help you to solve them.

STRATEGY	**Solving Trigonometric Equations**

1. Know the solutions to $\sin x = a$, $\cos x = a$, and $\tan x = a$.

2. Solve an equation involving only multiple angles as if the equation had a single variable.

3. Simplify complicated equations by using identities. Try to get an equation involving only one trigonometric function.

4. If possible, factor to get different trigonometric functions into separate factors.

5. For equations of quadratic type, solve by factoring or the quadratic formula.

6. Square each side of the equation, if necessary, so that identities involving squares can be applied. (Check for extraneous roots.)

Modeling Projectile Motion

The distance d (in feet) traveled by a projectile fired from the ground with an angle of elevation θ is related to the initial velocity v_0 (in feet per second) by the equation $v_0^2 \sin 2\theta = 32d$. If the projectile is pictured as being fired from the origin into the first quadrant, then the x- and y-coordinates (in feet) of the projectile at time t (in seconds) are given by $x = v_0 t \cos \theta$ and $y = -16t^2 + v_0 t \sin \theta$.

Example 🔟 The path of a projectile

A catapult is placed 100 feet from the castle wall, which is 35 feet high. The soldier wants the burning bale of hay to clear the top of the wall and land 50 feet inside the castle wall. If the initial velocity of the bale is 70 feet per second, then at what angle should the bale of hay be launched so that it will travel 150 feet and pass over the castle wall?

Solution

Use the equation $v_0^2 \sin 2\theta = 32d$ with $v_0 = 70$ feet per second and $d = 150$ feet to find θ:

$$70^2 \sin 2\theta = 32(150)$$

$$\sin 2\theta = \frac{32(150)}{70^2} \approx 0.97959$$

The launch angle θ must be in the interval $(0°, 90°)$, so we look for values of 2θ in the interval $(0°, 180°)$. Since $\sin^{-1}(0.97959) \approx 78.4°$, both $78.4°$ and $180° - 78.4° = 101.6°$ are possible values for 2θ. So possible values for θ are $39.2°$ and $50.8°$. See Fig. 3.101 on the next page.

▪ Foreshadowing Calculus

A standard calculus problem is to show that a projectile's maximum distance for a given velocity is achieved at 45°. As illustrated in Example 11, there are two angles that can be used to achieve a distance less than the maximum, one less than 45° and one greater than 45°.

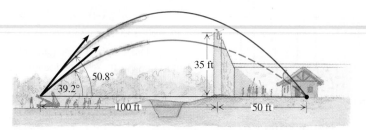

▪ **Figure 3.101**

Use the equation $x = v_0 t \cos \theta$ to find the time at which the bale is 100 feet from the catapult (measured horizontally) by using each of the possible values for θ:

$$100 = 70t \cos(39.2°) \qquad\qquad 100 = 70t \cos(50.8°)$$

$$t = \frac{100}{70 \cos(39.2°)} \approx 1.84 \text{ seconds} \qquad t = \frac{100}{70 \cos(50.8°)} \approx 2.26 \text{ seconds}$$

Use the equation $y = -16t^2 + v_0 t \sin \theta$ to find the altitude of the bale at time $t = 1.84$ seconds and $t = 2.26$ seconds:

$$y = -16(1.84)^2 + 70(1.84) \sin 39.2° \qquad y = -16(2.26)^2 + 70(2.26) \sin 50.8°$$

$$\approx 27.2 \text{ feet} \qquad\qquad\qquad \approx 40.9 \text{ feet}$$

If the burning bale is launched on a trajectory with an angle of 39.2°, then it will have an altitude of only 27.2 feet when it reaches the castle wall. If it is launched with an angle of 50.8°, then it will have an altitude of 40.9 feet when it reaches the castle wall. Since the castle wall is 35 feet tall, the 50.8° angle must be used for the bale to reach its intended target. ■

As illustrated in Example 10 there are two launch angles, one less than 45° and one greater than 45°, that can be used to propel a projectile a given distance with a given initial velocity. When you throw a ball you usually choose the smaller launch angle so that the ball reaches its target in less time. A standard calculus problem is to show that a projectile's maximum distance for a given initial velocity is achieved at a launch angle of 45°. If you want to throw a ball a fixed distance with a minimum of effort, then what launch angle would you use?

For Thought

True or False? Explain.

1. The only solutions to $\cos \alpha = 1/\sqrt{2}$ in [0°, 360°) are 45° and 135°.

2. The only solution to $\sin x = -0.55$ in $[0, \pi)$ is $\sin^{-1}(-0.55)$.

3. $\{x \mid x = -29° + k360°\} = \{x \mid x = 331° + k360°\}$, where k is any integer.

4. The solution set to $\tan x = -1$ is $\left\{x \mid x = \frac{7\pi}{4} + k\pi\right\}$, where k is any integer.

5. $2 \cos^2 x + \cos x - 1 = (2 \cos x - 1)(\cos x + 1)$ is an identity.

6. The equation $\sin^2 x = \sin x \cos x$ is equivalent to $\sin x = \cos x$.

7. One solution to $\sec x = 2$ is $\dfrac{1}{\cos^{-1}(2)}$.

8. The solution set to $\cot x = 3$ for x in $[0, \pi)$ is $\{\tan^{-1}(1/3)\}$.

9. The equation $\sin x = \cos x$ is equivalent to $\sin^2 x = \cos^2 x$.

10. $\left\{ x \mid 3x = \dfrac{\pi}{2} + 2k\pi \right\} = \left\{ x \mid x = \dfrac{\pi}{6} + 2k\pi \right\}$, where k is any integer.

3.8 Exercises

Find all real numbers that satisfy each equation. Do not use a calculator. See the summaries for solving $\cos x = a$, $\sin x = a$, and $\tan x = a$ on pages 276, 277, and 278.

1. $\cos x = -1$

2. $\cos x = 0$

3. $\sin x = 0$

4. $\sin x = 1$

5. $\sin x = -1$

6. $\cos x = 1$

7. $\cos x = 1/2$

8. $\cos x = \sqrt{2}/2$

9. $\sin x = \sqrt{2}/2$

10. $\sin x = \sqrt{3}/2$

11. $\tan x = 1$

12. $\tan x = \sqrt{3}/3$

13. $\cos x = -\sqrt{3}/2$

14. $\cos x = -\sqrt{2}/2$

15. $\sin x = -\sqrt{2}/2$

16. $\sin x = -\sqrt{3}/2$

17. $\tan x = -1$

18. $\tan x = -\sqrt{3}$

Find all angles in degrees that satisfy each equation. Round approximate answers to the nearest tenth of a degree.

19. $\cos \alpha = 0$ **20.** $\cos \alpha = -1$ **21.** $\sin \alpha = 1$

22. $\sin \alpha = -1$ **23.** $\tan \alpha = 0$ **24.** $\tan \alpha = -1$

25. $\cos \alpha = 0.873$ **26.** $\cos \alpha = -0.158$

27. $\sin \alpha = -0.244$ **28.** $\sin \alpha = 0.551$

29. $\tan \alpha = 5.42$ **30.** $\tan \alpha = -2.31$

Find all real numbers that satisfy each equation.

31. $\cos(x/2) = 1/2$ **32.** $2 \cos 2x = -\sqrt{2}$

33. $\cos 3x = 1$ **34.** $\cos 2x = 0$

35. $2 \sin(x/2) - 1 = 0$ **36.** $\sin 2x = 0$

37. $2 \sin 2x = -\sqrt{2}$ **38.** $\sin(x/3) + 1 = 0$

39. $\tan 2x = \sqrt{3}$

40. $\sqrt{3} \tan(3x) + 1 = 0$

41. $\tan 4x = 0$

42. $\tan 3x = -1$

43. $\sin(\pi x) = 1/2$

44. $\tan(\pi x/4) = 1$

45. $\cos(2\pi x) = 0$

46. $\sin(3\pi x) = 1$

Find all values of α in $[0°, 360°)$ that satisfy each equation.

47. $2 \sin \alpha = -\sqrt{3}$

48. $\tan \alpha = -\sqrt{3}$

49. $\sqrt{2} \cos 2\alpha - 1 = 0$

50. $\sin 6\alpha = 1$

51. $\sec 3\alpha = -\sqrt{2}$

52. $\csc(5\alpha) + 2 = 0$

53. $\cot(\alpha/2) = \sqrt{3}$

54. $\sec(\alpha/2) = \sqrt{2}$

Find all values of α in degrees that satisfy each equation. Round approximate answers to the nearest tenth of a degree.

55. $\sin 3\alpha = 0.34$

56. $\cos 2\alpha = -0.22$

57. $\sin 3\alpha = -0.6$

58. $\tan 4\alpha = -3.2$

59. $\sec 2\alpha = 4.5$

60. $\csc 3\alpha = -1.4$

61. $\csc(\alpha/2) = -2.3$

62. $\cot(\alpha/2) = 4.7$

Find all real numbers in the interval $[0, 2\pi)$ that satisfy each equation. Round approximate answers to the nearest tenth. See the strategy for solving trigonometric equations on page 283.

63. $3 \sin^2 x = \sin x$

64. $2 \tan^2 x = \tan x$

65. $2 \cos^2 x + 3 \cos x = -1$

66. $2 \sin^2 x + \sin x = 1$

67. $5 \sin^2 x - 2 \sin x = \cos^2 x$

68. $\sin^2 x - \cos^2 x = 0$

69. $\tan x = \sec x - \sqrt{3}$

70. $\csc x - \sqrt{3} = \cot x$

71. $\sin x + \sqrt{3} = 3\sqrt{3}\cos x$

72. $6\sin^2 x - 2\cos x = 5$

73. $\tan x \sin 2x = 0$

74. $3\sec^2 x \tan x = 4\tan x$

75. $\sin 2x - \sin x \cos x = \cos x$

76. $2\cos^2 2x - 8\sin^2 x \cos^2 x = -1$

77. $\sin x \cos(\pi/4) + \cos x \sin(\pi/4) = 1/2$

78. $\sin(\pi/6)\cos x - \cos(\pi/6)\sin x = -1/2$

79. $\sin 2x \cos x - \cos 2x \sin x = -1/2$

80. $\cos 2x \cos x - \sin 2x \sin x = 1/2$

Find all values of θ in the interval [0°, 360°) that satisfy each equation. Round approximate answers to the nearest tenth of a degree.

81. $\cos^2\left(\dfrac{\theta}{2}\right) = \sec\theta$

82. $2\sin^2\left(\dfrac{\theta}{2}\right) = \cos\theta$

83. $2\sin\theta = \cos\theta$

84. $3\sin 2\theta = \cos 2\theta$

85. $\sin 3\theta = \csc 3\theta$

86. $\tan^2\theta - \cot^2\theta = 0$

87. $\tan^2\theta - 2\tan\theta - 1 = 0$

88. $\cot^2\theta - 4\cot\theta + 2 = 0$

89. $9\sin^2\theta + 12\sin\theta + 4 = 0$

90. $12\cos^2\theta + \cos\theta - 6 = 0$

91. $\dfrac{\tan 3\theta - \tan\theta}{1 + \tan 3\theta \tan\theta} = \sqrt{3}$

92. $\dfrac{\tan 3\theta + \tan 2\theta}{1 - \tan 3\theta \tan 2\theta} = 1$

93. $8\cos^4\theta - 10\cos^2\theta + 3 = 0$

94. $4\sin^4\theta - 5\sin^2\theta + 1 = 0$

95. $\sec^4\theta - 5\sec^2\theta + 4 = 0$

96. $\cot^4\theta - 4\cot^2\theta + 3 = 0$

Solve each problem.

97. *Motion of a Spring*   A block is attached to a spring and set in motion on a frictionless plane. Its location on the surface at any time t in seconds is given in meters by

$$x = \sqrt{3}\sin 2t + \cos 2t.$$

For what values of t is the block at its resting position $x = 0$?

98. *Motion of a Spring*   A block is set in motion hanging from a spring and oscillates about its resting position $x = 0$ according to the function $x = -0.3\sin 3t + 0.5\cos 3t$. For what values of t is the block at its resting position $x = 0$?

99. *Firing an M-16*   A soldier is accused of breaking a window 3300 ft away during target practice. If the muzzle velocity for an M-16 is 325 ft/sec, then at what angle would it have to be aimed for the bullet to travel 3300 ft? The distance d (in feet) traveled by a projectile fired at an angle θ is related to the initial velocity v_0 (in feet per second) by the equation $v_0^2 \sin 2\theta = 32d$.

100. *Firing an M-16*   If you were accused of firing an M-16 into the air and breaking a window 4000 ft away, what would be your defense?

101. *Choosing the Right Angle*   Cincinnati Reds centerfielder Ken Griffey, Jr., fields a ground ball and attempts to make a 230-ft throw to home plate. Given that Griffey commonly makes long throws at 90 mph, find the two possible angles at which he can throw the ball to home plate. Find the time saved by choosing the smaller angle.

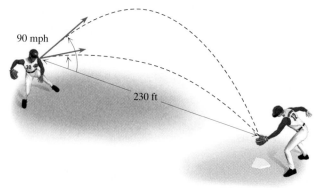

90 mph

230 ft

■ **Figure for Exercise 101**

102. *Muzzle Velocity*   The 8-in. (diameter) howitzer on the U.S. Army's M-110 can propel a projectile a distance of 18,500 yd. If the angle of elevation of the barrel is 45°, then what muzzle velocity (in feet per second) is required to achieve this distance?

Thinking Outside the Box XXVI

Two Common Triangles   An equilateral triangle with sides of length 1 and an isosceles right triangle with legs of length 1 are positioned as shown in the accompanying diagram. Find the exact area of the shaded triangle.

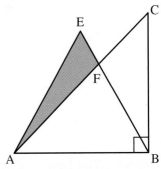

▪ **Figure for Thinking Outside the Box XXVI**

Pop Quiz

Find all angles α in degrees that satisfy each equation.

1. $\sin \alpha = \sqrt{2}/2$

2. $\cos \alpha = 1/2$

3. $\tan \alpha = -1$

Find all real numbers in $[0, 2\pi]$ that satisfy each equation.

4. $\sin(x/2) = 1/2$

5. $\cos(x) = 1$

6. $\tan(2x) = 1$

The Law of Sines and the Law of Cosines

In Section 3.6 we used trigonometry to solve right triangles. In this section and the next we will learn to solve any triangle for which we have enough information to determine the shape of the triangle.

Oblique Triangles

Any triangle without a right angle is called an **oblique triangle.** As usual, we use α, β, and γ for the angles of a triangle and a, b, and c, respectively, for the lengths of the sides opposite those angles, as shown in Fig. 3.102. The vertices at angles α, β, and γ are labeled A, B, and C, respectively. To solve an oblique triangle, we must know at least three parts of the triangle, at least one of which must be the length of a side. We can classify the different cases for the three known parts as follows:

1. One side and any two angles (ASA or AAS)

2. Two sides and a nonincluded angle (SSA)

3. Two sides and an included angle (SAS)

4. Three sides (SSS)

We can actually solve all of these cases by dividing the triangles into right triangles and using right triangle trigonometry. Since that method is quite tedious, we

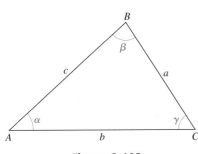

▪ **Figure 3.102**

develop the *law of sines* and the *law of cosines*. The first two cases can be handled with the law of sines. We will discuss the last two cases when we develop the law of cosines.

The Law of Sines

The **law of sines** says that *the ratio of the sine of an angle and the length of the side opposite the angle is the same for each angle of a triangle.*

Theorem: The Law of Sines

In any triangle,

$$\frac{\sin \alpha}{a} = \frac{\sin \beta}{b} = \frac{\sin \gamma}{c}.$$

PROOF Either the triangle is an acute triangle (all acute angles) or it is an obtuse triangle (one obtuse angle). We consider the case of the obtuse triangle here. The case of the acute triangle is proved similarly. Triangle ABC is shown in Fig. 3.103 with an altitude of length h_1 drawn from point C to the opposite side and an altitude of length h_2 drawn from point B to the extension of the opposite side.

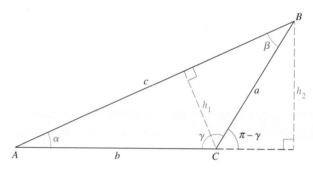

■ **Figure 3.103**

Since α and β are now in right triangles, we have

$$\sin \alpha = \frac{h_1}{b} \qquad \text{or} \qquad h_1 = b \sin \alpha$$

and

$$\sin \beta = \frac{h_1}{a} \qquad \text{or} \qquad h_1 = a \sin \beta.$$

Replace h_1 by $b \sin \alpha$ in the equation $h_1 = a \sin \beta$ to get

$$b \sin \alpha = a \sin \beta.$$

Dividing each side of the last equation by ab yields

$$\frac{\sin \alpha}{a} = \frac{\sin \beta}{b}.$$

Using the largest right triangle in Fig. 3.103, we have

$$\sin \alpha = \frac{h_2}{c} \quad \text{or} \quad h_2 = c \sin \alpha.$$

Using the identity for the sine of a difference we have

$$\sin(\pi - \gamma) = \sin \pi \cos \gamma - \cos \pi \sin \gamma$$

$$= \sin \gamma.$$

Since $\pi - \gamma$ is an angle of a right triangle in Fig. 3.103, we have

$$\sin \gamma = \sin(\pi - \gamma) = \frac{h_2}{a} \quad \text{or} \quad h_2 = a \sin \gamma.$$

From $h_2 = c \sin \alpha$ and $h_2 = a \sin \gamma$ we get $c \sin \alpha = a \sin \gamma$ or

$$\frac{\sin \alpha}{a} = \frac{\sin \gamma}{c}$$

and we have proved the law of sines.   ■

The law of sines can also be written in the form

$$\frac{a}{\sin \alpha} = \frac{b}{\sin \beta} = \frac{c}{\sin \gamma}.$$

In solving triangles, it is usually simplest to use the form in which the unknown quantity appears in the numerator.

In our first example we use the law of sines to solve a triangle for which we are given two angles and an included side.

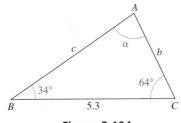

■ **Figure 3.104**

Example 1 Given two angles and an included side (ASA)

Given $\beta = 34°$, $\gamma = 64°$, and $a = 5.3$, solve the triangle.

Solution

To sketch the triangle, first draw side a, then draw angles of approximately 34° and 64° on opposite ends of a. Label all parts (see Fig. 3.104). Since the sum of the three angles of a triangle is 180°, the third angle α is 82°. By the law of sines,

$$\frac{5.3}{\sin 82°} = \frac{b}{\sin 34°}$$

$$b = \frac{5.3 \sin 34°}{\sin 82°} \approx 3.0.$$

Again, by the law of sines,

$$\frac{c}{\sin 64°} = \frac{5.3}{\sin 82°}$$

$$c = \frac{5.3 \sin 64°}{\sin 82°} \approx 4.8.$$

So $\alpha = 82°$, $b \approx 3.0$, and $c \approx 4.8$ solves the triangle.   ■

The AAS case is similar to the ASA case. If two angles are known, then the third can be found by using the fact that the sum of all angles is 180°. If we know all angles and any side, then we can proceed to find the remaining sides with the law of sines as in Example 1.

The Ambiguous Case (SSA)

In the AAS and ASA cases we are given any two angles with positive measures and the length of any side. If the total measure of the two angles is less than 180°, then a unique triangle is determined. However, for two sides and a *nonincluded* angle (SSA), there are several possibilities. So the SSA case is called the **ambiguous case.** Drawing the diagram in the proper order will help you in understanding the ambiguous case.

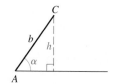

Draw α and b: find h

■ **Figure 3.105**

Suppose we are given an acute angle α $(0° < \alpha < 90°)$ and sides a and b. Side a is opposite the angle α, and side b is adjacent to it. Draw an angle of approximate size α in standard position with terminal side of length b, as shown in Fig. 3.105. Don't draw in side a yet. Let h be the distance from C to the initial side of α. Since $\sin \alpha = h/b$, we have $h = b \sin \alpha$. Now we are ready to draw in side a, but there are four possibilities for its location.

1. If $a < h$, then no triangle can be formed, because a cannot reach from point C to the initial side of α. This situation is shown in Fig. 3.106(a) .
2. If $a = h$, then exactly one right triangle is formed, as in Fig. 3.106(b).
3. If $h < a < b$, then exactly two triangles are formed, because a reaches to the initial side in two places, as in Fig. 3.106(c).
4. If $a \geq b$, then only one triangle is formed, as in Fig. 3.106(d).

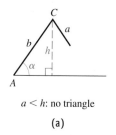
$a < h$: no triangle
(a)

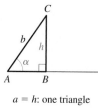

$a = h$: one triangle
(b)

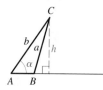
$h < a < b$: two triangles
(c)

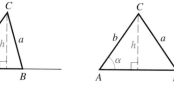
$a \geq b$: one triangle
(d)

■ **Figure 3.106**

If we start with α, a, and b, where $90° \leq \alpha < 180°$, then there are only two possibilities for the number of triangles determined. If $a \leq b$, then no triangle is formed, as in Fig. 3.107(a). If $a > b$, one triangle is formed, as in Fig. 3.107(b).

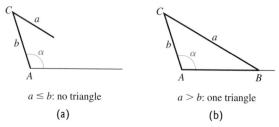
$a \leq b$: no triangle
(a)

$a > b$: one triangle
(b)

■ **Figure 3.107**

It is not necessary to memorize all of the SSA cases shown in Figs. 3.106 and 3.107. If you draw the triangle for a given problem in the order suggested, then it will be clear how many triangles are possible with the given parts. We must decide how many triangles are possible, before we can solve the triangle(s).

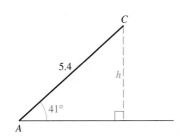

■ **Figure 3.108**

Example **2** SSA with no triangle

Given $\alpha = 41°$, $a = 3.3$, and $b = 5.4$, solve the triangle.

Solution

Draw the 41° angle and label its terminal side 5.4, as shown in Fig. 3.108. Side a must go opposite α, but do not put it in yet. Find the length of the altitude h from C. Using a trigonometric ratio, we get $\sin 41° = h/5.4$, or $h = 5.4 \sin 41° \approx 3.5$. Since $a = 3.3$, a is *shorter* than the altitude h, and side a will not reach from point C to the initial side of α. So there is no triangle with the given parts. ■

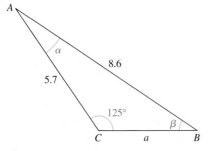

■ **Figure 3.109**

Example **3** SSA with one triangle

Given $\gamma = 125°$, $b = 5.7$, and $c = 8.6$, solve the triangle.

Solution

Draw the 125° angle and label its terminal side 5.7. Since $8.6 > 5.7$, side c will reach from point A to the initial side of γ, as shown in Fig. 3.109, and form a single triangle. To solve this triangle, we use the law of sines:

$$\frac{\sin 125°}{8.6} = \frac{\sin \beta}{5.7}$$

$$\sin \beta = \frac{5.7 \sin 125°}{8.6}$$

$$\beta = \sin^{-1}\left(\frac{5.7 \sin 125°}{8.6}\right) \approx 32.9°$$

Since the sum of α, β, and γ is 180°, $\alpha = 22.1°$. Now use α and the law of sines to find a:

$$\frac{a}{\sin 22.1°} = \frac{8.6}{\sin 125°}$$

$$a = \frac{8.6 \sin 22.1°}{\sin 125°} \approx 3.9$$ ■

Example **4** SSA with two triangles

Given $\beta = 56.3°$, $a = 8.3$, and $b = 7.6$, solve the triangle.

Solution

Draw an angle of approximately 56.3°, and label its terminal side 8.3. Side b must go opposite β, but do not put it in yet. Find the length of the altitude h from point C to the initial side of β, as shown in Fig. 3.110(a) on the next page. Since $\sin 56.3° = h/8.3$, we get $h = 8.3 \sin 56.3° \approx 6.9$. Because b is longer than h but

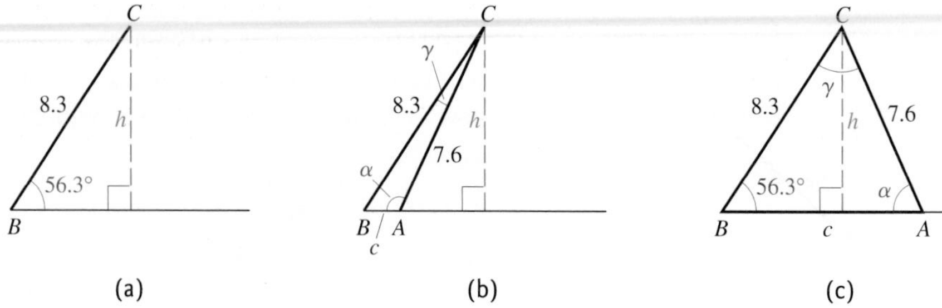

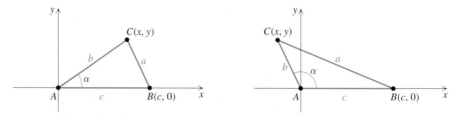

(a)   **(b)**   **(c)**

■ **Figure 3.110**

shorter than a, there are two triangles that satisfy the given conditions, as shown in parts (b) and (c) of Fig. 3.110. In either triangle we have

$$\frac{\sin \alpha}{8.3} = \frac{\sin 56.3°}{7.6}$$

$$\sin \alpha = 0.9086.$$

This equation has two solutions in $[0°, 180°]$. For part (c) we get $\alpha = \sin^{-1}(0.9086) = 65.3°$, and for part (b) we get $\alpha = 180° - 65.3° = 114.7°$. Using the law of sines, we get $\gamma = 9.0°$ and $c = 1.4$ for part (b), and we get $\gamma = 58.4°$ and $c = 7.8$ for part (c). ■

The Law of Cosines

The **law of cosines** gives a formula for the square of any side of an oblique triangle in terms of the other two sides and their included angle.

Theorem: Law of Cosines

If triangle ABC is an oblique triangle with sides a, b, and c and angles α, β, and γ, then

$$a^2 = b^2 + c^2 - 2bc \cos \alpha,$$

$$b^2 = a^2 + c^2 - 2ac \cos \beta,$$

$$c^2 = a^2 + b^2 - 2ab \cos \gamma.$$

PROOF   Given triangle ABC, position the triangle as shown in Fig. 3.111. The vertex C is in the first quadrant if α is acute and in the second if α is obtuse. Both cases are shown in Fig. 3.111.

■ **Figure 3.111**

In either case, the x-coordinate of C is $x = b \cos \alpha$, and the y-coordinate of C is $y = b \sin \alpha$. The distance from C to B is a, but we can also find that distance by using the distance formula:

$$a = \sqrt{(b \cos \alpha - c)^2 + (b \sin \alpha - 0)^2}$$

$$a^2 = (b \cos \alpha - c)^2 + (b \sin \alpha)^2$$

$$= b^2 \cos^2 \alpha - 2bc \cos \alpha + c^2 + b^2 \sin^2 \alpha$$

$$= b^2(\cos^2 \alpha + \sin^2 \alpha) + c^2 - 2bc \cos \alpha$$

Using $\cos^2 x + \sin^2 x = 1$, we get the first equation of the theorem:

$$a^2 = b^2 + c^2 - 2bc \cos \alpha$$

Similar arguments with B and C at $(0, 0)$ produce the other two equations. ▪

In any triangle with unequal sides, the largest angle is opposite the largest side, and the smallest angle is opposite the smallest side. We need this fact in the first example.

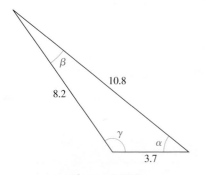

Figure 3.112

Example ▪5▪ Given three sides of a triangle (SSS)

Given $a = 8.2$, $b = 3.7$, and $c = 10.8$, solve the triangle.

Solution

Draw the triangle and label it as in Fig. 3.112. Since c is the longest side, use

$$c^2 = a^2 + b^2 - 2ab \cos \gamma$$

to find the largest angle γ:

$$-2ab \cos \gamma = c^2 - a^2 - b^2$$

$$\cos \gamma = \frac{c^2 - a^2 - b^2}{-2ab} = \frac{(10.8)^2 - (8.2)^2 - (3.7)^2}{-2(8.2)(3.7)} \approx -0.5885$$

$$\gamma = \cos^{-1}(-0.5885) \approx 126.1°$$

We could finish with the law of cosines, but in this case it is simpler to use the law of sines, which involves fewer computations:

$$\frac{\sin \beta}{b} = \frac{\sin \gamma}{c}$$

$$\frac{\sin \beta}{3.7} = \frac{\sin 126.1°}{10.8}$$

$$\sin \beta \approx 0.2768$$

There are two solutions to $\sin \beta = 0.2768$ in $[0°, 180°]$. However, β must be less than $90°$, because γ is $126.1°$. So $\beta = \sin^{-1}(0.2768) \approx 16.1°$. Finally, $\alpha = 180° - 126.1° - 16.1° = 37.8°$. ▪

In solving the SSS case, we always *find the largest angle first,* using the law of cosines. The remaining two angles must be acute angles. So when using the law of sines to find another angle, we need only find an acute solution to the equation.

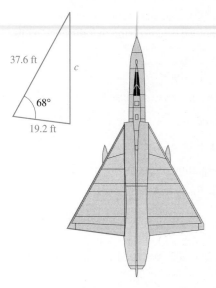

We have seen that two adjacent sides and a nonincluded angle (SSA) might not determine a triangle. To determine a triangle with three given sides (SSS), the sum of the lengths of any two must be greater than the length of the third side. This fact is called the **triangle inequality.** To understand the triangle inequality, try to draw a triangle with sides of lengths 1 in., 2 in., and 5 in. Two sides and the included angle (SAS) will determine a triangle provided the angle is between 0° and 180°.

Example **6** **Given two sides and an included angle (SAS)**

The wing of the F-106 Delta Dart is triangular in shape, with the dimensions given in Fig. 3.113. Find the length of the side labeled c in Fig. 3.113.

Solution

Using the law of cosines, we find c as follows:

$$c^2 = (19.2)^2 + (37.6)^2 - 2(19.2)(37.6)\cos 68° \approx 1241.5$$
$$c \approx \sqrt{1241.5} \approx 35.2 \text{ feet}$$

▪ **Figure 3.113**

PROCEDURE **Solving Triangles (In all cases draw pictures.)**

ASA (For example α, c, β)

1. Find γ using $\gamma = 180° - \alpha - \beta$.

2. Find a and c using the law of sines.

SSA (For example a, b, α)

1. Find $h = b\sin\alpha$. If $h > a$, then there is no triangle.

2. If $h = a$, then there is one right triangle ($\beta = 90°$ and b is the hypotenuse). Solve it using right triangle trigonometry.

3. If $h < a < b$ there are two triangles, one with β acute and one with β obtuse. Find the acute β using the law of sines. Subtract it from 180° to get the obtuse β. In each of the two triangles, find γ using $\gamma = 180° - \alpha - \beta$ and c using the law of sines.

4. If $a \geq b$ there is only one triangle and β is acute (α or γ might be obtuse). Find β using the law of sines. Then find γ using $\gamma = 180° - \alpha - \beta$. Find c using the law of sines.

SSS (For example a, b, c)

1. Find the largest angle using the law of cosines. The largest angle is opposite the largest side.

2. Find another angle using the law of sines.

3. Find the third angle by subtracting the first two from 180°.

SAS (For example b, α, c)

1. Find a using the law of cosines.

2. Use the law of sines to find β if $b < c$ or γ if $c < b$. If $b = c$, then find either β or γ.

3. Find the last angle by subtracting the first two from 180°.

Applications

When you reach for a light switch, your brain controls the angles at your elbow and your shoulder that put your hand at the location of the switch. An arm on a robot works in much the same manner. In the next example we see how the law of sines and the law of cosines are used to determine the proper angles at the joints so that a robot's hand can be moved to a position given in the coordinates of the workspace.

Example **7** Positioning a robotic arm

A robotic arm with a 0.5-meter segment and a 0.3-meter segment is attached at the origin, as shown in Fig. 3.114. The computer-controlled arm is positioned by rotating each segment through angles θ_1 and θ_2, as shown in Fig. 3.114. Given that we want to have the end of the arm at the point (0.7, 0.2), find θ_1 and θ_2 to the nearest tenth of a degree.

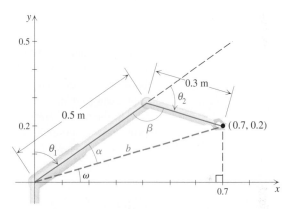

Figure 3.114

Historical Note

Archimedes of Syracuse (287 B.C.–212 B.C.) was a Greek mathematician, physicist, engineer, astronomer, and philosopher. Many consider him one of the greatest mathematicians in antiquity. Archimedes seems to have carried out the first theoretical calculation of π, concluding that $223/71 < \pi < 22/7$. He used a limiting process very similar to what you will do if you try the Concepts of Calculus at the end of this chapter.

Solution

In the right triangle shown in Fig. 3.114, we have $\tan \omega = 0.2/0.7$ and

$$\omega = \tan^{-1}\left(\frac{0.2}{0.7}\right) \approx 15.95°.$$

To get the angles to the nearest tenth, we use more accuracy along the way and round to the nearest tenth only on the final answer. Find b using the Pythagorean theorem:

$$b = \sqrt{0.7^2 + 0.2^2} \approx 0.73$$

Find β using the law of cosines:

$$(0.73)^2 = 0.5^2 + 0.3^2 - 2(0.5)(0.3) \cos \beta$$

$$\cos \beta = \frac{0.73^2 - 0.5^2 - 0.3^2}{-2(0.5)(0.3)} = -0.643$$

$$\beta = \cos^{-1}(-0.643) \approx 130.02°$$

Find α using the law of sines:

$$\frac{\sin \alpha}{0.3} = \frac{\sin 130.02°}{0.73}$$

$$\sin \alpha = \frac{0.3 \cdot \sin 130.02°}{0.73} = 0.3147$$

$$\alpha = \sin^{-1}(0.3147) \approx 18.34°$$

Since $\theta_1 = 90° - \alpha - \omega$,

$$\theta_1 = 90° - 18.34° - 15.95° = 55.71°.$$

Since β and θ_2 are supplementary,

$$\theta_2 = 180° - 130.02° = 49.98°.$$

So the longer segment of the arm is rotated about 55.7° and the shorter segment is rotated about 50.0°. ■

Since the angles in Example 7 describe a clockwise rotation, the angles could be given negative signs to indicate the direction of rotation. In robotics, the direction of rotation is important, because there may be more than one way to position a robotic arm at a desired location. In fact, Example 7 has infinitely many solutions. See if you can find another one.

In the next example we find the height of an object from a distance. This same problem was solved in Section 3.6, Example 6. Note how much easier the solution is here using the law of sines.

Example **8** **Finding the height of an object from a distance**

The angle of elevation of the top of a water tower from point A on the ground is 19.9°. From point B, 50.0 feet closer to the tower, the angle of elevation is 21.8°. What is the height of the tower?

Solution

Let y represent the height of the tower. All angles of triangle ABC can be determined as shown in Fig. 3.115.

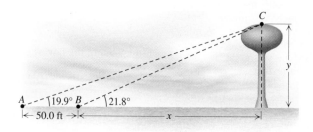

■ **Figure 3.115**

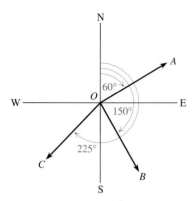

Figure 3.116

Apply the law of sines in triangle ABC:

$$\frac{a}{\sin 19.9°} = \frac{50}{\sin 1.9°}$$

$$a = \frac{50 \sin 19.9°}{\sin 1.9°} \approx 513.3 \text{ feet}$$

Now, using the smaller right triangle, we have $\sin 21.8° = y/513.3$ or $y = 513.3 \sin 21.8° \approx 191$. So the height of the tower is 191 feet. ■

The measure of an angle that describes the direction of a ray is called the **bearing** of the ray. In air navigation, bearing is given as a nonnegative angle less than 360° measured in a clockwise direction from a ray pointing due north. So in Fig. 3.116 the bearing of ray $\overrightarrow{OA}$ is 60°, the bearing of ray $\overrightarrow{OB}$ is 150°, and the bearing of $\overrightarrow{OC}$ is 225°.

Example 9 Using bearing in solving triangles

A bush pilot left the Fairbanks Airport in a light plane and flew 100 miles toward Fort Yukon in still air on a course with a bearing of 18°. She then flew due east (bearing 90°) for some time to drop supplies to a snowbound family. After the drop, her course to return to Fairbanks had a bearing of 225°. What was her maximum distance from Fairbanks?

Solution

Figure 3.117 shows the course of her flight. To change course from bearing 18° to bearing 90° at point B, the pilot must add 72° to the bearing. So $\angle ABC$ is 108°. A bearing of 225° at point C means that $\angle BCA$ is 45°. Finally, we obtain $\angle BAC = 27°$ by subtracting 18° and 45° from 90°.

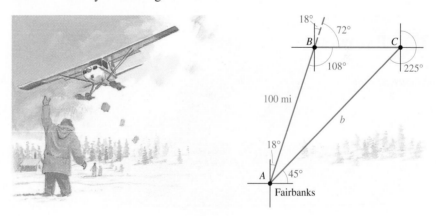

Figure 3.117

We can find the length of $\overline{AC}$ (the maximum distance from Fairbanks) by using the law of sines:

$$\frac{b}{\sin 108°} = \frac{100}{\sin 45°}$$

$$b = \frac{100 \cdot \sin 108°}{\sin 45°} \approx 134.5$$

So the pilot's maximum distance from Fairbanks was 134.5 miles. ■

In marine navigation and surveying, the bearing of a ray is the acute angle the ray makes with a ray pointing due north or due south. Along with the acute angle, directions are given that indicate in which quadrant the ray lies. For example, in Fig. 3.116, $\overrightarrow{OA}$ has a bearing 60° east of north (N60°E), $\overrightarrow{OB}$ has a bearing 30° east of south (S30°E), and $\overrightarrow{OC}$ has a bearing 45° west of south (S45°W).

For Thought

True or False? Explain.

1. If we know the measures of two angles of a triangle, then the measure of the third angle is determined.

2. If $\frac{\sin 9°}{a} = \frac{\sin 17°}{88}$, then $a = \frac{\sin 17°}{88 \sin 9°}$.

3. The equation $\frac{\sin \alpha}{5} = \frac{\sin 44°}{18}$ has exactly one solution in $[0, 180°]$.

4. One solution to $\frac{\sin \beta}{2.3} = \frac{\sin 39°}{1.6}$ is $\beta = \sin^{-1}\left(\frac{2.3 \sin 39°}{1.6}\right)$.

5. $\frac{\sin 60°}{\sqrt{3}} = \frac{\sin 30°}{1}$

6. No triangle exists with $\alpha = 60°$, $b = 10$ feet, and $a = 500$ feet.

7. If $\gamma = 90°$ in triangle ABC, then $c^2 = a^2 + b^2$.

8. If a, b, and c are the sides of a triangle, then $a = \sqrt{c^2 + b^2 - 2bc \cos \gamma}$.

9. If a, b, and c are the sides of any triangle, then $c^2 = a^2 + b^2$.

10. The smallest angle of a triangle lies opposite the shortest side.

3.9 Exercises

Solve each triangle.

1.

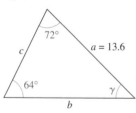

2.

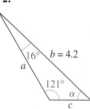

3.

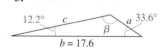

4.

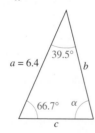

Solve each triangle with the given parts.

5. $\alpha = 10.3°$, $\gamma = 143.7°$, $c = 48.3$

6. $\beta = 94.7°$, $\alpha = 30.6°$, $b = 3.9$

7. $\beta = 120.7°$, $\gamma = 13.6°$, $a = 489.3$

8. $\alpha = 39.7°$, $\gamma = 91.6°$, $b = 16.4$

Determine the number of triangles with the given parts and solve each triangle.

9. $\alpha = 39.6°$, $c = 18.4$, $a = 3.7$

10. $\beta = 28.6°$, $a = 40.7$, $b = 52.5$

11. $\gamma = 60°$, $b = 20$, $c = 10\sqrt{3}$

12. $\alpha = 41.2°$, $a = 8.1$, $b = 10.6$

13. $\beta = 138.1°$, $c = 6.3$, $b = 15.6$

14. $\gamma = 128.6°, a = 9.6, c = 8.2$

15. $\beta = 32.7°, a = 37.5, b = 28.6$

16. $\alpha = 30°, c = 40, a = 20$

17. $\gamma = 99.6°, b = 10.3, c = 12.4$

18. $\alpha = 75.3, a = 12.4, b = 9.8$

Solve each triangle.

19.

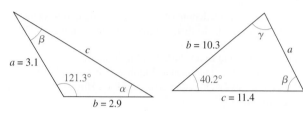

20.

21.

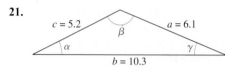

22.

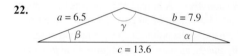

Solve each triangle with the given information. See the procedure for solving triangles on page 294.

23. $a = 6.8, c = 2.4, \beta = 10.5°$

24. $a = 1.3, b = 14.9, \gamma = 9.8°$

25. $a = 18.5, b = 12.2, c = 8.1$

26. $a = 30.4, b = 28.9, c = 31.6$

27. $b = 9.3, c = 12.2, \alpha = 30°$

28. $a = 10.3, c = 8.4, \beta = 88°$

29. $a = 6.3, b = 7.1, c = 6.8$

30. $a = 4.1, b = 9.8, c = 6.2$

31. $a = 7.2, \beta = 25°, \gamma = 35°$

32. $b = 12.3, \alpha = 20°, \gamma = 120°$

Determine the number of triangles with the given parts.

33. $a = 3, b = 4, c = 7$

34. $a = 2, b = 9, c = 5$

35. $a = 10, b = 5, c = 8$

36. $a = 3, b = 15, c = 16$

37. $c = 10, \alpha = 40°, \beta = 60°, \gamma = 90°$

38. $b = 6, \alpha = 62°, \gamma = 120°$

39. $b = 10, c = 1, \alpha = 179°$

40. $a = 10, c = 4, \beta = 2°$

41. $b = 8, c = 2, \gamma = 45°$

42. $a = \sqrt{3}/2, b = 1, \alpha = 60°$

Solve each problem.

43. *Observing Traffic* A traffic report helicopter left the WKPR studios on a course with a bearing of 210°. After flying 12 mi to reach interstate highway 20, the helicopter flew due east along I-20 for some time. The helicopter headed back to WKPR on a course with a bearing of 310° and reported no accidents along I-20. For how many miles did the helicopter fly along I-20?

44. *Course of a Fighter Plane* During an important NATO exercise, an F-14 Tomcat left the carrier Nimitz on a course with a bearing of 34° and flew 400 mi. Then the F-14 flew for some distance on a course with a bearing of 162°. Finally, the plane flew back to its starting point on a course with a bearing of 308°. What distance did the plane fly on the final leg of the journey?

45. *Cellular One* The angle of elevation of the top of a cellular telephone tower from point *A* on the ground is 18.1°. From point *B*, 32.5 ft closer to the tower, the angle of elevation is 19.3°. What is the height of the tower?

46. *Moving Back* A surveyor determines that the angle of elevation of the top of a building from a point on the ground is 30.4°. He then moves back 55.4 ft and determines that the angle of elevation is 23.2°. What is the height of the building?

47. *Designing an Addition* A 40-ft-wide house has a roof with a 6-12 pitch (the roof rises 6 ft for a run of 12 ft). The owner plans a 14-ft-wide addition that will have a 3-12 pitch to its roof. Find the lengths of $\overline{AB}$ and $\overline{BC}$ in the accompanying figure on the next page.

■ **Figure for Exercise 47**

48. *Shot Down* A cruise missile is traveling straight across the desert at 548 mph at an altitude of 1 mile, as shown in the figure. A gunner spots the missile coming in his direction and fires a projectile at the missile when the angle of elevation of the missile is 35°. If the speed of the projectile is 688 mph, then for what angle of elevation of the gun will the projectile hit the missile?

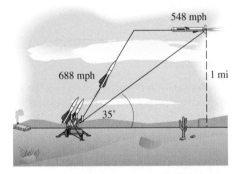

■ **Figure for Exercise 48**

49. *Length of a Chord* What is the length of the chord intercepted by a central angle of 19° in a circle of radius 30 ft?

50. *Boating* The boat shown in the accompanying figure is 3 mi from both lighthouses and the angle between the line of sight to the lighthouses is 20°. Find the distance between the lighthouses.

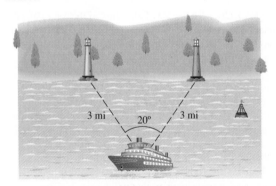

■ **Figure for Exercise 50**

51. *The Pentagon* The Pentagon in Washington, D.C. is 921 ft on each side, as shown in the accompanying figure. What is the distance from a vertex to the center of the Pentagon?

■ **Figure for Exercise 51**

52. *A Hexagon* If the length of each side of a regular hexagon is 10 ft, then what is the distance from a vertex to the center?

53. *Hiking* Jan and Dean started hiking from the same location at the same time. Jan hiked at 4 mph with bearing N12°E, and Dean hiked at 5 mph with bearing N31°W. How far apart were they after 6 hr?

54. *Flying* Andrea and Carlos left the airport at the same time. Andrea flew at 180 mph on a course with bearing 80°, and Carlos flew at 240 mph on a course with bearing 210°. How far apart were they after 3 hr?

55. *Positioning a Human Arm* A human arm consists of an upper arm of 30 cm and a lower arm of 30 cm, as shown in the figure. To move the hand to the point (36, 8), the human brain chooses angle θ_1 and θ_2, as shown in the figure. Find θ_1 and θ_2 to the nearest tenth of a degree.

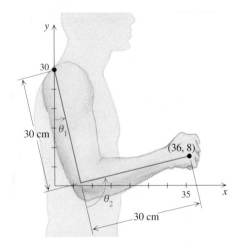

■ **Figure for Exercise 55**

56. *Attack of the Grizzly* A forest ranger is 150 ft above the ground in a fire tower when she spots an angry grizzly bear east of the tower with an angle of depression of 10°, as shown in the figure. Southeast of the tower she spots a hiker with an angle of depression of 15°. Find the distance between the hiker and the angry bear.

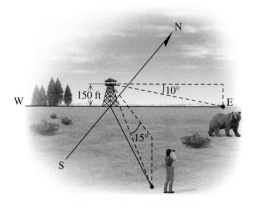

■ **Figure for Exercise 56**

Thinking Outside the Box XXVII

Watering the Lawn Josie places her lawn sprinklers at the vertices of a triangle that has sides of 9 m, 10 m, and 11 m. The sprinklers water in circular patterns with radii of 4, 5, and 6 m. No area is watered by more than one sprinkler. What amount of area inside the triangle is not watered by any of the three sprinklers? Give your answer to the nearest thousandth of a square meter.

1. If $\alpha = 80°$, $\beta = 121°$, and $c = 12$, then what is γ?

2. If $\alpha = 20.4°$, $\beta = 27.3°$, and $c = 38.5$, then what is a?

3. If $\alpha = 33.5°$, $a = 7.4$, and $b = 10.6$, then what is β?

4. If $\alpha = 12.3°$, $b = 10.4$, and $c = 8.1$, then what is a?

5. If $a = 6$, $b = 7$, and $c = 12$, then what is γ?

∎∎∎ Highlights

3.1 Angles and Their Measurements

Degree Measure	Divide the circumference of a circle into 360 equal arcs. The degree measure is the number of degrees through which the initial side of an angle is rotated to get to the terminal side (positive for counterclockwise, negative for clockwise).	90° is 1/4 of a circle. 180° is 1/2 of a circle.
Radian Measure	The length of the arc on the unit circle through which the initial side rotates to get to the terminal side	$\pi/2$ is 1/4 of the unit circle. π is 1/2 of the unit circle.
Converting	Use π radians $= 180°$ and cancellation of units.	$90° \cdot \dfrac{\pi \text{ rad}}{180°} = \dfrac{\pi}{2} \text{ rad}$
Arc Length	$s = \alpha r$ where s is the arc intercepted by a central angle of α radians on a circle of radius r	$\alpha = 90°$, $r = 10$ ft $s = \alpha r = \dfrac{\pi}{2} \cdot 10 \text{ ft} = 5\pi \text{ ft}$

3.2 The Sine and Cosine Functions

Sine and Cosine Functions	If α is an angle in standard position and (x, y) is the point of intersection of the terminal side and the unit circle, then $\sin \alpha = y$ and $\cos \alpha = x$.	$\sin(90°) = 1$ $\cos(\pi/2) = 0$ $\sin(0) = 0$
Domain	The domain for sine or cosine can be the set of angles in standard position, the measures of those angles, or the set of real numbers.	$\sin(30°) = 1/2$ $\sin(\pi/4) = \sqrt{2}/2$ $\sin(72.6) \approx -0.3367$
Fundamental Identity	For any angle α (or real number α), $\sin^2(\alpha) + \cos^2(\alpha) = 1$.	$\sin^2(30°) + \cos^2(30°) = 1$ $\sin^2\left(\sqrt[3]{2}\right) + \cos^2\left(\sqrt[3]{2}\right) = 1$

3.3 The Graphs of the Sine and Cosine Functions

Graphs	The graphs of $y = \sin x$ and $y = \cos x$ are sine waves, each with period 2π.	
General Sine and Cosine Functions	The graph of $y = A \sin[B(x - C)] + D$ or $y = A \cos[B(x - C)] + D$ is a sine wave with amplitude $|A|$, period $2\pi/|B|$, phase shift C, and vertical translation D.	$y = -3 \sin(2(x - \pi)) + 1$ amplitude 3, period π, phase shift π, vertical translation 1
Graphing a General Sine Function	Start with the five key points on $y = \sin x$: $(0, 0), \left(\frac{\pi}{2}, 1\right), (\pi, 0), \left(\frac{3\pi}{2}, -1\right), (2\pi, 0)$ Divide each x-coordinate by B and add C. Multiply each y-coordinate by A and add D. Sketch one cycle of the general function through the five new points. Graph a general cosine function in the same manner.	$y = 3 \sin\left(2\left(x - \frac{\pi}{2}\right)\right) + 1$ Five new points: $\left(\frac{\pi}{2}, 1\right), \left(\frac{3\pi}{4}, 4\right),$ $(\pi, 1), \left(\frac{5\pi}{4}, -2\right), \left(\frac{3\pi}{2}, 1\right)$ Draw one cycle through these points.
Frequency	$F = 1/P$, P is the period and F is the frequency.	$y = \sin(24\pi x), F = 12$

3.4 The Other Trigonometric Functions and Their Graphs

Tangent, Cotangent, Secant, and Cosecant	If (x, y) is the point of intersection of the unit circle and the terminal side of α, then $\tan \alpha = y/x$, $\cot \alpha = x/y$, $\sec \alpha = 1/x$, and $\csc \alpha = 1/y$. If 0 occurs in a denominator the function is undefined.	$\tan(\pi/4) = 1$ $\cot(\pi/2) = 0$ $\sec(30°) = 2/\sqrt{3}$ $\csc(0)$ is undefined.
Identities	$\tan \alpha = \dfrac{\sin \alpha}{\cos \alpha}, \quad \cot \alpha = \dfrac{\cos \alpha}{\sin \alpha}, \quad \sec \alpha = \dfrac{1}{\cos \alpha}, \quad \csc \alpha = \dfrac{1}{\sin \alpha}$	
Tangent	Period π, fundamental cycle on $(-\pi/2, \pi/2)$, vertical asymptotes $x = \pi/2 + k\pi$	
Cotangent	Period π, fundamental cycle on $(0, \pi)$, vertical asymptotes $x = k\pi$	
Secant	Period 2π, vertical asymptotes $x = \pi/2 + k\pi$	
Cosecant	Period 2π, vertical asymptotes $x = k\pi$	

3.5 **The Inverse Trigonometric Functions**

Inverse Sine If $y = \sin^{-1} x$ for x in $[-1, 1]$, then y is the real number in $[-\pi/2, \pi/2]$ such that $\sin y = x$.

$\sin^{-1}(1) = \pi/2$
$\sin^{-1}(-1/2) = -\pi/6$

Inverse Cosine If $y = \cos^{-1} x$ for x in $[-1, 1]$, then y is the real number in $[0, \pi]$ such that $\cos y = x$.

$\cos^{-1}(-1) = \pi$
$\cos^{-1}(-1/2) = 2\pi/3$

Inverse Tangent If $y = \tan^{-1} x$ for x in $(-\infty, \infty)$, then y is the real number in $(-\pi/2, \pi/2)$ such that $\tan y = x$.

$\tan^{-1}(1) = \pi/4$
$\tan^{-1}(-1) = -\pi/4$

3.6 **Right Triangle Trigonometry**

Trigonometric Ratios If (x, y) is any point other than $(0, 0)$ on the terminal side of α is standard position and $r = \sqrt{x^2 + y^2}$, then $\sin \alpha = y/r$, $\cos \alpha = x/r$, and $\tan \alpha = y/x$ $(x \neq 0)$.

Terminal side of α through $(3, 4)$, gives $r = 5$ and $\sin \alpha = 4/5$, $\cos \alpha = 3/5$, and $\tan \alpha = 4/3$.

Trigonometric Functions in a Right Triangle If α is an acute angle of a right triangle, then $\sin \alpha = $ opp/hyp, $\cos \alpha = $ adj/hyp, and $\tan \alpha = $ opp/adj.

$\sin \alpha = 5/13$, $\cos \alpha = 12/13$, $\tan \alpha = 5/12$

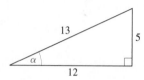

3.7 **Identities**

Identity An equation that is satisfied by every number for which both sides are defined.

$\sin^2 x + \cos^2 x = 1$

Disproving an Identity To show that an equation is not an identity, find a value for the variable that does not satisfy the equation.

$\sin(2x) = 2 \sin(x)$ is not an identity because $\sin(2 \cdot \pi/2) \neq 2 \sin(\pi/2)$.

Proving an Identity To verify or prove an identity, start with the expression on one side and use known identities and properties of algebra to convert it into the expression on the other side.

$$\frac{\sin x + \cos x}{\cos x} = \frac{\sin x}{\cos x} + \frac{\cos x}{\cos x}$$
$$= \tan x + 1$$

Reduction Formula If α is an angle in standard position whose terminal side contains (a, b) and x is a real number, then $a \sin x + b \cos x = \sqrt{a^2 + b^2} \sin(x + \alpha)$.

$\alpha = \tan^{-1}(4/3)$
$3 \sin x + 4 \cos x$
$= 5 \sin(x + \tan^{-1}(4/3))$

3.8 **Conditional Trigonometric Equations**

Solving $\cos x = a$ If $|a| \leq 1$, find all solutions in $[0, 2\pi]$, then add all multiples of 2π to them and simplify. If $|a| > 1$ there are no solutions.

$\cos x = 0, x = \dfrac{\pi}{2} + k\pi$

$\cos x = \dfrac{1}{2}, x = \dfrac{\pi}{3} + 2k\pi$

or $x = \dfrac{5\pi}{3} + 2k\pi$

Solving sin $x = a$	If $\|a\| \leq 1$, find all solutions in $[0, 2\pi]$, then add all multiples of 2π to them and simplify. If $\|a\| > 1$ there are no solutions.	$\sin x = 0, x = k\pi$ $\sin x = \dfrac{1}{2}, x = \dfrac{\pi}{6} + 2k\pi$ or $x = \dfrac{5\pi}{6} + 2k\pi$
Solving tan $x = a$	If a is a real number, find all solutions in $[-\pi/2, \pi/2]$, then add multiples of π to them and simplify.	$\tan x = 1, x = \dfrac{\pi}{4} + k\pi$

3.9 The Law of Sines and The Law of Cosines

Law of Sines	In any triangle $\dfrac{\sin \alpha}{a} = \dfrac{\sin \beta}{b} = \dfrac{\sin \gamma}{c}$.	$a = \sqrt{3}, b = 1, c = 2$ $\alpha = \pi/3, \beta = \pi/6, \gamma = \pi/2$ $\dfrac{\sin\left(\frac{\pi}{3}\right)}{\sqrt{3}} = \dfrac{\sin\left(\frac{\pi}{6}\right)}{1} = \dfrac{\sin\left(\frac{\pi}{2}\right)}{2}$
Law of Cosines	If a is the side opposite angle α in any triangle then $a^2 = b^2 + c^2 - 2bc \cos \alpha$.	$a = 4, b = 5, c = 6$ $4^2 = 5^2 + 6^2 - 60 \cos \alpha$

Function Gallery: **Some Basic Functions of Trigonometry**

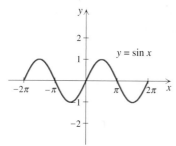

Period 2π
Amplitude 1

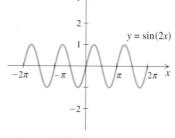

Period π
Amplitude 1

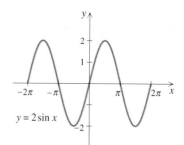

Period 2π
Amplitude 2

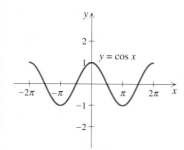

Period 2π
Amplitude 1

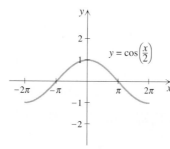

Period 4π
Amplitude 1

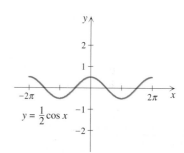

Period 2π
Amplitude $\frac{1}{2}$

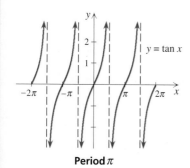

Period π

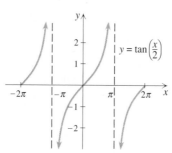

Period 2π

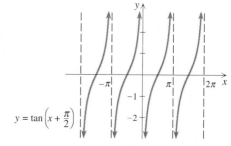

Period π

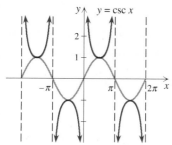

Period 2π

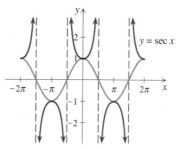

Period 2π

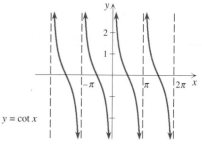

Period π

▪▪▪ Chapter 3 Review Exercises

Find the measure in degrees of the least positive angle that is coterminal with each given angle.

1. $388°$ **2.** $-840°$ **3.** $-153°14'27''$

4. $455°39'24''$ **5.** $-\pi$ **6.** $-35\pi/6$

7. $13\pi/5$ **8.** $29\pi/12$

Convert each radian measure to degree measure. Do not use a calculator.

9. $5\pi/3$ **10.** $-3\pi/4$ **11.** $3\pi/2$ **12.** $5\pi/6$

Convert each degree measure to radian measure. Do not use a calculator.

13. $330°$ **14.** $405°$ **15.** $-300°$ **16.** $-210°$

Fill in the tables. Do not use a calculator.

17.

θ deg	0	30	45	60	90	120	135	150	180
θ rad									
$\sin\theta$									
$\cos\theta$									
$\tan\theta$									

18.

θ rad	0	$\dfrac{\pi}{6}$	$\dfrac{\pi}{4}$	$\dfrac{\pi}{3}$	$\dfrac{\pi}{2}$	$\dfrac{2\pi}{3}$	$\dfrac{3\pi}{4}$	$\dfrac{5\pi}{6}$	π
θ deg									
$\sin\theta$									
$\cos\theta$									
$\tan\theta$									

Give the exact values of each of the following expressions. Do not use a calculator.

19. $\sin(-\pi/4)$ **20.** $\cos(-2\pi/3)$ **21.** $\tan(\pi/3)$

22. $\sec(\pi/6)$ **23.** $\csc(-120°)$ **24.** $\cot(135°)$

25. $\sin(180°)$ **26.** $\tan(0°)$ **27.** $\cos(3\pi/2)$

28. $\csc(5\pi/6)$ **29.** $\sec(-\pi)$ **30.** $\cot(-4\pi/3)$

31. $\cot(420°)$ **32.** $\sin(390°)$ **33.** $\cos(-135°)$

34. $\tan(225°)$ **35.** $\sec(2\pi/3)$ **36.** $\csc(-3\pi/4)$

37. $\tan(5\pi/6)$ **38.** $\sin(7\pi/6)$

For each triangle shown below, find the exact values of $\sin\alpha$, $\cos\alpha$, $\tan\alpha$, $\csc\alpha$, $\sec\alpha$, and $\cot\alpha$.

39. 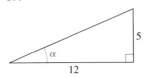   **40.**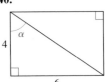

Find an approximate value for each expression. Round to four decimal places.

41. $\sin(44°)$ **42.** $\cos(-205°)$ **43.** $\cos(4.62)$

44. $\sin(3.14)$ **45.** $\tan(\pi/17)$ **46.** $\sec(2.33)$

47. $\csc(105°4')$ **48.** $\cot(55°3'12'')$

Find the exact value of each expression.

49. $\sin^{-1}(-0.5)$ **50.** $\cos^{-1}(-0.5)$

51. $\arctan(-1)$ **52.** $\operatorname{arccot}(1/\sqrt{3})$

53. $\sec^{-1}(\sqrt{2})$ **54.** $\csc^{-1}(\sec(\pi/3))$

55. $\sin^{-1}(\sin(5\pi/6))$ **56.** $\cos(\cos^{-1}(-\sqrt{3}/2))$

Find the exact value of each expression in degrees.

57. $\sin^{-1}(1)$ **58.** $\tan^{-1}(1)$

59. $\arccos(-1/\sqrt{2})$ **60.** $\operatorname{arcsec}(2)$

61. $\cot^{-1}(\sqrt{3})$ **62.** $\cot^{-1}(-\sqrt{3})$

63. $\operatorname{arccot}(0)$ **64.** $\operatorname{arccot}(-\sqrt{3}/3)$

Solve each right triangle with the given parts.

65. $a = 2, b = 3$ **66.** $a = 3, c = 7$

67. $a = 3.2, \alpha = 21.3°$ **68.** $\alpha = 34.6°, c = 9.4$

Sketch at least one cycle of the graph of each function and determine the period and range.

69. $f(x) = 2\sin(3x)$ **70.** $f(x) = 1 + \cos(x + \pi/4)$

71. $y = \tan(2x + \pi)$ **72.** $y = \cot(x - \pi/4)$

73. $y = \sec\left(\dfrac{1}{2}x\right)$ **74.** $y = \csc\left(\dfrac{\pi}{2}x\right)$

75. $y = \dfrac{1}{2}\cos(2x)$

76. $y = 1 - \sin(x - \pi/3)$

77. $y = \cot(2x + \pi/3)$

78. $y = 2\tan(x + \pi/4)$

79. $y = \dfrac{1}{3}\csc(2x + \pi)$

80. $y = 1 + 2\sec(x - \pi/4)$

For each of the following sine curves find an equation of the form
$y = A\sin[B(x - C)] + D$.

81.

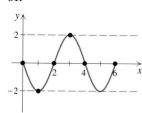

82.

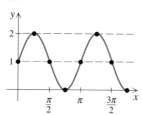

83.

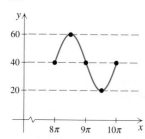

84.

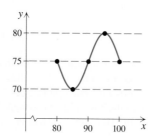

Solve each problem.

85. Find α if α is the angle between $90°$ and $180°$ whose sine is $1/2$.

86. Find β if β is the angle between $180°$ and $270°$ whose cosine is $-\sqrt{3}/2$.

87. Find $\sin \alpha$, given that $\cos \alpha = 1/5$ and α is in quadrant IV.

88. Find $\tan \alpha$, given that $\sin \alpha = 1/3$ and α is in quadrant II.

89. Find x if x is the length of the shortest side of a right triangle that has a $16°$ angle and a hypotenuse of 24 ft.

90. Find x if x is the length of the hypotenuse of a right triangle that has a $22°$ angle and a shortest side of length 12 m.

91. Find α if α is the largest acute angle of a right triangle that has legs with lengths 6 cm and 8 cm.

92. Find β if β is the largest acute angle of a right triangle that has a hypotenuse with length 19 yd and a leg with length 8 yd.

Simplify each expression.

93. $(1 - \sin \alpha)(1 + \sin \alpha)$

94. $\csc x \tan x + \sec(-x)$

95. $(1 - \csc x)(1 - \csc(-x))$

96. $\dfrac{\cos^2 x - \sin^2 x}{\sin 2x}$

97. $\dfrac{1}{1 + \sin \alpha} - \dfrac{\sin(-\alpha)}{\cos^2 \alpha}$

98. $2\sin\left(\dfrac{\pi}{2} - \alpha\right)\cos\left(\dfrac{\pi}{2} - \alpha\right)$

99. $\dfrac{2\tan 2s}{1 - \tan^2 2s}$

100. $\dfrac{\tan 2w - \tan 4w}{1 + \tan 2w \tan 4w}$

101. $\sin 3\theta \cos 6\theta - \cos 3\theta \sin 6\theta$

102. $\dfrac{\sin 2y}{1 + \cos 2y}$

103. $\dfrac{1 - \cos 2z}{\sin 2z}$

104. $\cos^2\left(\dfrac{x}{2}\right) - \sin^2\left(\dfrac{x}{2}\right)$

Prove that each of the following equations is an identity.

105. $\sec 2\theta = \dfrac{1 + \tan^2 \theta}{1 - \tan^2 \theta}$

106. $\tan^2 \theta = \dfrac{1 - \cos 2\theta}{1 + \cos 2\theta}$

107. $\sin^2\left(\dfrac{x}{2}\right) = \dfrac{\csc^2 x - \cot^2 x}{2\csc^2 x + 2\csc x \cot x}$

108. $\cot(-x) = \dfrac{1 - \sin^2 x}{\cos(-x)\sin(-x)}$

109. $\cot(\alpha - 45°) = \dfrac{1 + \tan \alpha}{\tan \alpha - 1}$

110. $\cos(\alpha + 45°) = \dfrac{\cos \alpha - \sin \alpha}{\sqrt{2}}$

111. $\dfrac{\sin 2\beta}{2\csc \beta} = \sin^2 \beta \cos \beta$

112. $\sin(45° - \beta) = \dfrac{\cos 2\beta}{\sqrt{2}(\cos \beta + \sin \beta)}$

113. $\dfrac{\cot^3 y - \tan^3 y}{\sec^2 y + \cot^2 y} = 2\cot 2y$

114. $\dfrac{\sin^3 y - \cos^3 y}{\sin y - \cos y} = \dfrac{2 + \sin 2y}{2}$

Find all real numbers that satisfy each equation.

115. $2\cos 2x + 1 = 0$

116. $2\sin 2x + \sqrt{3} = 0$

117. $\left(\sqrt{3}\csc x - 2\right)(\csc x - 2) = 0$

118. $\left(\sec x - \sqrt{2}\right)\left(\sqrt{3}\sec x + 2\right) = 0$

119. $2 \sin^2 x + 1 = 3 \sin x$ **120.** $4 \sin^2 x = \sin x + 3$

121. $-8\sqrt{3} \sin \frac{x}{2} = -12$ **122.** $-\cos \frac{x}{2} = \sqrt{2} + \cos \frac{x}{2}$

123. $\cos \frac{x}{2} - \sin x = 0$ **124.** $\sin 2x = \tan x$

125. $\cos 2x + \sin^2 x = 0$ **126.** $\tan \frac{x}{2} = \sin x$

Solve each triangle that exists with the given parts. If there is more than one triangle with the given parts, then solve each one.

127. $\gamma = 48°, a = 3.4, b = 2.6$

128. $a = 6, b = 8, c = 10$

129. $\alpha = 13°, \beta = 64°, c = 20$

130. $\alpha = 50°, a = 3.2, b = 8.4$

131. $a = 3.6, b = 10.2, c = 5.9$

132. $\beta = 36.2°, \gamma = 48.1°, a = 10.6$

133. $a = 30.6, b = 12.9, c = 24.1$

134. $\alpha = 30°, a = \sqrt{3}, b = 2\sqrt{3}$

135. $\beta = 22°, c = 4.9, b = 2.5$

136. $\beta = 121°, a = 5.2, c = 7.1$

Solve each problem.

137. *Broadcasting the Oldies*   If radio station Q92 is broadcasting its oldies at 92.3 FM, then it is broadcasting at a frequency of 92.3 megahertz, or 92.3×10^6 cycles per second. What is the period of a wave with this frequency? (FM stands for frequency modulation.)

138. *AM Radio*   If WLS in Chicago is broadcasting at 890 AM (amplitude modulation), then its signal has a frequency of 890 kilohertz, or 890×10^3 cycles per second. What is the period of a wave with this frequency?

139. *Crooked Man*   A man is standing 1000 ft from a surveyor. The surveyor measures an angle of 0.4° sighting from the man's feet to his head. Assume that the man can be represented by the arc intercepted by a central angle of 0.4° in a circle of radius 1000 ft. Find the height of the man.

140. *Straight Man*   Assume that the man of Exercise 139 can be represented by the side a of a right triangle with angle $\alpha = 0.4°$ and $b = 1000$ ft. Find the height of the man, and compare your answer with the answer of the previous exercise.

141. *Oscillating Depth*   The depth of water in a tank oscillates between 12 ft and 16 ft. It takes 10 min for the depth to go from 12 to 16 and 10 min for the depth to go from 16 to 12 ft.

Express the depth as a function of time in the form $y = A \sin[B(x - C)] + D$, where the depth is 16 ft at time 0. Graph one cycle of the function.

142. *Oscillating Temperature*   The temperature of the water in a tank oscillates between 100°F and 120°F. It takes 30 min for the temperature to go from 100° to 120° and 30 min for the temperature to go from 120° to 100°. Express the temperature as a function of time in the form $y = A \sin[B(x - C)] + D$, where the temperature is 100° at time 20 min. Graph one cycle of the function.

143. *Shooting a Target*   Judy is standing 200 ft from a circular target with a radius of 3 in. To hit the center of the circle, she must hold the gun perfectly level, as shown in the figure. Will she hit the target if her aim is off by one-tenth of a degree in any direction?

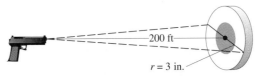

■ **Figure for Exercise 143**

144. *Height of Buildings*   Two buildings are 300 ft apart. From the top of the shorter building, the angle of elevation of the top of the taller building is 23°, and the angle of depression of the base of the taller building is 36°, as shown in the figure. How tall is each building?

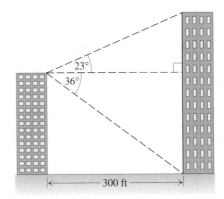

■ **Figure for Exercise 144**

145. *John Hancock*   From a point on the street the angle of elevation of the top of the John Hancock Building is 65.7°. From a point on the street that is 100 ft closer to the building the angle of elevation is 70.1°. Find the height of the building.

146. *Cloud Height*   Visual Flight Rules require that the height of the clouds be more than 1000 ft for a pilot to fly without instrumentation. At night, cloud height can be determined from the ground by using a searchlight and an observer, as shown in the figure. If the beam of light is aimed straight upward and the observer 500 ft away sights the cloud with an angle of elevation of 55°, then what is the height of the cloud cover?

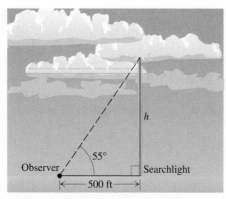

■ **Figure for Exercise 146**

147. *Radar Range* The radar antenna on a cargo ship is located 90 feet above the water, as shown in the accompanying figure. Find the distance to the horrizon from that height. Use 3950 miles as the radius of the earth.

148. *Increasing Visibility* To increase its radar visibility, a sailboat has a radar reflector mounted on the mast, 25 feet above the water. Find the distance to the horizon from that height. Assuming that radar waves travel in a straight line, at what distance can the cargo ship of the previous exercise get a radar image of the sailboat?

■ **Figure for Exercises 147 and 148**

Thinking Outside the Box XXVIII

Buckling Bridge A 100 ft bridge expands 1 in. during the heat of the day. Since the ends of the bridge are embedded in rock, the bridge buckles upward and forms an arc of a circle for which the original bridge is a chord. What is the approximate distance moved by the center of the bridge?

Concepts of Calculus

Area of a circle and π

The idea of a limiting value to a process was used long before it was formalized in calculus. Ancient mathematicians discovered the formula for the area of a circle and the value of π using the formula for the area of a triangle $A = \frac{1}{2}bh$, trigonometry, and the idea of limits. For the following exercises imagine that you are an ancient mathematician (with a modern calculator), but you have never heard of π or the formula $A = \pi r^2$.

Exercises

1. A regular pentagon is inscribed in a circle of radius r as shown in the following figure.

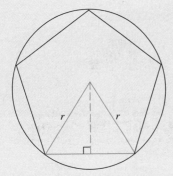

The pentagon is made up of five isosceles triangles. Show that the area of the pentagon is $\frac{5\sin(72°)}{2}r^2$.

2. A regular polygon of n sides is inscribed in a circle of radius r. Write the area of the n-gon in terms of r and n.

3. The area of a regular n-gon inscribed in a circle of radius r is a constant multiple of r^2. Find the constant for a decagon, kilogon, and megagon.

4. What happens to the shape of the inscribed n-gon as n increases?

5. What would you use as a formula for the area of a circle of radius r? (You have never heard of π.)

6. Find a formula for the perimeter of an n-gon inscribed in a circle of radius r.

7. Find a formula for the circumference of a circle of radius r. (You have never heard of π.)

8. Suppose that the π-key on your calculator is broken. What expression could you use to calculate π accurate to nine decimal places?

4

Exponential and Logarithmic Functions

The Sunshine Skyway Bridge

Location St. Petersburg and Bradenton, Florida

Completion Date 1987

Cost $244 million

Length 29,040 feet

Longest Single Span 1200 feet

Engineers Figg and Muller Engineering Group

Completed in 1987, the Sunshine Skyway is the world's longest cable-stayed concrete bridge. This bright yellow bridge is not the first bridge to span the mouth of Tampa Bay. In 1980 during a violent thunderstorm the freighter *Summit Venture* plowed into the 4-mile steel cantilever bridge that used to span this same location. Thirty-five people plunged to their death as 1000 feet of the bridge fell into the bay. Protecting the new skyway bridge from ships was high priority. So large concrete islands, called dolphins, were installed around each of the bridge's six piers to absorb any accidental impact from ships.

4.1   Exponential Functions and Their Applications

Functions that involve a finite combination of basic arithmetic operations, powers, and roots are called **algebraic functions.** Functions that transcend what can be described in this manner are called **transcendental functions.** In calculus it is shown that transcendental functions can be written as infinite combinations of basic arithmetic operations. The trigonometric functions from the last chapter and the exponential and logarithmic functions that we study in this chapter are transcendental functions. (See Concepts of Calculus at the end of this chapter.)

The Definition

In algebraic functions such as

$$j(x) = x^2, \qquad p(x) = x^5, \qquad \text{and} \qquad m(x) = x^{1/3}$$

the base is a variable and the exponent is constant. For the exponential functions the base is constant and the exponent is a variable. The functions

$$f(x) = 2^x, \qquad g(x) = 5^x, \qquad \text{and} \qquad h(x) = \left(\frac{1}{3}\right)^x$$

are exponential functions.

Definition:
Exponential Function

> An **exponential function** with **base a** is a function of the form
>
> $$f(x) = a^x$$
>
> where a and x are real numbers such that $a > 0$ and $a \neq 1$.

We rule out the base $a = 1$ in the definition because $f(x) = 1^x$ is the constant function $f(x) = 1$. Negative numbers and 0 are not used as bases because powers such as $(-4)^{1/2}$ and 0^{-2} are not real numbers or not defined. Other functions with variable exponents, such as $f(x) = 3^{2x+1}$ or $f(x) = 10 \cdot 7^{-x^2}$, may also be called exponential functions.

To evaluate an exponential function we use our knowledge of exponents. A review of exponents can be found in Section A.1 of the Appendix.

Example **1** **Evaluating exponential functions**

Let $f(x) = 4^x$, $g(x) = 5^{2-x}$, and $h(x) = \left(\frac{1}{3}\right)^x$. Find the following values. Check your answers with a calculator.

a. $f(3/2)$ **b.** $g(3)$ **c.** $h(-2)$

Solution

a. $f(3/2) = 4^{3/2} = \left(\sqrt{4}\right)^3 = 2^3 = 8$ Take the square root, then cube.

b. $g(3) = 5^{2-3} = 5^{-1} = \dfrac{1}{5}$ Exponent -1 means reciprocal.

c. $h(-2) = \left(\dfrac{1}{3}\right)^{-2} = 3^2 = 9$ Find the reciprocal, then square.

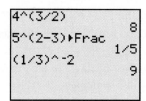

■ **Figure 4.1**

Evaluating these functions with a calculator, as in Fig. 4.1, yields the same results. ■

Domain of Exponential Functions

The domain of the exponential function $f(x) = 2^x$ is the set of all real numbers. To understand how this function can be evaluated for any *real* number, first note that the function is evaluated for any *rational* number using powers and roots. For example,

$$f(1.7) = f(17/10) = 2^{17/10} = \sqrt[10]{2^{17}} \approx 3.249009585.$$

Until now, we have not considered *irrational* exponents.

Consider the expression $2^{\sqrt{3}}$ and recall that the irrational number $\sqrt{3}$ is an infinite nonterminating nonrepeating decimal number:

$$\sqrt{3} = 1.7320508075 \ldots$$

If we use rational approximations to $\sqrt{3}$ as exponents, we see a pattern:

$$2^{1.7} = 3.249009585 \ldots$$

$$2^{1.73} = 3.317278183 \ldots$$

$$2^{1.732} = 3.321880096 \ldots$$

$$2^{1.73205} = 3.321995226 \ldots$$

$$2^{1.7320508} = 3.321997068 \ldots$$

As the exponents get closer and closer to $\sqrt{3}$ we get results that are approaching some number. We define $2^{\sqrt{3}}$ to be that number. Of course, it is impossible to write the exact value of $\sqrt{3}$ or $2^{\sqrt{3}}$ as a decimal, but you can use a calculator to get $2^{\sqrt{3}} \approx 3.321997085$. Since any irrational number can be approximated by rational numbers in this same manner, 2^x is defined similarly for any irrational number. This idea extends to any exponential function.

Domain of an Exponential Function

The domain of $f(x) = a^x$ for $a > 0$ and $a \neq 1$ is the set of all real numbers.

Graphing Exponential Functions

Even though the domain of an exponential function is the set of real numbers, for ease of computation, we generally choose only rational numbers for x to find ordered pairs on the graph of the function.

Example **2** Graphing an exponential function ($a > 1$)

Sketch the graph of each exponential function and state the domain and range.

a. $f(x) = 2^x$ **b.** $g(x) = 10^x$

Solution

a. Find some ordered pairs satisfying $f(x) = 2^x$ as follows:

x	-2	-1	0	1	2	3
$y = 2^x$	1/4	1/2	1	2	4	8

y-values increasing ⟶

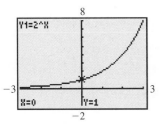

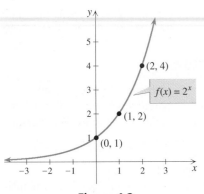

Figure 4.2

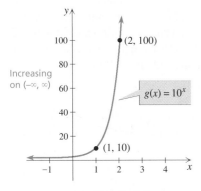

Figure 4.3

As $x \to \infty$, so does 2^x. Using limit notation, $\lim\limits_{x \to \infty} 2^x = \infty$. As $x \to -\infty$, 2^x approaches, but never reaches, 0. Using limit notation, $\lim\limits_{x \to -\infty} 2^x = 0$. So the x-axis is a horizontal asymptote for the curve. It can be shown that the graph of $f(x) = 2^x$ increases in a continuous manner, with no "jumps" or "breaks." So the graph is a smooth curve through the points shown in Fig. 4.2. The domain of $f(x) = 2^x$ is $(-\infty, \infty)$ and the range is $(0, \infty)$. The function is increasing on $(-\infty, \infty)$.

The calculator graph shown in Fig. 4.3 supports these conclusions. Because of the limited resolution of the calculator screen, the calculator graph appears to touch its horizontal asymptote in this window. You can have a mental picture of a curve that gets closer and closer but never touches its asymptote, but it is impossible to accurately draw such a picture. □

b. Find some ordered pairs satisfying $g(x) = 10^x$ as follows:

x	-2	-1	0	1	2	3
$y = 10^x$	0.01	0.1	1	10	100	1000

y-values increasing ⟶

These points indicate that the graph of $g(x) = 10^x$ behaves in the same manner as the graph of $f(x) = 2^x$. We have $\lim\limits_{x \to \infty} 10^x = \infty$ and $\lim\limits_{x \to -\infty} 10^x = 0$. So the x-axis is a horizontal asymptote. The domain of $g(x) = 10^x$ is $(-\infty, \infty)$ and the range is $(0, \infty)$. The function is increasing on $(-\infty, \infty)$. Plotting the points in the table yields the curve shown in Fig. 4.4. In theory the curve never touches the x-axis. So we usually try to keep our hand-drawn curve from touching the x-axis.

The calculator graph in Fig. 4.5 supports these conclusions. ■

Note the similarities between the graphs shown in Figs. 4.2 and 4.4. Both pass through the point (0, 1), both functions are increasing, and both have the x-axis as a horizontal asymptote. Since a horizontal line can cross these graphs only once, the functions $f(x) = 2^x$ and $g(x) = 10^x$ are one-to-one by the horizontal line test. All functions of the form $f(x) = a^x$ for $a > 1$ have graphs similar to those in Figs. 4.2 and 4.4.

The phrase "growing exponentially" is often used to describe a population or other quantity whose growth can be modeled with an increasing exponential function. For example, the earth's population is said to be growing exponentially because of the shape of the graph shown in Fig. 4.6.

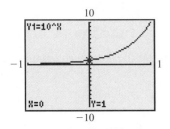

Figure 4.4

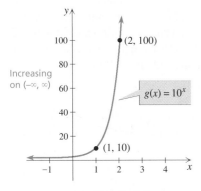

Figure 4.5

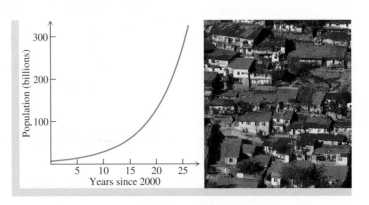

Figure 4.6

If $a > 1$, then the graph of $f(x) = a^x$ is increasing. In the next example we graph exponential functions in which $0 < a < 1$ and we will see that these functions are decreasing.

Example **3** **Graphing an exponential function** $(0 < a < 1)$

Sketch the graph of each function and state the domain and range of the function.

a. $f(x) = (1/2)^x$ **b.** $g(x) = 3^{-x}$

Solution

a. Find some ordered pairs satisfying $f(x) = (1/2)^x$ as follows:

x	-2	-1	0	1	2	3
$y = \left(\dfrac{1}{2}\right)^x$	4	2	1	1/2	1/4	1/8

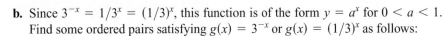
y-values decreasing

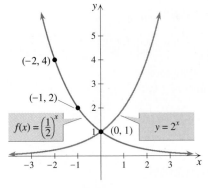

$(-2, 4)$

$(-1, 2)$

$f(x) = \left(\dfrac{1}{2}\right)^x$ $(0, 1)$ $y = 2^x$

■ **Figure 4.7**

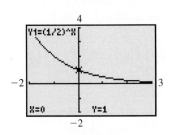

■ **Figure 4.8**

Plot these points and draw a smooth curve through them as in Fig. 4.7. Note that $\lim\limits_{x \to \infty} \left(\dfrac{1}{2}\right)^x = 0$ and $\lim\limits_{x \to -\infty} \left(\dfrac{1}{2}\right)^x = \infty$. The *x*-axis is a horizontal asymptote. The domain is $(-\infty, \infty)$ and the range is $(0, \infty)$. The function is decreasing on $(-\infty, \infty)$. Because $(1/2)^x = 2^{-x}$, the graph of $f(x) = (1/2)^x$ is a reflection in the *y*-axis of the graph of $y = 2^x$.

The calculator graph in Fig. 4.8 supports these conclusions. □

b. Since $3^{-x} = 1/3^x = (1/3)^x$, this function is of the form $y = a^x$ for $0 < a < 1$. Find some ordered pairs satisfying $g(x) = 3^{-x}$ or $g(x) = (1/3)^x$ as follows:

x	-2	-1	0	1	2	3
$y = 3^{-x}$	9	3	1	1/3	1/9	1/27

y-values decreasing

Plot these points and draw a smooth curve through them, as shown in Fig. 4.9. Note that $\lim\limits_{x \to \infty} 3^{-x} = 0$ and $\lim\limits_{x \to -\infty} 3^{-x} = \infty$. The domain is $(-\infty, \infty)$ and the range is $(0, \infty)$. The function is decreasing on its entire domain, $(-\infty, \infty)$. The *x*-axis is again a horizontal asymptote.

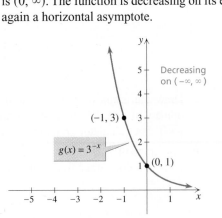

Decreasing on $(-\infty, \infty)$

$(-1, 3)$

$g(x) = 3^{-x}$

$(0, 1)$

■ **Figure 4.9**

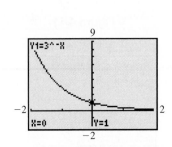

■ **Figure 4.10**

The calculator graph in Fig. 4.10 supports these conclusions. ■

Again note the similarities between the graphs in Figs. 4.7 and 4.9. Both pass through $(0, 1)$, both functions are decreasing, and both have the x-axis as a horizontal asymptote. By the horizontal line test, these functions are also one-to-one. The base determines whether the exponential function is increasing or decreasing. In general, we have the following properties of exponential functions.

Properties of Exponential Functions

The exponential function $f(x) = a^x$ has the following properties:

1. The function f is increasing for $a > 1$ and decreasing for $0 < a < 1$.
2. The y-intercept of the graph of f is $(0, 1)$.
3. The graph has the x-axis as a horizontal asymptote.
4. The domain of f is $(-\infty, \infty)$, and the range of f is $(0, \infty)$.
5. The function f is one-to-one.

The Exponential Family of Functions

A function of the form $f(x) = a^x$ is an exponential function. Any function of the form $g(x) = b \cdot a^{x-h} + k$ is a member of the **exponential family** of functions. The graph of g is a transformation of the graph of f as discussed in Section 1.7.

The graph of f moves to the left if $h < 0$ or to the right if $h > 0$.
The graph of f moves upward if $k > 0$ or downward if $k < 0$.
The graph of f is stretched if $b > 1$ and shrunk if $0 < b < 1$.
The graph of f is reflected in the x-axis if b is negative.

Example **4** Graphing members of the exponential family

Sketch the graph of each function and state its domain and range.

a. $y = 2^{x-3}$ **b.** $f(x) = -4 + 3^{x+2}$

Solution

a. The graph of $y = 2^{x-3}$ is a translation three units to the right of the graph of $y = 2^x$. For accuracy, find a few ordered pairs that satisfy $y = 2^{x-3}$:

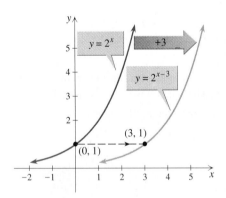

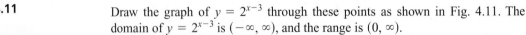

x	3	4	5
$y = 2^{x-3}$	1	2	4

■ **Figure 4.11**

Draw the graph of $y = 2^{x-3}$ through these points as shown in Fig. 4.11. The domain of $y = 2^{x-3}$ is $(-\infty, \infty)$, and the range is $(0, \infty)$.

△ The calculator graph in Fig. 4.12 supports these conclusions. □

b. The graph of $f(x) = -4 + 3^{x+2}$ is obtained by translating $y = 3^x$ to the left two units and downward four units. For accuracy, find a few points on the graph of $f(x) = -4 + 3^{x+2}$:

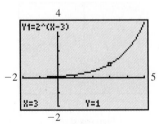

■ **Figure 4.12**

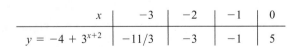

x	-3	-2	-1	0
$y = -4 + 3^{x+2}$	$-11/3$	-3	-1	5

Sketch a smooth curve through these points as shown in Fig. 4.13. The horizontal asymptote is the line $y = -4$. The domain of f is $(-\infty, \infty)$, and the range is $(-4, \infty)$.

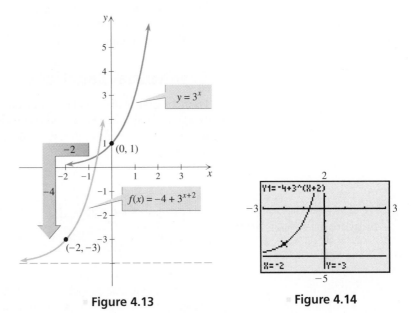

Figure 4.13   **Figure 4.14**

The calculator graph of $y = -4 + 3^{x+2}$ and the asymptote $y = -4$ in Fig. 4.14 supports these conclusions. ■

If you recognize that a function is a transformation of a simpler function, then you know what its graph looks like. This information along with a few ordered pairs will help you to make accurate graphs.

Example **5** **Graphing an exponential function with a reflection**

Sketch the graph of $f(x) = -\frac{1}{2} \cdot 3^{-x}$ and state the domain and range.

Solution

The graph of $y = 3^{-x}$ is shown in Fig. 4.9. Multiplying by $-1/2$ shrinks the graph and reflects it below the x-axis. The x-axis is the horizontal asymptote for the graph of f. Find a few ordered pairs as follows:

x	-2	-1	0	1	2
$y = -\frac{1}{2} \cdot 3^{-x}$	$-9/2$	$-3/2$	$-1/2$	$-1/6$	$-1/18$

y-values increasing →

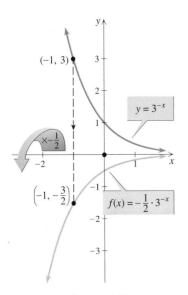

Figure 4.15

Sketch the curve through these points as in Fig. 4.15. The domain is $(-\infty, \infty)$ and the range is $(-\infty, 0)$. ■

The form $f(x) = b \cdot a^x$ is often used in applications. In this case b is the initial value of the function, because $f(0) = b \cdot a^0 = b$. For example, if you discover 3 ants in your cookie jar (time 0) and the number of ants doubles after each minute, then the number of ants x minutes after your initial discovery is $3 \cdot 2^x$. If it makes sense to start at $x = 1$, then use $x - 1$ for the exponent. For example, if you get \$2 in tips on your first day at work and your tips increase by 50% on each succeeding day, then the amount on day x is $2 \cdot 1.5^{x-1}$ dollars.

Exponential Equations

From the graphs of exponential functions, we observed that they are one-to-one. For an exponential function, one-to-one means that if two exponential expressions with the same base are equal, then the exponents are equal. For example, if $2^m = 2^n$, then $m = n$.

One-to-One Property of Exponential Functions

For $a > 0$ and $a \neq 1$,

$$\text{if} \quad a^{x_1} = a^{x_2}, \quad \text{then} \quad x_1 = x_2.$$

The one-to-one property is used in solving simple exponential equations. For example, to solve $2^x = 8$, we recall that $8 = 2^3$. So the equation becomes $2^x = 2^3$. By the one-to-one property, $x = 3$ is the only solution. The one-to-one property applies only to equations in which each side is a power of the same base.

Example **6** Solving exponential equations

Solve each exponential equation.

a. $4^x = \dfrac{1}{4}$ **b.** $\left(\dfrac{1}{10}\right)^x = 100$

Solution

a. Since $1/4 = 4^{-1}$, we can write the right-hand side as a power of 4 and use the one-to-one property:

$$4^x = \frac{1}{4} = 4^{-1} \qquad \text{Both sides are powers of 4.}$$

$$x = -1 \qquad \text{One-to-one property}$$

b. Since $(1/10)^x = 10^{-x}$ and $100 = 10^2$, we can write each side as a power of 10:

$$\left(\frac{1}{10}\right)^x = 100 \qquad \text{Original equation}$$

$$10^{-x} = 10^2 \qquad \text{Both sides are powers of 10.}$$

$$-x = 2 \qquad \text{One-to-one property}$$

$$x = -2 \qquad \blacksquare$$

The type of equation that we solved in Example 6 arises naturally when we try to find the first coordinate of an ordered pair of an exponential function when given the second coordinate, as shown in the next example.

Example **7** **Finding the first coordinate given the second**

Let $f(x) = 5^{2-x}$. Find x such that $f(x) = 125$.

Solution

To find x such that $f(x) = 125$, we must solve $5^{2-x} = 125$:

$$5^{2-x} = 5^3 \qquad \text{Since } 125 = 5^3$$

$$2 - x = 3 \qquad \text{One-to-one property}$$

$$x = -1 \qquad\qquad\quad ■$$

The Compound Interest Model

Exponential functions are used to model phenomena such as population growth, radioactive decay, and compound interest. Here we will show how these functions are used to determine the amount of an investment earning compound interest.

If P dollars are deposited in an account with a simple annual interest rate r for t years, then the amount A in the account at the end of t years is found by using the formula $A = P + Prt$ or $A = P(1 + rt)$. If \$1000 is deposited at 6% annual rate for one quarter of a year, then at the end of three months the amount is

$$A = 1000\left(1 + 0.06 \cdot \frac{1}{4}\right) = 1000(1.015) = \$1015.$$

If the account begins the next quarter with \$1015, then at the end of the second quarter we again multiply by 1.015 to get the amount

$$A = 1000(1.015)^2 \approx \$1030.23.$$

This process is referred to as compound interest because interest is put back into the account and the interest also earns interest. At the end of 20 years, or 80 quarters, the amount is

$$A = 1000(1.015)^{80} \approx \$3290.66.$$

Compound interest can be thought of as simple interest computed over and over. The general **compound interest formula** follows.

Compound Interest Formula

If a principal P is invested for t years at an annual rate r compounded n times per year, then the amount A, or ending balance, is given by

$$A = P\left(1 + \frac{r}{n}\right)^{nt}.$$

The principal P is also called **present value** and the amount A is called **future value**.

Figure 4.16

Leonhard Euler (1707–1783) was a Swiss mathematician and physicist. Euler introduced and popularized several notational conventions through his numerous textbooks. He introduced the concept of a function and was the first to write *f(x)*. He also introduced the modern notation for the trigonometric functions, the letter e (Euler's number) for the base of the natural logarithm, the Greek letter sigma for summations, and the letter *i* to denote the imaginary unit.

Example **8** **Using the compound interest formula**

Find the amount or future value when a principal of $20,000 is invested at 6% compounded daily for three years.

Solution

Use $P = \$20{,}000$, $r = 0.06$, $n = 365$, and $t = 3$ in the compound interest formula:

$$A = \$20{,}000\left(1 + \frac{0.06}{365}\right)^{365 \cdot 3} \approx \$23{,}943.99$$

Figure 4.16 shows this expression on a graphing calculator. ■

When interest is compounded daily, financial institutions usually use the exact number of days. To keep our discussion of interest simple, we assume that all years have 365 days and ignore leap years. When discussing months we assume that all months have 30 days. There may be other rules involved in computing interest that are specific to the institution doing the computing. One popular method is to compound quarterly and only give interest on money that is on deposit for full quarters. With this rule you could have money on deposit for nearly six months (say January 5 to June 28) and receive no interest.

Continuous Compounding and the Number *e*

The more often that interest is figured during the year, the more interest an investment will earn. The first five lines of Table 4.1 show the future value of $10,000 invested at 12% for one year for more and more frequent compounding. The last line shows the limiting amount $11,274.97, which we cannot exceed no matter how often we compound the interest on this investment for one year.

Table 4.1

Compounding	Future value in one year
Annually	$\$10{,}000\left(1 + \dfrac{0.12}{1}\right)^{1} = \$11{,}200$
Quarterly	$\$10{,}000\left(1 + \dfrac{0.12}{4}\right)^{4} \approx \$11{,}255.09$
Monthly	$\$10{,}000\left(1 + \dfrac{0.12}{12}\right)^{12} \approx \$11{,}268.25$
Daily	$\$10{,}000\left(1 + \dfrac{0.12}{365}\right)^{365} \approx \$11{,}274.75$
Hourly	$\$10{,}000\left(1 + \dfrac{0.12}{8760}\right)^{8760} \approx \$11{,}274.96$
Continuously	$\$10{,}000e^{0.12(1)} \approx \$11{,}274.97$

To better understand Table 4.1, we need to use a fact from calculus. The expression $[1 + r/n]^{nt}$ used in calculating the first five values in the table can be shown to approach e^{rt} as n goes to ∞. Using limit notation,

$$\lim_{n \to \infty}\left[1 + \frac{r}{n}\right]^{nt} = e^{rt}.$$

Figure 4.17

The number e (like π) is an irrational number that occurs in many areas of mathematics. Using $r = 1$ and $t = 1$ in this limit gives us $\lim\limits_{n \to \infty}[1 + 1/n]^n = e$, which can be used as the definition of the number e. You can find approximate values for powers of e using a calculator with an e^x-key. To find an approximation for e itself, find e^1 on a calculator:

$$e \approx 2.718281828459$$

Figure 4.17 shows the graphing calculator computation of the future value of $10,000 at 12% compounded continuously for one year and the value of e^1. □

So there is a limit to the results obtained by more and more frequent compounding. The limit is the product of the principal and e^{rt}. Using $A = P \cdot e^{rt}$ to find the amount is called **continuous compounding.**

Continuous Compounding Formula

If a principal P is invested for t years at an annual rate r compounded continuously, then the amount A, or ending balance, is given by

$$A = P \cdot e^{rt}.$$

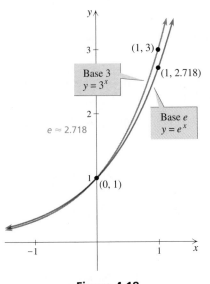

Figure 4.18

Example **9** **Interest compounded continuously**

Find the amount when a principal of $5600 is invested at $6\frac{1}{4}\%$ annual rate compounded continuously for 5 years and 9 months.

Solution

Convert 5 years and 9 months to 5.75 years. Use $r = 0.0625$, $t = 5.75$, and $P = \$5600$ in the continuous compounding formula:

$$A = 5600 \cdot e^{(0.0625)(5.75)} \approx \$8021.63 \qquad ■$$

The function $f(x) = e^x$ is called the **base-e exponential function.** Variations of this function are used to model many types of growth and decay. The graph of $y = e^x$ looks like the graph of $y = 3^x$, because the value of e is close to 3. Use your calculator to check that the graph of $y = e^x$ shown in Fig. 4.18 goes approximately through the points $(-1, 0.368)$, $(0, 1)$, $(1, 2.718)$, and $(2, 7.389)$.

For Thought

True or False? Explain.

1. The function $f(x) = (-2)^x$ is an exponential function.

2. The function $f(x) = 2^x$ is invertible.

3. If $2^x = \frac{1}{8}$, then $x = -3$.

4. If $f(x) = 3^x$, then $f(0.5) = \sqrt{3}$.

5. If $f(x) = e^x$, then $f(0) = 1$.

6. If $f(x) = e^x$ and $f(t) = e^2$, then $t = 2$.

7. The x-axis is a horizontal asymptote for the graph of $y = e^x$.

8. The function $f(x) = (0.5)^x$ is increasing.

9. The functions $f(x) = 4^{x-1}$ and $g(x) = (0.25)^{1-x}$ have the same graph.

10. $2^{1.73} = \sqrt[100]{2^{173}}$

4.1 Exercises

Let $f(x) = 3^x$, $g(x) = 2^{1-x}$, and $h(x) = (1/4)^x$. Find the following values.

1. $f(2)$

2. $f(4)$

3. $f(-2)$

4. $f(-3)$

5. $g(2)$

6. $g(1)$

7. $g(-2)$

8. $g(-3)$

9. $h(-1)$

10. $h(-2)$

11. $h(-1/2)$

12. $h(3/2)$

Sketch the graph of each function by finding at least three ordered pairs on the graph. State the domain, the range, and whether the function is increasing or decreasing.

13. $f(x) = 5^x$

14. $f(x) = 4^x$

15. $y = 10^{-x}$

16. $y = e^{-x}$

17. $f(x) = (1/4)^x$

18. $f(x) = (0.2)^x$

Use a graph or a table to find each limit.

19. $\lim\limits_{x \to \infty} 3^x$

20. $\lim\limits_{x \to -\infty} 3^x$

21. $\lim\limits_{x \to \infty} 5^{-x}$

22. $\lim\limits_{x \to -\infty} 5^{-x}$

23. $\lim\limits_{x \to \infty} \left(\frac{1}{3}\right)^x$

24. $\lim\limits_{x \to -\infty} \left(\frac{1}{3}\right)^x$

25. $\lim\limits_{x \to -\infty} e^{-x}$

26. $\lim\limits_{x \to \infty} e^{-x}$

Use transformations to help you graph each function. Identify the domain, range, and horizontal asymptote. Determine whether the function is increasing or decreasing.

27. $f(x) = 2^x - 3$

28. $f(x) = 3^{-x} + 1$

29. $f(x) = 2^{x+3} - 5$

30. $f(x) = 3^{1-x} - 4$

31. $y = -2^{-x}$

32. $y = -10^{-x}$

33. $y = 1 - 2^x$

34. $y = -1 - 2^{-x}$

35. $f(x) = 0.5 \cdot 3^{x-2}$

36. $f(x) = -0.1 \cdot 5^{x+4}$

37. $y = 500(0.5)^x$

38. $y = 100 \cdot 2^x$

Write the equation of each graph in its final position.

39. The graph of $y = 2^x$ is translated five units to the right and then two units downward.

40. The graph of $y = e^x$ is translated three units to the left and then one unit upward.

41. The graph of $y = (1/4)^x$ is translated one unit to the right, reflected in the *x*-axis, and then translated two units downward.

42. The graph of $y = 10^x$ is translated three units upward, two units to the left, and then reflected in the *x*-axis.

Solve each equation.

43. $2^x = 64$

44. $5^x = 1$

45. $10^x = 0.1$

46. $10^{2x} = 1000$

47. $-3^x = -27$

48. $-2^x = -\dfrac{1}{2}$

49. $3^{-x} = 9$

50. $2^x = \dfrac{1}{8}$

51. $8^x = 2$

52. $9^x = 3$

53. $e^x = \dfrac{1}{e^2}$

54. $e^{-x} = \dfrac{1}{e}$

55. $\left(\dfrac{1}{2}\right)^x = 8$

56. $\left(\dfrac{2}{3}\right)^x = \dfrac{9}{4}$

57. $10^{x-1} = 0.01$

58. $10^{|x|} = 1000$

Let $f(x) = 2^x$, $g(x) = (1/3)^x$, $h(x) = 10^x$, and $m(x) = e^x$. Find the value of x in each equation.

59. $f(x) = 4$

60. $f(x) = 32$

61. $f(x) = \dfrac{1}{2}$

62. $f(x) = 1$

63. $g(x) = 1$

64. $g(x) = 9$

65. $g(x) = 27$

66. $g(x) = \dfrac{1}{9}$

67. $h(x) = 1000$

68. $h(x) = 10^5$

69. $h(x) = 0.1$

70. $h(x) = 0.0001$

71. $m(x) = e$

72. $m(x) = e^3$

73. $m(x) = \dfrac{1}{e}$

74. $m(x) = 1$

Fill in the missing coordinate in each ordered pair so that the pair is a solution to the given equation.

75. $y = 3^x$ $(2,\ ), (\ ,3), (-1,\ ), (\ ,1/9)$

76. $y = 10^x$ $(3,\ ), (\ ,1), (-2,\ ), (\ ,0.01)$

77. $f(x) = 5^{-x}$ $(0,\ ), (\ ,25), (-1,\ ), (\ ,1/5)$

78. $f(x) = e^{-x}$ $(1,\ ), (\ ,e), (0,\ ), (\ ,e^2)$

79. $f(x) = -2^x$ $(4, \quad), (\quad, -1/4), (-1, \quad), (\quad, -32)$

80. $f(x) = -(1/4)^{x-1}$ $(3, \quad), (\quad, -4), (-1, \quad), (\quad, -1/16)$

Solve each problem. When needed, use 365 days per year and 30 days per month.

81. *Periodic Compounding*   A deposit of $5000 earns 8% annual interest. Find the amount in the account at the end of 6 years and the amount of interest earned during the 6 years if the interest is compounded

    a. annually

    b. quarterly

    c. monthly

    d. daily.

82. *Periodic Compounding*   Melinda invests her $80,000 winnings from Publishers Clearing House at a 9% annual percentage rate. Find the amount of the investment at the end of 20 years and the amount of interest earned during the 20 years if the interest is compounded.

    a. annually

    b. quarterly

    c. monthly

    d. daily.

83. *Compounded Continuously*   The Lakewood Savings Bank pays 8% annual interest compounded continuously. How much will a deposit of $5000 amount to for each time period?

    a. 6 years

    b. 8 years 3 months

    c. 5 years 4 months 22 days

    d. 20 years 321 days

84. *Compounding Continuously*   The Commercial Federal Credit Union pays $6\frac{3}{4}\%$ annual interest compounded continuously. How much will a deposit of $9000 amount to for each time period?

    a. 13 years

    b. 12 years 8 months

    c. 10 years 6 months 14 days

    d. 40 years 66 days

85. *Present Value Compounding Daily*   A credit union pays 6.5% annual interest compounded daily. What deposit today (present value) would amount to $3000 in 5 years and 4 months?

86. *Higher Yields*   Federal Savings and Loan offers $6\frac{3}{4}\%$ annual interest compounded daily on certificates of deposit. What dollar amount deposited now would amount to $40,000 in 3 years and 2 months?

87. *Present Value Compounding Continuously*   Peoples Bank offers 5.42% compounded continuously on CDs. What amount invested now would grow to $20,000 in 30 years.

88. *Saving for Retirement*   An investor wants to have a retirement nest egg of $100,000 and estimates that her investment now will grow at 3% compounded continuously for 40 years. What amount should she invest now to achieve this goal?

89. *Working by the Hour*   One million dollars is deposited in an account paying 6% compounded continuously.

    a. What amount of interest will it earn in its first hour on deposit?

    b. What amount of interest will it earn during its 500th hour on deposit?

90. *Big Debt*   The national debt is approximately $8 trillion. If Uncle Sam paid 5% annual interest compounded continuously on that debt, then what amount of interest would he pay for one day? How much could he save if he paid 5% annual interest compounded daily?

91. *Radioactive Decay*   The number of grams of a certain radioactive substance present at time t is given by the formula $A = 200e^{-0.001t}$, where t is the number of years. How many grams are present at time $t = 0$? How many grams are present at time $t = 500$?

92. *Population Growth*   The function $P = 2.4e^{0.03t}$ models the size of the population of a small country, where P is in millions of people in the year $2000 + t$.

    a. What was the population in 2000?

    b. Use the formula to estimate the population to the nearest tenth of a million in 2020.

93. *Position of a Football*   The football is on the 10-yard line. Several penalties in a row are given, and each penalty moves the ball half the distance to the closer goal line. Write a formula that gives the ball's position P after nth such penalty.

94. *Cost of a Parking Ticket*   The cost of a parking ticket on campus is $15 for the first offense. Given that the cost doubles for each additional offense, write a formula for the cost C as a function of the number of tickets n.

95. *Challenger Disaster*   Using data on O-ring damage from 24 previous space shuttle launches, Professor Edward R. Tufte of Yale University concluded that the number of O-rings damaged per launch is an exponential function of the temperature at the time of the launch. If NASA had used a model such as $n = 644e^{-0.15t}$, where t is the Fahrenheit temperature at the time of launch and n is the number of O-rings damaged, then the tragic end to the flight of the space shuttle Challenger might have been avoided. Using this model, find the number of O-rings that would be expected to fail at 31°F, the temperature at the time of the Challenger launch on January 28, 1986.

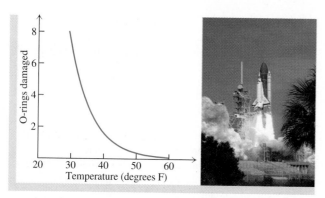

▪ **Figure for Exercise 95**

96. *Manufacturing Cost*   The cost in dollars for manufacturing x units of a certain drug is given by the function $C(x) = xe^{0.001x}$. Find the cost for manufacturing 500 units. Find the function $AC(x)$ that gives the average cost per unit for manufacturing x units. What happens to the average cost per unit as x gets larger and larger?

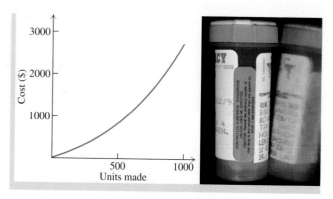

▪ **Figure for Exercise 96**

Thinking Outside the Box XXIX

Swim Meet   Two swimmers start out from opposite sides of a pool. Each swims the length of the pool and back at a constant rate. They pass each other for the first time 40 feet from one side of the pool and for the second time 45 feet from the other side of the pool. What is the length of the pool?

4.1   Pop Quiz

1. What is $f(4)$ if $f(x) = -2^x$?

2. Is $f(x) = 3^{-x}$ increasing or decreasing?

3. Find the domain and range for $y = e^{x-1} + 2$.

4. What is the horizontal asymptote for $y = 2^x - 1$?

5. Solve $(1/4)^x = 64$.

6. What is a, if $f(a) = 8$ and $f(x) = 2^{-x}$?

7. If $1000 earns 4% annual interest compounded quarterly, then what is the amount after 20 years?

8. Find the amount in the last problem if the interest is compounded continuously.

4.2 Logarithmic Functions and Their Applications

Since exponential functions are one-to-one functions (Section 4.1), they are invertible. In this section we will study the inverses of the exponential functions.

The Definition

The inverses of the exponential functions are called **logarithmic functions.** Since f^{-1} is a general name for an inverse function, we adopt a more descriptive notation for these inverses. If $f(x) = a^x$, then instead of $f^{-1}(x)$, we write $\log_a(x)$ for the inverse of the base-a exponential function. We read $\log_a(x)$ as "log of x with base a," and we call the expression $\log_a(x)$ a **logarithm.**

The meaning of $\log_a(x)$ will be clearer if we consider the exponential function

$$f(x) = 2^x$$

as an example. Since $f(3) = 2^3 = 8$, the base-2 exponential function pairs the exponent 3 with the value of the exponential expression 8. Since the function $\log_2(x)$ reverses that pairing, we have $\log_2(8) = 3$. So $\log_2(8)$ is the exponent that is used on the base 2 to obtain 8. *In general, $\log_a(x)$ is the exponent that is used on the base a to obtain the value x.*

■ **Foreshadowing Calculus**

In calculus the logarithmic functions can be defined geometrically (as area under a certain curve). Then exponential functions are defined as the inverses of the logarithmic functions.

Definition:
Logarithmic Function

> For $a > 0$ and $a \neq 1$, the **logarithmic function with base a** is denoted $f(x) = \log_a(x)$, where
>
> $$y = \log_a(x) \qquad \text{if and only if} \qquad a^y = x.$$

Example **1** **Evaluating logarithmic functions**

Find the indicated values of the logarithmic functions.

a. $\log_3(9)$ **b.** $\log_2\left(\dfrac{1}{4}\right)$ **c.** $\log_{1/2}(8)$

Solution

a. By the definition of logarithm, $\log_3(9)$ is the exponent that is used on the base 3 to obtain 9. Since $3^2 = 9$, we have $\log_3(9) = 2$.

b. Since $\log_2(1/4)$ is the exponent that is used on the base 2 to obtain $1/4$, we try various powers of 2 until we find $2^{-2} = 1/4$. So $\log_2(1/4) = -2$.

You can use the graph of $y = 2^x$ as in Fig. 4.19 to check that $2^{-2} = 1/4$. □

c. Since $\log_{1/2}(8)$ is the exponent that is used on a base of $1/2$ to obtain 8, we try various powers of $1/2$ until we find $(1/2)^{-3} = 8$. So $\log_{1/2}(8) = -3$. ■

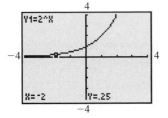

■ **Figure 4.19**

Since the exponential function $f(x) = a^x$ has domain $(-\infty, \infty)$ and range $(0, \infty)$, the logarithmic function $f(x) = \log_a(x)$ has domain $(0, \infty)$ and range $(-\infty, \infty)$. So there are no logarithms of negative numbers or zero. Expressions such as $\log_2(-4)$ and $\log_3(0)$ are undefined. Note that $\log_a(1) = 0$ for any base a, because $a^0 = 1$ for any base a.

There are two bases that are used more frequently than the others; they are 10 and e. The notation $\log_{10}(x)$ is abbreviated $\log(x)$, and $\log_e(x)$ is abbreviated $\ln(x)$. Most scientific calculators have function keys for the exponential functions 10^x and e^x and their inverses $\log(x)$ and $\ln(x)$, which are called **common logarithms** and **natural logarithms,** respectively. Natural logarithms are also called **Napierian logarithms** after John Napier (1550–1617).

Note that $\log(76)$ is approximately 1.8808 and $10^{1.8808}$ is approximately 76 as shown in Fig. 4.20. If you use more digits for $\log(76)$ as the power of 10, then the calculator gets closer to 76. □

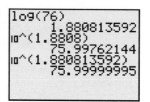

■ **Figure 4.20**

You can use a calculator to find common or natural logarithms, but you should know how to find the values of logarithms such as those in Examples 1 and 2 without using a calculator.

Example **2** **Evaluating logarithmic functions**

Find the indicated values of the logarithmic functions without a calculator. Use a calculator to check.

a. $\log(1000)$ **b.** $\ln(1)$ **c.** $\ln(-6)$

Solution

a. To find $\log(1000)$, we must find the exponent that is used on the base 10 to obtain 1000. Since $10^3 = 1000$, we have $\log(1000) = 3$.
b. Since $e^0 = 1$, $\ln(1) = 0$.
c. The expression $\ln(-6)$ is undefined because -6 is not in the domain of the natural logarithm function. There is no power of e that results in -6.

The calculator results for parts (a) and (b) are shown in Fig. 4.21. If you ask a calculator for $\ln(-6)$, it will give you an error message. ■

```
log(1000)
               3
ln(1)
               0
ln(-6)
```

■ **Figure 4.21**

Graphs of Logarithmic Functions

The functions of $y = a^x$ and $y = \log_a(x)$ for $a > 0$ and $a \neq 1$ are inverse functions. So the graph of $y = \log_a(x)$ is a reflection about the line $y = x$ of the graph of $y = a^x$. The graph of $y = a^x$ has the x-axis as its horizontal asymptote, while the graph of $y = \log_a(x)$ has the y-axis as its vertical asymptote.

Example **3** **Graph of a base-a logarithmic function with $a > 1$**

Sketch the graphs of $y = 2^x$ and $y = \log_2(x)$ on the same coordinate system. State the domain and range of each function.

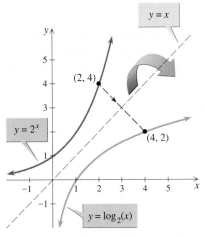

Figure 4.22

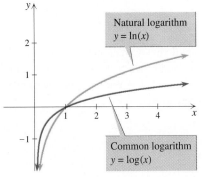

Figure 4.23

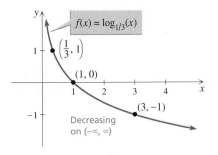

Decreasing
on $(-\infty, \infty)$

Figure 4.24

Solution

Since these two functions are inverses of each other, the graph $y = \log_2(x)$ is a reflection of the graph of $y = 2^x$ about the line $y = x$. Make a table of ordered pairs for each function:

x	-1	0	1	2
$y = 2^x$	$1/2$	1	2	4

x	$1/2$	1	2	4
$y = \log_2(x)$	-1	0	1	2

Sketch $y = 2^x$ and $y = \log_2(x)$ through the points given in the table, as shown in Fig. 4.22. Keep in mind that $y = \log_2(x)$ is a reflection of $y = 2^x$ about the line $y = x$. The important features of the two functions are given in the following table.

	Domain	Range	Asymptotes	Limits
$y = 2^x$	$(-\infty, \infty)$	$(0, \infty)$	x-axis	$\lim\limits_{x \to -\infty} 2^x = 0$, $\lim\limits_{x \to \infty} 2^x = \infty$
$y = \log_2(x)$	$(0, \infty)$	$(-\infty, \infty)$	y-axis	$\lim\limits_{x \to 0^+} \log_2(x) = -\infty$, $\lim\limits_{x \to \infty} \log_2(x) = \infty$

Note that the function $y = \log_2(x)$ graphed in Fig. 4.22 is one-to-one by the horizontal line test and is an increasing function. The graphs of the common logarithm function $y = \log(x)$ and the natural logarithm function $y = \ln(x)$ are shown in Fig. 4.23, and also are increasing functions. The function $y = \log_a(x)$ is increasing if $a > 1$. By contrast, if $0 < a < 1$, the function $y = \log_a(x)$ is decreasing, as illustrated in the next example.

Example **4** **Graph of a base-a logarithmic function with $0 < a < 1$**

Sketch the graph of $f(x) = \log_{1/3}(x)$ and state its domain and range.

Solution

Make a table of ordered pairs for the function:

x	$1/9$	$1/3$	1	3	9
$y = \log_{1/3}(x)$	2	1	0	-1	-2

Sketch the curve through these points as in Fig. 4.24. The y-axis is a vertical asymptote for the curve. In terms of limits, $\lim\limits_{x \to 0^+} \log_{1/3}(x) = \infty$ and $\lim\limits_{x \to \infty} \log_{1/3}(x) = -\infty$. The domain is $(0, \infty)$ and its range is $(-\infty, \infty)$. ■

The logarithmic functions have properties corresponding to the properties of the exponential functions stated in Section 4.1.

Properties of Logarithmic Functions

The logarithmic function $f(x) = \log_a(x)$ has the following properties:

1. The function f is increasing for $a > 1$ and decreasing for $0 < a < 1$.
2. The x-intercept of the graph of f is $(1, 0)$.
3. The graph has the y-axis as a vertical asymptote.
4. The domain of f is $(0, \infty)$, and the range of f is $(-\infty, \infty)$.
5. The function f is one-to-one.

The Logarithmic Family of Functions

A function of the form $f(x) = \log_a(x)$ is a logarithmic function. Any function of the form $g(x) = b \cdot \log_a(x - h) + k$ is a member of the **logarithmic family** of functions. The graph of g is a transformation of the graph of f as discussed in Section 1.7.

The graph of f moves to the left if $h < 0$ or to the right if $h > 0$.
The graph of f moves upward if $k > 0$ or downward if $k < 0$.
The graph of f is stretched if $b > 1$ and shrunk if $0 < b < 1$.
The graph of f is reflected in the x-axis if b is negative.

Example **5** **Graphing members of the logarithmic family**

Sketch the graph of each function and state its domain and range.

a. $y = \log_2(x - 1)$ **b.** $f(x) = -\dfrac{1}{2}\log_2(x + 3)$

Solution

a. The graph of $y = \log_2(x - 1)$ is obtained by translating the graph of $y = \log_2(x)$ to the right one unit. Since the domain of $y = \log_2(x)$ is $(0, \infty)$, the domain of $y = \log_2(x - 1)$ is $(1, \infty)$. The line $x = 1$ is the vertical asymptote. Calculate a few ordered pairs to get an accurate graph.

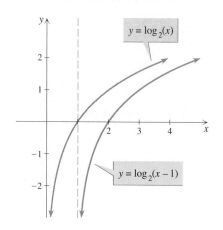

Figure 4.25

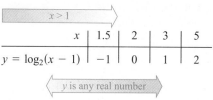

	x > 1			
x	1.5	2	3	5
$y = \log_2(x - 1)$	-1	0	1	2
	y is any real number			

Sketch the curve as shown in Fig. 4.25. The range is $(-\infty, \infty)$.

b. The graph of f is obtained by translating the graph of $y = \log_2(x)$ to the left three units. The vertical asymptote for f is the line $x = -3$. Multiplication by $-1/2$ shrinks the graph and reflects it in the x-axis. Calculate a few ordered pairs for accuracy.

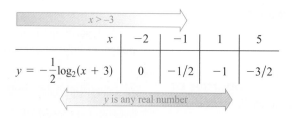

	x > -3			
x	-2	-1	1	5
$y = -\dfrac{1}{2}\log_2(x + 3)$	0	$-1/2$	-1	$-3/2$
	y is any real number			

Sketch the curve through these points as shown in Fig. 4.26. The domain of f is $(-3, \infty)$, and the range is $(-\infty, \infty)$.

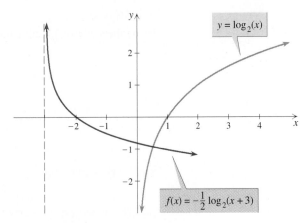

■ **Figure 4.26**   ■

◢◣ Note that logarithmic functions involving common or natural logarithms can be graphed with a graphing calculator because the calculator has keys for log and ln. To graph logarithmic functions involving other bases, we need the base-change formula that is discussed in Section 4.3.   □

Logarithmic and Exponential Equations

Some equations involving logarithms can be solved by writing an equivalent exponential equation, and some equations involving exponents can be solved by writing an equivalent logarithmic equation. Rewriting logarithmic and exponential equations is possible because of the definition of logarithms:

$$y = \log_a(x) \qquad \text{if and only if} \qquad a^y = x$$

Example **6** **Rewriting logarithmic and exponential equations**

Write each equation involving logarithms as an equivalent exponential equation, and write each equation involving exponents as an equivalent logarithmic equation.

a. $\log_5(625) = 4$ **b.** $\log_3(n) = 5$ **c.** $3^x = 50$ **d.** $e^{x-1} = 9$

Solution

a. $\log_5(625) = 4$ is equivalent to $5^4 = 625$.
b. $\log_3(n) = 5$ is equivalent to $3^5 = n$.
c. $3^x = 50$ is equivalent to $x = \log_3(50)$.
d. $e^{x-1} = 9$ is equivalent to $x - 1 = \ln(9)$.   ■

In the next example we use the definition of logarithms to find the inverse of an exponential function.

Example **7** **The inverse of an exponential function**

Find f^{-1} for $f(x) = \frac{1}{2} \cdot 6^{x-3}$.

Solution

Using the switch-and-solve method, we switch x and y then solve for y:

$$x = \frac{1}{2} \cdot 6^{y-3} \qquad \text{Switch } x \text{ and } y \text{ in } y = \frac{1}{2} \cdot 6^{x-3}.$$

$$2x = 6^{y-3} \qquad \text{Multiply each side by 2.}$$

$$y - 3 = \log_6(2x) \qquad \text{Definition of logarithm}$$

$$y = \log_6(2x) + 3 \qquad \text{Add 3 to each side.}$$

So $f^{-1}(x) = \log_6(2x) + 3$. ∎

The one-to-one property of exponential functions was used to solve exponential equations in Section 4.1. Likewise, the one-to-one property of logarithmic functions is used in solving logarithmic equations. The one-to-one property says that *if two quantities have the same base-a logarithm, then the quantities are equal.* For example, if $\log_2(m) = \log_2(n)$, then $m = n$.

One-to-One Property of Logarithms

> For $a > 0$ and $a \neq 1$,
>
> $$\text{if} \quad \log_a(x_1) = \log_a(x_2), \quad \text{then} \quad x_1 = x_2.$$

The one-to-one properties and the definition of logarithm are at present the only new tools that we have for solving equations. Later in this chapter we will develop more properties of logarithms and solve more complicated equations.

Example **8** Solving equations involving logarithms

Solve each equation.

a. $\log_3(x) = -2$ **b.** $\log_x(5) = 2$ **c.** $5^x = 9$ **d.** $\ln(x^2) = \ln(3x)$

Solution

a. Use the definition of logarithm to write the equivalent exponential equation.

$$\log_3(x) = -2 \qquad \text{Original equation}$$

$$x = 3^{-2} \qquad \text{Definition of logarithm}$$

$$= \frac{1}{9}$$

Since $\log_3(1/9) = -2$ is correct, the solution to the equation is $1/9$.

b. $\log_x(5) = 2$ Original equation

$$x^2 = 5 \qquad \text{Definition of logarithm}$$

$$x = \pm\sqrt{5}$$

Since the base of a logarithm is always nonnegative, the only solution is $\sqrt{5}$.

c. $5^x = 9$ Original equation

$$x = \log_5(9) \qquad \text{Definition of logarithm}$$

The exact solution to $5^x = 9$ is the irrational number $\log_5(9)$. In the next section we will learn how to find a rational approximation for $\log_5(9)$.

d. $\quad \ln(x^2) = \ln(3x)$ $\qquad$ Original equation

$\qquad x^2 = 3x$ $\qquad$ One-to-one property of logarithms

$\quad x^2 - 3x = 0$

$\quad x(x - 3) = 0$

$\quad x = 0 \quad$ or $\quad x = 3$

Checking $x = 0$ in the original equation, we get the undefined expression $\ln(0)$. So the only solution to the equation is 3. ■

Example **9** **Equations involving common and natural logarithms**

Use a calculator to find the value of x rounded to four decimal places.

a. $10^x = 50$ $\quad$ **b.** $2e^{0.5x} = 6$

Solution

a. $10^x = 50$ $\qquad$ Original equation

$\quad x = \log(50)$ $\qquad$ Definition of logarithm

$\quad x \approx 1.6990$

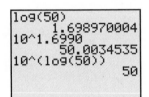 Figure 4.27 shows how to find the approximate value of $\log(50)$ and how to check the approximate answer and exact answer. □

b. $2e^{0.5x} = 6$ $\qquad$ Original equation

$\quad e^{0.5x} = 3$ $\qquad$ Divide by 2 to get the form $a^x = y$.

$\quad 0.5x = \ln(3)$ $\qquad$ Definition of logarithm

$\quad x = \dfrac{\ln(3)}{0.5}$

$\quad x \approx 2.1972$

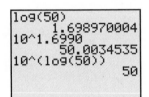 Figure 4.28 shows how to find the approximate value of x and how to check it in the original equation. ■

Figure 4.27

Figure 4.28

```
log(50)
        1.698970004
10^1.6990
        50.0034535
10^(log(50))
               50
```

```
ln(3)/.5
        2.197224577
2e^(.5Ans)
               6
```

Applications

We saw in Section 4.1 that if a principal of P dollars earns interest at an annual rate r compounded continuously, then the amount after t years is given by

$$A = Pe^{rt}.$$

If either the rate or the time is the only unknown in this formula, then the definition of logarithms can be used to solve for the rate or the time.

Example **10** **Finding the time in a continuous compounding problem**

If $8000 is invested at 9% compounded continuously, then how long will it take for the investment to grow to $20,000?

Function Gallery: Exponential and Logarithmic Functions

Exponential: $f(x) = a^x$, domain $(-\infty, \infty)$, range $(0, \infty)$

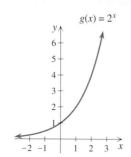

Increasing on $(-\infty, \infty)$
y-intercept $(0, 1)$

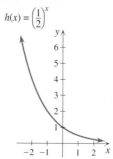

Decreasing on $(-\infty, \infty)$
y-intercept $(0, 1)$

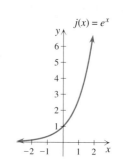

Increasing on $(-\infty, \infty)$
y-intercept $(0, 1)$

Logarithmic: $f^{-1}(x) = \log_a(x)$, domain $(0, \infty)$, range $(-\infty, \infty)$

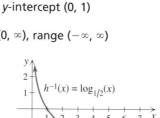

Increasing on $(0, \infty)$
x-intercept $(1, 0)$

Decreasing on $(0, \infty)$
x-intercept $(1, 0)$

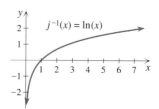

Increasing on $(0, \infty)$
x-intercept $(1, 0)$

Solution

Use the formula $A = Pe^{rt}$ with $A = \$20{,}000$, $P = \$8000$, and $r = 0.09$:

$$20{,}000 = 8000e^{0.09t}$$

$$2.5 = e^{0.09t} \qquad \text{Divide by 8000 to get the form } y = a^x.$$

$$0.09t = \ln(2.5) \qquad \text{Definition of logarithm: } y = a^x \text{ if and only if } x = \log_a(y).$$

$$t = \frac{\ln(2.5)}{0.09}$$

$$\approx 10.181 \text{ years}$$

We can multiply 365 by 0.181 to get approximately 66 days. So the investment grows to \$20,000 in approximately 10 years and 66 days.

You can use a graphing calculator to check as shown in Fig. 4.29.

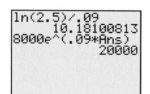

Figure 4.29

The formula $A = Pe^{rt}$ was introduced to model the continuous growth of money. However, this type of formula is used in a wide variety of applications. In the following exercises you will find problems involving population growth, declining forests, and global warming.

For Thought

True or False? Explain.

1. The first coordinate of an ordered pair in an exponential function is a logarithm.

2. $\log_{100}(10) = 2$

3. If $f(x) = \log_3(x)$, then $f^{-1}(x) = 3^x$.

4. $10^{\log(1000)} = 1000$

5. The domain of $f(x) = \ln(x)$ is $(-\infty, \infty)$.

6. $\ln(e^{2.451}) = 2.451$

7. For any positive real number x, $e^{\ln(x)} = x$.

8. For any base a, where $a > 0$ and $a \neq 1$, $\log_a(0) = 1$.

9. $\log(10^3) + \log(10^5) = \log(10^8)$

10. $\log_2(32) - \log_2(8) = \log_2(4)$

4.2 Exercises

Determine the number that can be used in place of the question mark to make the equation true.

1. $2^? = 64$

2. $2^? = 16$

3. $3^? = \dfrac{1}{81}$

4. $3^? = 1$

5. $16^? = 2$

6. $16^? = 16$

7. $\left(\dfrac{1}{5}\right)^? = 125$

8. $\left(\dfrac{1}{5}\right)^? = \dfrac{1}{125}$

Find the indicated value of the logarithmic functions.

9. $\log_2(64)$

10. $\log_2(16)$

11. $\log_3(1/81)$

12. $\log_3(1)$

13. $\log_{16}(2)$

14. $\log_{16}(16)$

15. $\log_{1/5}(125)$

16. $\log_{1/5}(1/125)$

17. $\log(0.1)$

18. $\log(10^6)$

19. $\log(1)$

20. $\log(10)$

21. $\ln(e)$

22. $\ln(0)$

23. $\ln(e^{-5})$

24. $\ln(e^9)$

Sketch the graph of each function, and state the domain and range of each function.

25. $y = \log_3(x)$

26. $y = \log_4(x)$

27. $y = \log_{1/2}(x)$

28. $y = \log_{1/4}(x)$

29. $f(x) = \ln(x - 1)$

30. $f(x) = \log_3(x + 2)$

31. $f(x) = -3 + \log(x + 2)$

32. $f(x) = 4 - \log(x + 6)$

33. $f(x) = -\dfrac{1}{2}\log(x - 1)$

34. $f(x) = -2 \cdot \log_2(x + 2)$

Use a graph or a table to find each limit.

35. $\lim\limits_{x \to \infty} \log_3(x)$

36. $\lim\limits_{x \to 0^+} \log_3(x)$

37. $\lim\limits_{x \to 0^+} \log_{1/2}(x)$

38. $\lim\limits_{x \to \infty} \log_{1/2}(x)$

39. $\lim\limits_{x \to 0^+} \ln(x)$

40. $\lim\limits_{x \to \infty} \ln(x)$

41. $\lim\limits_{x \to \infty} \log(x)$

42. $\lim\limits_{x \to 0^+} \log(x)$

Write the equation of each graph in its final position.

43. The graph of $y = \ln(x)$ is translated three units to the right and then four units downward.

44. The graph of $y = \log(x)$ is translated five units to the left and then seven units upward.

45. The graph of $y = \log_2(x)$ is translated five units to the right, reflected in the x-axis, and then translated one unit downward.

46. The graph of $y = \log_3(x)$ is translated four units upward, six units to the left, and then reflected in the x-axis.

Write each equation as an equivalent exponential equation.

47. $\log_2(32) = 5$

48. $\log_3(81) = 4$

49. $\log_5(x) = y$

50. $\log_4(a) = b$

51. $\log(1000) = z$

52. $\ln(y) = 3$

Write each equation as an equivalent logarithmic equation.

53. $5^3 = 125$

54. $2^7 = 128$

55. $e^3 = y$

56. $10^5 = w$

57. $y = 10^m$

58. $p = e^x$

For each function, find f^{-1}.

59. $f(x) = 2^x$ **60.** $f(x) = 5^x$

61. $f(x) = \log_7(x)$ **62.** $f(x) = \log(x)$

63. $f(x) = \ln(x - 1)$ **64.** $f(x) = \log(x + 4)$

65. $f(x) = 3^{x+2}$ **66.** $f(x) = 6^{x-1}$

67. $f(x) = \dfrac{1}{2} \cdot 10^{x-1} + 5$ **68.** $f(x) = 2^{3x+1} - 6$

Solve each equation. Find the exact solutions.

69. $\log_2(x) = 8$ **70.** $\log_5(x) = 3$

71. $\log_3(x) = \dfrac{1}{2}$ **72.** $\log_4(x) = \dfrac{1}{3}$

73. $\log_x(16) = 2$ **74.** $\log_x(16) = 4$

75. $3^x = 77$ **76.** $\dfrac{1}{2^x} = 5$

77. $\ln(x - 3) = \ln(2x - 9)$ **78.** $\log_2(4x) = \log_2(x + 6)$

79. $\log_x(18) = 2$ **80.** $\log_x(9) = \dfrac{1}{2}$

81. $3^{x+1} = 7$ **82.** $5^{3-x} = 12$

83. $\log(x) = \log(6 - x^2)$ **84.** $\log_3(2x) = \log_3(24 - x^2)$

Find the approximate solution to each equation. Round to four decimal places.

85. $10^x = 25$ **86.** $e^x = 2$

87. $e^{2x} = 3$ **88.** $10^{3x} = 5$

89. $5e^x = 4$ **90.** $10^x - 3 = 5$

91. $\dfrac{1}{10^x} = 2$ **92.** $\dfrac{1}{e^{x-1}} = 5$

Solve each problem.

93. *Finding Time* Find the amount of time to the nearest tenth of a year that it would take for $10 to grow to $20 at each of the following annual rates compounded continuously.
a. 2%   **b.** 4%

c. 8%   **d.** 16%

94. *Finding Time* Find the amount of time, to the nearest tenth of a year that it would take for $10 to grow to $40 at each of the following annual rates compounded continuously.
a. 1%   **b.** 2%

c. 8.327%   **d.** $7\dfrac{2}{3}$%

95. *Finding Rate* Find the annual percentage rate compounded continuously to the nearest tenth of a percent for which $10 would grow to $30 for each of the following time periods.
a. 5 years   **b.** 10 years

c. 20 years   **d.** 40 years

96. *Finding Rate* Find the annual percentage rate to the nearest tenth of a percent for which $10 would grow to $50 for each of the following time periods.
a. 4 years   **b.** 8 years

c. 16 years   **d.** 32 years

97. *Becoming a Millionaire* Find the amount of time to the nearest day that it would take for a deposit of $1000 to grow to $1 million at 14% compounded continuously.

98. *Doubling Your Money* How long does it take for a deposit of $1000 to double at 8% compounded continuously?

99. *Finding the Rate* Solve the equation $A = Pe^{rt}$ for r, then find the rate at which a deposit of $1000 would double in 3 years compounded continuously.

100. *Finding the Rate* At what interest rate would a deposit of $30,000 grow to $2,540,689 in 40 years with continuous compounding?

101. *Traffic Jam* The amount of U.S. long-distance traffic for data transmission is expected to grow exponentially in the coming years. See the accompanying figure. The expected growth can be modeled by the function

$$d = 0.1e^{0.46t},$$

where t is the number of years since 1994 and d is measured in billions of gigabits per year.
a. What will be the amount of long-distance data transmission in 2009?

b. In what year will long-distance data transmission reach 14 billion gigabits per year?

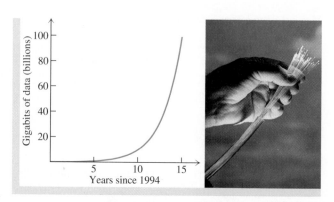

Figure for Exercise 101

102. *Global Warming*   The increasing global temperature can be modeled by the function

$$I = 0.1e^{0.02t},$$

where I is the increase in global temperature in degrees Celsius since 1900, and t is the number of years since 1900 (NASA, www.science.nasa.gov).

   a. How much warmer will it be in 2010 than it was in 1950?

   b. In what year will the global temperature be $4°$ greater than the global temperature in 2000?

Thinking Outside the Box XXX

Seven-Eleven   A convenience store sells a gallon of milk for $7 and a loaf of bread for $11. You are allowed to buy any combination of milk and bread, including only milk or only bread. Your total bill is always a whole number of dollars, but there are many whole numbers that cannot be the total. For example, the total cannot be $15. What is the largest whole number of dollars that cannot be the total?

4.2 Pop Quiz

1. What is x if $2^x = 32$?

2. Find $\log_2(32)$.

3. Is $y = \log_3(x)$ increasing or decreasing?

4. Find all asymptotes for $f(x) = \ln(x - 1)$.

5. Write $3^a = b$ as a logarithmic equation.

6. Find f^{-1} if $f(x) = \log(x + 3)$

7. Solve $\log_5(x) = 3$.

8. Solve $\log_x(36) = 2$.

9. How long (to the nearest day) does it take for $2 to grow to $4 at 5% compounded continuously?

4.3   Rules of Logarithms

The rules of logarithms are closely related to the rules of exponents, because logarithms are exponents. In this section we use the rules of exponents to develop some rules of logarithms. With these rules of logarithms we will be able to solve more equations involving exponents and logarithms. The rules of exponents are discussed in Section A.1 of the Appendix.

The Inverse Rules

The definition of logarithms leads to two rules that are useful in solving equations. If $f(x) = a^x$ and $g(x) = \log_a(x)$, then

$$g(f(x)) = g(a^x) = \log_a(a^x) = x$$

for any real number x. The result of this composition is x because the functions are inverses. If we compose in the opposite order we get

$$f(g(x)) = f(\log_a(x)) = a^{\log_a(x)} = x$$

for any positive real number x. The results are called the **inverse rules.**

Inverse Rules

If $a > 0$ and $a \neq 1$, then

1. $\log_a(a^x) = x$ for any real number x
2. $a^{\log_a(x)} = x$ for $x > 0$.

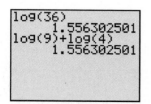

```
e^(ln(17.2))
              17.2
ln(e^(6))
              6
10^(log(98.6))
              98.6
```

Figure 4.30

The inverse rules are easy to use if you remember that $\log_a(x)$ is the power of a that produces x. For example, $\log_2(67)$ is the power of 2 that produces 67. So $2^{\log_2(67)} = 67$. Similarly, $\ln(e^{99})$ is the power of e that produces e^{99}. So $\ln(e^{99}) = 99$. You can illustrate the inverse rules with a calculator, as shown in Fig. 4.30. □

Example 1 Using the inverse rules

Simplify each expression.

a. $e^{\ln(x^2)}$ **b.** $\log_7(7^{2x-1})$

Solution

By the inverse rules, $e^{\ln(x^2)} = x^2$ and $\log_7(7^{2x-1}) = 2x - 1$. ■

The Logarithm of a Product

By the product rule for exponents, we add exponents when multiplying exponential expressions having the same base. To find a corresponding rule for logarithms, let's examine the equation $2^3 \cdot 2^2 = 2^5$. Notice that the exponents 3, 2, and 5 are the base-2 logarithms of 8, 4, and 32, respectively.

$$\begin{array}{ccc} \log_2(8) & \log_2(4) & \log_2(32) \\ \searrow & \downarrow & \swarrow \\ 2^3 & \cdot \; 2^2 = & 2^5 \end{array}$$

When we add the exponents 3 and 2 to get 5, we are adding logarithms and getting another logarithm as the result. So the base-2 logarithm of 32 (the product of 8 and 4) is the sum of the base-2 logarithms of 8 and 4:

$$\log_2(8 \cdot 4) = \log_2(8) + \log_2(4)$$

This example suggests the **product rule for logarithms.**

Product Rule for Logarithms

For $M > 0$ and $N > 0$,

$$\log_a(MN) = \log_a(M) + \log_a(N).$$

PROOF

$$a^{\log_a M + \log_a N} = a^{\log_a M} \cdot a^{\log_a N} \qquad \text{Product rule for exponents}$$
$$= M \cdot N \qquad \text{Inverse rule}$$

Now by the definition of logarithm ($a^x = y \Leftrightarrow \log_a y = x$) we have

$$\log_a(MN) = \log_a(M) + \log_a(N). \qquad ■$$

```
log(36)
        1.556302501
log(9)+log(4)
        1.556302501
```

Figure 4.31

The product rule for logarithms says that *the logarithm of a product of two numbers is equal to the sum of their logarithms*, provided that all of the logarithms are defined and all have the same base. There is no rule about the logarithm of a sum, and the logarithm of a sum is generally *not* equal to the sum of the logarithms. For example, $\log_2(8 + 8) \neq \log_2(8) + \log_2(8)$ because $\log_2(8 + 8) = 4$ while $\log_2(8) + \log_2(8) = 6$. You can use a calculator to illustrate the product rule, as in Fig. 4.31. □

Example **2** **Using the product rule for logarithms**

Write each expression as a single logarithm. All variables represent positive real numbers.

a. $\log_3(x) + \log_3(6)$ **b.** $\ln(3) + \ln(x^2) + \ln(y)$

Solution

a. By the product rule for logarithms, $\log_3(x) + \log_3(6) = \log_3(6x)$.
b. By the product rule for logarithms, $\ln(3) + \ln(x^2) + \ln(y) = \ln(3x^2y)$.   ■

The Logarithm of a Quotient

By the quotient rule for exponents, we subtract the exponents when dividing exponential expressions having the same base. To find a corresponding rule for logarithms, examine the equation $2^5/2^2 = 2^3$. Notice that the exponents 5, 2, and 3 are the base-2 logarithms of 32, 4, and 8, respectively:

$$\overset{\log_2(32)}{\underset{\log_2(4)}{\frac{2^5}{2^2}}} = \overset{\log_2(8)}{2^3}$$

When we subtract the exponents 5 and 2 to get 3, we are subtracting logarithms and getting another logarithm as the result. So the base-2 logarithm of 8 (the quotient of 32 and 4) is the difference of the base-2 logarithms of 32 and 4:

$$\log_2\left(\frac{32}{4}\right) = \log_2(32) - \log_2(4)$$

This example suggests the **quotient rule for logarithms.**

Quotient Rule for Logarithms

For $M > 0$ and $N > 0$,

$$\log_a\left(\frac{M}{N}\right) = \log_a(M) - \log_a(N).$$

The quotient rule for logarithms says that *the logarithm of a quotient of two numbers is equal to the difference of their logarithms*, provided that all logarithms are defined and all have the same base. (The proof of the quotient rule is similar to that of the product rule and so it is left for the reading.) Note that the quotient rule does not apply to division of logarithms. For example,

$$\frac{\log_2(32)}{\log_2(4)} \neq \log_2(32) - \log_2(4),$$

because $\log_2(32)/\log_2(4) = 5/2$, while $\log_2(32) - \log_2(4) = 3$.

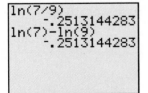

Figure 4.32

You can use a calculator to illustrate the quotient rule, as shown in Fig. 4.32.   □

Example **3** **Using the quotient rule for logarithms**

Write each expression as a single logarithm. All variables represent positive real numbers.

a. $\log_3(24) - \log_3(4)$ **b.** $\ln(x^6) - \ln(x^2)$

Solution

a. By the quotient rule, $\log_3(24) - \log_3(4) = \log_3(24/4) = \log_3(6)$.

b. $\ln(x^6) - \ln(x^2) = \ln\!\left(\dfrac{x^6}{x^2}\right)$ By the quotient rule for logarithms

$\qquad\qquad\qquad\quad = \ln(x^4)$ By the quotient rule for exponents   ■

The Logarithm of a Power

By the power rule for exponents, we multiply the exponents when finding a power of an exponential expression. For example, $(2^3)^2 = 2^6$. Notice that the exponents 3 and 6 are the base-2 logarithms of 8 and 64, respectively.

$$\overset{\log_2(8)}{\underset{\downarrow}{}}\quad\overset{\log_2(64)}{\underset{\downarrow}{}}$$
$$(2^3)^2 = 2^6$$

So the base-2 logarithm of 64 (the second power of 8) is twice the base-2 logarithm of 8:

$$\log_2(8^2) = 2 \cdot \log_2(8)$$

This example suggests the **power rule for logarithms.**

Power Rule for Logarithms

For $M > 0$ and any real number N,
$$\log_a(M^N) = N \cdot \log_a(M).$$

```
ln(17^3)
        8.499640032
3ln(17)
        8.499640032
```

■ **Figure 4.33**

The power rule for logarithms says that *the logarithm of a power of a number is equal to the power times the logarithm of the number*, provided that all logarithms are defined and have the same base. The proof is left for the reader.

◺◿ You can illustrate the power rule on a calculator, as shown in Fig. 4.33.   □

Example **4** **Using the power rule for logarithms**

Rewrite each expression in terms of $\log(3)$.

a. $\log(3^8)$ **b.** $\log(\sqrt{3})$ **c.** $\log\!\left(\dfrac{1}{3}\right)$

Solution

a. $\log(3^8) = 8 \cdot \log(3)$ By the power rule for logarithms

b. $\log\left(\sqrt{3}\right) = \log(3^{1/2}) = \dfrac{1}{2}\log(3)$ By the power rule for logarithms

c. $\log\left(\dfrac{1}{3}\right) = \log(3^{-1}) = -\log(3)$ By the power rule for logarithms   ■

Using the Rules

When simplifying or rewriting expressions, we often apply several rules. In the following box we list all of the available rules of logarithms.

**Rules of Logarithms
with Base *a***

If M, N, and a are positive real numbers with $a \neq 1$, and x is any real number, then

1. $\log_a(a) = 1$ **2.** $\log_a(1) = 0$
3. $\log_a(a^x) = x$ **4.** $a^{\log_a(N)} = N$
5. $\log_a(MN) = \log_a(M) + \log_a(N)$ **6.** $\log_a(M/N) = \log_a(M) - \log_a(N)$
7. $\log_a(M^x) = x \cdot \log_a(M)$ **8.** $\log_a(1/N) = -\log_a(N)$

Note that rule 1 is a special case of rule 3 with $x = 1$, rule 2 follows from the fact that $a^0 = 1$ for any nonzero base a, and rule 8 is a special case of rule 6 with $M = 1$.

The rules for logarithms with base a in the preceding box apply to all logarithms, including common logarithms (base 10) and natural logarithms (base e). Since natural logarithms are very popular, we list the rules of logarithms again for base e in the following box for easy reference.

Rules of Natural Logarithms

If M and N are positive real numbers and x is any real number, then

1. $\ln(e) = 1$ **2.** $\ln(1) = 0$
3. $\ln(e^x) = x$ **4.** $e^{\ln(N)} = N$
5. $\ln(MN) = \ln(M) + \ln(N)$ **6.** $\ln(M/N) = \ln(M) - \ln(N)$
7. $\ln(M^x) = x \cdot \ln(M)$ **8.** $\ln(1/N) = -\ln(N)$

Example **5** **Using the rules of logarithms**

Rewrite each expression in terms of $\log(2)$ and $\log(3)$.

a. $\log(6)$ **b.** $\log\left(\dfrac{16}{3}\right)$ **c.** $\log\left(\dfrac{1}{3}\right)$

Solution

a. $\log(6) = \log(2 \cdot 3)$ Factor.

$\qquad = \log(2) + \log(3)$ Rule 5

b. $\log\left(\dfrac{16}{3}\right) = \log(16) - \log(3)$ Rule 6

$\qquad = \log(2^4) - \log(3)$ Write 16 as a power of 2.

$\qquad = 4 \cdot \log(2) - \log(3)$ Rule 7

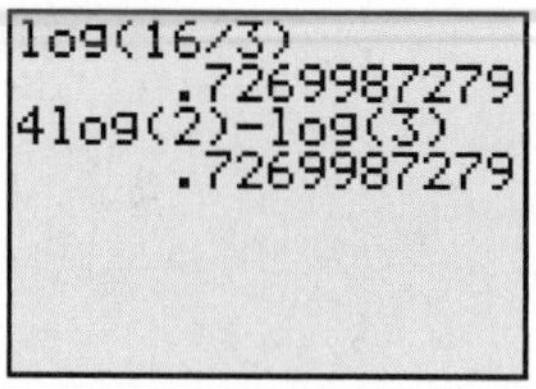

```
log(16/3)
      .7269987279
4log(2)-log(3)
      .7269987279
```

■ **Figure 4.34**

⬚ You can check this answer with a calculator, as shown in Fig. 4.34. ☐

c. $\log\left(\dfrac{1}{3}\right) = -\log(3)$ Rule 8   ■

Be careful to use the rules of logarithms exactly as stated. For example, the logarithm of a product is the sum of the logarithms. The logarithm of a product is generally *not* equal to the product of the logarithms. That is,

$$\log(2 \cdot 3) = \log(2) + \log(3) \quad \text{but} \quad \log(2 \cdot 3) \neq \log(2) \cdot \log(3).$$

Example **6** **Rewriting a logarithmic expression**

Rewrite each expression using a sum or difference of multiples of logarithms.

a. $\ln\left(\dfrac{3x^2}{yz}\right)$ **b.** $\log_3\left(\dfrac{(x-1)^2}{z^{3/2}}\right)$

Solution

a. $\ln\left(\dfrac{3x^2}{yz}\right) = \ln(3x^2) - \ln(yz)$ Quotient rule for logarithms

$$= \ln(3) + \ln(x^2) - [\ln(y) + \ln(z)] \quad \text{Product rule for logarithms}$$

$$= \ln(3) + 2 \cdot \ln(x) - \ln(y) - \ln(z) \quad \text{Power rule for logarithms}$$

Note that $\ln(y) + \ln(z)$ must be in brackets (or parentheses) because of the subtraction symbol preceding it.

b. $\log_3\left(\dfrac{(x-1)^2}{z^{3/2}}\right) = \log_3((x-1)^2) - \log_3(z^{3/2})$ Quotient rule for logarithms

$$= 2 \cdot \log_3(x-1) - \frac{3}{2}\log_3(z) \quad \text{Power rule for logarithms} \quad ■$$

In Example 7 we use the rules "in reverse" of the way we did in Example 6.

Example **7** **Rewriting as a single logarithm**

Rewrite each expression as a single logarithm.

a. $\ln(x-1) + \ln(3) - 3 \cdot \ln(x)$ **b.** $\dfrac{1}{2}\log(y) - \dfrac{1}{3}\log(z)$

Solution

a. Use the product rule, the power rule, and then the quotient rule for logarithms:

$$\ln(x-1) + \ln(3) - 3 \cdot \ln(x) = \ln[3(x-1)] - \ln(x^3)$$

$$= \ln[3x - 3] - \ln(x^3)$$

$$= \ln\left(\frac{3x-3}{x^3}\right)$$

b. $\dfrac{1}{2}\log(y) - \dfrac{1}{3}\log(z) = \log(y^{1/2}) - \log(z^{1/3})$

$$= \log(\sqrt{y}) - \log(\sqrt[3]{z})$$

$$= \log\left(\frac{\sqrt{y}}{\sqrt[3]{z}}\right) \quad ■$$

The Base-Change Formula

The exact solution to an exponential equation is often expressed in terms of logarithms. For example, the exact solution to $(1.03)^x = 5$ is $x = \log_{1.03}(5)$. But how do we calculate $\log_{1.03}(5)$? The next example shows how to find a rational approximation for $\log_{1.03}(5)$ using rules of logarithms. In this example we also introduce a new idea in solving equations, *taking the logarithm of each side.* The base-*a* logarithms of two equal quantities are equal because logarithm is a function.

Example 8 A rational approximation for a logarithm

Find an approximate rational solution to $(1.03)^x = 5$. Round to four decimal places.

Solution

Take the logarithm of each side using one of the bases available on a calculator.

$$(1.03)^x = 5$$

$$\log((1.03)^x) = \log(5) \qquad \text{Take the logarithm of each side.}$$

$$x \cdot \log(1.03) = \log(5) \qquad \text{Power rule for logarithms}$$

$$x = \frac{\log(5)}{\log(1.03)} \qquad \text{Divide each side by } \log(1.03).$$

$$\approx 54.4487$$

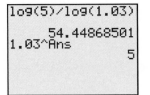 Divide the logarithms and check with a calculator, as shown in Fig. 4.35. ■

■ **Figure 4.35**

The solution to the equation of Example 8 is a base-1.03 logarithm, but we obtained a rational approximation for it using base-10 logarithms. We can use the procedure of Example 8 to write a base-*a* logarithm in terms of a base-*b* logarithm for any bases *a* and *b*:

$$a^x = M \qquad \text{Equivalent equation: } x = \log_a(M)$$

$$\log_b(a^x) = \log_b(M) \qquad \text{Take the base-} b \text{ logarithm of each side.}$$

$$x \cdot \log_b(a) = \log_b(M) \qquad \text{Power rule for logarithms}$$

$$x = \frac{\log_b(M)}{\log_b(a)} \qquad \text{Divide each side by } \log_b(a).$$

Since $x = \log_a(M)$, we have the following **base-change formula.**

Base-Change Formula

> If $a > 0$, $b > 0$, $a \neq 1$, $b \neq 1$, and $M > 0$, then
>
> $$\log_a(M) = \frac{\log_b(M)}{\log_b(a)}.$$

The base-change formula says that the logarithm of a number in one base is equal to the logarithm of the number in the new base divided by the logarithm of the old base. Using this formula, a logarithm such as $\log_3(7)$ can be easily found with a calculator. Let $a = 3$, $b = 10$, and $M = 7$ in the formula to get

$$\log_3(7) = \frac{\log(7)}{\log(3)} \approx 1.7712.$$

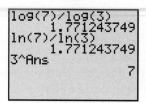

◪ Note that you get the same value for $\log_3(7)$ using natural logarithms, as shown in Fig. 4.36. □

Example ⑨ **Using the base-change formula with compound interest**

If $2500 is invested at 6% compounded daily, then how long (to the nearest day) would it take for the investment to double in value?

Solution

We want the number of years for $2500 to grow to $5000 at 6% compounded daily. Use $P = \$2500$, $A = \$5000$, $n = 365$, and $r = 0.06$ in the formula for compound interest:

$$A = P\left(1 + \frac{r}{n}\right)^{nt}$$

$$5000 = 2500\left(1 + \frac{0.06}{365}\right)^{365t}$$

$$2 = \left(1 + \frac{0.06}{365}\right)^{365t} \qquad \text{Divide by 2500 to get the form } y = a^x.$$

$$2 \approx (1.000164384)^{365t} \qquad \text{Approximate the base with a decimal number.}$$

$$365t \approx \log_{1.000164384}(2) \qquad \text{Definition of logarithm}$$

$$t \approx \frac{1}{365} \cdot \frac{\ln(2)}{\ln(1.000164384)} \qquad \text{Base-change formula (Use either ln or log.)}$$

$$\approx 11.553 \text{ years}$$

The investment of $2500 will double in about 11 years, 202 days. ■

Note that Example 9 can also be solved by taking the log (or ln) of each side and applying the power rule for logarithms.

$$\log(2) = \log((1 + 0.06/365)^{365t})$$

$$\log(2) = 365t \cdot \log(1 + 0.06/365)$$

$$\frac{\log(2)}{365 \cdot \log(1 + 0.06/365)} = t$$

$$11.553 \approx t$$

If a variable is a function of time, we are often interested in finding the rate of growth or decay. In the continuous model, $y = ae^{rt}$, the rate r appears as an exponent. If we know the values of the other variables, we find the rate by solving for r using natural logarithms. If $y = ab^t$ is used instead of $y = ae^{rt}$, the rate does not appear in the formula. However, we can write $y = ab^t$ as $y = ae^{\ln(b)\cdot t}$, because $b = e^{\ln(b)}$. Comparing $e^{\ln(b)\cdot t}$ to e^{rt}, we see that the rate is $\ln(b)$.

For Thought

True or False? Explain.

1. $\dfrac{\log(8)}{\log(3)} = \log(8) - \log(3)$

2. $\ln\left(\sqrt{3}\right) = \dfrac{\ln(3)}{2}$

3. $\dfrac{\log_{19}(8)}{\log_{19}(2)} = \log_3(27)$

4. $\dfrac{\log_2(7)}{\log_2(5)} = \dfrac{\log_3(7)}{\log_3(5)}$

5. $e^{\ln(x)} = x$ for any real number x.

6. The equations $\log(x - 2) = 4$ and $\log(x) - \log(2) = 4$ are equivalent.

7. The equations $x + 1 = 2x + 3$ and $\log(x + 1) = \log(2x + 3)$ are equivalent.

8. $\ln(e^x) = x$ for any real number x.

9. If $30 = x^{50}$, then x is between 1 and 2.

10. If $20 = a^4$, then $\ln(a) = \frac{1}{4}\ln(20)$.

4.3 Exercises

Simplify each expression.

1. $e^{\ln(\sqrt{y})}$ 2. $10^{\log(3x+1)}$ 3. $\log(10^{y+1})$

4. $\ln(e^{2k})$ 5. $7^{\log_7(999)}$ 6. $\log_4(2^{300})$

Rewrite each expression as a single logarithm.

7. $\log(5) + \log(3)$ 8. $\ln(6) + \ln(2)$

9. $\log_2(x - 1) + \log_2(x)$

10. $\log_3(x + 2) + \log_3(x - 1)$

11. $\log_4(12) - \log_4(2)$ 12. $\log_2(25) - \log_2(5)$

13. $\ln(x^8) - \ln(x^3)$ 14. $\log(x^2 - 4) - \log(x - 2)$

Rewrite each expression as a sum or difference of logarithms.

15. $\log_2(3x)$ 16. $\log_3(xy)$

17. $\log\left(\dfrac{x}{2}\right)$ 18. $\log\left(\dfrac{a}{b}\right)$

19. $\log(x^2 - 1)$ 20. $\log(a^2 - 9)$

Rewrite each expression in terms of $\log_a(5)$.

21. $\log_a(5^3)$ 22. $\log_a(25)$ 23. $\log_a\left(\sqrt{5}\right)$

24. $\log_a\left(\sqrt[3]{5}\right)$ 25. $\log_a\left(\dfrac{1}{5}\right)$ 26. $\log_a\left(\dfrac{1}{125}\right)$

Rewrite each expression in terms of $\log_a(2)$ and $\log_a(5)$.

27. $\log_a(10)$ 28. $\log_a(0.4)$

29. $\log_a(2.5)$ 30. $\log_a(250)$

31. $\log_a\left(\sqrt{20}\right)$ 32. $\log_a(0.0005)$

Rewrite each expression as a sum or difference of multiples of logarithms.

33. $\log_3(5x)$ 34. $\log_2(xyz)$ 35. $\log_2\left(\dfrac{5}{2y}\right)$

36. $\log\left(\dfrac{4a}{3b}\right)$ 37. $\log(3\sqrt{x})$ 38. $\ln\left(\sqrt{x/4}\right)$

39. $\log(3 \cdot 2^{x-1})$ 40. $\ln\left(\dfrac{5^{-x}}{2}\right)$

41. $\ln\left(\dfrac{\sqrt[3]{xy}}{t^{4/3}}\right)$ 42. $\log\left(\dfrac{3x^2}{(ab)^{2/3}}\right)$

Rewrite each expression as a single logarithm.

43. $\log_2(5) + 3 \cdot \log_2(x)$

44. $\log(x) + 5 \cdot \log(x)$

45. $\log_7(x^5) - 4 \cdot \log_7(x^2)$

46. $\dfrac{1}{3}\ln(6) - \dfrac{1}{3}\ln(2)$

47. $\log(2) + \log(x) + \log(y) - \log(z)$

48. $\ln(2) + \ln(3) + \ln(5) - \ln(7)$

49. $\frac{1}{2}\log(x) - \log(y) + \log(z) - \frac{1}{3}\log(w)$

50. $\frac{5}{6}\log_2(x) + \frac{2}{3}\log_2(y) - \frac{1}{2}\log_2(x) - \log_2(y)$

Find an approximate rational solution to each equation. Round answers to four decimal places.

51. $2^x = 9$

52. $3^x = 12$

53. $(0.56)^x = 8$

54. $(0.23)^x = 18.4$

55. $(1.06)^x = 2$

56. $(1.09)^x = 3$

Use a calculator and the base-change formula to find each logarithm to four decimal places.

57. $\log_4(9)$

58. $\log_3(4.78)$

59. $\log_{9.1}(2.3)$

60. $\log_{1.2}(13.7)$

61. $\log_{1/2}(12)$

62. $\log_{1.05}(3.66)$

Solve each equation. Round answers to four decimal places.

63. $(1.02)^{4t} = 3$

64. $(1.025)^{12t} = 3$

65. $(1.0001)^{365t} = 3.5$

66. $(1.00012)^{365t} = 2.4$

Solve each equation. Round answers to four decimal places.

67. $\log_x(33.4) = 5$

68. $\log_x(12.33) = 2.3$

69. $\log_x(0.546) = -1.3$

70. $\log_x(0.915) = -3.2$

For Exercises 71–74 find the time required for each investment given in the table to grow to the specified amount. The letter W represents an unknown principal.

	Principal	Ending Balance	Rate	Compounded	Time
71.	$800	$2000	8%	Daily	?
72.	$10,000	$1,000,000	7.75%	Annually	?
73.	$W	$3W	10%	Quarterly	?
74.	$W	$2W	12%	Monthly	?

75. *Ben's Gift to Boston* Ben Franklin's gift of $4000 to Boston grew to $4.5 million in 200 years. At what interest rate compounded annually would this growth occur?

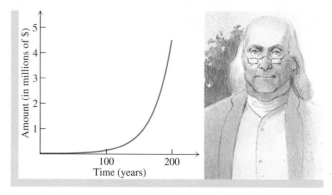

■ **Figure for Exercise 75**

76. *Ben's Gift to Philadelphia* Ben Franklin's gift of $4000 to Philadelphia grew to $2 million in 200 years. At what interest rate compounded monthly would this growth occur?

77. *Richter Scale* The common logarithm is used to measure the intensity of an earthquake on the Richter scale. The Richter scale rating of an earthquake of intensity I is given by $\log(I) - \log(I_0)$, where I_0 is the intensity of a small "benchmark" earthquake. Write the Richter scale rating as a single logarithm. What is the Richter scale rating of an earthquake for which $I = 1000 \cdot I_0$?

78. *Colombian Earthquake of 1906* At 8.6 on the Richter scale, the Colombian earthquake of January 31, 1906, was one of the strongest earthquakes on record. Use the formula from Exercise 77 to write I as a multiple of I_0 for this earthquake.

79. *Time for Growth* The time in years for a population of size P_0 to grow to size P at the annual growth rate r is given by $t = \ln((P/P_0)^{1/r})$. Use the rules of logarithms to express t in terms of $\ln(P)$ and $\ln(P_0)$.

80. *Rollover Time* The probability that a $1 ticket does not win the Louisiana Lottery is $\frac{7,059,051}{7,059,052}$. The probability p that n independently sold tickets are all losers and the lottery rolls over is given by

$$p = \left(\frac{7,059,051}{7,059,052}\right)^n.$$

a. As n increases is p increasing or decreasing?

b. For what number of tickets sold is the probability of a rollover greater than 50%?

Thinking Outside the Box XXXI

Unit Fractions The fractions $\frac{1}{2}, \frac{1}{3}, \frac{1}{4}, \frac{1}{5}, \frac{1}{6}, \frac{1}{7}$, etc., are called *unit fractions*. Some rational numbers can be expressed as sums of unit fractions. For example,

$$\frac{4}{7} = \frac{1}{2} + \frac{1}{14} \quad \text{and} \quad \frac{11}{18} = \frac{1}{3} + \frac{1}{6} + \frac{1}{9}.$$

Write each of the following fractions as a sum of unit fractions using the fewest number of unit fractions, and keep the total of all of the denominators in the unit fractions as small as possible.

a. $\dfrac{6}{23}$ **b.** $\dfrac{14}{15}$ **c.** $\dfrac{7}{11}$

4.3 Pop Quiz

1. Write $\log(9) + \log(3)$ as a single logarithm.

2. Write $\log(9) - \log(3)$ as a single logarithm.

3. Write $3\ln(x) + \ln(y)$ as a single logarithm.

4. Write $\ln(5x)$ as a sum of logarithms.

5. Write $\ln\left(\sqrt{18}\right)$ in terms of $\ln(2)$ and $\ln(3)$.

Solve each equation. Round answers to four decimal places.

6. $3^x = 11$ 7. $\log_x(22.5) = 3$

4.4 More Equations and Applications

The rules of Section 4.3 combined with the techniques that we have already used in Sections 4.1 and 4.2 allow us to solve several new types of equations involving exponents and logarithms.

Logarithmic Equations

An equation involving a single logarithm can usually be solved by using the definition of logarithm as we did in Section 4.2.

Example **1** **An equation involving a single logarithm**

Solve the equation $\log(x - 3) = 4$.

Solution

Write the equivalent equation using the definition of logarithm:

$$\log(x - 3) = 4 \qquad \text{Original equation } \log_a(y) = x \Leftrightarrow y = a^x$$
$$x - 3 = 10^4$$
$$x = 10{,}003$$

Check this number in the original equation. The solution is 10,003. ∎

When more than one logarithm is present, we can use the one-to-one property as in Section 4.2 or use the other rules of logarithms to combine logarithms.

Example 2 Equations involving more than one logarithm

Solve each equation.

a. $\log_2(x) + \log_2(x + 2) = \log_2(6x + 1)$ **b.** $\log(x) - \log(x - 1) = 2$
c. $2 \cdot \ln(x) = \ln(x + 3) + \ln(x - 1)$

Solution

a. Since the sum of two logarithms is equal to the logarithm of a product, we can rewrite the left-hand side of the equation.

$$\log_2(x) + \log_2(x + 2) = \log_2(6x + 1)$$

$$\log_2(x(x + 2)) = \log_2(6x + 1) \quad \text{Product rule for logarithms}$$

$$x^2 + 2x = 6x + 1 \quad \text{One-to-one property of logarithms}$$

$$x^2 - 4x - 1 = 0 \quad \text{Solve quadratic equation.}$$

$$x = \frac{4 \pm \sqrt{16 - 4(-1)}}{2} = 2 \pm \sqrt{5}$$

Since $2 - \sqrt{5}$ is a negative number, $\log_2(2 - \sqrt{5})$ is undefined and $2 - \sqrt{5}$ is not a solution. The only solution to the equation is $2 + \sqrt{5}$. Check this solution by using a calculator and the base-change formula.

⊞ The check is shown with a graphing calculator in Fig. 4.37. □

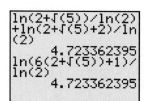

■ **Figure 4.37**

b. Since the difference of two logarithms is equal to the logarithm of a quotient, we can rewrite the left-hand side of the equation:

$$\log(x) - \log(x - 1) = 2$$

$$\log\left(\frac{x}{x - 1}\right) = 2 \quad \text{Quotient rule for logarithms}$$

$$\frac{x}{x - 1} = 10^2 \quad \text{Definition of logarithm}$$

$$x = 100x - 100 \quad \text{Solve for } x.$$

$$-99x = -100$$

$$x = \frac{100}{99}$$

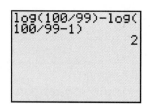

■ **Figure 4.38**

⊞ Check 100/99 in the original equation, as shown in Fig. 4.38. □

c. $2 \cdot \ln(x) = \ln(x + 3) + \ln(x - 1)$

$$\ln(x^2) = \ln(x^2 + 2x - 3) \quad \text{Power rule, product rule}$$

$$x^2 = x^2 + 2x - 3 \quad \text{One-to-one property}$$

$$0 = 2x - 3$$

$$\frac{3}{2} = x$$

⊞ Use a calculator to check 3/2 in the original equation. The only solution to the equation is 3/2. ∎

All solutions to logarithmic and exponential equations should be checked in the original equation, because extraneous roots can occur, as in Example 2(a). Use your calculator to check every solution and you will increase your proficiency with your calculator.

Exponential Equations

An exponential equation with a single exponential expression can usually be solved by using the definition of logarithm, as in Section 4.2.

Example 3 Equations involving a single exponential expression

Solve the equation $(1.02)^{4t-1} = 5$.

Solution

Write an equivalent equation, using the definition of logarithm.

$$4t - 1 = \log_{1.02}(5)$$

$$t = \frac{1 + \log_{1.02}(5)}{4} \qquad \text{The exact solution}$$

$$= \frac{1 + \dfrac{\ln(5)}{\ln(1.02)}}{4} \qquad \text{Base-change formula}$$

$$\approx 20.5685$$

The approximate solution is 20.5685.

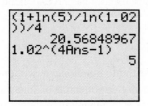

■ **Figure 4.39**

Check with a graphing calculator, as shown in Fig. 4.39.   ■

Note that Example 3 can also be solved by taking the ln (or log) of each side and applying the power rule for logarithms:

$$\ln\left(1.02^{4t-1}\right) = \ln(5)$$

$$(4t - 1)\ln(1.02) = \ln(5)$$

$$4t \cdot \ln(1.02) - \ln(1.02) = \ln(5)$$

$$t = \frac{\ln(5) + \ln(1.02)}{4 \cdot \ln(1.02)} \approx 20.5685$$

If an equation has an exponential expression on each side, as in Example 4, then it is best to take the log or ln of each side to solve it.

Example 4 Equations involving two exponential expressions

Find the exact and approximate solutions to $3^{2x-1} = 5^x$.

Solution

$$\ln(3^{2x-1}) = \ln(5^x) \qquad \text{Take the natural logarithm of each side.}$$

$$(2x - 1)\ln(3) = x \cdot \ln(5) \qquad \text{Power rule for logarithms}$$

$$2x \cdot \ln(3) - \ln(3) = x \cdot \ln(5) \qquad \text{Distributive property}$$

$$2x \cdot \ln(3) - x \cdot \ln(5) = \ln(3)$$

$$x[2 \cdot \ln(3) - \ln(5)] = \ln(3)$$

$$x = \frac{\ln(3)}{2 \cdot \ln(3) - \ln(5)} \qquad \text{Exact solution}$$

$$\approx 1.8691$$

Had we used common logarithms, similar steps would give

$$x = \frac{\log(3)}{2 \cdot \log(3) - \log(5)}$$

$$\approx 1.8691.$$

■ **Figure 4.40**

Check the solution in the original equation, as shown in Fig. 4.40. ■

The technique of Example 4 can be used on any equation of the form $a^M = b^N$. Take the natural logarithm of each side and apply the power rule, to get an equation of the form $M \cdot \ln(a) = N \cdot \ln(b)$. This last equation usually has no exponents and is relatively easy to solve.

Strategy for Solving Equations

We solved equations involving exponential and logarithmic functions in Sections 4.1 through 4.4. There is no formula that will solve every exponential or logarithmic equation, but the following strategy will help you solve exponential and logarithmic equations.

STRATEGY **Solving Exponential and Logarithmic Equations**

1. If the equation involves a single logarithm or a single exponential expression, then use the definition of logarithm: $y = \log_a(x)$ if and only if $a^y = x$.

2. Use the one-to-one properties when applicable:
   a) if $a^M = a^N$, then $M = N$.
   b) if $\log_a(M) = \log_a(N)$, then $M = N$.

3. If an equation has several logarithms with the same base, then combine them using the product and quotient rules:
   a) $\log_a(M) + \log_a(N) = \log_a(MN)$
   b) $\log_a(M) - \log_a(N) = \log_a(M/N)$

4. If an equation has only exponential expressions with different bases on each side, then take the natural or common logarithm of each side and use the power rule: $a^M = b^N$ is equivalent to $\ln(a^M) = \ln(b^N)$ or $M \cdot \ln(a) = N \cdot \ln(b)$.

Radioactive Dating

In Section 4.1, we stated that the amount A of a radioactive substance remaining after t years is given by

$$A = A_0 e^{rt},$$

where A_0 is the initial amount present and r is the annual rate of decay for that particular substance. A standard measurement of the speed at which a radioactive substance decays is its **half-life.** The half-life of a radioactive substance is the amount of time that it takes for one-half of the substance to decay. Of course, when one-half has decayed, one-half remains.

Now that we have studied logarithms, we can use the formula for radioactive decay to determine the age of an ancient object that contains a radioactive substance. One such substance is potassium-40, which is found in rocks. Once the rock is formed, the potassium-40 begins to decay. The amount of time that has passed since the formation of the rock can be determined by measuring the amount of potassium-40 that has decayed into argon-40. Dating rocks using potassium-40 is known as **potassium-argon dating.**

Example **5** **Finding the age of *Deinonychus* ("terrible claw")**

Our chapter-opening case described the 1964 find of *Deinonychus* (*National Geographic*, August 1978). Since dinosaur bones are too old to contain enough organic material for radiocarbon dating, paleontologists often estimate the age of bones by dating volcanic debris in the surrounding rock. The age of *Deinonychus* was determined from the age of surrounding rocks by using potassium-argon dating. The half-life of potassium-40 is 1.31 billion years. If 94.5% of the original amount of potassium-40 is still present in the rock, then how old are the bones of *Deinonychus*?

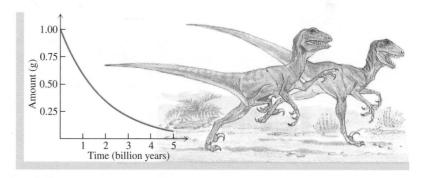

Solution

The half-life is the amount of time that it takes for 1 gram to decay to 0.5 gram. Use $A_0 = 1$, $A = 0.5$, and $t = 1.31 \times 10^9$ in the formula $A = A_0 e^{rt}$ to find r:

$$0.5 = 1 \cdot e^{(1.31 \times 10^9)(r)}$$

$$(1.31 \times 10^9)(r) = \ln(0.5) \qquad \text{Definition of logarithm}$$

$$r = \frac{\ln(0.5)}{1.31 \times 10^9}$$

$$\approx -5.29 \times 10^{-10}$$

Now we can find the amount of time that it takes for 1 gram to decay to 0.945 gram. Use $r \approx -5.29 \times 10^{-10}$, $A_0 = 1$, and $A = 0.945$ in the formula.

$$0.945 = 1 \cdot e^{(-5.29 \times 10^{-10})(t)}$$

$$(-5.29 \times 10^{-10})(t) = \ln(0.945) \qquad \text{Definition of logarithm}$$

$$t = \frac{\ln(0.945)}{-5.29 \times 10^{-10}}$$

$$\approx 107 \text{ million years}$$

The dinosaur *Deinonychus* lived about 107 million years ago. ■

Newton's Model for Cooling

Newton's law of cooling states that when a warm object is placed in colder surroundings or a cold object is placed in warmer surroundings, then the difference between the two temperatures decreases in an exponential manner. If D_0 is the initial difference in temperature, then the difference D at time t is modeled by the formula

$$D = D_0 e^{kt},$$

where k is a constant that depends on the object and the surroundings. In the next example we use Newton's law of cooling to answer a question that you might have asked yourself as the appetizers were running low.

Example **6** **Using Newton's law of cooling**

A turkey with a temperature of 40°F is moved to a 350° oven. After 4 hours the internal temperature of the turkey is 170°F. If the turkey is done when its temperature reaches 185°, then how much longer must it cook?

■ **Figure 4.41**

Solution

The initial difference of 310° (350° − 40°) has dropped to a difference of 180° (350° − 170°) after 4 hours. See Fig. 4.41. With this information we can find k:

$$180 = 310 e^{4k}$$

$$e^{4k} = \frac{180}{310} \qquad \text{Isolate } e^{4k} \text{ by dividing by 310.}$$

$$4k = \ln(18/31) \qquad \text{Definition of logarithm: } e^x = y \Leftrightarrow x = \ln(y)$$

$$k = \frac{\ln(18/31)}{4} \approx -0.1359$$

The turkey is done when the difference in temperature between the turkey and the oven is 165° (the oven temperature 350° minus the turkey temperature 185°). Now find the time for which the difference will be 165°:

$$165 = 310e^{-0.1359t}$$

$$e^{-0.1359t} = \frac{165}{310}$$

$$-0.1359t = \ln(165/310) \qquad \text{Definition of logarithm: } e^x = y \Leftrightarrow x = \ln(y)$$

$$t = \frac{\ln(165/310)}{-0.1359} \approx 4.6404$$

The difference in temperature between the turkey and the oven will be 165° when the turkey has cooked approximately 4.6404 hours. So the turkey must cook about 0.6404 hour (38.4 minutes) longer.   ■

Paying off a Loan

If n is the number of periods per year, r is the annual percentage rate (APR), t is the number of years, and i is the interest rate per period ($i = r/n$), then the periodic payment R that will pay off a loan of P dollars is given by

$$R = P\frac{i}{1 - (1 + i)^{-nt}}.$$

Homeowners are often interested in how the time will change if they increase the monthly payment. Solving for t requires logarithms.

Example **7** Finding the time

A couple still owes $90,000 on a house that is financed at 8% annual percentage rate compounded monthly. If they start paying $1200 per month, then when will the loan be paid off?

Solution

Use $R = 1200$, $P = 90{,}000$, $n = 12$, and $i = 0.08/12$ in the formula for the monthly payment:

$$1200 = 90{,}000\frac{0.08/12}{1 - (1 + 0.08/12)^{-12t}}$$

$$1200\left(1 - (1 + 0.08/12)^{-12t}\right) = 600$$

$$1 - (1 + 0.08/12)^{-12t} = 0.5$$

$$0.5 = (1 + 0.08/12)^{-12t}$$

$$\ln(0.5) = -12t \cdot \ln(1 + 0.08/12)$$

$$\frac{\ln(0.5)}{-12 \cdot \ln(1 + 0.08/12)} = t$$

$$8.6932 \approx t$$

So in approximately 8.6932 years or about 8 years and 8 months the loan will be paid off.   ■

For Thought

True or False? Explain.

1. The equation $3(1.02)^x = 21$ is equivalent to $x = \log_{1.02}(7)$.

2. If $x - x \cdot \ln(3) = 8$, then $x = \dfrac{8}{1 - \ln(3)}$.

3. The solution to $\ln(x) - \ln(x - 1) = 6$ is $1 - \sqrt{6}$.

4. If $2^{x-3} = 3^{2x+1}$, then $x - 3 = \log_2(3^{2x+1})$.

5. The exact solution to $3^x = 17$ is 2.5789.

6. The equation $\log(x) + \log(x - 3) = 1$ is equivalent to $\log(x^2 - 3x) = 1$.

7. The equation $4^x = 2^{x-1}$ is equivalent to $2x = x - 1$.

8. The equation $(1.09)^x = 2.3$ is equivalent to $x \cdot \ln(1.09) = \ln(2.3)$.

9. $\ln(2) \cdot \log(7) = \log(2) \cdot \ln(7)$

10. $\log(e) \cdot \ln(10) = 1$

4.4 Exercises

Each of these equations involves a single logarithm. Solve each equation. See the strategy for solving exponential and logarithmic equations on page 348.

1. $\log_2(x) = 3$

2. $\log_3(x) = 0$

3. $\log(x + 20) = 2$

4. $\log(x - 6) = 1$

5. $\log(x^2 - 15) = 1$

6. $\log(x^2 - 5x + 16) = 1$

7. $\log_x(9) = 2$

8. $\log_x(16) = 4$

9. $-2 = \log_x(4)$

10. $-\dfrac{1}{2} = \log_x(9)$

11. $\log_x(10) = 3$

12. $\log_x(5) = 2$

13. $\log_8(x) = -\dfrac{2}{3}$

14. $\log_4(x) = -\dfrac{5}{2}$

Each of these equations involves more than one logarithm. Solve each equation. Give exact solutions.

15. $\log_2(x + 2) + \log_2(x - 2) = 5$

16. $\log(x + 1) - \log(x) = 3$

17. $\log(5) = 2 - \log(x)$

18. $\log(4) = 1 + \log(x - 1)$

19. $\ln(x) + \ln(x + 2) = \ln(8)$

20. $\log_3(x) = \log_3(2) - \log_3(x - 2)$

21. $\log(4) + \log(x) = \log(5) - \log(x)$

22. $\ln(x) - \ln(x + 1) = \ln(x + 3) - \ln(x + 5)$

23. $\log_2(x) - \log_2(3x - 1) = 0$

24. $\log_3(x) + \log_3(1/x) = 0$

25. $x \cdot \ln(3) = 2 - x \cdot \ln(2)$

26. $x \cdot \log(5) + x \cdot \log(7) = \log(9)$

Each of these equations involves a single exponential expression. Solve each equation. Round approximate solutions to four decimal places.

27. $2^{x-1} = 7$

28. $5^{3x} = 29$

29. $(1.09)^{4x} = 3.4$

30. $(1.04)^{2x} = 2.5$

31. $3^{-x} = 30$

32. $10^{-x+3} = 102$

33. $9 = e^{-3x^2}$

34. $25 = 10^{-2x}$

Each of these equations involves more than one exponential expression. Solve each equation. Round approximate solutions to four decimal places.

35. $6^x = 3^{x+1}$

36. $2^x = 3^{x-1}$

37. $e^{x+1} = 10^x$

38. $e^x = 2^{x+1}$

39. $2^{x-1} = 4^{3x}$

40. $3^{3x-4} = 9^x$

41. $6^{x+1} = 12^x$

42. $2^x \cdot 2^{x+1} = 4^{x^2+x}$

Solve each equation. Round approximate solutions to four decimal places.

43. $e^{-\ln(w)} = 3$

44. $10^{2 \cdot \log(y)} = 4$

45. $(\log(z))^2 = \log(z^2)$

46. $\ln(e^x) - \ln(e^6) = \ln(e^2)$

47. $4(1.02)^x = 3(1.03)^x$

48. $500(1.06)^x = 400(1.02)^{4x}$

49. $e^{3 \cdot \ln(x^2) - 2 \cdot \ln(x)} = \ln(e^{16})$

50. $\sqrt{\log(x) - 3} = \log(x) - 3$

51. $\left(\dfrac{1}{2}\right)^{2x-1} = \left(\dfrac{1}{4}\right)^{3x+2}$ **52.** $\left(\dfrac{2}{3}\right)^{x+1} = \left(\dfrac{9}{4}\right)^{x+2}$

Find the approximate solution to each equation by graphing an appropriate function on a graphing calculator and locating the x-intercept. Note that these equations cannot be solved by the techniques that we have learned in this chapter.

53. $2^x = 3^{x-1} + 5^{-x}$ **54.** $2^x = \log(x + 4)$

55. $\ln(x + 51) = \log(-48 - x)$ **56.** $2^x = 5 - 3^{x+1}$

57. $x^2 = 2^x$ **58.** $x^3 = e^x$

Solve each problem.

59. *Finding the Rate* If the half-life of a radioactive substance is 10,000 years, then at what rate is it decaying?

60. *Finding the Rate* If the half-life of a drug is 12 hours, then at what rate is it being eliminated from the body?

61. *Dating a Bone* A piece of bone from an organism is found to contain 10% of the carbon-14 that it contained when the organism was living. If the half-life of carbon-14 is 5730 years, then how long ago was the organism alive?

62. *Old Clothes* If only 15% of the carbon-14 in a remnant of cloth has decayed, then how old is the cloth?

63. *Dating a Tree* How long does it take for 12 g of carbon-14 in a tree trunk to be reduced to 10 g of carbon-14 by radioactive decay?

64. *Carbon-14 Dating* How long does it take for 2.4 g of carbon-14 to be reduced to 1.3 g of carbon-14 by radioactive decay?

65. *Radioactive Waste* If 25 g of radioactive waste reduces to 20 g of radioactive waste after 8000 years, then what is the half-life for this radioactive element?

66. *Finding the Half-Life* If 80% of a radioactive element remains radioactive after 250 million years, then what percent remains radioactive after 600 million years? What is the half-life of this element?

67. *Lorazepam* The drug lorazepam, used to relieve anxiety and nervousness, has a half-life of 14 hours. Its chemical structure is shown in the accompanying figure. If a doctor prescribes one 2.5 milligram tablet every 24 hours, then what percentage of the last dosage remains in the patient's body when the next dosage is taken?

Lorazepam

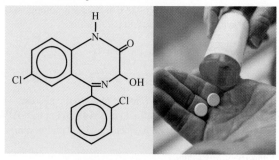

■ **Figure for Exercise 67**

68. *Drug Build-Up* The level of a prescription drug in the human body over time can be found using the formula

$$L = \frac{D}{1 - (0.5)^{n/h}},$$

where D is the amount taken every n hours and h is the drug's half-life in hours.

  a. If 2.5 milligrams of lorazepam with a half-life of 14 hours is taken every 24 hours, then to what level does the drug build up over time?

  b. If a doctor wants the level of lorazepam to build up to a level of 5.58 milligrams in a patient taking 2.5 milligram doses, then how often should the doses be taken?

  c. What is the difference between taking 2.5 milligrams of lorazepam every 12 hours and taking 5 milligrams every 24 hours?

69. *Cooking a Roast* James knows that to get well-done beef, it should be brought to a temperature of 170°F. He placed a sirloin tip roast with a temperature of 35°F in an oven with a temperature of 325°, and after 3 hr the temperature of the roast was 140°. How much longer must the roast be in the oven to get it well done? If the oven temperature is set at 170°, how long will it take to get the roast well done?

70. *Room Temperature* Marlene brought a can of polyurethane varnish that was stored at 40°F into her shop, where the temperature was 74°. After 2 hr the temperature of the varnish was 58°. If the varnish must be 68° for best results, then how much longer must Marlene wait until she uses the varnish?

71. *Time of Death* A detective discovered a body in a vacant lot at 7 A.M. and found that the body temperature was 80°F. The county coroner examined the body at 8 A.M. and found that the body temperature was 72°. Assuming that the body temperature was 98° when the person died and that the air temperature was a constant 40° all night, what was the approximate time of death?

72. *Cooling Hot Steel* A blacksmith immersed a piece of steel at 600°F into a large bucket of water with a temperature of 65°. After 1 min the temperature of the steel was 200°. How much longer must the steel remain immersed to bring its temperature down to 100°?

73. *Paying off a Loan* Find the time (to the nearest month) that it takes to pay off a loan of $100,000 at 9% APR compounded monthly with payments of $1250 per month.

74. *Solving for Time* Solve the formula

$$R = P\frac{i}{1 - (1 + i)^{-nt}}$$

for t. Then use the result to find the time (to the nearest month) that it takes to pay off a loan of $48,265 at $8\frac{3}{4}$% APR compounded monthly with payments of $700 per month.

75. *Equality of Investments* Fiona invested $1000 at 6% compounded continuously. At the same time, Maria invested $1100 at 6% compounded daily. How long will it take (to the nearest day) for their investments to be equal in value?

76. *Depreciation and Inflation* Boris won a $35,000 luxury car on *Wheel of Fortune*. He plans to keep it until he can trade it evenly for a new compact car that currently costs $10,000. If the value of the luxury car decreases by 8% each year and the cost of the compact car increases by 5% each year, then in how many years will he be able to make the trade?

77. *Habitat Destruction* Biologists use the species-area curve $n = k \log(A)$ to estimate the number of species n that live in a region of area A, where k is a constant.
  a. If 2500 species live in a rain forest of 400 square kilometers, then how many species will be left when half of this rain forest is destroyed by logging?
  b. A rain forest of 1200 square kilometers supported 3500 species in 1950. Due to intensive logging, what remains of this forest supported only 1000 species in 2000. What percent of this rain forest has been destroyed?

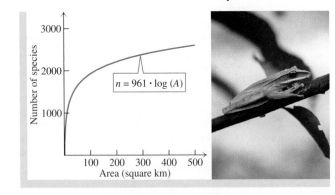

Figure for Exercise 77

78. *Extinction of Species* In 1980 an Amazon rain forest of area A contained n species. In the year 2000, after intensive logging, this rain forest was reduced to an area that contained only half as many species. Use the species-area formula from the previous exercise to find the area in 2000.

79. *Noise Pollution* The level of a sound in decibels (db) is determined by the formula

$$\text{sound level} = 10 \cdot \log(I \times 10^{12}) \text{ db},$$

where I is the intensity of the sound in watts per square meter. To combat noise pollution, a city has an ordinance prohibiting sounds above 90 db on a city street. What value of I gives a sound of 90 db?

80. *Doubling the Sound Level* A small stereo amplifier produces a sound of 50 db at a distance of 20 ft from the speakers. Use the formula from Exercise 79 to find the intensity of the sound at this point in the room. If the intensity just found is doubled, what happens to the sound level? What must the intensity be to double the sound level to 100 db?

81. *Present Value of a CD* What amount (present value) must be deposited today in a certificate of deposit so that the investment will grow to $20,000 in 18 years at 6% compounded continuously.

82. *Present Value of a Bond* A $50 U.S. Savings Bond paying 6.22% compounded monthly matures in 11 years 2 months. What is the present value of the bond?

Thinking Outside the Box XXXII

Pile of Pipes Six pipes are placed in a pile as shown in the diagram. The three pipes on the bottom each have a radius of 2 feet. The two pipes on top of those each have a radius of 1 foot. At the top of the pile is a pipe that is tangent to the two smaller pipes and one of the larger pipes. What is its exact radius?

■ Figure for Thinking Outside the Box XXXII

4.4 Pop Quiz

Solve each equation. Round approximate solutions to four decimal places.

1. $\log_5(x) = 4$

2. $\log(x + 1) = 3$

3. $\ln(x) + \ln(x - 1) = \ln(12)$

4. $3^{x-5} = 10$

5. $8^x = 3^{x+5}$

■■■ Highlights

4.1 Exponential Functions and Their Applications

Exponential Function

$f(x) = a^x$ for $a > 0$ and $a \neq 1$
Domain: $(-\infty, \infty)$, Range: $(0, \infty)$
Horizontal asymptote: $y = 0$

$f(x) = 2^x, g(x) = e^x$

Increasing and Decreasing

$f(x) = a^x$ is increasing on $(-\infty, \infty)$ if $a > 1$,
decreasing on $(-\infty, \infty)$ if $0 < a < 1$.

$f(x) = 3^x$ is increasing
$g(x) = (0.4)^x$ is decreasing

One-to-One

If $a^{x_1} = a^{x_2}$, then $x_1 = x_2$.

$2^{x-1} = 2^3 \Rightarrow x - 1 = 3$

Amount Formula

P dollars, annual interest rate r, for t years
Compounded periodically: n periods per year,

$$A = P\left(1 + \frac{r}{n}\right)^{nt}$$

Compounded continuously:
$A = Pe^{rt} (e \approx 2.718)$

$P = \$1000, r = 5\%, t = 10$ yr
Compounded monthly:

$$A = 1000\left(1 + \frac{0.05}{12}\right)^{120}$$

Compounded continuously:
$A = 1000e^{0.05(10)}$

4.2 Logarithmic Functions and Their Applications

Logarithmic Function

$f(x) = \log_a(x)$ for $a > 0$ and $a \neq 1$
$y = \log_a(x) \Leftrightarrow a^y = x$
Domain: $(0, \infty)$, Range: $(-\infty, \infty)$
Vertical asymptote: $x = 0$
Common log: base 10, $f(x) = \log(x)$
Natural log: base e, $f(x) = \ln(x)$

$f(x) = \log_2(x)$
$f(32) = \log_2(32) = 5$
$f(1) = \log_2(1) = 0$
$f(1/4) = \log_2(1/4) = -2$

Increasing and Decreasing

$f(x) = \log_a(x)$ is increasing on $(0, \infty)$ if $a > 1$,
decreasing on $(0, \infty)$ if $0 < a < 1$.

$f(x) = \ln(x)$ is increasing.
$g(x) = \log_{1/2}(x)$ is decreasing.

One-to-One

If $\log_a(x_1) = \log_a(x_2)$, then $x_1 = x_2$.

$\log_2(3x) = \log_2(4) \Rightarrow 3x = 4$

4.3 Rules of Logarithms

Inverse Rules

$\log_a(a^x) = x$ and $a^{\log_a(x)} = x$

$\ln(e^5) = 5, 10^{\log(7)} = 7$

Product Rule

$\log_a(MN) = \log_a(M) + \log_a(N)$

$\log_2(8x) = 3 + \log_2(x)$

Quotient Rule

$\log_a(M/N) = \log_a(M) - \log_a(N)$

$\log(1/2) = \log(1) - \log(2)$

Power Rule

$\log_a(M^N) = N \cdot \log_a(M)$

$\ln(e^3) = 3 \cdot \ln(e)$

Base-Change

$$\log_a(M) = \frac{\log_b(M)}{\log_b(a)}$$

$$\log_3(7) = \frac{\ln(7)}{\ln(3)} = \frac{\log(7)}{\log(3)}$$

4.4 More Equations and Applications

Equation-Solving Strategy

1. Use $y = \log_a(x) \Leftrightarrow a^y = x$ on equations with a single logarithm or single exponential.
2. Use the one-to-one properties to eliminate logarithms or exponential expressions.
3. Combine logarithms using the product and quotient rules.
4. Take a logarithm of each side and use the power rule.

$\log_2(x) = 3 \Leftrightarrow x = 2^3$

$\log(x^2) = \log(5) \Leftrightarrow x^2 = 5$
$e^x = e^{2x-1} \Leftrightarrow x = 2x - 1$
$\ln(x) + \ln(2) = 3$
$\qquad \ln(2x) = 3$
$3^x = 4^{2x}$
$x \cdot \ln(3) = 2x \cdot \ln(4)$

Function Gallery: Some Basic Functions of Algebra with Transformations

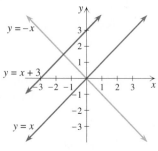

Linear

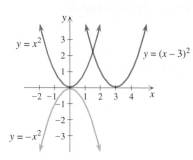

Quadratic

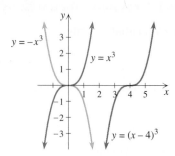

Cubic

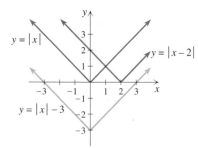

Absolute value

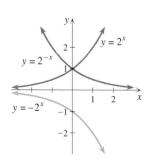

Exponential

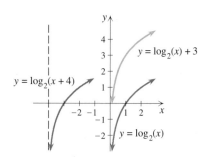

Logarithmic

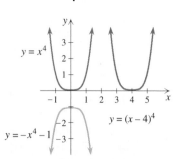

Square root

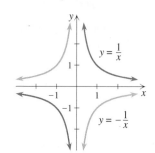

Reciprocal

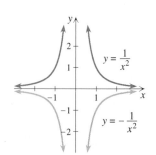

Rational

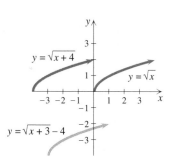

Fourth degree

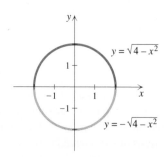

Semicircle

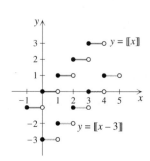

Greatest integer

▪▪▪Chapter 4 Review Exercises

Simplify each expression.

1. 2^6

2. $\ln(e^2)$

3. $\log_2(64)$

4. $3 + 2 \cdot \log(10)$

5. $\log_9(1)$

6. $5^{\log_5(99)}$

7. $\log_2(2^{17})$

8. $\log_2(\log_2(16))$

Let $f(x) = 2^x$, $g(x) = 10^x$, and $h(x) = \log_2(x)$. Simplify each expression.

9. $f(5)$

10. $g(-1)$

11. $\log(g(3))$

12. $g(\log(5))$

13. $(h \circ f)(9)$

14. $(f \circ h)(7)$

15. $g^{-1}(1000)$

16. $g^{-1}(1)$

17. $h(1/8)$

18. $f(1/2)$

19. $f^{-1}(8)$

20. $(f^{-1} \circ f)(13)$

Rewrite each expression as a single logarithm.

21. $\log(x - 3) + \log(x)$

22. $(1/2)\ln(x) - 2 \cdot \ln(y)$

23. $2 \cdot \ln(x) + \ln(y) + \ln(3)$

24. $3 \cdot \log_2(x) - 2 \cdot \log_2(y) + \log_2(z)$

Rewrite each expression as a sum or difference of multiples of logarithms.

25. $\log(3x^4)$

26. $\ln\left(\dfrac{x^5}{y^3}\right)$

27. $\log_3\left(\dfrac{5\sqrt{x}}{y^4}\right)$

28. $\log_2\left(\sqrt{xy^3}\right)$

Rewrite each expression in terms of $\ln(2)$ and $\ln(5)$.

29. $\ln(10)$

30. $\ln(0.4)$

31. $\ln(50)$

32. $\ln\left(\sqrt{20}\right)$

Find the exact solution to each equation.

33. $\log(x) = 10$

34. $\log_3(x + 1) = -1$

35. $\log_x(81) = 4$

36. $\log_x(1) = 0$

37. $\log_{1/3}(27) = x + 2$

38. $\log_{1/2}(4) = x - 1$

39. $3^{x+2} = \dfrac{1}{9}$

40. $2^{x-1} = \dfrac{1}{4}$

41. $e^{x-2} = 9$

42. $\dfrac{1}{2^{1-x}} = 3$

43. $4^{x+3} = \dfrac{1}{2^x}$

44. $3^{2x-1} \cdot 9^x = 1$

45. $\log(x) + \log(2x) = 5$

46. $\log(x + 90) - \log(x) = 1$

47. $\log_2(x) + \log_2(x - 4) = \log_2(x + 24)$

48. $\log_5(x + 18) + \log_5(x - 6) = 2 \cdot \log_5(x)$

49. $2 \cdot \ln(x + 2) = 3 \cdot \ln(4)$

50. $x \cdot \log_2(12) = x \cdot \log_2(3) + 1$

51. $x \cdot \log(4) = 6 - x \cdot \log(25)$

52. $\log(\log(x)) = 1$

Find the missing coordinate so that each ordered pair satisfies the given equation.

53. $y = \left(\dfrac{1}{3}\right)^x$: $(-1,\ \ )$, $(\ \ , 27)$, $(-1/2,\ \ )$, $(\ \ , 1)$

54. $y = \log_9(x - 1)$: $(2,\ \ )$, $(4,\ \ )$, $(\ \ , 3/2)$, $(\ \ , -1)$

Match each equation to one of the graphs (a)–(h).

55. $y = 2^x$

56. $y = 2^{-x}$

57. $y = \log_2(x)$

58. $y = \log_{1/2}(x)$

59. $y = 2^{x+2}$

60. $y = \log_2(x + 2)$

61. $y = 2^x + 2$

62. $y = 2 + \log_2(x)$

a.

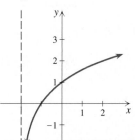

b.

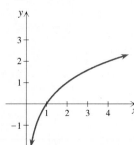

c.

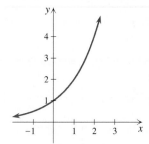

d.

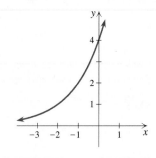

e.

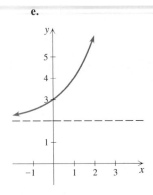

f.

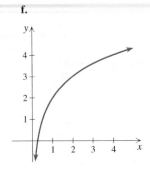

g.

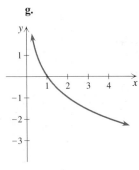

h.

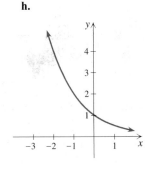

Sketch the graph of each function. State the domain, the range, and whether the function is increasing or decreasing. Identify any asymptotes.

63. $f(x) = 5^x$

64. $f(x) = e^x$

65. $f(x) = 10^{-x}$

66. $f(x) = (1/2)^x$

67. $y = \log_3(x)$

68. $y = \log_5(x)$

69. $y = 1 + \ln(x + 3)$

70. $y = 3 - \log_2(x)$

71. $f(x) = 1 + 2^{x-1}$

72. $f(x) = 3 - 2^{x+1}$

73. $y = \log_3(-x + 2)$

74. $y = 1 + \log_3(x + 2)$

For each function f, find f^{-1}.

75. $f(x) = 7^x$

76. $f(x) = 3^x$

77. $f(x) = \log_5(x)$

78. $f(x) = \log_8(x)$

79. $f(x) = 3 \cdot \log(x - 1)$

80. $f(x) = \log_2(x + 3) - 5$

81. $f(x) = e^{x+2} - 3$

82. $f(x) = 2 \cdot 3^x + 1$

Use a calculator to find an approximate solution to each equation. Round answers to four decimal places.

83. $3^x = 10$

84. $4^{2x} = 12$

85. $\log_3(x) = 1.876$

86. $\log_5(x + 2) = 2.7$

87. $5^x = 8^{x+1}$

88. $3^x = e^{x+1}$

Use the rules of logarithms to determine whether each equation is true or false. Explain your answer.

89. $\log_3(81) = \log_3(9) \cdot \log_3(9)$

90. $\log(81) = \log(9) \cdot \log(9)$

91. $\ln(3^2) = (\ln(3))^2$

92. $\ln\left(\dfrac{5}{8}\right) = \dfrac{\ln(5)}{\ln(8)}$

93. $\log_2(8^4) = 12$

94. $\log(8.2 \times 10^{-9}) = -9 + \log(8.2)$

95. $\log(1006) = 3 + \log(6)$

96. $\dfrac{\log_2(16)}{\log_2(4)} = \log_2(16) - \log_2(4)$

97. $\dfrac{\log_2(8)}{\log_2(16)} = \log_2(8) - \log_2(16)$

98. $\ln(e^{(e^x)}) = e^x$

99. $\log_2(25) = 2 \cdot \log(5)$

100. $\log_3\left(\dfrac{5}{7}\right) = \log(5) - \log(7)$

101. $\log_2(7) = \dfrac{\log(7)}{\log(2)}$

102. $\dfrac{\ln(17)}{\ln(3)} = \dfrac{\log(17)}{\log(3)}$

Solve each problem.

103. *Solving for r*   Solve the formula $A = Pe^{rt}$ for r.

104. *Solving a Formula*   Solve the formula $A = P + Ce^{-kt}$ for t.

105. *Future Value*   If $50,000 is deposited in a bank account paying 5% compounded quarterly, then what will be the value of the account at the end of 18 years?

106. *Future Value*   If $30,000 is deposited in First American Savings and Loan in an account paying 6.18% compounded continuously, then what will be the value of the account after 12 years and 3 months?

107. *Doubling Time with Quarterly Compounding*   How long (to the nearest quarter) will it take for the investment of Exercise 105 to double?

108. *Doubling Time with Continuous Compounding*   How long (to the nearest day) will it take for the investment of Exercise 106 to double?

109. *Finding the Half-Life*   The number of grams A of a certain radioactive substance present at time t is given by the formula $A = 25e^{-0.00032t}$, where t is the number of years from the present. How many grams are present initially? How many grams are present after 1000 years? What is the half-life of this substance?

110. *Comparing Investments* Shinichi invested $800 at 6% compounded continuously. At the same time, Toshio invested $1000 at 5% compounded monthly. How long (to the nearest month) will it take for their investments to be equal in value?

111. *Learning Curve* According to an educational psychologist, the number of words learned by a student of a foreign language after t hours in the language laboratory is given by $f(t) = 40,000(1 - e^{-0.0001t})$. How many hours would it take to learn 10,000 words?

112. *Pediatric Tuberculosis* Health researchers divided the state of Maryland into regions according to per capita income and found a relationship between per capita income i (in thousands) and the percentage of pediatric TB cases P. The equation $P = 31.5(0.935)^i$ can be used to model the data.

  a. What percentage of TB cases would you expect to find in the region with a per capita income of $8000?

  b. What per capita income would you expect in the region where only 2% of the TB cases occurred?

113. *Sea Worthiness* Three measures of a boat's sea worthiness are given in the accompanying table, where A is the sail area in square feet, d is the displacement in pounds, L is the length at the water line in feet, and b is the beam in feet. The Freedom 40 is a $291,560 sailboat with a sail area of 1026 ft², a displacement of 25,005 lb, a beam of 13 ft 6 in., and a length at the water line of 35 ft 1 in. Its sail area-displacement ratio is 19.20, its displacement-length ratio is 258.51, and its capsize screening value is 1.85. Find x, y, and z.

■ **Table for Exercise 113**

Measure	Expression	
Sail area-displacement ratio	$A\left(\dfrac{d}{64}\right)^x$	
Displacement-length ratio	$\dfrac{d}{2240}\left(\dfrac{L}{100}\right)^y$	
Capsize-screening value	$b\left(\dfrac{d}{64}\right)^z$	

114. *Doubling the Bet* A strategy used by some gamblers when they lose is to play again and double the bet, assuming that the amount won is equal to the amount bet. So if a gambler loses $2, then the gambler plays again and bets $4, then $8, and so on.

  a. Why would the gambler use this strategy?

  b. On the nth bet, the gambler risks 2^n dollars. Estimate the first value of n for which the amount bet is more than $1 million using the accompanying graph of $y = 2^n$.

  c. Use logarithms to find the answer to part (b).

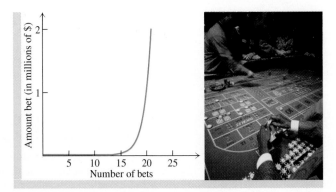

■ **Figure for Exercise 114**

Thinking Outside the Box XXXIII

Crescent City Three semicircles are drawn so that their diameters are the three sides of a right triangle as shown in the diagram. What is the ratio of the total area of the two crescents A and B to the area of the triangle T?

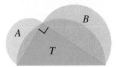

■ **Figure for Thinking Outside the Box XXXIII**

Concepts of Calculus

Evaluating transcendental functions

If we want the value of sin(5), e^3, *or* ln(6) *we simply press a few buttons on our calculators and instantly get the answer. But there are no algebraic formulas that will produce values of transcendental functions like the trigonometric, exponential, and logarithmic functions. So how does a calculator do it? A calculator uses programs and formulas, but not the usual kind of formulas. A calculator uses infinite algebraic formulas called infinite series. The following infinite series are studied in calculus:*

$$e^x = 1 + x + \frac{x^2}{2!} + \frac{x^3}{3!} + \frac{x^4}{4!} + \cdots \text{ where } -\infty < x < \infty$$

$$\ln(1 + x) = x - \frac{x^2}{2} + \frac{x^3}{3} - \frac{x^4}{4} + \cdots \text{ where } -1 < x \leq 1$$

$$\sin(x) = x - \frac{x^3}{3!} + \frac{x^5}{5!} - \frac{x^7}{7!} + \cdots \text{ where } -\infty < x < \infty$$

$$\cos(x) = 1 - \frac{x^2}{2!} + \frac{x^4}{4!} - \frac{x^6}{6!} + \cdots \text{ where } -\infty < x < \infty$$

If the value of one of these functions is an irrational number, we cannot calculate it exactly. The more terms that we use from the infinite formula, the closer we get to the true value of the function.

Exercises

1. Use the first five terms of the formula for e^x to calculate $e^{0.1}$ and compare your result to the value of $e^{0.1}$ obtained using the e^x-key on your calculator.

2. Find ln(2) by using the first 65 terms of the series and compare your result to the calculator value for ln(2).

3. Why can't we use the series for $\ln(1 + x)$ with $x = 4$ to calculate ln(5)? Assuming that we know ln(2) correct to nine decimal places, find ln(5) by using the infinite series and the equation $\ln(5) = \ln(2^2 \cdot 1.25)$.

4. Find ln(9) using the method of the previous exercise.

5. Let $x = \pi/6$ and evaluate

$$x,$$
$$x = x^3/3!,$$
$$x = x^3/3! + x^5/5!,$$
$$x - x^3/3! + x^5/5! - x^7/7!,$$

and so on. Which is the first expression to give 0.5 on your calculator. What is $\sin(\pi/6)$?

6. Repeat the previous exercise with $x = 13\pi/6$.

7. Let $x = \pi/4$ and evaluate.

$$1 - x^2/2!,$$
$$1 - x^2/2! + x^4/4!,$$
$$1 - x^2/2! + x^4/4! - x^6/6!,$$

and so on. Which is the first expression to agree with your calculator's value for cos $(\pi/4)$?

8. Repeat the previous exercise with $x = 9\pi/4$.

9. Make a conjecture about the size of x and the number of terms needed to give an accurate value for $\sin(x)$ or $\cos(x)$.

5 Conic Sections, Polar Coordinates, and Parametric Equations

The Akashi Kaikyo Bridge

Location Kobe and Awaji-shima, Japan

Completion Date 1998

Cost $4.3 billion

Length 12,828 feet

Longest Single Span 6527 feet

Engineers Honshu-Shikoku Bridge Authority

In 1998, Japanese engineers stretched the limits of bridge building when they completed the Akashi Kaikyo Bridge, currently the longest suspension bridge in the world. It would take four Brooklyn Bridges to span the same distance. Its 928-foot towers are higher than any other bridge towers in the world. The Akashi Kaikyo was designed to withstand 180-mile-per-hour hurricane winds, tsunamis, and earthquakes up to 8.5 on the Richter scale. Such events occur almost annually in that location. A unique feature of this bridge is its 20 tuned mass dampers (TMDs) in each tower. The TMDs swing in the opposite direction of the wind sway, balancing the bridge and canceling out the sway.

The parabola, circle, ellipse, and hyperbola can be defined as the four curves that are obtained by intersecting a double right circular cone and a plane, as shown in Fig. 5.1. That is why these curves are known as **conic sections.** If the plane passes through the vertex of the cone, then the intersection of the cone and the plane could be a single point, a line, or a pair of intersecting lines. These are called the **degenerate conics.** The conic sections can also be defined as the graphs of certain equations (as we did for the parabola in Section 2.1). However, the useful properties of the curves are not apparent in either of these approaches. In this chapter we give geometric definitions from which we derive equations for the conic sections. This approach allows us to better understand the properties of the conic sections.

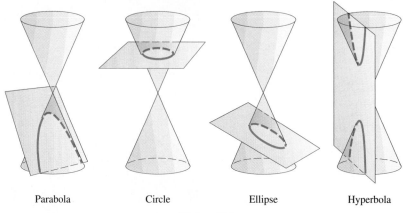

| Parabola | Circle | Ellipse | Hyperbola |

▪ **Figure 5.1**

Definition

Previously in Section 2.1, we defined a parabola algebraically as the graph of $y = ax^2 + bx + c$ for $a \neq 0$. This equation can be expressed also in the form $y = a(x - h)^2 + k$. No equation is mentioned in our geometric definition of a parabola.

Definition: Parabola

> A **parabola** is the set of all points in the plane that are equidistant from a fixed line (the **directrix**) and a fixed point not on the line (the **focus**).

Figure 5.2 shows a parabola with its directrix, focus, axis of symmetry, and vertex. In terms of the directrix and focus, the **axis of symmetry** can be described as the line perpendicular to the directrix and containing the focus. The **vertex** is the point on the axis of symmetry that is equidistant from the focus and directrix. If we position the directrix and focus in a coordinate plane with the directrix horizontal, we can find an equation that is satisfied by all points of the parabola.

Parabola

Axis of symmetry

Focus

Vertex

Directrix

▪ **Figure 5.2**

Developing the Equation

Start with the focus at $(0, p)$ and the directrix $y = -p$ for $p > 0$, as shown in Fig. 5.3. Since the vertex is halfway between the focus and directrix, the vertex is $(0, 0)$. Note that p (the **focal length**) is the directed distance from the vertex to the

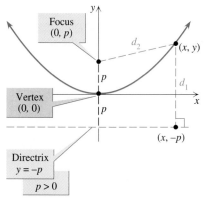

Figure 5.3

focus. If the focus is above the vertex, then $p > 0$, and if the focus is below the vertex, then $p < 0$. The distance from the vertex to the focus or the vertex to the directrix is $|p|$.

Now the distance d_1 from an arbitrary point (x, y) on the parabola to the directrix is the distance from (x, y) to $(x, -p)$, as shown in Fig. 5.3. Using the distance formula from Section 1.3 we have

$$d_1 = \sqrt{(x - x)^2 + (y - (-p))^2} = \sqrt{(y + p)^2} = y + p.$$

Now we find the distance d_2 between (x, y) and the focus $(0, p)$:

$$d_2 = \sqrt{(x - 0)^2 + (y - p)^2} = \sqrt{x^2 + y^2 - 2yp + p^2}$$

Since $d_1 = d_2$ for every point (x, y) on the parabola, we have the following:

$$y + p = \sqrt{x^2 + y^2 - 2yp + p^2}$$

$$y^2 + 2yp + p^2 = x^2 + y^2 - 2yp + p^2 \qquad \text{Square each side.}$$

$$y = \frac{1}{4p}x^2 \qquad \text{Solve for } y.$$

Translating the focus and directrix h units horizontally and k units vertically yields a focus at $(h, k + p)$, a directrix $y = k - p$, a vertex at (h, k), and the new equation

$$y = \frac{1}{4p}(x - h)^2 + k.$$

This equation is of the form $y = a(x - h)^2 + k$ where $a = 1/(4p)$. So the curve determined by the geometric definition has an equation that is an equation of a parabola according to the algebraic definition. We state this result as follows.

Theorem: The Equation of a Parabola

> The equation of a parabola with focus $(h, k + p)$ and directrix $y = k - p$ is
>
> $$y = a(x - h)^2 + k,$$
>
> where $a = 1/(4p)$ and (h, k) is the vertex.

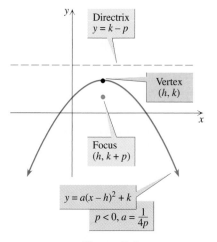

Figure 5.4

The link between the geometric definition and the equation of a parabola is

$$a = \frac{1}{4p}.$$

For any particular parabola, a and p have the same sign. If they are both positive, the parabola opens upward and the focus is above the directrix. If they are both negative, the parabola opens downward and the focus is below the directrix. Figure 5.4 shows the positions of the focus, directrix, and vertex for a parabola with $p < 0$. Since a is inversely proportional to p, smaller values of $|p|$ correspond to larger values of $|a|$ and to "narrower" parabolas.

Example **1** **Writing the equation from the focus and directrix**

Find the equation of the parabola with focus $(-1, 3)$ and directrix $y = 2$.

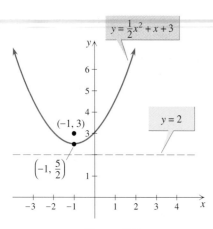

Figure 5.5

Solution

The focus is one unit above the directrix, as shown in Fig. 5.5. So $p = 1/2$. Therefore $a = 1/(4p) = 1/2$. Since the vertex is halfway between the focus and directrix, the y-coordinate of the vertex is $(3 + 2)/2$ and the vertex is $(-1, 5/2)$. Use $a = 1/2, h = -1$, and $k = 5/2$ in the formula $y = a(x - h)^2 + k$ to get the equation

$$y = \frac{1}{2}(x - (-1))^2 + \frac{5}{2}.$$

Simplify to get the equation $y = \frac{1}{2}x^2 + x + 3$. ■

The Standard Equation of a Parabola

If we start with the standard equation of a parabola, $y = ax^2 + bx + c$, we can identify the vertex, focus, and directrix by rewriting it in the form $y = a(x - h)^2 + k$.

Example **2** Finding the vertex, focus, and directrix

Find the vertex, focus, and directrix of the graph of $y = -3x^2 - 6x + 2$.

Solution

Use completing the square to write the equation in the form $y = a(x - h)^2 + k$:

$$y = -3(x^2 + 2x) + 2$$
$$= -3(x^2 + 2x + 1 - 1) + 2$$
$$= -3(x^2 + 2x + 1) + 2 + 3$$
$$= -3(x + 1)^2 + 5$$

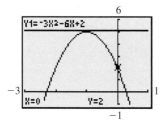

Figure 5.6

The vertex is $(-1, 5)$, and the parabola opens downward because $a = -3$. Since $a = 1/(4p)$, we have $1/(4p) = -3$, or $p = -1/12$. Because the parabola opens downward, the focus is $1/12$ unit below the vertex $(-1, 5)$ and the directrix is a horizontal line $1/12$ unit above the vertex. The focus is $(-1, 59/12)$, and the directrix is $y = 61/12$.

The graphs of $y_1 = -3x^2 - 6x + 2$ and $y_2 = 61/12$ in Fig. 5.6 show how close the directrix is to the vertex for this parabola. ■

In Section 2.1 we learned that the x-coordinate of the vertex of the parabola $y = ax^2 + bx + c$ is $-b/(2a)$. We can use $x = -b/(2a)$ and $a = 1/(4p)$ to determine the focus and directrix without completing the square.

Example **3** Finding the vertex, focus, and directrix

Find the vertex, focus, and directrix of the parabola $y = 2x^2 + 6x - 7$ without completing the square, and determine whether the parabola opens upward or downward.

Solution

First use $x = -b/(2a)$ to find the x-coordinate of the vertex:

$$x = \frac{-b}{2a} = \frac{-6}{2 \cdot 2} = -\frac{3}{2}$$

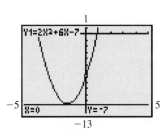

Figure 5.7

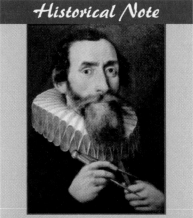
To find the y-coordinate of the vertex, let $x = -3/2$ in $y = 2x^2 + 6x - 7$:

$$y = 2\left(-\frac{3}{2}\right)^2 + 6\left(-\frac{3}{2}\right) - 7 = \frac{9}{2} - 9 - 7 = -\frac{23}{2}$$

The vertex is $(-3/2, -23/2)$. Since $a = 2$, the parabola opens upward. Use $2 = 1/(4p)$ to get $p = 1/8$. Since the parabola opens upward, the directrix is $1/8$ unit below the vertex and the focus is $1/8$ unit above the vertex. The directrix is $y = -93/8$, and the focus is $(-3/2, -91/8)$.

You can graph $y_1 = -93/8$ and $y_2 = 2x^2 + 6x - 7$, as shown in Fig. 5.7, to check the position of the parabola and its directrix. ■

Graphing a Parabola

According to the geometric definition of a parabola, every point on the parabola is equidistant from its focus and directrix. However, it is not easy to find points satisfying that condition. The German mathematician Johannes Kepler (1571–1630) devised a method for drawing a parabola with a given focus and directrix: A piece of string, with length equal to the length of the T-square, is attached to the end of the T-square and the focus, as shown in Fig. 5.8. A pencil is moved down the edge of the T-square, holding the string against it, while the T-square is moved along the directrix toward the focus. While the pencil is aligning the string with the edge of the T-square, it remains equidistant from the focus and directrix.

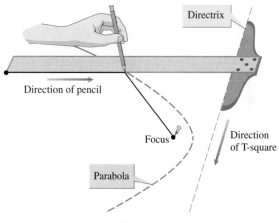

Figure 5.8

Although Kepler's method does give the graph of the parabola from the focus and directrix, the most accurate graphs are now drawn with the equation and a computer. In the next example we start with the focus and directrix, find the equation, and then draw the graph.

Example **4** Graphing a parabola given its focus and directrix

Find the vertex, axis of symmetry, x-intercepts, and y-intercept of the parabola that has focus $(3, 7/4)$ and directrix $y = 9/4$. Sketch the graph, showing the focus and directrix.

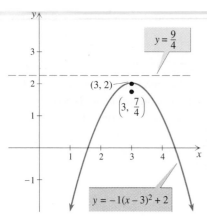

■ **Figure 5.9**

Solution

First draw the focus and directrix on the graph, as shown in Fig. 5.9. Since the vertex is midway between the focus and directrix, the vertex is $(3, 2)$. The distance between the focus and vertex is $1/4$. Since the directrix is above the focus, the parabola opens downward, and we have $p = -1/4$. Use $a = 1/(4p)$ to get $a = -1$. Use $a = -1$ and the vertex $(3, 2)$ in the equation $y = a(x - h)^2 + k$ to get

$$y = -1(x - 3)^2 + 2.$$

The axis of symmetry is the vertical line $x = 3$. If $x = 0$, then $y = -1(0 - 3)^2 + 2 = -7$. So the y-intercept is $(0, -7)$. Find the x-intercepts by setting y equal to 0 in the equation:

$$-1(x - 3)^2 + 2 = 0$$
$$(x - 3)^2 = 2$$
$$x - 3 = \pm\sqrt{2}$$
$$x = 3 \pm \sqrt{2}$$

The x-intercepts are $\left(3 - \sqrt{2}, 0\right)$ and $\left(3 + \sqrt{2}, 0\right)$. Two additional points that satisfy $y = -1(x - 3)^2 + 2$ are $(1, -2)$ and $(4, 1)$. Using all of this information, we get the graph shown in Fig. 5.9. ■

Parabolas Opening to the Left or Right

The graphs of $y = 2x^2$ and $x = 2y^2$ are both parabolas. Interchanging the variables simply changes the roles of the x- and y-axes. The parabola $y = 2x^2$ opens upward, whereas the parabola $x = 2y^2$ opens to the right. For parabolas opening right or left, the directrix is a vertical line. If the focus is to the right of the directrix, then the parabola opens to the right, and if the focus is to the left of the directrix, then the parabola opens to the left. Figure 5.10 shows the relative locations of the vertex, focus, and directrix for parabolas of the form $x = a(y - k)^2 + h$.

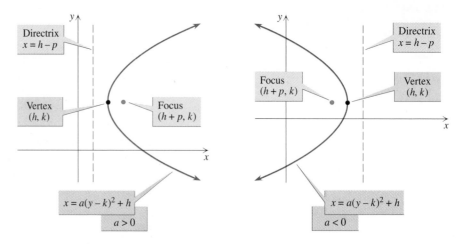

■ **Figure 5.10**

The equation $x = a(y - k)^2 + h$ can be written as $x = ay^2 + by + c$. So the graph of $x = ay^2 + by + c$ is also a parabola opening to the right for $a > 0$ and to the left for $a < 0$. Because the roles of x and y are interchanged, $-b/(2a)$ is now the y-coordinate of the vertex, and the axis of symmetry is the horizontal line $y = -b/(2a)$.

Example 5 Graphing a parabola with a vertical directrix

Find the vertex, axis of symmetry, y-intercepts, focus, and directrix for the parabola $x = y^2 - 2y$. Find several other points on the parabola and sketch the graph.

Solution

Because $a = 1$, the parabola opens to the right. The y-coordinate of the vertex is

$$y = \frac{-b}{2a} = \frac{-(-2)}{2(1)} = 1.$$

If $y = 1$, then $x = (1)^2 - 2(1) = -1$ and the vertex is $(-1, 1)$. The axis of symmetry is the horizontal line $y = 1$. To find the y-intercepts, we solve $y^2 - 2y = 0$ by factoring:

$$y(y - 2) = 0$$
$$y = 0 \quad \text{or} \quad y = 2$$

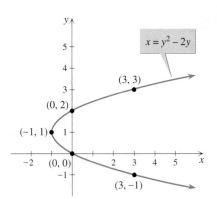

$x = y^2 - 2y$

(3, 3)

(0, 2)

(−1, 1)

(0, 0)

(3, −1)

Figure 5.11

The y-intercepts are $(0, 0)$ and $(0, 2)$. Using all of this information and the additional points $(3, 3)$ and $(3, -1)$, we get the graph shown in Fig. 5.11. Because $a = 1$ and $a = 1/(4p)$, we have $p = 1/4$. So the focus is $1/4$ unit to the right of the vertex at $(-3/4, 1)$. The directrix is the vertical line $1/4$ unit to the left of the vertex, $x = -5/4$. ∎

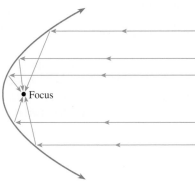

Focus

Figure 5.12

Applications

In Section 2.1 we saw an important application of a parabola. Because of the shape of a parabola, a quadratic function has a maximum value or a minimum value at the vertex of the parabola. However, parabolas are important for another totally different reason. When a ray of light, traveling parallel to the axis of symmetry, hits a parabolic reflector, it is reflected toward the focus of the parabola. See Fig. 5.12. This property is used in telescopes to magnify the light from distant stars. For spotlights, in which the light source is at the focus, the reflecting property is used in reverse. Light originating at the focus is reflected off the parabolic reflector and projected outward in a narrow beam. The reflecting property is used also in telephoto camera lenses, radio antennas, satellite dishes, eavesdropping devices, and flashlights.

Example 6 The Hubble telescope

The Hubble space telescope uses a glass mirror with a parabolic cross section and a diameter of 2.4 meters, as shown in Fig. 5.13 on the next page. If the focus of the parabola is 57.6 meters from the vertex, then what is the equation of the parabola used for the mirror? How much thicker is the mirror at the edge than at its center?

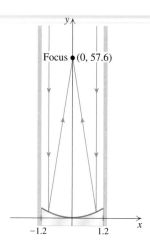

■ **Figure 5.13**

Solution

If we position the parabola with the vertex at the origin and opening upward, as shown in Fig. 5.13, then its equation is of the form $y = ax^2$. Since the distance between the focus and vertex is 57.6 meters, $p = 57.6$. Using $a = 1/(4p)$, we get $a = 1/(4 \cdot 57.6) = 1/230.4$. The equation of the parabola is

$$y = \frac{1}{230.4}x^2.$$

To find the difference in thickness at the edge, let $x = 1.2$ in the equation of the parabola:

$$y = \frac{1}{230.4}(1.2)^2 = 0.006250 \text{ meter}$$

The mirror is 0.006250 meter thicker at the edge than at the center.   ■

For a telescope to work properly, the glass mirror must be ground with great precision. Prior to the 1993 repair, the Hubble space telescope did not work as well as hoped (Space Telescope Science Institute, www.stsci.edu), because the mirror was actually ground to be only 0.006248 meter thicker at the outside edge, two millionths of a meter smaller than it should have been!

For Thought

True or False? Explain.

1. A parabola with focus (0, 0) and directrix $y = -1$ has vertex (0, 1).

2. A parabola with focus (3, 0) and directrix $x = -1$ opens to the right.

3. A parabola with focus (4, 5) and directrix $x = 1$ has vertex (5/2, 5).

4. For $y = x^2$, the focus is $(0, -1/4)$.

5. For $x = y^2$, the focus is $(1/4, 0)$.

6. A parabola with focus (2, 3) and directrix $y = -5$ has no x-intercepts.

7. The parabola $y = 4x^2 - 9x$ has no y-intercept.

8. The vertex of the parabola $x = 2(y + 3)^2 - 5$ is $(-5, -3)$.

9. The parabola $y = (x - 5)^2 + 4$ has its focus at (5, 4).

10. The parabola $x = -3y^2 + 7y - 5$ opens downward.

5.1 Exercises

Each of the following graphs shows a parabola along with its vertex, focus, and directrix. Determine the coordinates of the vertex and focus, and the equation of the directrix.

1.

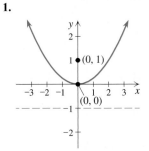

2.

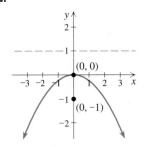

3.

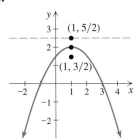

4.

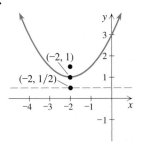

5.

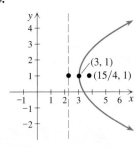

6.

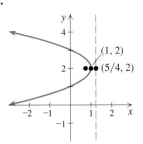

Find the equation of the parabola with the given focus and directrix.

7. Focus $(0, 2)$, directrix $y = -2$

8. Focus $(0, 1)$, directrix $y = -1$

9. Focus $(0, -3)$, directrix $y = 3$

10. Focus $(0, -1)$, directrix $y = 1$

11. Focus $(3, 5)$, directrix $y = 2$

12. Focus $(-1, 5)$, directrix $y = 3$

13. Focus $(1, -3)$, directrix $y = 2$

14. Focus $(1, -4)$, directrix $y = 0$

15. Focus $(-2, 1.2)$, directrix $y = 0.8$

16. Focus $(3, 9/8)$, directrix $y = 7/8$

Find the equation of the parabola with the given focus and directrix.

17.

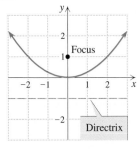

18.

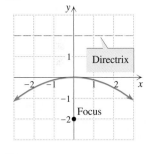

19.

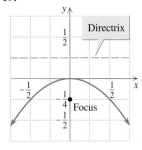

20.

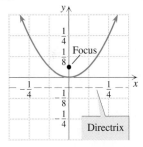

Determine the vertex, focus, and directrix for each parabola.

21. $y = (x - 1)^2$

22. $y = (x + 2)^2$

23. $y = \dfrac{1}{4}(x - 3)^2$

24. $y = \dfrac{1}{2}(x + 5)^2$

25. $y = -2(x - 3)^2 + 4$

26. $y = -4(x - 1)^2 + 3$

Use completing the square to write each equation in the form $y = a(x - h)^2 + k$. Identify the vertex, focus, and directrix.

27. $y = x^2 - 8x + 3$

28. $y = x^2 + 2x - 5$

29. $y = 2x^2 + 12x + 5$

30. $y = 3x^2 + 12x + 1$

31. $y = -2x^2 + 6x + 1$ **32.** $y = -3x^2 - 6x + 5$

33. $y = 5x^2 + 30x$ **34.** $y = -2x^2 + 12x$

35. $y = \dfrac{1}{8}x^2 - \dfrac{1}{2}x + \dfrac{9}{2}$ **36.** $y = \dfrac{1}{4}x^2 + \dfrac{1}{2}x - \dfrac{7}{4}$

Find the vertex, focus, and directrix of each parabola without completing the square, and determine whether the parabola opens upward or downward.

37. $y = x^2 - 4x + 3$ **38.** $y = x^2 - 6x - 7$

39. $y = -x^2 + 2x - 5$ **40.** $y = -x^2 + 4x + 3$

41. $y = 3x^2 - 6x + 1$ **42.** $y = 2x^2 + 4x - 1$

43. $y = -\dfrac{1}{2}x^2 - 3x + 2$ **44.** $y = -\dfrac{1}{2}x^2 + 3x - 1$

45. $y = \dfrac{1}{4}x^2 + 5$ **46.** $y = -\dfrac{1}{8}x^2 - 6$

Find the vertex, axis of symmetry, x-intercepts, and y-intercept of the parabola that has the given focus and directrix. Sketch the graph, showing the focus and directrix.

47. Focus $(1/2, -2)$, directrix $y = -5/2$

48. Focus $(1, -35/4)$, directrix $y = -37/4$

49. Focus $(-1/2, 6)$, directrix $y = 13/2$

50. Focus $(-1, 35/4)$, directrix $y = 37/4$

Find the vertex, axis of symmetry, x-intercepts, y-intercept, focus, and directrix for each parabola. Sketch the graph, showing the focus and directrix.

51. $y = \dfrac{1}{2}(x + 2)^2 + 2$ **52.** $y = \dfrac{1}{2}(x - 4)^2 + 1$

53. $y = -\dfrac{1}{4}(x + 4)^2 + 2$ **54.** $y = -\dfrac{1}{4}(x - 2)^2 + 4$

55. $y = \dfrac{1}{2}x^2 - 2$ **56.** $y = -\dfrac{1}{4}x^2 + 4$

57. $y = x^2 - 4x + 4$ **58.** $y = (x - 4)^2$

59. $y = \dfrac{1}{3}x^2 - x$ **60.** $y = \dfrac{1}{5}x^2 + x$

Find the vertex, axis of symmetry, x-intercept, y-intercepts, focus, and directrix for each parabola. Sketch the graph, showing the focus and directrix.

61. $x = -y^2$ **62.** $x = y^2 - 2$

63. $x = -\dfrac{1}{4}y^2 + 1$ **64.** $x = \dfrac{1}{2}(y - 1)^2$

65. $x = y^2 + y - 6$ **66.** $x = y^2 + y - 2$

67. $x = -\dfrac{1}{2}y^2 - y - 4$ **68.** $x = -\dfrac{1}{2}y^2 + 3y + 4$

69. $x = 2(y - 1)^2 + 3$ **70.** $x = 3(y + 1)^2 - 2$

71. $x = -\dfrac{1}{2}(y + 2)^2 + 1$ **72.** $x = -\dfrac{1}{4}(y - 2)^2 - 1$

Find the equation of the parabola determined by the given information.

73. Focus $(1, 5)$, vertex $(1, 4)$

74. Directrix $y = 5$, vertex $(2, 3)$

75. Vertex $(0, 0)$, directrix $x = -2$

76. Focus $(-2, 3)$, vertex $(-9/4, 3)$

Solve each problem.

77. *The Hale Telescope* The focus of the Hale telescope, in the accompanying figure, on Palomar Mountain in California is 55 ft above the mirror (at the vertex). The Pyrex glass mirror is 200 in. in diameter and 23 in. thick at the center. Find the equation for the parabola that was used to shape the glass. How thick is the glass on the outside edge?

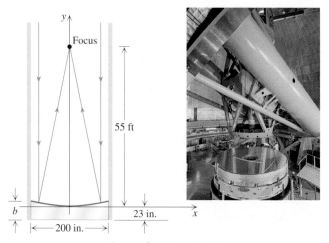

Figure for Exercise 77

78. *Eavesdropping*   From the Edmund Scientific catalog you can buy a device that will "pull in voices up to three-quarters of a mile away with our electronic parabolic microphone." The 18.75-in.-diameter plastic shield reflects sound waves to a microphone located at the focus. Given that the microphone is located 6 in. from the vertex of the parabolic shield, find the equation for a cross section of the shield. What is the depth of the shield?

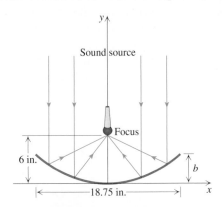

Figure for Exercise 78

79. *Bridge Cables*   A cable on the Akashi Kaikyo bridge has a parabolic shape as shown in the accompanying figure. Find the equation of the parabola that has vertex at (0, 0) and passes through the point (3264, 675). What are the coordinates of the focus?

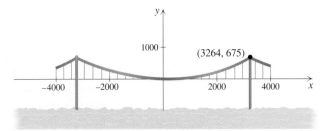

Figure for Exercise 79

Thinking Outside the Box XXXIV

Stacking Pipes   Pipes with radii of 2 ft and 3ft are placed next to each other and anchored so that they cannot move, as shown in the diagram. What is the radius of the largest pipe that can be placed on top of these two pipes?

Figure for Thinking Outside the Box XXXIV

5.1 Pop Quiz

1. Find the equation of the parabola with focus (0, 3) and directrix $y = 1$.

2. Find the vertex, focus, and directrix for

$$y = -\frac{1}{16}(x - 2)^2 + 3.$$

3. Find the vertex, axis of symmetry, x-intercept, y-intercepts, focus, and directrix for $x = y^2 - 4y + 3$.

5.2 The Ellipse and the Circle

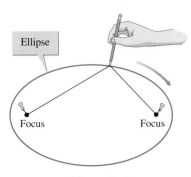

Figure 5.14

Ellipses and circles can be obtained by intersecting planes and cones, as shown in Fig. 5.1. Here we will develop general equations for ellipses, and then circles, by starting with geometric definitions, as we did with the parabola.

Definition of Ellipse

An easy way to draw an ellipse is illustrated in Fig. 5.14. A string is attached at two fixed points, and a pencil is used to take up the slack. As the pencil is moved around the paper, the sum of the distances of the pencil point from the two fixed points remains constant, because the length of the string is constant. This method of drawing an ellipse is used in construction, and it illustrates the geometric definition of an ellipse.

Definition: Ellipse

> An **ellipse** is the set of points in a plane such that the sum of their distances from two fixed points is a constant. Each fixed point is called a **focus** (plural: foci) of the ellipse.

The ellipse, like the parabola, has interesting reflecting properties. All light or sound waves emitted from one focus are reflected off the ellipse to concentrate at the other focus, as shown in Fig. 5.15. This property is used in light fixtures such as a dentist's light, for which a concentration of light at a point is desired, and in a whispering gallery like the one in the U.S. Capitol Building. In a whispering gallery, a whisper emitted at one focus is reflected off the elliptical ceiling and is amplified so that it can be heard at the other focus, but not anywhere in between.

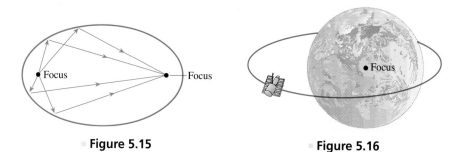

Figure 5.15   **Figure 5.16**

The orbits of the planets around the sun and satellites around Earth are elliptical. For the orbit of Earth, the sun is at one focus of the elliptical path. For the orbit of a satellite such as the Hubble space telescope, the center of Earth is one focus. See Fig. 5.16.

The Equation of the Ellipse

The ellipse shown in Fig. 5.17 has foci at $(c, 0)$ and $(-c, 0)$, and y-intercepts $(0, b)$ and $(0, -b)$, where $c > 0$ and $b > 0$. The line segment $\overline{V_1 V_2}$ is the **major axis,** and the line segment $\overline{B_1 B_2}$ is the **minor axis.** For any ellipse, the major axis is longer than the minor axis, and the foci are on the major axis. The **center** of an ellipse is the midpoint of the major (or minor) axis. The ellipse in Fig. 5.17 is centered at the origin.

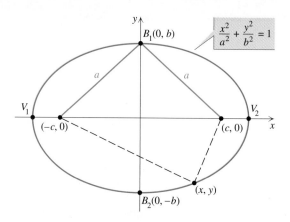

Figure 5.17

The **vertices** of an ellipse are the endpoints of the major axis. The vertices of the ellipse in Fig. 5.17 are the x-intercepts.

Let a be the distance between $(c, 0)$ and the y-intercept $(0, b)$, as shown in Fig. 5.17. The sum of the distances from the two foci to $(0, b)$ is $2a$. So for any point (x, y) on the ellipse, the distance from (x, y) to $(c, 0)$ plus the distance from (x, y) to $(-c, 0)$ is equal to $2a$. Writing this last statement as an equation (using the distance formula) gives the equation of the ellipse shown in Fig. 5.17:

$$\sqrt{(x - c)^2 + (y - 0)^2} + \sqrt{(x - (-c))^2 + (y - 0)^2} = 2a$$

With some effort, this equation can be greatly simplified. We will provide the major steps in simplifying it, and leave the details as an exercise. First simplify inside the radicals and isolate them to get

$$\sqrt{x^2 - 2xc + c^2 + y^2} = 2a - \sqrt{x^2 + 2xc + c^2 + y^2}.$$

Next, square each side and simplify again to get

$$a\sqrt{x^2 + 2xc + c^2 + y^2} = a^2 + xc.$$

Squaring each side again yields

$$a^2x^2 - c^2x^2 + a^2y^2 = a^4 - a^2c^2$$

$$(a^2 - c^2)x^2 + a^2y^2 = a^2(a^2 - c^2) \quad \text{Factor.}$$

Since $a^2 = b^2 + c^2$, or $a^2 - c^2 = b^2$ (from Fig. 5.17), replace $a^2 - c^2$ by b^2:

$$b^2x^2 + a^2y^2 = a^2b^2$$

$$\frac{x^2}{a^2} + \frac{y^2}{b^2} = 1 \quad \text{Divide each side by } a^2b^2.$$

We have proved the following theorem.

Theorem: Equation of an Ellipse with Center (0, 0) and Horizontal Major Axis

> The equation of an ellipse centered at the origin with foci $(c, 0)$ and $(-c, 0)$ and y-intercepts $(0, b)$ and $(0, -b)$ is
>
> $$\frac{x^2}{a^2} + \frac{y^2}{b^2} = 1, \quad \text{where } a^2 = b^2 + c^2.$$

If we start with the foci on the y-axis and x-intercepts $(b, 0)$ and $(-b, 0)$, then we can develop a similar equation, which is stated in the following theorem. If the foci are on the y-axis, then the y-intercepts are the vertices and the major axis is vertical.

Theorem: Equation of an Ellipse with Center (0, 0) and Vertical Major Axis

> The equation of an ellipse centered at the origin with foci $(0, c)$ and $(0, -c)$ and x-intercepts $(b, 0)$ and $(-b, 0)$ is
>
> $$\frac{x^2}{b^2} + \frac{y^2}{a^2} = 1, \quad \text{where } a^2 = b^2 + c^2.$$

Consider the ellipse in Fig. 5.18 on the next page with foci $(c, 0)$ and $(-c, 0)$ and equation

$$\frac{x^2}{a^2} + \frac{y^2}{b^2} = 1,$$

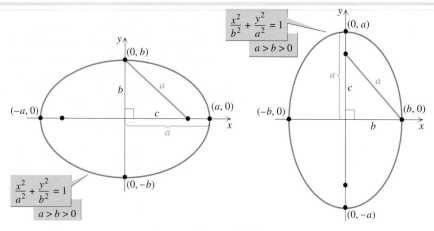

Figure 5.18

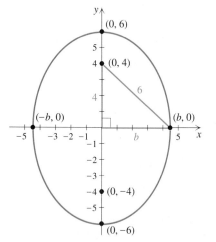

Figure 5.19

where $a > b > 0$. If $y = 0$ in this equation, then $x = \pm a$. So the vertices (or x-intercepts) of the ellipse are $(a, 0)$ and $(-a, 0)$, and a is the distance from the center to a vertex. The distance from a focus to an endpoint of the minor axis is a also. *So in any ellipse the distance from the focus to an endpoint of the minor axis is the same as the distance from the center to a vertex.* The graphs in Fig. 5.18 will help you remember the relationship between a, b, and c.

Example **1** **Writing the equation of an ellipse**

Sketch an ellipse with foci at $(0, 4)$ and $(0, -4)$ and vertices $(0, 6)$ and $(0, -6)$, and find the equation for this ellipse.

Solution

Since the vertices are on the y-axis, the major axis is vertical and the ellipse is elongated in the direction of the y-axis. A sketch of the ellipse appears in Fig. 5.19. Because a is the distance from the center to a vertex, $a = 6$. Because c is the distance from the center to a focus, $c = 4$. To write the equation of the ellipse, we need the value of b^2. Use $a = 6$ and $c = 4$ in $a^2 = b^2 + c^2$ to get

$$6^2 = b^2 + 4^2$$

$$b^2 = 20.$$

So the x-intercepts are $\left(\sqrt{20}, 0\right)$ and $\left(-\sqrt{20}, 0\right)$, and the equation of the ellipse is

$$\frac{x^2}{20} + \frac{y^2}{36} = 1. \qquad ■$$

Graphing an Ellipse Centered at the Origin

For $a > b > 0$, the graph of

$$\frac{x^2}{a^2} + \frac{y^2}{b^2} = 1$$

is an ellipse centered at the origin with a horizontal major axis, x-intercepts $(a, 0)$ and $(-a, 0)$, and y-intercepts $(0, b)$ and $(0, -b)$, as shown in Fig. 5.18. The foci are on the major axis and are determined by $a^2 = b^2 + c^2$ or $c^2 = a^2 - b^2$. Remember that when the denominator for x^2 is larger than the denominator for y^2, the major axis is horizontal. *To sketch the graph of an ellipse centered at the origin, simply locate the four intercepts and draw an ellipse through them.*

Example **2** **Graphing an ellipse with foci on the x-axis**

Sketch the graph and identify the foci of the ellipse

$$\frac{x^2}{9} + \frac{y^2}{4} = 1.$$

Solution

To sketch the ellipse, we find the x-intercepts and the y-intercepts. If $x = 0$, then $y^2 = 4$ or $y = \pm 2$. So the y-intercepts are $(0, 2)$ and $(0, -2)$. If $y = 0$, then $x = \pm 3$. So the x-intercepts are $(3, 0)$ and $(-3, 0)$. To make a rough sketch of an ellipse, plot only the intercepts and draw an ellipse through them, as shown in Fig. 5.20. Since this ellipse is elongated in the direction of the x-axis, the foci are on the x-axis. Use $a = 3$ and $b = 2$ in $c^2 = a^2 - b^2$, to get $c^2 = 9 - 4 = 5$. So $c = \pm\sqrt{5}$, and the foci are $\left(\sqrt{5}, 0\right)$ and $\left(-\sqrt{5}, 0\right)$.

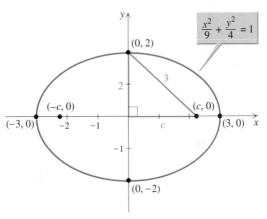

Figure 5.20

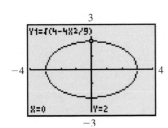

Figure 5.21

To check, solve for y to get $y = \pm\sqrt{4 - 4x^2/9}$. Then graph $y_1 = \sqrt{4 - 4x^2/9}$ and $y_2 = -y_1$, as shown in Fig. 5.21. ■

For $a > b > 0$, the graph of

$$\frac{x^2}{b^2} + \frac{y^2}{a^2} = 1$$

is an ellipse with a vertical major axis, x-intercepts $(b, 0)$ and $(-b, 0)$, and y-intercepts $(0, a)$ and $(0, -a)$, as shown in Fig. 5.18. When the denominator for y^2 is larger than the denominator for x^2, the major axis is vertical. The foci are always on the major axis and are determined by $a^2 = b^2 + c^2$ or $c^2 = a^2 - b^2$. Remember that this relationship between a, b, and c is determined by the Pythagorean theorem, because a, b, and c are the lengths of sides of a right triangle, as shown in Fig. 5.18.

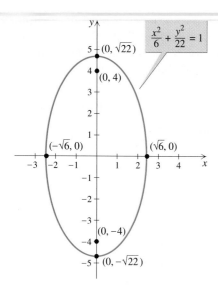

Figure 5.22

Example **3** Graphing an ellipse with foci on the y-axis

Sketch the graph of $11x^2 + 3y^2 = 66$ and identify the foci.

Solution

We first divide each side of the equation by 66 to get the standard equation:

$$\frac{x^2}{6} + \frac{y^2}{22} = 1$$

If $x = 0$, then $y^2 = 22$ or $y = \pm\sqrt{22}$. So the y-intercepts are $\left(0, \sqrt{22}\right)$ and $\left(0, -\sqrt{22}\right)$. If $y = 0$, then $x = \pm\sqrt{6}$ and the x-intercepts are $\left(\sqrt{6}, 0\right)$ and $\left(-\sqrt{6}, 0\right)$. Plot these four points and draw an ellipse through them, as shown in Fig. 5.22. Since this ellipse is elongated in the direction of the y-axis, the foci are on the y-axis. Use $a^2 = 22$ and $b^2 = 6$ in $c^2 = a^2 - b^2$ to get $c^2 = 22 - 6 = 16$. So $c = \pm 4$, and the foci are $(0, 4)$ and $(0, -4)$. ∎

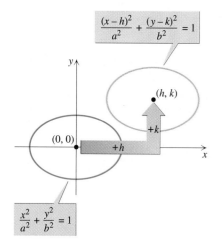

Figure 5.23

Translations of Ellipses

Although an ellipse is not the graph of a function, its graph can be translated in the same manner. Figure 5.23 shows the graphs of

$$\frac{(x - h)^2}{a^2} + \frac{(y - k)^2}{b^2} = 1 \quad \text{and} \quad \frac{x^2}{a^2} + \frac{y^2}{b^2} = 1.$$

They have the same size and shape, but the graph of the first equation is centered at (h, k) rather than at the origin. So the graph of the first equation is obtained by translating the graph of the second horizontally h units and vertically k units.

Example **4** Graphing an ellipse centered at (h, k)

Sketch the graph and identify the foci of the ellipse

$$\frac{(x - 3)^2}{25} + \frac{(y + 1)^2}{9} = 1.$$

Solution

The graph of this equation is a translation of the graph of

$$\frac{x^2}{25} + \frac{y^2}{9} = 1,$$

three units to the right and one unit downward. The center of the ellipse is $(3, -1)$. Since $a^2 = 25$, the vertices lie five units to the right and five units to the left of $(3, -1)$. So the ellipse goes through $(8, -1)$ and $(-2, -1)$. Since $b^2 = 9$, the graph includes points three units above and three units below the center. So the ellipse goes through $(3, 2)$ and $(3, -4)$, as shown in Fig. 5.24. Since $c^2 = 25 - 9 = 16$, $c = \pm 4$. The major axis is on the horizontal line $y = -1$. So the foci are found four units to the right and four units to the left of the center at $(7, -1)$ and $(-1, -1)$.

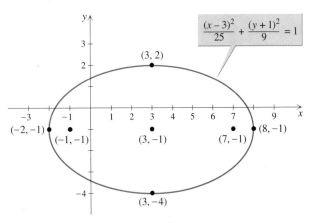

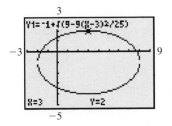

■ Figure 5.25

■ **Figure 5.24**

To check the location of the ellipse, graph

$$y_1 = -1 + \sqrt{9 - 9(x - 3)^2/25} \quad \text{and} \quad y_2 = -1 - \sqrt{9 - 9(x - 3)^2/25}$$

on a graphing calculator, as in Fig. 5.25.   ■

The Circle

The circle is the simplest curve of the four conic sections, and it is a special case of the ellipse. In keeping with our approach to the conic sections, we give a geometric definition of a circle and then use the distance formula to derive the standard equation for a circle. The standard equation was derived in this way in Section 1.3. For completeness, we now repeat the definition and derivation.

Definition: Circle

> A **circle** is a set of all points in a plane such that their distance from a fixed point (the **center**) is a constant (the **radius**).

As shown in Fig. 5.26, a point (x, y) is on a circle with center (h, k) and radius r if and only if

$$\sqrt{(x - h)^2 + (y - k)^2} = r.$$

If we square both sides of this equation, we get the standard equation of a circle.

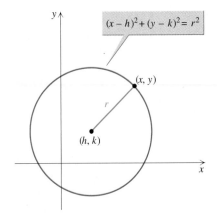

■ **Figure 5.26**

Theorem: Standard Equation of a Circle

The standard equation of a circle with center (h, k) and radius r $(r > 0)$ is

$$(x - h)^2 + (y - k)^2 = r^2.$$

Historical Note

Appollonius of Perga (262–190 B.C.), who was known as the Great Geometer, consolidated and extended previous results of conics into a monograph *Conic Sections,* consisting of eight books with 487 propositions. Appollonius' *Conic Sections* and Euclid's *Elements* may represent the quintessence of Greek mathematics. Later contributors to the subject include Kepler, Descartes, Fermat, Pascal, and Newton.

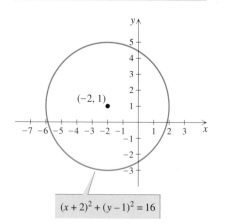

$(x + 2)^2 + (y - 1)^2 = 16$

Figure 5.27

The equation $(x - h)^2 + (y - k)^2 = a$ is a circle of radius $\sqrt{a}$ if $a > 0$. If $a = 0$, only (h, k) satisfies $(x - h)^2 + (y - k)^2 = 0$ and the point (h, k) is a degenerate circle. If $a < 0$, then no ordered pair satisfies $(x - h)^2 + (y - k)^2 = a$. If h and k are zero, then we get the standard equation of a circle centered at the origin, $x^2 + y^2 = r^2$.

A circle is an ellipse in which the two foci coincide at the center. If the foci are identical, then $a = b$ and the equation for an ellipse becomes the equation for a circle with radius a.

Example **5** Finding the equation for a circle

Write the equation for the circle that has center $(4, 5)$ and passes through $(-1, 2)$.

Solution

The radius is the distance from $(4, 5)$ to $(-1, 2)$:

$$r = \sqrt{(4 - (-1))^2 + (5 - 2)^2} = \sqrt{25 + 9} = \sqrt{34}$$

Use $h = 4$, $k = 5$, and $r = \sqrt{34}$ in $(x - h)^2 + (y - k)^2 = r^2$ to get the equation

$$(x - 4)^2 + (y - 5)^2 = 34.$$ ■

To graph a circle, we must know the center and radius. A compass or a string can be used to keep the pencil at a fixed distance from the center.

Example **6** Finding the center and radius

Determine the center and radius, and sketch the graph of $x^2 + 4x + y^2 - 2y = 11$.

Solution

Use completing the square to get the equation into the standard form:

$$x^2 + 4x + 4 + y^2 - 2y + 1 = 11 + 4 + 1$$
$$(x + 2)^2 + (y - 1)^2 = 16$$

From the standard form, we recognize that the center is $(-2, 1)$ and the radius is 4. The graph is shown in Fig. 5.27. ■

Applications

The **eccentricity** e of an ellipse is defined by $e = c/a$, where c is the distance from the center to a focus and a is one-half the length of the major axis. Since $0 < c < a$, we have $0 < e < 1$. For an ellipse that appears circular, the foci are close to the center and c is small compared with a. So its eccentricity is near 0. For an ellipse that is very elongated, the foci are close to the vertices and c is nearly equal to a. So its eccentricity is near 1. See Fig. 5.28. A satellite that orbits the earth has an elliptical orbit that is nearly circular, with eccentricity close to 0. On the other hand, Halley's comet is in a very elongated elliptical orbit around the sun, with eccentricity close to 1.

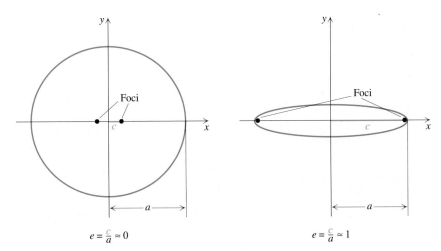

$e = \dfrac{c}{a} \approx 0$ $\qquad\qquad\qquad$ $e = \dfrac{c}{a} \approx 1$

Figure 5.28

Example **7** **Eccentricity of an orbit**

The first artificial satellite to orbit Earth was Sputnik I, launched by the Soviet Union in 1957. The altitude of Sputnik varied from 132 miles to 583 miles above the surface of Earth. If the center of Earth is one focus of its elliptical orbit and the radius of Earth is 3950 miles, then what was the eccentricity of the orbit?

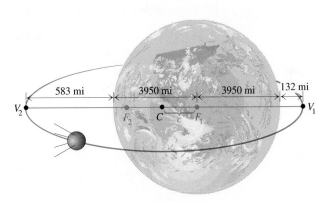

Figure 5.29

Solution

The center of Earth is one focus F_1 of the elliptical orbit, as shown in Fig. 5.29. When Sputnik I was 132 miles above Earth, it was at a vertex V_1, 4082 miles from F_1 ($4082 = 3950 + 132$). When Sputnik I was 583 miles above Earth, it was at the other vertex V_2, 4533 miles from F_1 ($4533 = 3950 + 583$). So the length of the major axis is $4082 + 4533$, or 8615 miles. Since the length of the major axis is $2a$, we get $a = 4307.5$ miles. Since c is the distance from the center of the ellipse C to F_1, we get $c = 4307.5 - 4082 = 225.5$. Use $e = c/a$ to get

$$e = \frac{225.5}{4307.5} \approx 0.052.$$

So the eccentricity of the orbit was approximately 0.052. ■

For Thought

True or False? Explain.

1. The x-intercepts for $\dfrac{x^2}{9} + \dfrac{y^2}{4} = 1$ are $(9, 0)$ and $(-9, 0)$.

2. The graph of $2x^2 + y^2 = 1$ is an ellipse.

3. The ellipse $\dfrac{x^2}{16} + \dfrac{y^2}{25} = 1$ has a major axis of length 10.

4. The x-intercepts for $0.5x^2 + y^2 = 1$ are $(\sqrt{2}, 0)$ and $(-\sqrt{2}, 0)$.

5. The y-intercepts for $x^2 + \dfrac{y^2}{3} = 1$ are $(0, \sqrt{3})$ and $(0, -\sqrt{3})$.

6. A circle is a set of points, and the center is one of those points.

7. If the foci of an ellipse coincide, then the ellipse is a circle.

8. No ordered pair satisfies the equation $(x - 3)^2 + (y + 1)^2 = 0$.

9. The graph of $(x - 1)^2 + (y + 2)^2 + 9 = 0$ is a circle of radius 3.

10. The radius of the circle $x^2 - 4x + y^2 + y = 9$ is 3.

5.2 Exercises

Each of the following graphs shows an ellipse along with its foci, vertices, and center. Determine the coordinates of the foci, vertices, and center.

1.

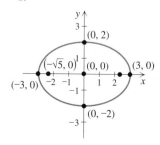

2.

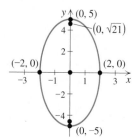

3.

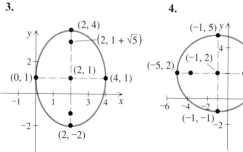

4.

Find the equation of each ellipse described below and sketch its graph.

5. Foci $(-2, 0)$ and $(2, 0)$, and y-intercepts $(0, -3)$ and $(0, 3)$.

6. Foci $(-3, 0)$ and $(3, 0)$, and y-intercepts $(0, -4)$ and $(0, 4)$.

7. Foci $(-4, 0)$ and $(4, 0)$, and x-intercepts $(-5, 0)$ and $(5, 0)$.

8. Foci $(-1, 0)$ and $(1, 0)$, and x-intercepts $(-3, 0)$ and $(3, 0)$.

9. Foci $(0, 2)$ and $(0, -2)$, and x-intercepts $(2, 0)$ and $(-2, 0)$.

10. Foci $(0, 6)$ and $(0, -6)$, and x-intercepts $(2, 0)$ and $(-2, 0)$.

11. Foci $(0, 4)$ and $(0, -4)$, and y-intercepts $(0, 7)$ and $(0, -7)$.

12. Foci $(0, 3)$ and $(0, -3)$, and y-intercepts $(0, 4)$ and $(0, -4)$.

Sketch the graph of each ellipse and identify the foci.

13. $\dfrac{x^2}{16} + \dfrac{y^2}{4} = 1$

14. $\dfrac{x^2}{16} + \dfrac{y^2}{9} = 1$

15. $\dfrac{x^2}{9} + \dfrac{y^2}{36} = 1$

16. $x^2 + \dfrac{y^2}{4} = 1$

17. $\dfrac{x^2}{25} + y^2 = 1$

18. $\dfrac{x^2}{6} + \dfrac{y^2}{10} = 1$

19. $\dfrac{y^2}{25} + \dfrac{x^2}{9} = 1$ **20.** $\dfrac{y^2}{9} + \dfrac{x^2}{4} = 1$

21. $9x^2 + y^2 = 9$ **22.** $x^2 + 4y^2 = 4$

23. $4x^2 + 9y^2 = 36$ **24.** $9x^2 + 25y^2 = 225$

Sketch the graph of each ellipse and identify the foci.

25. $\dfrac{(x-1)^2}{16} + \dfrac{(y+3)^2}{9} = 1$ **26.** $\dfrac{(x+2)^2}{16} + \dfrac{(y+1)^2}{4} = 1$

27. $\dfrac{(x-3)^2}{9} + \dfrac{(y+2)^2}{25} = 1$ **28.** $(x-5)^2 + \dfrac{(y-3)^2}{9} = 1$

29. $(x+4)^2 + 36(y+3)^2 = 36$

30. $9(x-1)^2 + 4(y+3)^2 = 36$

31. $9x^2 - 18x + 4y^2 + 16y = 11$

32. $4x^2 + 16x + y^2 - 6y = -21$

33. $9x^2 - 54x + 4y^2 + 16y = -61$

34. $9x^2 + 90x + 25y^2 - 50y = -25$

Find the equation of each ellipse and identify its foci.

35.

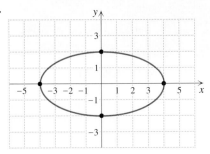

36.

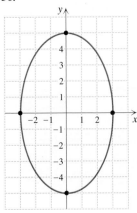

37.

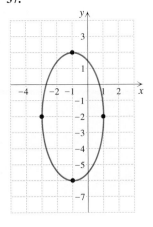

38.

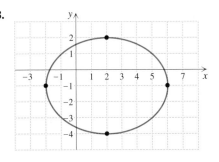

Write the equation for each circle described.

39. Center $(0, 0)$ and passing through $(4, 5)$

40. Center $(0, 0)$ and passing through $(-3, -4)$

41. Center $(2, -3)$ and passing through $(4, 1)$

42. Center $(-2, -4)$ and passing through $(1, -1)$

43. Diameter has endpoints $(3, 4)$ and $(-1, 2)$.

44. Diameter has endpoints $(3, -1)$ and $(-4, 2)$.

Determine the center and radius of each circle and sketch its graph.

45. $x^2 + y^2 = 100$ **46.** $x^2 + y^2 = 25$

47. $(x-1)^2 + (y-2)^2 = 4$ **48.** $(x+2)^2 + (y-3)^2 = 9$

Find the center and radius of each circle.

49. $x^2 + y^2 + 2y = 8$ **50.** $x^2 - 6x + y^2 = 1$

51. $x^2 + 8x + y^2 = 10y$ **52.** $x^2 + y^2 = 12x - 12y$

53. $x^2 + 4x + y^2 = 5$ **54.** $x^2 + y^2 - 6y = 0$

55. $x^2 - x + y^2 + y = \dfrac{1}{2}$ **56.** $x^2 + 5x + y^2 + 3y = \dfrac{1}{2}$

Write each of the following equations in one of the forms:

$y = a(x-h)^2 + k, \quad x = a(y-h)^2 + k,$

$\dfrac{(x-h)^2}{a^2} + \dfrac{(y-k)^2}{b^2} = 1, \quad \text{or} \quad (x-h)^2 + (y-k)^2 = r^2.$

Then identify each equation as the equation of a parabola, an ellipse, or a circle.

57. $y = x^2 + y^2$ **58.** $4x = x^2 + y^2$

59. $4x^2 + 12y^2 = 4$ **60.** $2x^2 + 2y^2 = 4 - y$

61. $2x^2 + 4x = 4 - y$

62. $2x^2 + 4y^2 = 4 - y$

63. $2 - x = (2 - y)^2$

64. $3(3 - x)^2 = 9 - y$

65. $2(4 - x)^2 = 4 - y^2$

66. $\dfrac{x^2}{4} + \dfrac{y^2}{4} = 1$

67. $9x^2 = 1 - 9y^2$

68. $9x^2 = 1 - 9y$

Solve each problem.

69. *Foci and a Point*   Find the equation of the ellipse that goes through the point $(2, 3)$ and has foci $(2, 0)$ and $(-2, 0)$.

70. *Foci and Eccentricity*   Find the equation of an ellipse with foci $(\pm 5, 0)$ and eccentricity $1/2$.

71. *Focus of Elliptical Reflector*   An elliptical reflector is 10 in. in diameter and 3 in. deep. A light source is at one focus $(0, 0)$, 3 in. from the vertex $(-3, 0)$, as shown in the accompanying figure. Find the location of the other focus.

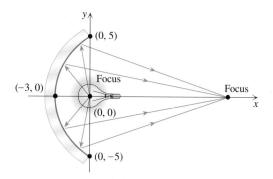

■ **Figure for Exercise 71**

72. *Constructing an Elliptical Arch*   A mason is constructing an elliptical form for an arch out of a 4-ft by 12-ft sheet of plywood, as shown in the accompanying figure. What length string is needed to draw the ellipse on the plywood (by attaching the ends at the foci as discussed at the beginning of this section)? Where are the foci for this ellipse?

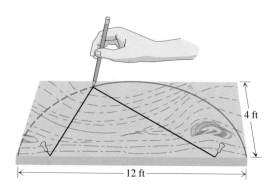

■ **Figure for Exercise 72**

73. *Comet Hale-Bopp*   Comet Hale-Bopp, which was clearly seen in April of 1997, orbits the sun in an elliptical orbit every 4200 years. At *perihelion,* the closest point to the sun, the comet is approximately 1 AU from the sun, as shown in the accompanying figure (*Sky and Telescope,* April 1997, used with permission. Guy Ottewell, Astronomical Calendar, 1997, p. 59). At *aphelion,* the farthest point from the sun, the comet is 520 AU from the sun. Find the equation of the ellipse. What is the eccentricity of the orbit?

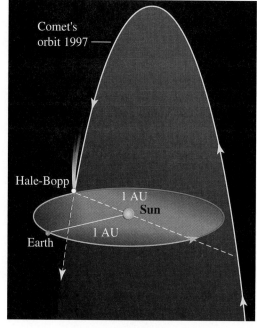

■ **Figure for Exercise 73**

74. *Orbit of the Moon*   The moon travels on an elliptical path with Earth at one focus. If the maximum distance from the moon to Earth is 405,500 km and the minimum distance is 363,300 km, then what is the eccentricity of the orbit?

75. *Halley's Comet*   The comet Halley, last seen in 1986, travels in an elliptical orbit with the sun at one focus. Its minimum distance to the sun is 8×10^7 km, and the eccentricity of its orbit is 0.97. What is its maximum distance from the sun? Comet Halley will next be visible from Earth in 2062.

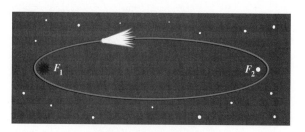

■ **Figure for Exercise 75**

76. *Adjacent Circles*   A 13-in.-diameter mag wheel and a 16-in.-diameter mag wheel are placed in the first quadrant, as shown in the figure. Write an equation for the circular boundary of each wheel.

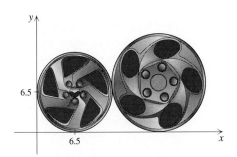

6.5

6.5

▪ **Figure for Exercise 76**

Thinking Outside the Box XXXV

Three Circles   A circle of radius 1 is centered at the origin and a circle of radius 1/2 is centered at (1/2, 0). A third circle is

positioned so that it is tangent to the other two circles and the *y*-axis as shown in the figure. Find the center and radius of the third circle.

$\frac{1}{2}$

$-\frac{1}{2}$ $\frac{1}{2}$

$-\frac{1}{2}$

▪ **Figure for Thinking Outside the Box XXXV**

5.2   Pop Quiz

1. Find the equation of the ellipse with foci $(\pm 3, 0)$ and *y*-intercepts $(0, \pm 5)$.

2. Find the foci for $\dfrac{(x - 1)^2}{9} + \dfrac{(y - 3)^2}{25} = 1$.

3. Find the center and radius for the circle $x^2 + 4x + y^2 - 10y = 0$.

5.3   The Hyperbola

The last of the four conic sections, the hyperbola, has two branches and each one has a focus, as shown in Fig. 5.30. The hyperbola also has a useful reflecting property. A light ray aimed at one focus is reflected toward the other focus, as shown in Fig. 5.30. This reflecting property is used in telescopes, as shown in Fig. 5.31.

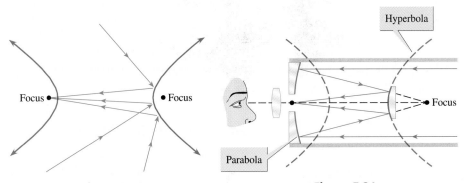

Focus   • Focus

Hyperbola

• Focus

Parabola

▪ **Figure 5.30**   ▪ **Figure 5.31**

Within the telescope, a small hyperbolic mirror with the same focus as the large parabolic mirror reflects the light to a more convenient location for viewing.

Hyperbolas also occur in the context of supersonic noise pollution. The sudden change in air pressure from an aircraft traveling at supersonic speed creates a cone-shaped wave through the air. The plane of the ground and this cone intersect along one branch of a hyperbola, as shown in Fig. 5.32. A sonic boom is heard on the ground along this branch of the hyperbola. Since this curve where the sonic boom is heard travels along the ground with the aircraft, supersonic planes are restricted from flying across the continental United States.

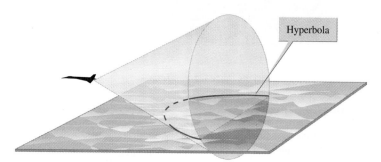

Hyperbola

■ **Figure 5.32**

A hyperbola may also occur as the path of a moving object such as a spacecraft traveling past the moon on its way toward Venus, a comet passing in the neighborhood of the sun, or an alpha particle passing by the nucleus of an atom.

The Definition

A hyperbola can be defined as the intersection of a cone and a plane, as shown in Fig. 5.1. As we did for the other conic sections, we will give a geometric definition of a hyperbola and use the distance formula to derive its equation.

Definition: Hyperbola

> A **hyperbola** is the set of points in a plane such that the difference between the distances from two fixed points (foci) is constant.

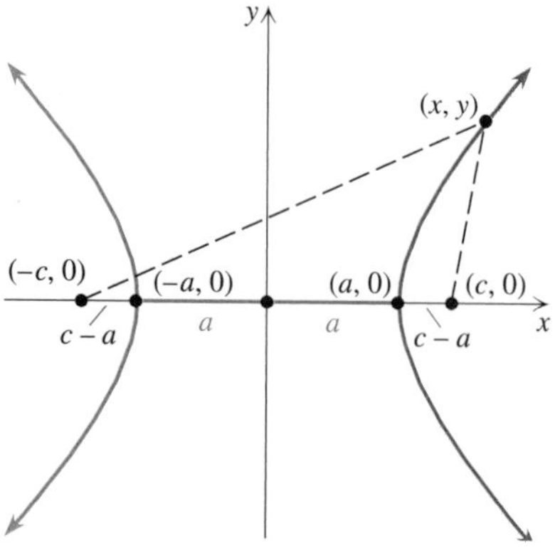

For a point on a hyperbola, the *difference* between the distances from two fixed points is constant, and for a point on an ellipse, the *sum* of the distances from two fixed points is constant. The definitions of a hyperbola and an ellipse are similar, and we will see their equations are similar also. Their graphs, however, are very different. In the hyperbola shown in Fig. 5.33, the branches look like parabolas, *but they are not parabolas* because they do not satisfy the geometric definition of a parabola. A hyperbola with foci on the x-axis, as in Fig. 5.33, is said to open to the left and right.

■ **Figure 5.33**

Developing the Equation

The hyperbola shown in Fig. 5.33 has foci at $(c, 0)$ and $(-c, 0)$ and x-intercepts or **vertices** at $(a, 0)$ and $(-a, 0)$, where $a > 0$ and $c > 0$. The line segment between the vertices is the **transverse axis.** The point $(0, 0)$, halfway between the foci, is the

center. The point $(a, 0)$ is on the hyperbola. The distance from $(a, 0)$ to the focus $(c, 0)$ is $c - a$. The distance from $(a, 0)$ to $(-c, 0)$ is $c + a$. So the constant difference is $2a$. For an arbitrary point (x, y), the distance to $(c, 0)$ is subtracted from the distance to $(-c, 0)$ to get the constant $2a$:

$$\sqrt{(x - (-c))^2 + (y - 0)^2} - \sqrt{(x - c)^2 + (y - 0)^2} = 2a$$

For (x, y) on the other branch, we would subtract in the opposite order, but we get the same simplified form. The simplified form for this equation is similar to that for the ellipse. We will provide the major steps for simplifying it and leave the details as an exercise. First simplify inside the radicals and isolate them to get

$$\sqrt{x^2 + 2xc + c^2 + y^2} = 2a + \sqrt{x^2 - 2xc + c^2 + y^2}.$$

Squaring each side and simplifying yields

$$xc - a^2 = a\sqrt{x^2 - 2xc + c^2 + y^2}.$$

Square each side again and simplify to get

$$c^2x^2 - a^2x^2 - a^2y^2 = c^2a^2 - a^4$$
$$(c^2 - a^2)x^2 - a^2y^2 = a^2(c^2 - a^2) \quad \text{Factor.}$$

Now $c^2 - a^2$ is positive, because $c > a$. So let $b^2 = c^2 - a^2$ and substitute:

$$b^2x^2 - a^2y^2 = a^2b^2$$
$$\frac{x^2}{a^2} - \frac{y^2}{b^2} = 1 \quad \text{Divide each side by } a^2b^2.$$

We have proved the following theorem.

Theorem: Equation of a Hyperbola Centered at (0, 0) Opening Left and Right

The equation of a hyperbola centered at the origin with foci $(c, 0)$ and $(-c, 0)$ and x-intercepts $(a, 0)$ and $(-a, 0)$ is

$$\frac{x^2}{a^2} - \frac{y^2}{b^2} = 1,$$

where $b^2 = c^2 - a^2$.

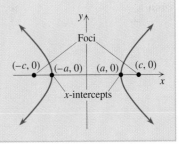

If the foci are positioned on the y-axis, then we say that the hyperbola opens up and down. For hyperbolas that open up and down, we have the following theorem.

Theorem: Equation of a Hyperbola Centered at (0, 0) Opening Up and Down

The equation of a hyperbola centered at the origin with foci $(0, c)$ and $(0, -c)$ and y-intercepts $(0, a)$ and $(0, -a)$ is

$$\frac{y^2}{a^2} - \frac{x^2}{b^2} = 1,$$

where $b^2 = c^2 - a^2$.

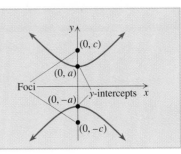

Graphing a Hyperbola Centered at (0, 0)

If we solve the equation $\dfrac{x^2}{a^2} - \dfrac{y^2}{b^2} = 1$, for y, we get the following:

$$\frac{y^2}{b^2} = \frac{x^2}{a^2} - 1$$

$$y^2 = \frac{b^2 x^2}{a^2} - b^2 \qquad \text{Multiply each side by } b^2.$$

$$y^2 = \frac{b^2 x^2}{a^2} - \frac{a^2 b^2 x^2}{a^2 x^2} \qquad \text{Write } b^2 \text{ as } \frac{a^2 b^2 x^2}{a^2 x^2}.$$

$$y^2 = \frac{b^2 x^2}{a^2}\left(1 - \frac{a^2}{x^2}\right) \qquad \text{Factor out } \frac{b^2 x^2}{a^2}.$$

$$y = \pm \frac{b}{a} x \sqrt{1 - \frac{a^2}{x^2}}$$

As $x \to \infty$, the value of $(a^2/x^2) \to 0$ and the value of y can be approximated by $y = \pm(b/a)x$. So the lines

$$y = \frac{b}{a}x \qquad \text{and} \qquad y = -\frac{b}{a}x$$

are oblique asymptotes for the graph of the hyperbola. The graph of this hyperbola is shown with its asymptotes in Fig. 5.34. The asymptotes are essential for determining the proper shape of the hyperbola. The asymptotes go through the points (a, b), $(a, -b)$, $(-a, b)$, and $(-a, -b)$. The rectangle with these four points as vertices is called the **fundamental rectangle.** The line segment with endpoints $(0, b)$ and $(0, -b)$ is called the **conjugate axis.** The location of the fundamental rectangle is determined by the conjugate axis and the transverse axis.

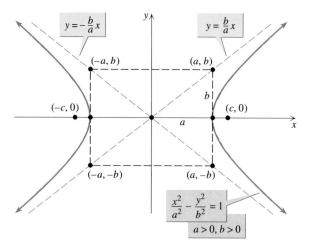

Figure 5.34

Note that since $c^2 = a^2 + b^2$, the distance c from the center to a focus is equal to the distance from the center to (a, b), as shown in Fig. 5.35.

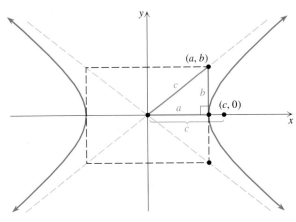

■ **Figure 5.35**

The following steps will help you graph hyperbolas opening left and right.

PROCEDURE   **Graphing the Hyperbola $\dfrac{x^2}{a^2} - \dfrac{y^2}{b^2} = 1$**

To graph $\dfrac{x^2}{a^2} - \dfrac{y^2}{b^2} = 1$ for $a > 0$ and $b > 0$, do the following:

1. Locate the x-intercepts $(a, 0)$ and $(-a, 0)$.
2. Draw the rectangle through $(\pm a, 0)$ and through $(0, \pm b)$.
3. Extend the diagonals of the rectangle to get the asymptotes.
4. Draw a hyperbola opening to the left and right from the x-intercepts approaching the asymptotes.

Example ❶ **Graphing a hyperbola opening left and right**

Determine the foci and the equations of the asymptotes, and sketch the graph of

$$\frac{x^2}{36} - \frac{y^2}{9} = 1.$$

Solution

If $y = 0$, then $x = \pm 6$. The x-intercepts are $(6, 0)$ and $(-6, 0)$. Since $b^2 = 9$, the fundamental rectangle goes through $(0, 3)$ and $(0, -3)$ and through the x-intercepts. Next draw the fundamental rectangle and extend its diagonals to get the asymptotes. Draw the hyperbola opening to the left and right, as shown in Fig. 5.36. To find the foci, use $c^2 = a^2 + b^2$:

$$c^2 = 36 + 9 = 45$$
$$c = \pm\sqrt{45}$$

So the foci are $\left(\sqrt{45}, 0\right)$ and $\left(-\sqrt{45}, 0\right)$. From the graph we get the asymptotes

$$y = \frac{1}{2}x \qquad \text{and} \qquad y = -\frac{1}{2}x.$$ ■

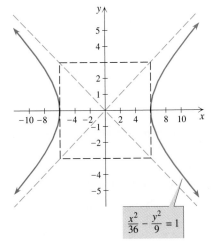

■ **Figure 5.36**

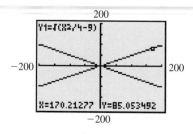

▪ **Figure 5.37**

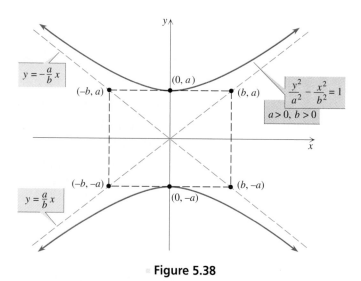 The graph of a hyperbola gets closer and closer to its asymptotes. So in a large viewing window you cannot tell the difference between the hyperbola and its asymptotes. For example, the hyperbola of Example 1, $y = \pm\sqrt{x^2/4 - 9}$, looks like its asymptotes $y = \pm(1/2)x$ in Fig. 5.37. □

We can show that hyperbolas opening up and down have asymptotes just as hyperbolas opening right and left have. The lines

$$y = \frac{a}{b}x \qquad \text{and} \qquad y = -\frac{a}{b}x$$

are asymptotes for the graph of

$$\frac{y^2}{a^2} - \frac{x^2}{b^2} = 1,$$

as shown in Fig. 5.38. Note that the asymptotes are essential for determining the shape of the hyperbola.

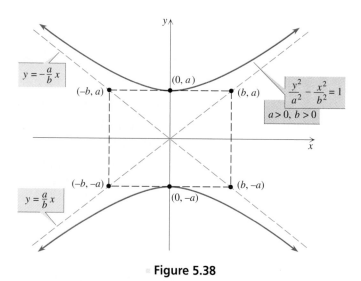

▪ **Figure 5.38**

The following steps will help you graph hyperbolas opening up and down.

PROCEDURE   **Graphing the Hyperbola $\dfrac{y^2}{a^2} - \dfrac{x^2}{b^2} = 1$**

To graph $\dfrac{y^2}{a^2} - \dfrac{x^2}{b^2} = 1$ for $a > 0$ and $b > 0$, do the following:

1. Locate the y-intercepts $(0, a)$ and $(0, -a)$.
2. Draw the rectangle through $(0, \pm a)$ and through $(b, 0)$ and $(-b, 0)$.
3. Extend the diagonals of the rectangle to get the asymptotes.
4. Draw a hyperbola opening up and down from the y-intercepts approaching the asymptotes.

Example **2** Graphing a hyperbola opening up and down

Determine the foci and the equations of the asymptotes, and sketch the graph of

$$4y^2 - 9x^2 = 36.$$

Solution

Divide each side of the equation by 36 to get the equation into the standard form for the equation of the hyperbola:

$$\frac{y^2}{9} - \frac{x^2}{4} = 1$$

If $x = 0$, then $y = \pm 3$. The y-intercepts are $(0, 3)$ and $(0, -3)$. Since $b^2 = 4$, the fundamental rectangle goes through the y-intercepts and through $(2, 0)$ and $(-2, 0)$. Draw the fundamental rectangle and extend its diagonals to get the asymptotes. Draw a hyperbola opening up and down from the y-intercepts approaching the asymptotes, as shown in Fig. 5.39. To find the foci, use $c^2 = a^2 + b^2$:

$$c^2 = 9 + 4 = 13$$

$$c = \pm\sqrt{13}$$

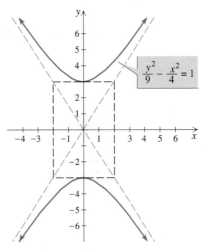

Figure 5.39

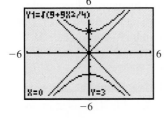

Figure 5.40

The foci are $\left(0, \sqrt{13}\right)$ and $\left(0, -\sqrt{13}\right)$. From the fundamental rectangle, we can see that the equations of the asymptotes are

$$y = \frac{3}{2}x \qquad \text{and} \qquad y = -\frac{3}{2}x.$$

Check by graphing $y = \pm\sqrt{9 + 9x^2/4}$ and $y = \pm 1.5x$ with a calculator, as shown in Fig. 5.40. ■

Hyperbolas Centered at (h, k)

The graph of a hyperbola can be translated horizontally and vertically by replacing x by $x - h$ and y by $y - k$.

Theorem: Hyperbolas Centered at (h, k)

A hyperbola centered at (h, k), opening left and right, has a horizontal transverse axis and equation

$$\frac{(x - h)^2}{a^2} - \frac{(y - k)^2}{b^2} = 1.$$

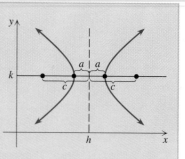

A hyperbola centered at (h, k), opening up and down, has a vertical transverse axis and equation

$$\frac{(y - k)^2}{a^2} - \frac{(x - h)^2}{b^2} = 1.$$

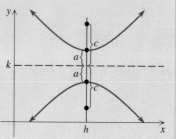

The foci and vertices are on the transverse axis. The distance from center to the vertices is a and the distance from center to foci is c where $c^2 = a^2 + b^2$, $a > 0$, and $b > 0$.

Example **3** **Graphing a hyperbola centered at (h, k)**

Determine the foci and equations of the asymptotes, and sketch the graph of

$$\frac{(x - 2)^2}{9} - \frac{(y + 1)^2}{4} = 1.$$

Solution

The graph that we seek is the graph of

$$\frac{x^2}{9} - \frac{y^2}{4} = 1$$

translated so that its center is $(2, -1)$. Since $a = 3$, the vertices are three units from $(2, -1)$ at $(5, -1)$ and $(-1, -1)$. Because $b = 2$, the fundamental rectangle passes through the vertices and points that are two units above and below $(2, -1)$. Draw the fundamental rectangle through the vertices and through $(2, 1)$ and $(2, -3)$. Extend the diagonals of the fundamental rectangle for the asymptotes, and draw the hyperbola opening to the left and right, as shown in Fig. 5.41. Use $c^2 = a^2 + b^2$ to get $c = \sqrt{13}$. The foci are on the transverse axis, $\sqrt{13}$ units from the center $(2, -1)$. So the foci are $\left(2 + \sqrt{13}, -1\right)$ and $\left(2 - \sqrt{13}, -1\right)$. The asymptotes have slopes $\pm 2/3$ and pass through $(2, -1)$. Using the point-slope form for the equation of the line, we get the equations

$$y = \frac{2}{3}x - \frac{7}{3} \quad \text{and} \quad y = -\frac{2}{3}x + \frac{1}{3}$$

as the equations of the asymptotes.

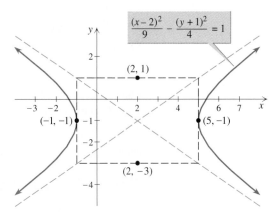

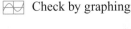

Figure 5.41

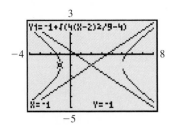

Figure 5.42

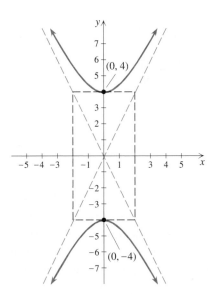

Figure 5.43

Check by graphing

$$y_1 = -1 + \sqrt{4(x - 2)^2/9 - 4},$$

$$y_2 = -1 - \sqrt{4(x - 2)^2/9 - 4},$$

$$y_3 = (2x - 7)/3, \text{ and}$$

$$y_4 = (-2x + 1)/3$$

on a graphing calculator, as in Fig. 5.42.   ■

Finding the Equation of a Hyperbola

The equation of a hyperbola depends on the location of the foci, center, vertices, transverse axis, conjugate axis, and asymptotes. However, it is not necessary to have all of this information to write the equation. In the next example, we find the equation of a hyperbola given only its transverse axis and conjugate axis.

Example **4** **Writing the equation of a hyperbola**

Find the equation of a hyperbola whose transverse axis has endpoints $(0, \pm 4)$ and whose conjugate axis has endpoints $(\pm 2, 0)$.

Solution

Since the vertices are the endpoints of the transverse axis, the vertices are $(0, \pm 4)$ and the hyperbola opens up and down. Since the fundamental rectangle goes through the endpoints of the transverse axis and the endpoints of the conjugate axis, we can sketch the hyperbola shown in Fig. 5.43. Since the hyperbola opens up and down and is centered at the origin, its equation is of the form

$$\frac{y^2}{a^2} - \frac{x^2}{b^2} = 1,$$

for $a > 0$ and $b > 0$. From the fundamental rectangle we get $a = 4$ and $b = 2$. So the equation of the hyperbola is

$$\frac{y^2}{16} - \frac{x^2}{4} = 1.$$ ■

Example **5** Writing the equation of a hyperbola

Find the equation of the hyperbola with asymptotes $y = \pm 4x$ and vertices $(\pm 3, 0)$.

Solution

The vertices of the hyperbola and the vertices of the fundamental rectangle have the same x-coordinates, ± 3. Since the asymptotes go through the vertices of the fundamental rectangle, the vertices of the fundamental rectangle are $(\pm 3, \pm 12)$. Since the parabola opens right and left, its equation is $\dfrac{x^2}{3^2} - \dfrac{y^2}{12^2} = 1$ or $\dfrac{x^2}{9} - \dfrac{y^2}{144} = 1$. ■

Classifying the Conics

We have developed the standard equations for the circle, ellipse, parabola, and hyperbola from the definitions of these curves. We now start with an equation of the form $Ax^2 + Cy^2 + Dx + Ey + F = 0$ and complete the squares to identify the curve.

Example **6** Identifying a conic by completing the squares

Identify the conic $4x^2 + 2y^2 + 8x - y + 1 = 0$.

Solution

Complete the squares for x and y as follows:

$$4x^2 + 8x + 2y^2 - y + 1 = 0$$

$$4(x^2 + 2x + 1) + 2\left(y^2 - \frac{1}{2}y + \frac{1}{16}\right) = -1 + 4 + \frac{2}{16}$$

$$4(x + 1)^2 + 2\left(y - \frac{1}{4}\right)^2 = \frac{25}{8}$$

$$\frac{(x + 1)^2}{\dfrac{25}{32}} + \frac{\left(y - \dfrac{1}{4}\right)^2}{\dfrac{25}{16}} = 1$$

The last equation is that of an ellipse centered at $\left(-1, \frac{1}{4}\right)$. ■

The process used in Example 6 will work on any equation of the form $Ax^2 + Cy^2 + Dx + Ey + F = 0$ in which A and C are not both equal to zero. However, for certain values of F we will get a pair of lines, a single point, or no graph at all, but we will not be concerned with those cases. Assuming that the equation represents a conic section, we can actually determine which one from the value AC. Because A and C had the same sign in Example 6 we got an ellipse. (If $A = C$ we get a circle.) If A and C have opposite signs the conic is a hyperbola and if either A or C is 0 the conic is a parabola. The following table summarizes these ideas.

Value of AC	Type of conic
Positive	Ellipse (or circle)
Negative	Hyperbola
Zero	Parabola

Example **7** Using *AC* to determine the type of conic

Use the value of *AC* to identify the conic.

a. $x^2 - 2y^2 + 3y + 6x = 0$
b. $-3x^2 - 6y^2 + 5y + 7x - 2 = 0$
c. $x^2 + 3y + 6x - 5 = 0$

Solution

a. Because $AC < 0$, the conic is a hyperbola.
b. Because $AC > 0$, the conic is an ellipse.
c. Because $AC = 0$, the conic is a parabola.

■

For Thought

True or False? Explain.

1. The graph of $\frac{x^2}{4} - \frac{y}{9} = 1$ is a hyperbola.

2. The graph of $\frac{x^2}{16} - \frac{y^2}{9} = 1$ has y-intercepts $(0, 3)$ and $(0, -3)$.

3. The hyperbola $y^2 - x^2 = 1$ opens up and down.

4. The graph of $y^2 = 4 + 16x^2$ is a hyperbola.

5. Every point that satisfies $y = \frac{b}{a}x$ must satisfy $\frac{x^2}{a^2} - \frac{y^2}{b^2} = 1$.

6. The asymptotes for $x^2 - \frac{y^2}{4} = 1$ are $y = 2x$ and $y = -2x$.

7. The foci for $\frac{x^2}{16} - \frac{y^2}{9} = 1$ are $(5, 0)$ and $(-5, 0)$.

8. The points $(0, \sqrt{8})$ and $(0, -\sqrt{8})$ are the foci for $\frac{y^2}{3} - \frac{x^2}{5} = 1$.

9. The graph of $\frac{x^2}{9} - \frac{y^2}{4} = 1$ intersects the line $y = \frac{2}{3}x$.

10. The graph of $y^2 = 1 - x^2$ is a hyperbola centered at the origin.

5.3 Exercises

Each of the following graphs shows a hyperbola along with its vertices, foci, and asymptotes. Determine the coordinates of the vertices and foci, and the equations of the asymptotes.

1.

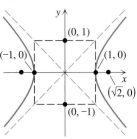

2.

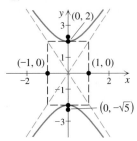

3.

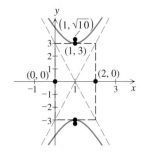

4.

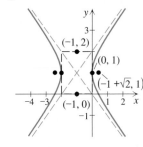

Determine the foci and the equations of the asymptotes, and sketch the graph of each hyperbola. See the procedures for graphing hyperbolas on pages 387 and 388.

5. $\dfrac{x^2}{4} - \dfrac{y^2}{9} = 1$

6. $\dfrac{x^2}{16} - \dfrac{y^2}{9} = 1$

7. $\dfrac{y^2}{4} - \dfrac{x^2}{25} = 1$

8. $\dfrac{y^2}{9} - \dfrac{x^2}{16} = 1$

9. $\dfrac{x^2}{4} - y^2 = 1$

10. $x^2 - \dfrac{y^2}{4} = 1$

11. $x^2 - \dfrac{y^2}{9} = 1$

12. $\dfrac{x^2}{9} - y^2 = 1$

13. $16x^2 - 9y^2 = 144$

14. $9x^2 - 25y^2 = 225$

15. $x^2 - y^2 = 1$

16. $y^2 - x^2 = 1$

Sketch the graph of each hyperbola. Determine the foci and the equations of the asymptotes.

17. $\dfrac{(x+1)^2}{4} - \dfrac{(y-2)^2}{9} = 1$

18. $\dfrac{(x+3)^2}{16} - \dfrac{(y+2)^2}{25} = 1$

19. $\dfrac{(y-1)^2}{4} - (x+2)^2 = 1$

20. $\dfrac{(y-2)^2}{4} - \dfrac{(x-1)^2}{9} = 1$

21. $\dfrac{(x+2)^2}{16} - \dfrac{(y-3)^2}{9} = 1$

22. $\dfrac{(x+1)^2}{16} - \dfrac{(y+2)^2}{25} = 1$

23. $(y-3)^2 - (x-3)^2 = 1$

24. $(y+2)^2 - (x+2)^2 = 1$

Find the equation of each hyperbola described below.

25. Asymptotes $y = \frac{1}{2}x$ and $y = -\frac{1}{2}x$, and x-intercepts $(6, 0)$ and $(-6, 0)$

26. Asymptotes $y = x$ and $y = -x$, and y-intercepts $(0, 2)$ and $(0, -2)$

27. Foci $(5, 0)$ and $(-5, 0)$ and x-intercepts $(3, 0)$ and $(-3, 0)$

28. Foci $(0, 5)$ and $(0, -5)$, and y-intercepts $(0, 4)$ and $(0, -4)$

29. Vertices of the fundamental rectangle $(3, \pm 5)$ and $(-3, \pm 5)$, and opening left and right

30. Vertices of the fundamental rectangle $(1, \pm 7)$ and $(-1, \pm 7)$, and opening up and down

Find the equation of each hyperbola.

31.

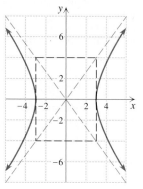

32.

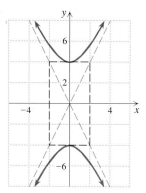

33.

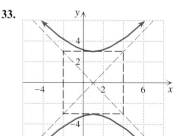

34.

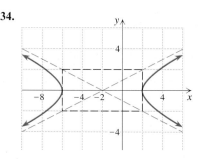

Rewrite each equation in one of the standard forms of the conic sections and identify the conic section.

35. $y^2 - x^2 + 2x = 2$

36. $4y^2 + x^2 - 2x = 15$

37. $y - x^2 = 2x$

38. $y^2 + x^2 - 2x = 0$

39. $25x^2 = 2500 - 25y^2$

40. $100x^2 = 25y^2 + 2500$

41. $100y^2 = 2500 - 25x$

42. $100y^2 = 2500 - 25x^2$

43. $2x^2 - 4x + 2y^2 - 8y = -9$

44. $2x^2 + 4x + y = -7$

45. $2x^2 + 4x + y^2 + 6y = -7$

46. $9x^2 - 18x + 4y^2 + 16y = 11$

47. $25x^2 - 150x - 8y = 4y^2 - 121$

48. $100y^2 + 4x = x^2 + 104$

Use the value of AC to identify the conic.

49. $4x^2 + 3y^2 + 5x - 7y - 1 = 0$

50. $-3x^2 - 8y^2 - 3x + 5y + 2 = 0$

51. $2x^2 - 2x - 3y - 9 = 0$

52. $5y^2 + 7x - 9y - 3 = 0$

53. $9x^2 - 5y^2 - 4x + 3y - 8 = 0$

54. $-8x^2 + y^2 - 2x + 5y = 0$

Solve each problem.

55. *Telephoto Lens*   The focus of the main parabolic mirror of a telephoto lens is 10 in. above the vertex, as shown in the drawing. A hyperbolic mirror is placed so that its vertex is at $(0, 8)$. If the hyperbola has center $(0, 0)$ and foci $(0, \pm 10)$, then what is the equation of the hyperbola?

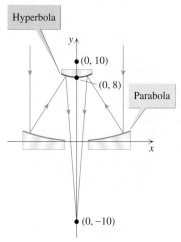

■ **Figure for Exercise 55**

56. *Parabolic Mirror*   Find the equation of the cross section of the parabolic mirror described in Exercise 55.

57. *Marine Navigation*   In 1990 the loran (long range navigation) system had about 500,000 users (International Loran Association, www.loran.org). A loran unit measures the difference in time that it takes for radio signals from pairs of fixed points to reach a ship. The unit then finds the equations of two hyperbolas that pass through the location of the ship and determines the location of the ship. Suppose that the hyperbolas $9x^2 - 4y^2 = 36$

and $16y^2 - x^2 = 16$ pass through the location of a ship in the first quadrant. Find the exact location of the ship.

58. *Air Navigation*   A pilot is flying in the coordinate system shown in the accompanying figure. Using radio signals emitted from $(4, 0)$ and $(-4, 0)$, he knows that he is four units closer to $(4, 0)$ than he is to $(-4, 0)$. Find the equation of the hyperbola with foci $(\pm 4, 0)$ that passes through his location. He also knows that he is two units closer to $(0, 4)$ than he is to $(0, -4)$. Find the equation of the hyperbola with foci $(0, \pm 4)$ that passes through his location. If the pilot is in the first quadrant, then what is his exact location?

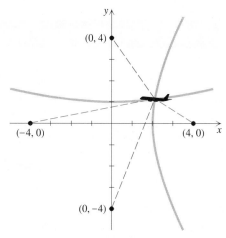

■ **Figure for Exercise 58**

59. *Points on a Hyperbola*   Find all points of the hyperbola $x^2 - y^2 = 1$ that are twice as far from one focus as they are from the other focus.

60. *Perpendicular Asymptotes*   For what values of a and b are the asymptotes of the hyperbola $x^2/a^2 - y^2/b^2 = 1$ perpendicular?

🖩   *Graph each hyperbola on a graphing calculator along with its asymptotes. Observe how close the hyperbola gets to its asymptotes as x gets larger and larger. If x = 50, what is the difference between the y-value on the asymptotes and the y-value on the hyperbola?*

61. $x^2 - y^2 = 1$ **62.** $\dfrac{y^2}{4} - \dfrac{x^2}{9} = 1$

Thinking Outside the Box XXXVI

A Little Lagniappe   A regulation basketball in the NBA has a circumference of 30 in. A manufacturer packages a basketball in a cubic box so that it just fits. As lagniappe, the manufacturer includes a small rubber ball in each corner of the box. The small rubber balls also fit exactly into the corners. Find the exact radius of each of the small rubber balls.

5.3 Pop Quiz

1. Find the foci and the equations of the asymptotes for the hyperbola $\dfrac{x^2}{9} - \dfrac{y^2}{4} = 1$.

2. Find the equation of the hyperbola that has vertices $(0, \pm 6)$ and asymptotes $y = \pm 3x$.

5.4 Polar Coordinates

The Cartesian coordinate system is a system for describing the location of points in a plane. It is not the only coordinate system available. In this section we will study the **polar coordinate system.** In this coordinate system, a point is located by using a *directed distance* and an *angle*.

Polar Coordinates

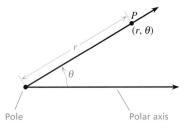

Figure 5.44

In the rectangular coordinate system, points are named according to their position with respect to an x-axis and a y-axis. In the polar coordinate system, we have a fixed point called the **pole** and a fixed ray called the **polar axis.** A point P has coordinates (r, θ), where r is the directed distance from the pole to P and θ is an angle whose initial side is the polar axis and whose terminal side contains the point. See Fig. 5.44.

Since we are so familiar with rectangular coordinates, we retain the x- and y-axes when using polar coordinates. The pole is placed at the origin, and the polar axis is placed along the positive x-axis. The angle θ is any angle (in degrees or radians) in standard position whose terminal side contains the point. As usual, θ is positive for a counterclockwise rotation and negative for a clockwise rotation. In polar coordinates, r can be any real number. For example, the ordered pair $(2, \pi/4)$ represents the point that lies two units from the origin on the terminal side of the angle $\pi/4$. The point $(0, \pi/4)$ is at the origin. The point $(-2, \pi/4)$ lies two units from the origin on the line through the terminal side of $\pi/4$ but in the direction opposite to $(2, \pi/4)$. See Fig. 5.45. Any ordered pair in polar coordinates names a single point, but the coordinates of a point in polar coordinates are not unique. For example, $(-2, \pi/4)$, $(2, 5\pi/4)$, and $(2, -3\pi/4)$ all name the same point.

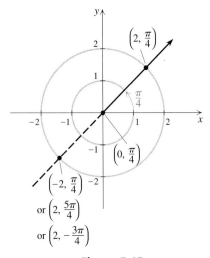

Figure 5.45

Example **1** Plotting points in polar coordinates

Plot the points whose polar coordinates are $(2, 5\pi/6)$, $(-3, \pi)$, $(1, -\pi/2)$, and $(-1, 450°)$.

Solution

The terminal side of $5\pi/6$ lies in the second quadrant, so $(2, 5\pi/6)$ is two units from the origin along this ray. The terminal side of the angle π points in the direction of the negative x-axis, but since the first coordinate of $(-3, \pi)$ is negative, the point

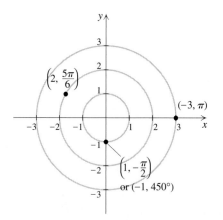

Figure 5.46

is located three units in the direction opposite to the direction of the ray. So $(-3, \pi)$ lies three units from the origin on the positive *x*-axis. The point $(-3, \pi)$ is the same point as $(3, 0)$. The terminal side of $-\pi/2$ lies on the negative *y*-axis. So $(1, -\pi/2)$ lies one unit from the origin on the negative *y*-axis. The terminal side of 450° lies on the positive *y*-axis. Since the first coordinate of $(-1, 450°)$ is negative, the point is located one unit in the opposite direction. The point $(-1, 450°)$ is the same as $(1, -\pi/2)$. All points are shown in Fig. 5.46. ■

Polar-Rectangular Conversions

To graph equations in polar coordinates, we must be able to switch the coordinates of a point in either system to the other. Suppose that (r, θ) is a point in the first quadrant with $r > 0$ and θ acute, as shown in Fig. 5.47. If (x, y) is the same point in rectangular coordinates, then *x* and *y* are the lengths of the legs of the right triangle shown in Fig. 5.47. Using the Pythagorean theorem and trigonometric ratios, we have

$$x^2 + y^2 = r^2, \qquad x = r\cos\theta, \qquad y = r\sin\theta, \qquad \text{and} \qquad \tan\theta = \frac{y}{x}.$$

It can be shown that these equations hold for any point (r, θ), except that $\tan\theta$ is undefined if $x = 0$. The following rules for converting from one system to the other follow from these relationships.

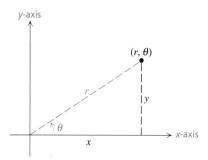

Figure 5.47

Polar-Rectangular Conversion Rules

To convert (r, θ) to rectangular coordinates (x, y), use

$$x = r\cos\theta \qquad \text{and} \qquad y = r\sin\theta.$$

To convert (x, y) to polar coordinates (r, θ), use

$$r = \sqrt{x^2 + y^2}$$

and any angle θ in standard position whose terminal side contains (x, y).

There are several ways to find an angle θ whose terminal side contains (x, y). One way to find θ (provided $x \neq 0$) is to find an angle that satisfies $\tan\theta = y/x$ and goes through (x, y). Remember that the angle $\tan^{-1}(y/x)$ is between $-\pi/2$ and $\pi/2$ and will go through (x, y) only if $x > 0$. If $x = 0$, then the point (x, y) is on the *y*-axis, and you can use $\theta = \pi/2$ or $\theta = -\pi/2$ as appropriate.

Example **2** Polar-rectangular conversion

a. Convert $(6, 210°)$ to rectangular coordinates.

b. Convert the rectangular coordinates $(-3, 6)$ to polar coordinates.

Solution

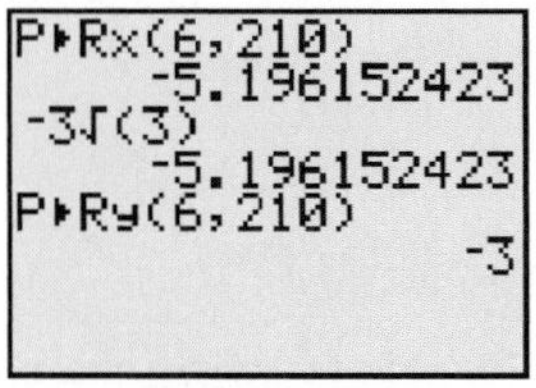

```
P▶Rx(6,210)
         -5.196152423
-3√(3)
         -5.196152423
P▶Ry(6,210)
                   -3
```

■ **Figure 5.48**

a. Use $r = 6$, $\cos 210° = -\sqrt{3}/2$, and $\sin 210° = -1/2$ in the formulas $x = r \cos \theta$ and $y = r \sin \theta$:

$$x = 6\left(-\frac{\sqrt{3}}{2}\right) = -3\sqrt{3} \qquad \text{and} \qquad y = 6\left(-\frac{1}{2}\right) = -3$$

So $(6, 210°)$ in polar coordinates is $\left(-3\sqrt{3}, -3\right)$ in rectangular coordinates.

⌐⌐ You can check this result with a graphing calculator, as shown in Fig. 5.48. □

b. To convert $(-3, 6)$ to polar coordinates, find r:

$$r = \sqrt{(-3)^2 + 6^2} = 3\sqrt{5}$$

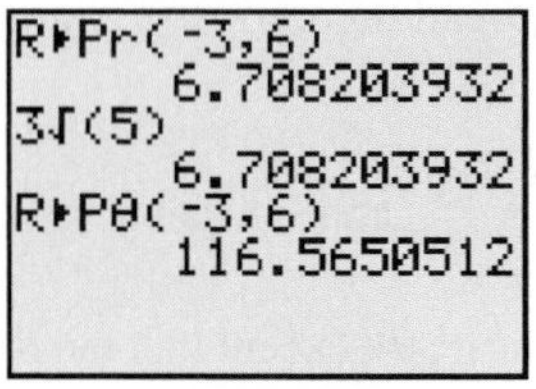

```
R▶Pr(-3,6)
         6.708203932
3√(5)
         6.708203932
R▶Pθ(-3,6)
          116.5650512
```

■ **Figure 5.49**

Use a calculator to find $\tan^{-1}(-2) \approx -63.4°$. To get an angle whose terminal side contains $(-3, 6)$, use $\theta = 180° - 63.4° = 116.6°$. So $(-3, 6)$ in polar coordinates is $\left(3\sqrt{5}, 116.6°\right)$. Since there are infinitely many representations for any point in polar coordinates, this answer is not unique. In fact, another possibility is $\left(-3\sqrt{5}, -63.4°\right)$.

⌐⌐ You can check this result with a graphing calculator, as shown in Fig. 5.49. ■

Polar Equations

An equation in two variables (typically x and y) that is graphed in the rectangular coordinate system is called a **rectangular** or **Cartesian equation.** An equation in two variables (typically r and θ) that is graphed in the polar coordinate system is called a **polar equation.** Certain polar equations are easier to graph than the equivalent Cartesian equations. We can graph a polar equation in the same way that we graph a rectangular equation, that is, we can simply plot enough points to get the shape of the graph. However, since most of our polar equations involve trigonometric functions, finding points on these curves can be tedious. A graphing calculator can be used to great advantage here.

Example **3** Graphing a polar equation

Sketch the graph of the polar equation $r = 2 \cos \theta$.

Solution

If $\theta = 0°$, then $r = 2 \cos 0° = 2$. So, the ordered pair $(2, 0°)$ is on the graph. If $\theta = 30°$, then $r = 2 \cos 30° = \sqrt{3}$, and $\left(\sqrt{3}, 30°\right)$ is on the graph. These ordered pairs and several others that satisfy the equation are listed in the following table:

θ	0°	30°	45°	60°	90°	120°	135°	150°	180°
r	2	$\sqrt{3}$	$\sqrt{2}$	1	0	-1	$-\sqrt{2}$	$-\sqrt{3}$	-2

Plot these points and draw a smooth curve through them to get the graph shown in Fig. 5.50. If θ is larger than 180° or smaller than 0°, we get different ordered pairs, but they are all located on the curve drawn in Fig. 5.50. For example, $\left(-\sqrt{3}, 210°\right)$ satisfies $r = 2 \cos \theta$, but it has the same location as $\left(\sqrt{3}, 30°\right)$.

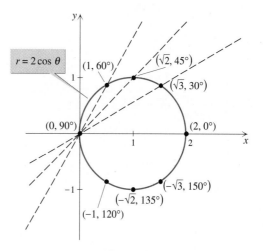

■ **Figure 5.50**

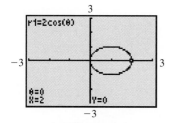

■ **Figure 5.51**

To check this graph with a calculator, set the mode to polar and enter $r = 2 \cos \theta$. The graph in Fig. 5.51 supports the graph in Fig. 5.50. ■

The graph $r = 2 \cos \theta$ in Fig. 5.50 looks like a circle. To verify that it is a circle, we can convert the polar equation to an equivalent rectangular equation because we know the form of the equation of a circle in rectangular coordinates. This conversion is done in Example 6.

In the next example, a simple polar equation produces a curve that is not usually graphed when studying rectangular equations because the equivalent rectangular equation is quite complicated.

Example **4** Graphing a polar equation

Sketch the graph of the polar equation $r = 3 \sin 2\theta$.

Solution

The ordered pairs in the following table satisfy the equation $r = 3 \sin 2\theta$. The values of r are rounded to the nearest tenth.

θ	0°	15°	30°	45°	60°	90°	135°	180°	225°	270°	315°	360°
r	0	1.5	2.6	3	2.6	0	−3	0	3	0	−3	0

As θ varies from 0° to 90°, the value of r goes from 0 to 3, then back to 0, creating a loop in quadrant I. As θ varies from 90° to 180°, the value of r goes from 0 to −3 then back to 0, creating a loop in quadrant IV. As θ varies from 180° to 270°, a loop in quadrant III is formed; and as θ varies from 270° to 360°, a loop in quadrant II is formed. If θ is chosen greater than 360° or less than 0°, we get the same points over and over because the sine function is periodic. The graph of $r = 3 \sin 2\theta$ is shown in Fig. 5.52. The graph is called a **four-leaf rose.**

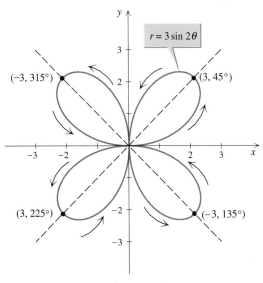

Figure 5.52

With a graphing calculator you can make a table of ordered pairs, as shown in Fig. 5.53. Make this table yourself and scroll through the table to see how the radius oscillates between 3 and -3 as the angle varies. The calculator graph of $r = 3 \sin(2\theta)$ in polar mode with θ between $0°$ and $360°$ is shown in Fig. 5.54. This graph supports the graph shown in Fig. 5.52.

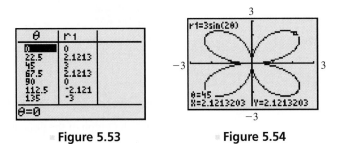

Figure 5.53   **Figure 5.54**   ■

Example 5 Graphing a polar equation

Sketch the graph of the polar equation $r = \theta$, where θ is in radians and $\theta \geq 0$.

Solution

The ordered pairs in the following table satisfy $r = \theta$. The values of r are rounded to the nearest tenth.

θ	0	$\pi/4$	$\pi/2$	$3\pi/4$	π	$3\pi/2$	2π	3π
r	0	0.8	1.6	2.4	3.1	4.7	6.3	9.4

The graph of $r = \theta$, called **the spiral of Archimedes,** is shown as a smooth curve in Fig. 5.55. As the value of θ increases, the value of r increases, causing the graph to

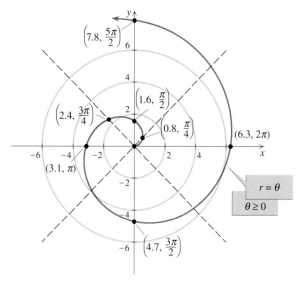

Figure 5.55

spiral out from the pole. There is no repetition of points as there was in Examples 3 and 4, because no periodic function is involved.

With a graphing calculator you can make a table of ordered pairs for $r = \theta$, as shown in Fig. 5.56. The calculator graph of $r = \theta$ in polar coordinates for θ ranging from 0 through 40 radians is shown in Fig. 5.57. This graph shows much more of the spiral than Fig. 5.55. It supports the conclusion that the graph of $r = \theta$ is a spiral.

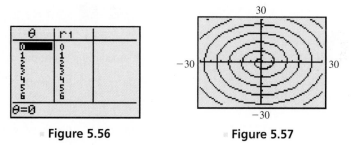

Figure 5.56   **Figure 5.57**   ▪

The Function Gallery on page 403 shows the graphs of several types of polar equations.

Converting Equations

▪ **Foreshadowing Calculus**

In calculus we often switch to polar coordinates to find areas and volumes of figures that are easier to understand and draw in polar coordinates.

We know that certain types of rectangular equations have graphs that are particular geometric shapes, such as lines, circles, and parabolas. We can use our knowledge of equations in rectangular coordinates with equations in polar coordinates (and vice versa) by converting the equations from one system to the other. For example, we can determine whether the graph of $r = 2\cos\theta$ in Example 3 is a circle by finding the equivalent Cartesian equation and deciding whether it is the equation of a circle. When converting from one system to another, we use the relationships

$$x^2 + y^2 = r^2, \qquad x = r\cos\theta, \qquad \text{and} \qquad y = r\sin\theta.$$

Example **6** **Converting a polar equation to a rectangular equation**

Write an equivalent rectangular equation for the polar equation $r = 2\cos\theta$.

Solution

First multiply each side of $r = 2\cos\theta$ by r to get $r^2 = 2r\cos\theta$. Now eliminate r and θ by making substitutions using $r^2 = x^2 + y^2$ and $x = r\cos\theta$:

$$r^2 = 2r\cos\theta$$
$$x^2 + y^2 = 2x$$

From Section 1.3, the standard equation of a circle with center (h, k) and radius r is $(x - h)^2 + (y - k)^2 = r^2$. Complete the square to get $x^2 + y^2 = 2x$ into the standard form of the equation of a circle:

$$x^2 - 2x + y^2 = 0$$
$$x^2 - 2x + 1 + y^2 = 0 + 1$$
$$(x - 1)^2 + y^2 = 1$$

Since we recognize the rectangular equation as the equation of a circle centered at $(1, 0)$ with radius 1, the graph of $r = 2\cos\theta$ shown in Fig. 5.58 is a circle centered at $(1, 0)$ with radius 1. ■

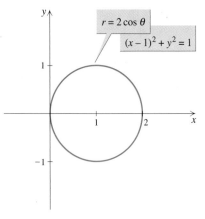

$r = 2\cos\theta$

$(x - 1)^2 + y^2 = 1$

■ **Figure 5.58**

In the next example we convert the rectangular equation of a line and circle into polar coordinates.

Example **7** **Converting a rectangular equation to a polar equation**

For each rectangular equation, write an equivalent polar equation.

a. $y = 3x - 2$ **b.** $x^2 + y^2 = 9$

Solution

a. Substitute $x = r\cos\theta$ and $y = r\sin\theta$, and then solve for r:

$$y = 3x - 2$$
$$r\sin\theta = 3r\cos\theta - 2$$
$$r\sin\theta - 3r\cos\theta = -2$$
$$r(\sin\theta - 3\cos\theta) = -2$$
$$r = \frac{-2}{\sin\theta - 3\cos\theta}$$

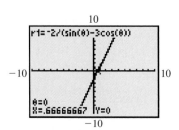

■ **Figure 5.59**

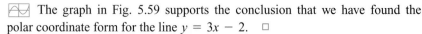 The graph in Fig. 5.59 supports the conclusion that we have found the polar coordinate form for the line $y = 3x - 2$. □

b. Substitute $r^2 = x^2 + y^2$ to get polar coordinates:

$$x^2 + y^2 = 9$$
$$r^2 = 9$$
$$r = \pm 3$$

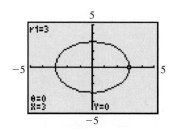

Figure 5.60

A polar equation for a circle of radius 3 centered at the origin is $r = 3$. The equation $r = -3$ is also a polar equation for the same circle.

The graph in Fig. 5.60 supports the conclusion that $r = 3$ is a circle in polar coordinates. Graph $r = 3$ on your calculator, using a viewing window that makes the circle look round. ▪

In Example 7 we saw that a straight line has a rather simple equation in rectangular coordinates but a more complicated equation in polar coordinates. A circle centered at the origin has a very simple equation in polar coordinates but a more complicated equation in rectangular coordinates. The graphs of simple polar equations are typically circular or somehow "centered" at the origin.

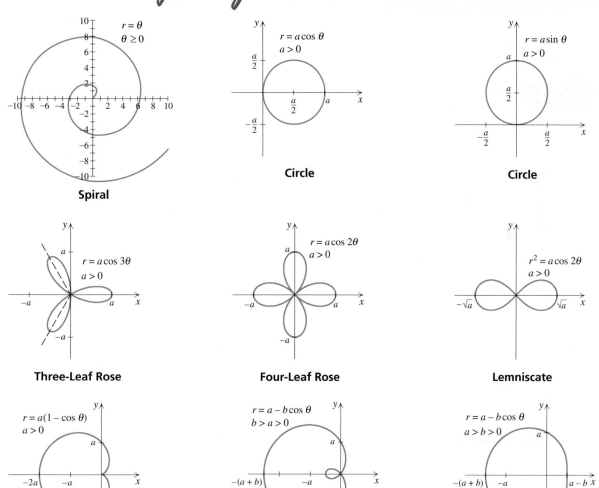

Function Gallery: **Functions in Polar Coordinates**

$r = \theta$
$\theta \geq 0$

Spiral

$r = a \cos \theta$
$a > 0$

Circle

$r = a \sin \theta$
$a > 0$

Circle

$r = a \cos 3\theta$
$a > 0$

Three-Leaf Rose

$r = a \cos 2\theta$
$a > 0$

Four-Leaf Rose

$r^2 = a \cos 2\theta$
$a > 0$

Lemniscate

$r = a(1 - \cos \theta)$
$a > 0$

Cardioid

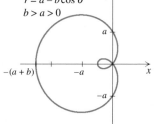

$r = a - b \cos \theta$
$b > a > 0$

Limaçon

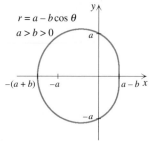

$r = a - b \cos \theta$
$a > b > 0$

Limaçon

For Thought

True or False? Explain.

1. The distance of the point (r, θ) from the origin depends only on r.

2. The distance of the point (r, θ) from the origin is r.

3. The ordered pairs $(2, \pi/4)$, $(2, -3\pi/4)$, and $(-2, 5\pi/4)$ all represent the same point in polar coordinates.

4. The equations relating rectangular and polar coordinates are $x = r \sin\theta$, $y = r \cos\theta$, and $x^2 + y^2 = r^2$.

5. The point $(-4, 225°)$ in polar coordinates is $\left(2\sqrt{2}, 2\sqrt{2}\right)$ in rectangular coordinates.

6. The graph of $0 \cdot r + \theta = \pi/4$ in polar coordinates is a straight line.

7. The graphs of $r = 5$ and $r = -5$ are identical.

8. The ordered pairs $\left(-\sqrt{2}/2, \pi/3\right)$ and $\left(\sqrt{2}/2, \pi/3\right)$ satisfy $r^2 = \cos 2\theta$.

9. The graph of $r = 1/\sin\theta$ is a vertical line.

10. The graphs of $r = \theta$ and $r = -\theta$ are identical.

5.4 Exercises

Find polar coordinates for each given point, using radian measure for the angle.

1.

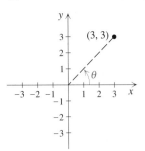

2.

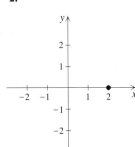

3.

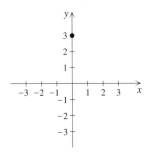

4.

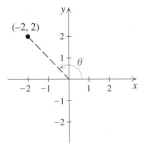

Plot the points whose polar coordinates are given.

5. $(2, 0°)$

6. $(-3, 0°)$

7. $(0, 35°)$

8. $(0, 90°)$

9. $(3, \pi/6)$

10. $(2, \pi/4)$

11. $(-2, 2\pi/3)$

12. $(-1, \pi/6)$

13. $(2, -\pi/4)$

14. $(1, -2\pi/3)$

15. $(3, -225°)$

16. $(2, -180°)$

17. $(-2, 45°)$

18. $(-3, 30°)$

19. $(4, 390°)$

20. $(3, 13\pi/6)$

Convert the polar coordinates of each point to rectangular coordinates.

21. $(4, 0°)$

22. $(-5, 0°)$

23. $(0, \pi/4)$

24. $(0, -\pi)$

25. $(1, \pi/6)$

26. $(2, \pi/4)$

27. $(-3, 3\pi/2)$

28. $(-2, 2\pi)$

29. $\left(\sqrt{2}, 135°\right)$

30. $\left(\sqrt{3}, 150°\right)$

31. $\left(-\sqrt{6}, -60°\right)$

32. $\left(-\sqrt{2}/2, -45°\right)$

Convert the rectangular coordinates of each point to polar coordinates. Use degrees for θ.

33. $\left(\sqrt{3}, 3\right)$

34. $(4, 4)$

35. $(-2, 2)$

36. $\left(-2, 2\sqrt{3}\right)$

37. $(0, 2)$

38. $(-2, 0)$

39. $(-3, -3)$

40. $(2, -2)$

41. $(1, 4)$

42. $(-2, 3)$

43. $\left(\sqrt{2}, -2\right)$

44. $\left(-2, -\sqrt{3}\right)$

Convert the polar coordinates of each point to rectangular coordinates rounded to the nearest hundredth.

45. $(4, 26°)$

46. $(-5, 33°)$

47. $(2, \pi/7)$

48. $(3, 2\pi/9)$

49. $(-2, 1.1)$

50. $(6, 2.3)$

Convert the rectangular coordinates of each point to polar coordinates. Round r to the nearest tenth and θ to the nearest tenth of a degree.

51. $(4, 5)$

52. $(-5, 3)$

53. $(-2, -7)$

54. $(3, -8)$

Sketch the graph of each polar equation.

55. $r = 2 \sin \theta$

56. $r = 3 \cos \theta$

57. $r = 3 \cos 2\theta$

58. $r = -2 \sin 2\theta$

59. $r = 2\theta$ for θ in radians

60. $r = \theta$ for $\theta \leq 0$ and θ in radians

61. $r = 1 + \cos \theta$ (cardioid)

62. $r = 1 - \cos \theta$ (cardioid)

63. $r^2 = 9 \cos 2\theta$ (lemniscate)

64. $r^2 = 4 \sin 2\theta$ (lemniscate)

65. $r = 4 \cos 2\theta$ (four-leaf rose)

66. $r = 3 \sin 2\theta$ (four-leaf rose)

67. $r = 2 \sin 3\theta$ (three-leaf rose)

68. $r = 4 \cos 3\theta$ (three-leaf rose)

69. $r = 1 + 2 \cos \theta$ (limaçon)

70. $r = 2 + \cos \theta$ (limaçon)

71. $r = 3.5$

72. $r = -5$

73. $\theta = 30°$

74. $\theta = 3\pi/4$

For each polar equation, write an equivalent rectangular equation.

75. $r = 4 \cos \theta$

76. $r = 2 \sin \theta$

77. $r = \dfrac{3}{\sin \theta}$

78. $r = \dfrac{-2}{\cos \theta}$

79. $r = 3 \sec \theta$

80. $r = 2 \csc \theta$

81. $r = 5$

82. $r = -3$

83. $\theta = \dfrac{\pi}{4}$

84. $\theta = 0$

85. $r = \dfrac{2}{1 - \sin \theta}$

86. $r = \dfrac{3}{1 + \cos \theta}$

For each rectangular equation, write an equivalent polar equation.

87. $x = 4$

88. $y = -6$

89. $y = -x$

90. $y = x\sqrt{3}$

91. $x^2 = 4y$

92. $y^2 = 2x$

93. $x^2 + y^2 = 4$

94. $2x^2 + y^2 = 1$

95. $y = 2x - 1$

96. $y = -3x + 5$

97. $x^2 + (y - 1)^2 = 1$

98. $(x + 1)^2 + y^2 = 4$

Thinking Outside the Box XXXVII

Laying Pipe A circular pipe with radius 1 is placed in a v-shaped trench whose sides form an angle of θ radians. In the cross section shown here, the pipe touches the sides of the trench at points A and B.

a. Find the area inside the circle and below the line segment AB in terms of θ (the blue area).

b. Find the area below the circle and inside the trench in terms of θ (the red area).

c. If θ is nearly equal to π, then the blue area and the red area are both very small. If θ is equal to π both areas are zero. As θ approaches π there is a limit to the ratio of the blue area to the red area. Use a calculator to determine this limit.

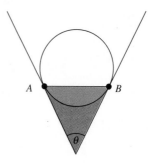

■ **Figure for Thinking Outside the Box XXXVII**

5.4 Pop Quiz

1. If the polar coordinates of a point are $(-1, 15\pi/4)$, then in which quadrant does the point lie?

2. Convert $(4, 150°)$ to rectangular coordinates.

3. Convert the rectangular coordinates $(-2, 2)$ to polar coordinates using radians for the angle.

4. Convert the polar equation $r = 4 \cos \theta$ into a rectangular equation.

5. Find the center and radius of the circle $r = -16 \sin \theta$.

6. Convert $y = x + 1$ into polar coordinates.

5.5   Polar Equations of the Conics

At the beginning of this chapter we gave definitions for each of the conic sections separately. In this section we give one definition that covers them all.

Alternative Definition of the Conics

In the following definition we will use a positive number e (the eccentricity), a fixed line D (the directrix), and a fixed point F (the focus) not on the line. Let $d(P, F)$ represent the distance from P to F and $d(P, D)$ represent the distance from P to D.

Alternative Definition: Conic

A **conic** is the set of points P in the plane such that the ratio of the distance from P to the focus to the distance from P to the directrix is a fixed positive number e. In symbols,

$$\frac{d(P, F)}{d(P, D)} = e.$$

The conic is called a **parabola** if $e = 1$, an **ellipse** if $e < 1$, and a **hyperbola** if $e > 1$.

Note that if $e = 1$ then the distance from P to the focus equals the distance from P to the directrix, which is exactly the definition of a parabola given in Section 5.1. Since we did not use the directrix when we defined the ellipse and hyperbola in Sections 5.2 and 5.3, it is not obvious that the ellipse and hyperbola defined here are the same as the ellipse and hyperbola defined earlier. So we will derive equations for the conics from this new definition and then see that the equations are equivalent to the equations derived earlier in this chapter.

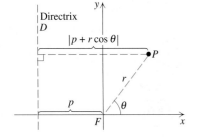

Figure 5.61

Polar Equations of the Conics

For simplicity we use polar coordinates and position the focus at the pole $(0, 0)$. Choose a vertical directrix D that lies p units to the left of the origin as shown in Fig. 5.61. If P is a point on the conic we have $d(P, F) = r$. P might lie to the left or right of D, but in either case $d(P, D) = |p + r\cos\theta|$. Since the ratio of these distances is e we have

$$\frac{r}{|p + r\cos\theta|} = e \quad \text{or} \quad \frac{\pm r}{p + r\cos\theta} = e.$$

Now if (r_1, θ_1) satisfies $\frac{-r}{p + r\cos\theta} = e$ then $(-r_1, \theta_1 + \pi)$ satisfies $\frac{r}{p + r\cos\theta} = e$. So we lose no points on the conic by simply writing its equation as $\frac{r}{p + r\cos\theta} = e$, which is equivalent to

$$r = \frac{ep}{1 - e\cos\theta}.$$

Likewise, if D lies p units to the right of the pole we get the equation

$$r = \frac{ep}{1 + e\cos\theta}.$$

With a horizontal directrix we get similar equations. These results are summarized as follows.

Theorem: Polar Equations of the Conics

The conics defined by the alternative definition have the following equations:

$$r = \frac{ep}{1 - e\cos\theta}$$ If the directrix is perpendicular to the polar axis p units to the left of the pole

$$r = \frac{ep}{1 + e\cos\theta}$$ If the directrix is perpendicular to the polar axis p units to the right of the pole

$$r = \frac{ep}{1 + e\sin\theta}$$ If the directrix is parallel to the polar axis p units above the pole

$$r = \frac{ep}{1 - e\sin\theta}$$ If the directrix is parallel to the polar axis p units below the pole

The conic is a parabola if $e = 1$, an ellipse if $e < 1$, and a hyperbola if $e > 1$.

Figure 5.62 illustrates the three types of conics and shows the location of the directrix in each case.

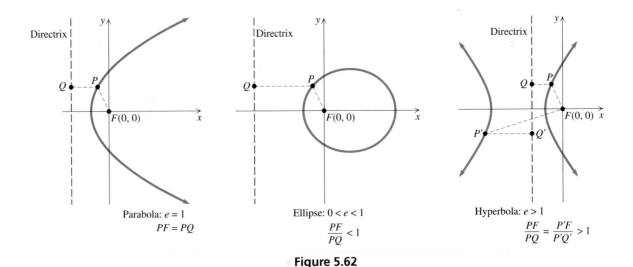

Parabola: $e = 1$
$PF = PQ$

Ellipse: $0 < e < 1$
$\dfrac{PF}{PQ} < 1$

Hyperbola: $e > 1$
$\dfrac{PF}{PQ} = \dfrac{P'F}{P'Q'} > 1$

■ **Figure 5.62**

Example **1** Identifying and graphing a conic from a polar equation

Identify the conic $r = \dfrac{2}{2 - \cos\theta}$ and graph it with a graphing calculator.

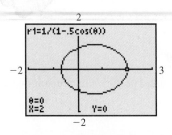

Figure 5.63

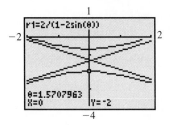

Figure 5.64

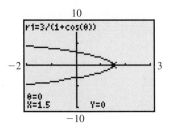

Figure 5.65

Solution

The given equation is equivalent to $r = \dfrac{1}{1 - \frac{1}{2}\cos\theta}$. So from the previous theorem we have $e = \frac{1}{2}$, $ep = 1$, and $p = 2$. So the vertical directrix is 2 units to the left of the pole and the conic is an ellipse. To graph the conic use a graphing calculator that graphs in polar coordinates as shown in Fig. 5.63. ■

Example **2** **Identifying and graphing a conic from a polar equation**

Identify the conic $r = \dfrac{2}{1 - 2\sin\theta}$ and graph it with a graphing calculator.

Solution

From the previous theorem we have $e = 2$, $ep = 2$, and $p = 1$. So the directrix is 1 unit below the pole and the conic is a hyperbola. To graph the conic use a graphing calculator that graphs in polar coordinates as shown in Fig. 5.64. ■

Example **3** **Finding the polar equation of a conic**

Find the equation of the conic that has its focus at the pole, directrix $x = 3$, and eccentricity 1. Graph it with a graphing calculator.

Solution

From the previous theorem we have $e = 1$, $p = 3$, and $r = \dfrac{3}{1 + \cos\theta}$. The conic is a parabola opening to the left. To graph the conic we use a graphing calculator that graphs in polar coordinates as shown in Fig. 5.65. ■

Equivalency of the Definitions

To see that $r = \dfrac{ep}{1 - e\cos\theta}$ is a conic according to the original definition, we use $x = r\cos\theta$ and $r^2 = x^2 + y^2$ to convert to rectangular coordinates:

$$r = \frac{ep}{1 - e\cos\theta}$$

$$r - er\cos\theta = ep$$

$$\sqrt{x^2 + y^2} = ep + ex$$

$$x^2 + y^2 = e^2(p + x)^2$$

$$(1 - e^2)x^2 + y^2 - 2pe^2x - e^2p^2 = 0$$

The last equation is in the form of the general equation of a conic from Section 5.3. Because e^2p^2 is positive we will not get a degenerate conic from this equation. From Section 5.3 we can identify the conic from the value of AC, which is $1 - e^2$. If $e = 1$, then $AC = 0$ and the conic is a parabola. If $e > 1$, then $AC < 0$ and the conic is a hyperbola. If $e < 1$, then $AC > 0$ and the conic is an ellipse. Similarly we can convert the other three polar forms of the conics into conics in rectangular coordinates. So a parabola, ellipse, or hyperbola by the alternative definition is a parabola, ellipse, or hyperbola by the original definition.

Example **4** **Converting a polar conic to rectangular coordinates**

Convert $r = \dfrac{4}{2 - 2\sin\theta}$ to rectangular coordinates and identify the conic. Sketch the graph.

Solution

Convert to rectangular coordinates as follows:

$$r = \frac{4}{2 - 2\sin\theta}$$

$$2r - 2r\sin\theta = 4$$

$$r = 2 + r\sin\theta$$

$$\sqrt{x^2 + y^2} = 2 + y$$

$$x^2 + y^2 = 4 + 4y + y^2$$

$$x^2 - 4y - 4 = 0$$

$$y = \frac{1}{4}x^2 - 1$$

The last equation is the equation for a parabola opening upward as shown in Fig. 5.66. ■

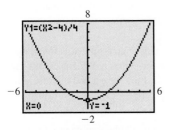

■**Figure 5.66**

For Thought

True or False? Explain.

1. The eccentricity for a parabola is 1.

2. The eccentricity for an ellipse is greater than 1.

3. Every conic section has a directrix.

4. For the conic $r = \dfrac{ep}{1 - e\cos\theta}$ the distance between the focus and directrix is p.

5. The directrix for $r = \dfrac{ep}{1 + e\sin\theta}$ is a horizontal line.

6. The directrix for $r = \dfrac{4}{3 - \cos\theta}$ is the line $x = 4$.

7. The focus for $r = \dfrac{5}{6 + \sin\theta}$ is the pole.

8. The graph of $2x^2 - 3y^2 + 4x - 5y - 1 = 0$ is a hyperbola.

9. The graph of $2x^2 - 4x + 8y - 2 = 0$ is a parabola.

10. The graph of $2x^2 + 5y^2 - x + y - 9 = 0$ is an ellipse.

5.5   Exercises

Find the eccentricity and use it to identify each conic as a parabola, ellipse, or hyperbola. Also find the distance between the focus and directrix.

1. $r = \dfrac{6}{1 - 2\cos\theta}$

2. $r = \dfrac{3}{1 - 5\sin\theta}$

3. $r = \dfrac{3}{4 - 4\sin\theta}$

4. $r = \dfrac{6}{5 - 5\cos\theta}$

5. $r = \dfrac{3}{3 + 4\sin\theta}$

6. $r = \dfrac{6}{2 + 5\cos\theta}$

Identify each conic and sketch its graph. Give the equation of the directrix in rectangular coordinates.

7. $r = \dfrac{2}{1 - \sin\theta}$

8. $r = \dfrac{3}{1 - \cos\theta}$

9. $r = \dfrac{5}{3 + 2\cos\theta}$

10. $r = \dfrac{6}{5 + 3\sin\theta}$

11. $r = \dfrac{1}{2 - 6\cos\theta}$

12. $r = \dfrac{1}{5 - \sin\theta}$

13. $r = \dfrac{6}{2 + \sin\theta}$

14. $r = \dfrac{6}{3 + 2\cos\theta}$

Write the polar equation for a conic with focus at the origin and the given eccentricity and directrix.

15. $e = 1, y = 2$

16. $e = 1, y = -3$

17. $e = 2, x = 5$

18. $e = 3, x = 6$

19. $e = \dfrac{1}{2}, y = 4$

20. $e = \dfrac{1}{3}, y = 6$

21. $e = \dfrac{3}{4}, x = -8$

22. $e = \dfrac{1}{5}, x = -10$

Convert each conic into rectangular coordinates and identify the conic.

23. $r = \dfrac{3}{1 + \sin\theta}$

24. $r = \dfrac{6}{1 + \cos\theta}$

25. $r = \dfrac{3}{4 - \cos\theta}$

26. $r = \dfrac{9}{4 - 3\sin\theta}$

27. $r = \dfrac{1}{3 + 9\cos\theta}$

28. $r = \dfrac{1}{5 - 10\sin\theta}$

29. $r = \dfrac{2}{6 - \sin\theta}$

30. $r = \dfrac{6}{3 - 9\cos\theta}$

Solve each problem.

31. Find the polar coordinates of both foci for the conic
$r = \dfrac{12}{3 + \cos\theta}.$

32. Find the polar coordinates of both foci for the conic
$r = \dfrac{12}{1 + 3\cos\theta}.$

33. In Section 5.2 the eccentricity of an ellipse was found by dividing the distance from the center to a focus by one-half the length of the major axis. Find the eccentricity using this definition for the ellipse $r = \dfrac{2}{2 - \cos\theta}.$

34. In Section 5.3 the eccentricity of a hyperbola was found by dividing the distance from the center to a focus by one-half the length of the transverse axis. Find the eccentricity using this definition for the hyperbola $r = \dfrac{6}{1 - 2\cos\theta}.$

Thinking Outside the Box XXXVIII

Double-Boxed   A rectangular box contains a delicate statue. The shipping department places the box containing the statue inside a 3 ft by 4 ft rectangular box as shown from above in the accompanying figure. If the box containing the statue is 1 ft wide, then what is its length? Find a four-decimal place approximation.

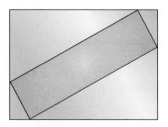

∎ **Figure for Thinking Outside the Box XXXVIII**

5.5 Pop Quiz

1. What is the eccentricity for the conic $r = \dfrac{4}{3 - 2\cos\theta}$?

2. What is the equation of the directrix for the conic $r = \dfrac{4}{2 + 2\sin\theta}$?

3. Write the polar equation of the conic with focus $(0, 0)$, directrix $y = -4$, and $e = 2$.

4. Convert $r = \dfrac{5}{2 - \cos\theta}$ into rectangular coordinates and identify the conic.

5.6 Parametric Equations

We know how to graph points in the plane using the rectangular coordinate system and using polar coordinates. Points in the plane can also be located using parametric equations.

Graphs of Parametric Equations

If $f(t)$ and $g(t)$ are functions of t, where t is in some interval of real numbers, then the equations $x = f(t)$ and $y = g(t)$ are called **parametric equations.** The variable t is called the **parameter** and the graph of the parametric equations is said to be defined **parametrically.** If the parameter is thought of as time, then we know when each point of the graph is plotted. If no interval is specified for t, then t is assumed to be any real number for which both $f(t)$ and $g(t)$ are defined.

Example 1 Graphing a line segment

Graph the parametric equations $x = 3t - 2$ and $y = t + 1$ for t in the interval $[0, 3]$. Determine the domain (the set of x-coordinates) and the range (the set of y-coordinates) for the function or relation that you graphed.

Solution

Make a table of ordered pairs corresponding to values of t between 0 and 3:

t	x	y
0	-2	1
1	1	2
2	4	3
3	7	4

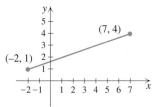

Figure 5.67

Since $x = 3t - 2$, we have $t = \dfrac{x + 2}{3}$. Since t is in the interval $[0, 3]$, we have $0 \le \dfrac{x + 2}{3} \le 3$. Solving for x, we get $-2 \le x \le 7$. So x is in the interval $[-2, 7]$. The graph is not the whole line. It is the line segment with endpoints $(-2, 1)$ and $(7, 4)$ shown in Fig. 5.67. We use solid dots at the end points to show that they are included in the graph. The domain is the interval $[-2, 7]$ and the range is the interval $[1, 4]$.

To check Example 1 with a graphing calculator, set your calculator to parametric mode and enter the parametric equations, as shown in Fig. 5.68(a). Set the limits on the viewing window and the parameter as in Fig. 5.68(b). The graph is shown in Fig. 5.68(c). □

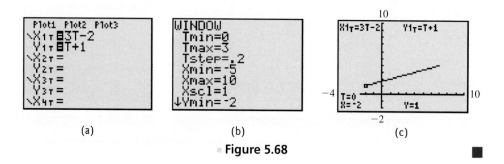

(a)   (b)   (c)

■ **Figure 5.68**   ■

Eliminating the Parameter

In rectangular coordinates, we know that $y = mx + b$ is a line, $(x - h)^2 + (y - k)^2 = r^2$ is a circle, and $y = ax^2 + bx + c$ is a parabola. Since we have little experience with parametric equations, it may not be obvious when a system of parametric equations has a familiar graph. However, it is often possible to identify a graph by eliminating the parameter and writing an equation involving only x and y.

Example **2** **Eliminating the parameter**

Eliminate the parameter and identify the graph of the parametric equations. Determine the domain (the set of x-coordinates) and the range (the set of y-coordinates).

a. $x = 3t - 2, y = t + 1, -\infty < t < \infty$
b. $x = 7 \sin t, y = 7 \cos t, -\infty < t < \infty$

Solution

a. Solve $y = t + 1$ for t to get $t = y - 1$. Now replace t in the other equation by $y - 1$:

$$x = 3(y - 1) - 2$$
$$x = 3y - 5$$
$$3y = x + 5$$
$$y = \frac{1}{3}x + \frac{5}{3}$$

After eliminating the parameter, we get $y = \frac{1}{3}x + \frac{5}{3}$, which is the equation of a line with slope 1/3 and y-intercept (0, 5/3). Because $-\infty < t < \infty$ and both x and y are linear functions of t, the doman of $y = \frac{1}{3}x + \frac{5}{3}$ is $(-\infty, \infty)$ and its range is $(-\infty, \infty)$. So the graph is the entire line. Note that in Example 1 these same parametric equations with a different interval for t determined a line segment.

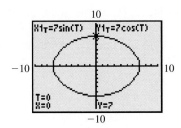

Figure 5.69

b. The simplest way to eliminate the parameter in this case is to use the trigonometric identity $\sin^2(\theta) + \cos^2(\theta) = 1$. Because $\sin t = x/7$ and $\cos t = y/7$, we have $(x/7)^2 + (y/7)^2 = 1$ or $x^2 + y^2 = 49$. So the graph is a circle centered at the origin with radius 7. The domain of this relation is $[-7, 7]$ and the range is $[-7, 7]$.

You can check this conclusion with a graphing calculator, as shown in Fig. 5.69. ■

Writing Parametric Equations

Because a nonvertical straight line has a unique slope and y-intercept, it has a unique equation in slope-intercept form. However, a polar equation for a curve is not unique and neither are parametric equations for a curve. For example, consider the line $y = 2x + 1$. For parametric equations we could let $x = t$ and $y = 2t + 1$. We could also let $x = 4t$ and $y = 8t + 1$. We could even write $x = t^3 + 7$ and $y = 2t^3 + 15$. Each of these pairs of parametric equations produces the line $y = 2x + 1$.

Example **3** **Writing parametric equations for a line segment**

Write parametric equations for the line segment between $(1, 3)$ and $(5, 8)$ for t in the interval $[0, 2]$, with $t = 0$ corresponding to $(1, 3)$ and $t = 2$ corresponding to $(5, 8)$.

Solution

We can make both parametric equations linear functions of t. If $x = mt + b$ and $t = 0$ corresponds to $x = 1$, then $1 = m \cdot 0 + b$ and $b = 1$. So $x = mt + 1$. If $t = 2$ corresponds to $x = 5$, then $5 = m \cdot 2 + 1$ or $m = 2$. So we have $x = 2t + 1$. Using similar reasoning for the y-coordinates we get $y = 2.5t + 3$.

You can use a graphing calculator to check, as shown in Fig. 5.70. ■

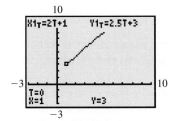

Figure 5.70

While it may not seem obvious how to write parametric equations for a particular rectangular equation, there is a simple way to find parametric equations for a polar equation $r = f(\theta)$. Because $x = r \cos \theta$ and $y = r \sin \theta$, we can substitute $f(\theta)$ for r and write $x = f(\theta) \cos \theta$ and $y = f(\theta) \sin \theta$. In this case the parameter is θ and we have parametric equations for the polar curve. The parametric equations for a polar curve can be used to graph a polar curve on a calculator that is capable of handling parametric equations but not polar equations.

Example **4** **Converting a polar equation to parametric equations**

Write parametric equations for the polar equation $r = 1 - \cos \theta$.

Solution

Replace r by $1 - \cos \theta$ in the equations $x = r \cos \theta$ and $y = r \sin \theta$ to get $x = (1 - \cos \theta) \cos \theta$ and $y = (1 - \cos \theta) \sin \theta$. We know from Section 5.4 that the graph of $r = 1 - \cos\theta$ is the cardioid shown in Fig. 5.71. So the graph of these parametric equations is the same cardioid.

The calculator graph of the parametric equations in Fig. 5.72 appears to be a cardioid and supports the graph shown in Fig. 5.71.

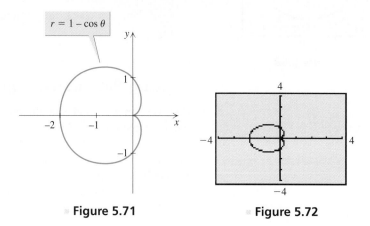

| Figure 5.71 | Figure 5.72 |

Shooting Baskets with Parametric Equations

With x and y in feet and t in seconds, the parametric equations for the path of a projectile are

$$x = v_0(\cos\theta)t \quad \text{and} \quad y = -16t^2 + v_0(\sin\theta)t + h_0$$

where θ is the angle of inclination in degrees, v_0 is the initial velocity in feet per second, and h_0 is the initial height in feet. With these equations and a graphing calculator you can illustrate the flight of a basketball.

To enter these equations into a graphing calculator first set the MODE as in Fig. 5.73. We will throw the ball from the location $(0, 6)$ to a basket at $(20, 10)$. To get these locations to appear on the screen we use STAT PLOT. Use STAT EDIT to enter these ordered pairs in L1 and L2, as shown in Fig. 5.74. Then use STAT PLOT to turn on the feature that plots a square mark at these locations as in Fig. 5.75.

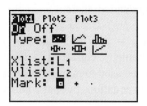

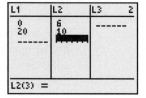

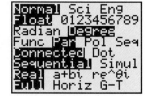

| Figure 5.73 | Figure 5.74 | Figure 5.75 |

Using A for the initial velocity, B for the angle of inclination of the throw, and $h_0 = 6$ feet, enter the parametric equations using Y= as in Fig. 5.76. To get the "ball," move the cursor to the left of X_{1T} and press Enter until a small circle appears to the left of X_{1T} as in Fig. 5.76. Set the window so that the time T ranges from 0 to 2 seconds in steps of 0.1 second as in Figure 5.77. Set the window so that $-2 \le x \le 25$ and $0 \le y \le 20$. Press QUIT to enter the initial values for A and B on the home screen using the STO button as in Figure 5.78.

Figure 5.76

Figure 5.77

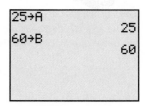

Figure 5.78

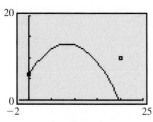

Figure 5.79

Now press GRAPH to see the ball tossed toward the basket. After the toss, the path of the ball will appear as in Fig. 5.79.

With the initial velocity of 25 feet per second and the angle 60°, the ball falls short of the basket, as seen in Fig. 5.79. Enter a new angle and velocity on the home screen until you find a combination for which the path of the ball goes through the basket at (20, 10). Before pressing GRAPH, press DRAW and ClrDraw to clear the old path from the graph screen.

For Thought

True or False? Explain.

1. If $x = 3t + 1$ and $y = 4t - 2$, then t is the variable and x and y are the parameters.

2. Parametric equations are graphed in the rectangular coordinate system.

3. The graph of $x = 0.5t$ and $y = 2t + 1$ is a straight line with slope 4.

4. The graph of $x = \cos t$ and $y = \sin t$ is a sine wave.

5. The graph of $x = 3t + 1$ and $y = 6t - 1$ for $0 \le t \le 3$ includes the point (2, 1).

6. The graph of $x = w^2 - 3$ and $y = w + 5$ for $-2 < w < 2$ includes the point (1, 7).

7. The graph of $x = e^t$ and $y = e^t$ lies entirely within the first quadrant.

8. The graph of $x = -\sin t$ and $y = \cos t$ for $0 < t < \pi/2$ lies entirely within the second quadrant.

9. The parametric equations $x = e^t$ and $y = e^t$ have the same graph as $x = \ln t$ and $y = \ln t$.

10. The polar equation $r = \cos \theta$ can be graphed using the parametric equations $x = \cos^2 \theta$ and $y = \cos \theta \sin \theta$.

5.6 Exercises

Complete the table that accompanies each pair of parametric equations.

1. $x = 4t + 1$, $y = t - 2$, for $0 \le t \le 3$

t	x	y
0		
1		
	7	
		1

2. $x = 3 - t$, $y = 2t + 5$, for $2 \le t \le 7$

t	x	y
2		
3		
	-2	
		19

3. $x = t^2$, $y = 3t - 1$, for $1 \le t \le 5$

t	x	y
1		
2.5		
	5	
		11
	25	

4. $x = \sqrt{t}$, $y = t + 4$, for $0 \le t \le 9$

t	x	y
0		
2		
4		
		12
	3	

Graph each pair of parametric equations in the rectangular coordinate system. Determine the domain (the set of x-coordinates) and the range (the set of y-coordinates).

5. $x = 3t - 2, y = t + 3$, for $0 \le t \le 4$

6. $x = 4 - 3t, y = 3 - t$, for $1 \le t \le 3$

7. $x = t - 1, y = t^2$, for t in $(-\infty, \infty)$

8. $x = t - 3, y = 1/t$, for t in $(-\infty, \infty)$

9. $x = \sqrt{w}, y = \sqrt{1 - w}$, for $0 < w < 1$

10. $x = \ln t, y = t + 3$, for $-2 < t < 2$

11. $x = \cos t, y = \sin t$

12. $x = 0.5t, y = \sin t$

Eliminate the parameter and identify the graph of each pair of parametric equations. Determine the domain (the set of x-coordinates) and the range (the set of y-coordinates).

13. $x = 4t - 5, y = 3 - 4t$

14. $x = 5t - 1, y = 4t + 6$

15. $x = -4 \sin 3t, y = 4 \cos 3t$

16. $x = 2 \sin t \cos t, y = 3 \sin 2t$

17. $x = t/4, y = e^t$

18. $x = t - 5, y = t^2 - 10t + 25$

19. $x = \tan t, y = 2 \tan t + 3$

20. $x = \tan t, y = -\tan^2 t + 3$

Write a pair of parametric equations that will produce the indicated graph. Answers may vary.

21. The line segment starting at $(2, 3)$ with $t = 0$ and ending at $(5, 9)$ with $t = 2$

22. The line segment starting at $(-2, 4)$ with $t = 3$ and ending at $(5, -9)$ with $t = 7$

23. That portion of the circle $x^2 + y^2 = 4$ that lies in the third quadrant

24. That portion of the circle $x^2 + y^2 = 9$ that lies below the x-axis.

25. The vertical line through $(3, 1)$

26. The horizontal line through $(5, 2)$

27. The circle whose polar equation is $r = 2 \sin \theta$

28. The four-leaf rose whose polar equation is $r = 5 \sin(2\theta)$

Graph the following pairs of parametric equations with the aid of a graphing calculator. These are uncommon curves that would be difficult to describe in rectangular or polar coordinates.

29. $x = \cos 3t, y = \sin t$

30. $x = \sin t, y = t^2$

31. $x = t - \sin t, y = 1 - \cos t$ (cycloid)

32. $x = t - \sin t, y = -1 + \cos t$ (inverted cycloid)

33. $x = 4 \cos t - \cos 4t, y = 4 \sin t - \sin 4t$ (epicycloid)

34. $x = \sin^3 t, y = \cos^3 t$ (hypocycloid)

The following problems involve the parametric equations for the path of a projectile

$$x = v_0(\cos \theta)t \qquad \text{and} \qquad y = -16t^2 + v_0(\sin \theta)t + h_0,$$

where θ is the angle of inclination of the projectile at the launch, v_0 is the initial velocity of the projectile in feet per second, and h_0 is the initial height of the projectile in feet.

35. An archer shoots an arrow from a height of 5 ft at an angle of inclination of 30° with a velocity of 300 ft/sec. Write the parametric equations for the path of the projectile and sketch the graph of the parametric equations.

36. If the arrow of Exercise 35 strikes a target at a height of 5 ft, then how far is the target from the archer?

37. For how many seconds is the arrow of Exercise 35 in flight?

38. What is the maximum height reached by the arrow in Exercise 35?

Thinking Outside the Box XXXIX

Lakefront Property A man-made lake in the shape of a triangle is bounded on each of its sides by a square lot as shown in the figure. The square lots are 8, 13, and 17 acres, respectively. What is the area of the lake in square feet?

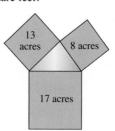

■ **Figure for Thinking Outside the Box XXXIX**

5.6 Pop Quiz

1. The graph of $x = 2t + 5$ and $y = 3t - 7$ for t in $[3, 5]$ is a line segment. What are the endpoints?

2. Eliminate the parameter and identify the graph of $x = 3\cos t$ and $y = 3\sin t$ for $-\infty < t < \infty$.

3. Write parametric equations for the line segment between $(0, 1)$ and $(3, 5)$, where $t = 0$ corresponds to $(0, 1)$ and $t = 4$ corresponds to $(3, 5)$.

▪▪▪ Highlights

5.1 The Parabola

Parabola: Geometric Definition	The set of all points in the plane that are equidistant from a fixed line (directrix) and a fixed point (focus) not on the line	All points on $y = x^2$ are equidistant from $y = -1/4$ and $(0, 1/4)$.
Parabolas Opening Up and Down	$y = ax^2 + bx + c$ opens up if $a > 0$ or down if $a < 0$. The x-coordinate of the vertex is $-b/(2a)$.	$y = 2x^2 + 4x - 1$ Opens up, vertex $(-1, -3)$
	$y = a(x - h)^2 + k$ opens up if $a > 0$ or down if $a < 0$, with vertex (h, k), focus $(h, k + p)$, and directrix $y = k - p$, where $a = 1/(4p)$.	$y = 2(x + 1)^2 - 3$ Opens up, vertex $(-1, -3)$, focus $(-1, -23/8)$, directrix $y = -25/8$
Parabolas Opening Left and Right	$x = ay^2 + by + c$ opens right if $a > 0$ or left if $a < 0$. The y-coordinate of the vertex is $-b/(2a)$.	$x = y^2 - 4y + 3$ Opens right, vertex $(-1, 2)$
	$x = a(y - k)^2 + h$ opens right if $a > 0$ or left if $a < 0$, with vertex (h, k), focus $(h + p, k)$, and directrix $x = h - p$, where $a = 1/(4p)$.	$x = (y - 2)^2 - 1$ Opens right, vertex $(-1, 2)$, focus $(-3/4, 2)$, directrix $y = -5/4$

5.2 The Ellipse and the Circle

Ellipse: Geometric Definition	The set of points in the plane such that the sum of their distances from two fixed points (foci) is constant	
Horizontal Major Axis	If $a > b > 0$, the ellipse $\dfrac{x^2}{a^2} + \dfrac{y^2}{b^2} = 1$ has intercepts $(\pm a, 0)$ and $(0, \pm b)$ and foci $(\pm c, 0)$ where $c^2 = a^2 - b^2$.	$\dfrac{x^2}{25} + \dfrac{y^2}{9} = 1$ intercepts $(\pm 5, 0)$, $(0, \pm 3)$, foci $(\pm 4, 0)$
Vertical Major Axis	If $a > b > 0$, and ellipse $\dfrac{x^2}{b^2} + \dfrac{y^2}{a^2} = 1$ has intercepts $(\pm b, 0)$ and $(0, \pm a)$ and foci $(0, \pm c)$ where $c^2 = a^2 - b^2$.	$\dfrac{x^2}{9} + \dfrac{y^2}{25} = 1$ intercepts $(\pm 3, 0)$, $(0, \pm 5)$, foci $(0, \pm 4)$

Circle: Geometric Definition	The set of points in a plane such that their distance from a fixed point (the center) is constant (the radius)	
Centered at Origin	$x^2 + y^2 = r^2$ for $r > 0$ has center $(0, 0)$ and radius r.	$x^2 + y^2 = 9$ center $(0, 0)$, radius 3
Centered at (h, k)	$(x - h)^2 + (y - k)^2 = r^2$ has center (h, k) and radius r.	$(x - 2)^2 + (y + 3)^2 = 25$ center $(2, -3)$, radius 5

5.3 The Hyperbola

Hyperbola: Geometric Definition	The set of points in the plane such that the difference between the distances from two fixed points (foci) is constant	
Centered at Origin, Opening Left and Right	$\dfrac{x^2}{a^2} - \dfrac{y^2}{b^2} = 1$ opens left and right, x-intercepts $(\pm a, 0)$ and foci $(\pm c, 0)$ where $c^2 = a^2 + b^2$.	$\dfrac{x^2}{16} - \dfrac{y^2}{9} = 1$ x-intercepts $(\pm 4, 0)$, foci $(\pm 5, 0)$
Fundamental Rectangle	Goes through $(\pm a, 0)$ and $(0, \pm b)$, asymptotes $y = \pm (b/a)x$	Asymptotes $y = \pm \dfrac{3}{4} x$
Centered at Origin, Opening Up and Down	$\dfrac{y^2}{a^2} - \dfrac{x^2}{b^2} = 1$ opens up and down, intercepts $(0, \pm a)$ and foci $(0, \pm c)$ where $c^2 = a^2 + b^2$.	$\dfrac{y^2}{16} - \dfrac{x^2}{9} = 1$ y-intercepts $(\pm 4, 0)$, foci $(0, \pm 5)$
Fundamental Rectangle	Goes through $(0, \pm a)$ and $(\pm b, 0)$, asymptotes $y = \pm (a/b)x$	Asymptotes $y = \pm \dfrac{4}{3} x$
Centered at (h, k)	$\dfrac{(x - h)^2}{a^2} - \dfrac{(y - k)^2}{b^2} = 1$ opens left and right, center (h, k), foci $(h \pm c, k)$ where $c^2 = a^2 + b^2$.	$\dfrac{(x - 1)^2}{16} - \dfrac{(y + 3)^2}{9} = 1$ center $(1, -3)$, foci $(6, -3)$, $(-4, -3)$
	$\dfrac{(y - k)^2}{a^2} - \dfrac{(x - h)^2}{b^2} = 1$ opens up and down, center (h, k), foci $(h, k \pm c)$ where $c^2 = a^2 + b^2$.	$\dfrac{(y - 5)^2}{16} - \dfrac{(x + 2)^2}{9} = 1$ center $(-2, 5)$, foci $(-2, 0)$, $(-2, 10)$

5.4 Polar Coordinates

Polar Coordinates	If $r > 0$, then (r, θ) is r units from the origin on the terminal side of θ in standard position. If $r < 0$, then (r, θ) is $|r|$ units from the origin on the extension of the terminal side of θ.	$(2, \pi/4)$ $(-2, 5\pi/4)$
Converting	Polar to rectangular: $x = r \cos \theta$, $y = r \sin \theta$. Rectangular to polar: $r = \sqrt{x^2 + y^2}$ and the terminal side of θ contains (x, y).	Polar: $(2, \pi/4)$ Rectangular: $\left(\sqrt{2}, \sqrt{2} \right)$

5.5 Polar Equations of the Conics

Alternate Definition of Conics

A conic is the set of points in the plane such that the ratio of the distance from a point P to the focus to the distance from P to the directrix is a fixed positive number e. The conic is a parabola if $e = 1$, an ellipse if $e < 1$, and a hyperbola if $e > 1$.

Polar Equations with Focus at (0, 0)

The distance from the focus to directrix is p.

Vertical directrix: $r = \dfrac{ep}{1 \pm e \cos \theta}$

Horizontal directrix: $r = \dfrac{ep}{1 \pm e \sin \theta}$

Parabola $e = 1, p = 1$

$$r = \dfrac{1}{1 + \cos \theta}$$

Ellipse $e = \dfrac{1}{2}, p = 1$

$$r = \dfrac{1}{2 + \sin \theta}$$

Hyperbola $e = 2, p = 1$

$$r = \dfrac{2}{1 + 2 \cos \theta}$$

5.6 Parametric Equations

Parametric Equations

$x = f(t)$ and $y = g(t)$ where t is a parameter in some interval of real numbers.

$x = 2t, y = t^2$
for t in $(0, 5)$

Converting to Rectangular

Eliminate the parameter.

$y = (x/2)^2$
for x in $(0, 10)$

■■■ Chapter 5 Review Exercises

For Exercises 1–6, sketch the graph of each parabola. Determine the x- and y-intercepts, vertex, axis of symmetry, focus, and directrix for each.

1. $y = x^2 + 4x - 12$

2. $y = 4x - x^2$

3. $y = 6x - 2x^2$

4. $y = 2x^2 - 4x + 2$

5. $x = y^2 + 4y - 6$

6. $x = -y^2 + 6y - 9$

Sketch the graph of each ellipse, and determine its foci.

7. $\dfrac{x^2}{16} + \dfrac{y^2}{36} = 1$

8. $\dfrac{x^2}{64} + \dfrac{y^2}{16} = 1$

9. $\dfrac{(x - 1)^2}{8} + \dfrac{(y - 1)^2}{24} = 1$

10. $\dfrac{(x + 2)^2}{16} + \dfrac{(y - 1)^2}{7} = 1$

11. $5x^2 - 10x + 4y^2 + 24y = -1$

12. $16x^2 - 16x + 4y^2 + 4y = 59$

Determine the center and radius of each circle, and sketch its graph.

13. $x^2 + y^2 = 81$

14. $6x^2 + 6y^2 = 36$

15. $(x + 1)^2 + y^2 = 4$

16. $(x - 2)^2 + (y + 3)^2 = 9$

17. $x^2 + 5x + y^2 + \dfrac{1}{4} = 0$

18. $x^2 + 3x + y^2 + 5y = \dfrac{1}{2}$

Write the standard equation for each circle with the given center and radius.

19. Center $(0, -4)$, radius 3

20. Center $(-2, -5)$, radius 1

21. Center $(-2, -7)$, radius $\sqrt{6}$

22. Center $\left(\dfrac{1}{2}, -\dfrac{1}{4}\right)$, radius $\dfrac{\sqrt{2}}{2}$

Sketch the graph of each hyperbola. Determine the foci and the equations of the asymptotes.

23. $\dfrac{x^2}{64} - \dfrac{y^2}{36} = 1$ **24.** $\dfrac{y^2}{100} - \dfrac{x^2}{64} = 1$

25. $\dfrac{(y-2)^2}{64} - \dfrac{(x-4)^2}{16} = 1$

26. $\dfrac{(x-5)^2}{100} - \dfrac{(y-10)^2}{225} = 1$

27. $x^2 - 4x - 4y^2 + 32y = 64$

28. $y^2 - 6y - 4x^2 + 48x = 279$

Identify each equation as the equation of a parabola, ellipse, circle, or hyperbola. Try to do these problems without rewriting the equations.

29. $x^2 = y^2 + 1$ **30.** $x^2 + y^2 = 1$

31. $x^2 = 1 - 4y^2$ **32.** $x^2 + 4x + y^2 = 0$

33. $x^2 + y = 1$ **34.** $y^2 = 1 - x$

35. $x^2 + 4x = y^2$ **36.** $9x^2 + 7y^2 = 63$

Write each equation in standard form, then sketch the graph of the equation.

37. $x^2 = 4 - y^2$ **38.** $x^2 = 4y^2 + 4$

39. $x^2 = 4y + 4$ **40.** $y^2 = 4x - 4$

41. $x^2 = 4 - 4y^2$ **42.** $x^2 = 4y - y^2$

43. $4y^2 = 4x - x^2$ **44.** $x^2 - 4x = y^2 + 4y + 4$

Determine the equation of each conic section described below.

45. A parabola with focus $(1, 3)$ and directrix $x = 1/2$

46. A circle centered at the origin and passing through $(2, 8)$

47. An ellipse with foci $(\pm 4, 0)$ and vertices $(\pm 6, 0)$

48. A hyperbola with asymptotes $y = \pm 3x$ and y-intercepts $(0, \pm 3)$

49. A circle with center $(1, 3)$ and passing through $(-1, -1)$

50. An ellipse with foci $(3, 2)$ and $(-1, 2)$, and vertices $(5, 2)$ and $(-3, 2)$

51. A hyperbola with foci $(\pm 3, 0)$ and x-intercepts $(\pm 2, 0)$

52. A parabola with focus $(0, 3)$ and directrix $y = 1$

Assume that each of the following graphs is the graph of a parabola, ellipse, circle, or hyperbola. Find the equation for each graph.

53.
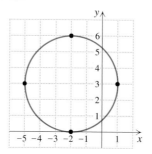

54.
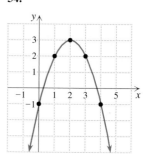

55.
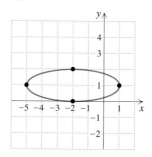

56.
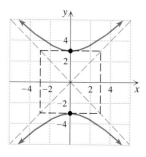

57.
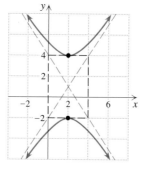

58.
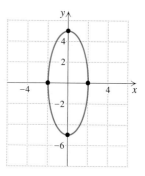

Convert the polar coordinates of each point to rectangular coordinates.

59. $(5, 60°)$ **60.** $(-4, 30°)$

61. $\left(\sqrt{3}, 100°\right)$ **62.** $\left(\sqrt{5}, 230°\right)$

Convert the rectangular coordinates of each point to polar coordinates. Use radians for θ.

63. $\left(-2, -2\sqrt{3}\right)$ **64.** $\left(-3\sqrt{2}, 3\sqrt{2}\right)$

65. $(2, -3)$ **66.** $(-4, -5)$

Sketch the graph of each polar equation.

67. $r = -2\sin\theta$ **68.** $r = 5\sin 3\theta$

69. $r = 2\cos 2\theta$ **70.** $r = 1.1 - \cos\theta$

71. $r = 500 + \cos\theta$ **72.** $r = 500$

73. $r = \dfrac{1}{\sin\theta}$ **74.** $r = \dfrac{-2}{\cos\theta}$

For each polar equation, write an equivalent rectangular equation.

75. $r = \dfrac{1}{\sin\theta + \cos\theta}$ **76.** $r = -6\cos\theta$

77. $r = -5$ **78.** $r = \dfrac{1}{1 + \sin\theta}$

For each rectangular equation, write an equivalent polar equation.

79. $y = 3$ **80.** $x^2 + (y + 1)^2 = 1$

81. $x^2 + y^2 = 49$ **82.** $2x + 3y = 6$

Identify each conic and sketch its graph. Give the equation of the directrix in rectangular coordinates.

83. $r = \dfrac{3}{1 + \sin\theta}$ **84.** $r = \dfrac{5}{1 + \cos\theta}$

85. $r = \dfrac{4}{2 + \cos\theta}$ **86.** $r = \dfrac{6}{3 + \sin\theta}$

87. $r = \dfrac{1}{3 - 6\cos\theta}$ **88.** $r = \dfrac{1}{4 - \sin\theta}$

Write the polar equation for a conic with focus at the origin and the given eccentricity and directrix.

89. $e = 1, y = 3$ **90.** $e = 1, y = -4$

91. $e = 3, x = -6$ **92.** $e = 4, x = 8$

93. $e = \dfrac{1}{3}, y = 9$ **94.** $e = \dfrac{1}{4}, y = -12$

Sketch the graph of each pair of parametric equations.

95. $x = 3t, y = 3 - t$, for t in $(0, 1)$

96. $x = t - 3, y = t^2$, for t in $(-\infty, \infty)$

97. $x = -\sin t, y = -\cos t$, for t in $[0, \pi/2]$

98. $x = -\cos t, y = \sin t$, for t in $[0, \pi]$

Set each problem.

99. *Nuclear Power*   A cooling tower for a nuclear power plant has a hyperbolic cross section, as shown in the accompanying figure. The diameter of the tower at the top and bottom is 240 ft, while the diameter at the middle is 200 ft. The height of the tower is $48\sqrt{11}$ ft. Find the equation of the hyperbola, using the coordinate system shown in the figure.

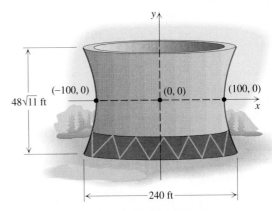

■ Figure for Exercise 99

100. *Searchlight*   The bulb in a searchlight is positioned 10 in. above the vertex of its parabolic reflector, as shown in the accompanying figure. The width of the reflector is 30 in. Using the coordinate system given in the figure, find the equation of the parabola and the thickness t of the reflector at its outside edge.

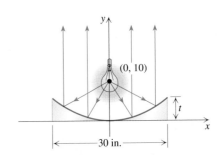

■ Figure for Exercise 100

101. *Whispering Gallery*   In the whispering gallery shown in the figure on the next page, the foci of the ellipse are 60 ft apart. Each focus is 4 ft from the vertex of an elliptical reflector. Using the coordinate system given in the figure, find the equation of the ellipse that is used to make the elliptical reflectors and the dimension marked h in the figure.

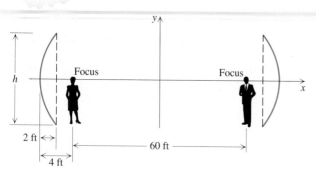

Figure for Exercise 101

102. A lithotripter is used to disintegrate kidney stones by bombarding them with high-energy shock waves generated at one focus of an elliptical reflector. The lithotripter is positioned so that the kidney stone is at the other focus of the reflector. If the equation $81x^2 + 225y^2 = 18{,}225$ is used for the cross section of the elliptical reflector, with centimeters as the unit of measurement, then how far from the point of generation will the waves be focused?

Thinking Outside the Box XL

Falling Painter   A painter has her brush positioned a ft from the top of the ladder and b ft from the bottom as shown in the figure. Unfortunately, the ladder was placed on a frictionless surface and the bottom starts sliding away from the wall on which the ladder is leaning. As the painter falls to the ground, she keeps the brush in the same position so that it paints an arc of a curve on the adjacent wall.
a. What kind of curve is it?

b. Find an equation for the curve in an appropriate coordinate system.

Figure for Thinking Outside the Box XL

Concepts of Calculus

The reflection property of a parabola

Parabolic antennas are used to gather light waves, radar waves, and sound waves to a focus where they are amplified. But how does a ray of light reflect off a parabola and pass through its focus? If a light ray is reflected off a flat mirror, the angle of incidence is equal to the angle of reflection as shown in Fig 5.80. Pool players use that same principle when hitting a ball off the rail on a pool table. The principle is not quite as simple for a curve. For a curve, the angle of incidence and the angle of reflection are the angles between the ray and the tangent line to the curve as shown in Fig. 5.81. We generally use techniques from calculus to find the equation of a tangent line, but for a parabola we can find it using only algebra. In the following exercises we will show that the parabola $y = x^2$ really does have a reflecting property that sends incoming rays to its focus.

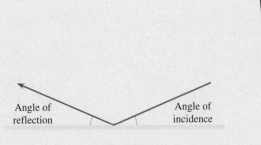

Figure 5.80

Angle of incidence

Angle of reflection

Tangent line

Figure 5.81

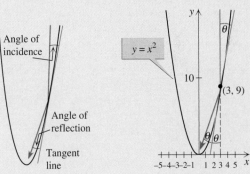

Figure 5.82

Exercises

1. The tangent line to $y = x^2$ is a line $y = mx + b$ that intersects $y = x^2$ exactly once. By substitution $x^2 = mx + b$. Show that this quadratic has exactly one solution only if $b = -\frac{1}{4}m^2$.

2. Show that the slope of the tangent line at a point (x, y) on $y = x^2$ is twice the x-coordinate by solving $x^2 = mx - \frac{1}{4}m^2$ for m.

3. Find the equation of the tangent line to $y = x^2$ at $(3, 9)$.

4. Let θ be the angle of incidence for the light ray traveling parallel to the y-axis and heading to $(3, 9)$ as shown in Fig. 5.82. Find $\tan \theta$.

5. Since that angle of reflection is also θ, the light ray has changed course from its original vertical path by the angle 2θ. Use a double-angle identity to find $\tan 2\theta$.

6. Now find $\tan 2\theta$ by using a right triangle for which the endpoints of the hypotenuse are $(3, 9)$ and the focus of $y = x^2$. If this answer agrees with $\tan 2\theta$ found by the identity, then the light ray passes through the focus.

7. Repeat Exercises 3 through 6 using an arbitrary point (w, w^2) on $y = x^2$ to show that a vertical ray reflects off any point on $y = x^2$ and then passes through the focus.

APPENDIX

A Basic Algebra Review

A.1 Exponents and Radicals

In this section we review the rules for exponential and radical expressions.

Exponential Expressions

We use positive integral exponents to indicate the number of times a number occurs in a product. For example, $2 \cdot 2 \cdot 2 \cdot 2$ is written as 2^4. We read 2^4 as "the fourth power of 2" or "2 to the fourth power."

Definition: Positive Integral Exponents

For any positive integer n:

$$a^n = \underbrace{a \cdot a \cdot a \cdot \cdots \cdot a}_{n \text{ factors of } a}$$

We call a the **base**, n the **exponent** or **power**, and a^n an **exponential expression.**

We read a^n as "a to the nth power." For a^1 we usually omit the exponent and just write a. We refer to the exponents 2 and 3 as squares and cubes. For example, 3^2 is read "3 squared," 2^3 is read "2 cubed," x^4 is read "x to the fourth," b^5 is read "b to the fifth," and so on. To evaluate an expression such as -3^2 we square 3 first, then take the opposite. So $-3^2 = -9$ and $(-3)^2 = (-3)(-3) = 9$.

Example **1** Evaluating exponential expressions

Evaluate.

a. 4^3 **b.** $(-2)^4$ **c.** -2^4

Solution

a. $4^3 = 4 \cdot 4 \cdot 4 = 16 \cdot 4 = 64$
b. $(-2)^4 = (-2)(-2)(-2)(-2) = 16$
c. $-2^4 = -(2 \cdot 2 \cdot 2 \cdot 2) = -16$ ∎

Negative Integral Exponents

We use a negative sign in an exponent to represent multiplicative inverses or reciprocals. For negative exponents we do not allow the base to be zero because zero does not have a reciprocal.

AA-1

Definition: Negative Integral Exponents

If a is a nonzero real number and n is a positive integer, then $a^{-n} = \dfrac{1}{a^n}$.

Example **2** Evaluating expressions that have negative exponents

Simplify each expression without using a calculator, then check with a calculator.

a. $4 \cdot 2^{-3}$ **b.** $\left(\dfrac{2}{3}\right)^{-3}$ **c.** $\dfrac{6^{-2}}{2^{-3}}$

Solution

a. $4 \cdot 2^{-3} = 4 \cdot \dfrac{1}{2^3} = 4 \cdot \dfrac{1}{8} = \dfrac{1}{2}$

b. $\left(\dfrac{2}{3}\right)^{-3} = \dfrac{1}{\left(\dfrac{2}{3}\right)^3} = \dfrac{1}{\dfrac{2}{3} \cdot \dfrac{2}{3} \cdot \dfrac{2}{3}} = \dfrac{1}{\dfrac{8}{27}} = \dfrac{27}{8}$ Note that $\left(\dfrac{2}{3}\right)^{-3} = \left(\dfrac{3}{2}\right)^3$.

c. $\dfrac{6^{-2}}{2^{-3}} = \dfrac{\dfrac{1}{6^2}}{\dfrac{1}{2^3}} = \dfrac{1}{6^2} \cdot \dfrac{2^3}{1} = \dfrac{8}{36} = \dfrac{2}{9}$ Note that $\dfrac{6^{-2}}{2^{-3}} = \dfrac{2^3}{6^2}$. ■

Example 2(b) illustrates the fact that a fractional base can be inverted, if the sign of the exponent is changed. Example 2(c) illustrates the fact that a factor of the numerator or denominator can be moved from the numerator to the denominator or vice versa as long as we change the sign of the exponent.

Rules for Negative Exponents and Fractions

If a and b are nonzero real numbers and m and n are integers, then

$$\left(\dfrac{a}{b}\right)^{-m} = \left(\dfrac{b}{a}\right)^{m} \quad \text{and} \quad \dfrac{a^{-m}}{b^{-n}} = \dfrac{b^n}{a^m}.$$

Be careful with the rules of exponents when sums or differences are involved. For example, $\dfrac{2^{-1} - 1}{3^{-1}} \neq \dfrac{3^1 - 1}{2^1}$ because

$$\dfrac{2^{-1} - 1}{3^{-1}} = \dfrac{\dfrac{1}{2} - 1}{\dfrac{1}{3}} = -\dfrac{3}{2} \quad \text{and} \quad \dfrac{3^1 - 1}{2^1} = 1.$$

Rules of Exponents

So far we have defined positive and negative integral exponents. Zero as an exponent is defined as follows.

Definition: Zero Exponent

If a is a nonzero real number, then $a^0 = 1$.

Using the definitions of positive, negative, and zero exponents, we can show that the following rules hold for any integral exponents.

Rules for Integral Exponents

If a and b are nonzero real numbers and m and n are integers, then

1. $a^m a^n = a^{m+n}$ **Product Rule**
2. $\dfrac{a^m}{a^n} = a^{m-n}$ **Quotient Rule**
3. $(a^m)^n = a^{mn}$ **Power of a Power Rule**
4. $(ab)^n = a^n b^n$ **Power of a Product Rule**
5. $\left(\dfrac{a}{b}\right)^n = \dfrac{a^n}{b^n}$ **Power of a Quotient Rule**

The rules for integral exponents are used to simplify expressions in the next example.

Example **3** **Simplifying expressions with integral exponents**

Simplify each expression. Write your answer without negative exponents. Assume that all variables represent nonzero real numbers.

a. $(3x^2)^3(-2x^{-2})$ **b.** $\left(\dfrac{a^5 b^{-1}}{a^7}\right)^4$

Solution

a. $(3x^2)^3(-2x^{-2}) = 3^3(x^2)^3(-2x^{-2})$ Power of a product rule

$\phantom{(3x^2)^3(-2x^{-2})} = 27x^6(-2x^{-2})$ Power of a power rule

$\phantom{(3x^2)^3(-2x^{-2})} = -54x^4$ Product rule

b. $\left(\dfrac{a^5 b^{-1}}{a^7}\right)^4 = \dfrac{(a^5 b^{-1})^4}{(a^7)^4}$ Power of a quotient rule

$\phantom{\left(\dfrac{a^5 b^{-1}}{a^7}\right)^4} = \dfrac{(a^5)^4(b^{-1})^4}{(a^7)^4}$ Power of a product rule

$\phantom{\left(\dfrac{a^5 b^{-1}}{a^7}\right)^4} = \dfrac{a^{20} b^{-4}}{a^{28}}$ Power of a power rule

$\phantom{\left(\dfrac{a^5 b^{-1}}{a^7}\right)^4} = a^{-8} b^{-4}$ Quotient rule

$\phantom{\left(\dfrac{a^5 b^{-1}}{a^7}\right)^4} = \dfrac{1}{a^8 b^4}$ Definition of negative exponent ■

Roots

Since $2^4 = 16$ and $(-2)^4 = 16$, both 2 and -2 are fourth roots of 16. The nth root of a number is defined in terms of the nth power.

Definition: *n*th Roots

> If n is a positive integer and $a^n = b$, then a is called an ***n*th root** of b.
> If $a^2 = b$, then a is a **square root** of b. If $a^3 = b$, then a is the **cube root** of b.

If n is even and a is an nth root of b, then a is called an **even root** of b. If n is odd and a is an nth root of b, then a is called an **odd root** of b. Every positive real number has *two* real even roots, a positive root and a negative root. For example, both 5 and -5 are square roots of 25 because $5^2 = 25$ and $(-5)^2 = 25$. Moreover, every real number has exactly *one* real odd root. For example, because $2^3 = 8$ and 3 is odd, 2 is the only real cube root of 8. Because $(-2)^3 = -8$ and 3 is odd, -2 is the only real cube root of -8.

Finding an nth root is the reverse of finding an nth power, so we use the notation $a^{1/n}$ for the nth root of a. For example, since the positive square root of 25 is 5, we write $25^{1/2} = 5$.

Definition: Exponent 1/*n*

> If n is a positive even integer and a is positive, then $a^{1/n}$ denotes the **positive real nth root of a** and is called the **principal nth root of a**.
>
> If n is a positive odd integer and a is any real number, then $a^{1/n}$ denotes the real nth root of a.
>
> If n is a positive integer, then $0^{1/n} = 0$.

Example **4** **Evaluating expressions involving exponent 1/*n***

Evaluate each expression.

a. $4^{1/2}$ **b.** $(-8)^{1/3}$ **c.** $(-4)^{1/2}$

Solution

a. Because the positive real square root of 4 is 2, $4^{1/2} = 2$.
b. Because the real cube root of -8 is -2, $(-8)^{1/3} = -2$.
c. Since the definition of nth root does not include an even root of a negative number, $(-4)^{1/2}$ has not yet been defined. Even roots of negative numbers do exist in the complex number system, but an even root of a negative number is not a real number. ■

Rational Exponents

A rational exponent indicates both a root and a power. The expression $a^{m/n}$ is defined as the mth power of the nth root of a.

Definition: Rational Exponents

> If m and n are positive integers, then
> $$a^{m/n} = (a^{1/n})^m$$
> provided that $a^{1/n}$ is a real number.

Note that $a^{1/n}$ is not real when a is negative and n is even. According to the definition of rational exponents, expressions such as $(-25)^{-3/2}$, $(-43)^{1/4}$, and $(-1)^{2/2}$ are not defined because each of them involves an even root of a negative number. Note that some authors define $a^{m/n}$ only for m/n in lowest terms. In that case the fourth power of the square root of three could *not* be written as $3^{4/2}$. This author prefers the more general definition given above.

A negative rational exponent indicates a reciprocal just as a negative integral exponent does. So $7^{-2/3} = 1/7^{2/3}$. The root or the power indicated by a rational exponent can be evaluated in either order. That is, $(a^{1/n})^m = (a^m)^{1/n}$ provided $a^{1/n}$ is real. However, for mental evaluation the following order is best.

Procedure: Evaluating $a^{-m/n}$

To evaluate $a^{-m/n}$ mentally,

1. find the nth root of a,   **2.** raise it to the m power,   **3.** find the reciprocal.

Example **5** **Evaluating expressions with rational exponents**

Evaluate each expression.

a. $(-8)^{2/3}$ **b.** $27^{-2/3}$ **c.** $100^{6/4}$

Solution

a. Mentally, the cube root of -8 is -2 and the square of -2 is 4. In symbols:

$$(-8)^{2/3} = ((-8)^{1/3})^2 = (-2)^2 = 4$$

b. Mentally, the cube root of 27 is 3, the square of 3 is 9, and the reciprocal of 9 is $\frac{1}{9}$. In symbols:

$$27^{-2/3} = \frac{1}{(27^{1/3})^2} = \frac{1}{3^2} = \frac{1}{9}$$

c. $100^{6/4} = 100^{3/2} = 10^3 = 1000$ ■

Note how we reduced the exponent in Example 5(c). However, exponents can be reduced only on expressions that are defined. For example, $(-1)^{2/2} \neq (-1)^1$ because $(-1)^{2/2}$ is an undefined expression.

The rules for integral exponents stated previously in this section also hold for rational exponents. Note that the power of a power rule can fail if the base is negative. For example, $(x^2)^{1/2} = x$ according to the rule. However, if x is a negative number the left side of this equation is positive and the right side is negative. So $(x^2)^{1/2} = x$ is only correct if x is nonnegative. The equation $(x^2)^{1/2} = |x|$ is correct for any x.

Example **6** **Simplifying expressions with rational exponents**

Use the rules of exponents to simplify each expression. Assume that the variables represent positive real numbers. Write answers without negative exponents.

a. $x^{2/3}x^{4/3}$ **b.** $(x^4y^{1/2})^{1/4}$ **c.** $\left(\dfrac{a^{3/2}b^{2/3}}{a^2}\right)^3$

Solution

a. $x^{2/3}x^{4/3} = x^{6/3}$ Product rule

$= x^2$ Simplify the exponent.

b. $(x^4y^{1/2})^{1/4} = (x^4)^{1/4}(y^{1/2})^{1/4}$ Power of a product rule

$= xy^{1/8}$ Power of a power rule

c. $\left(\dfrac{a^{3/2}b^{2/3}}{a^2}\right)^3 = \dfrac{(a^{3/2})^3(b^{2/3})^3}{(a^2)^3}$ Power of a quotient rule

$= \dfrac{a^{9/2}b^2}{a^6}$ Power of power rule

$= a^{-3/2}b^2$ Quotient rule $\left(\dfrac{9}{2} - 6 = -\dfrac{3}{2}\right)$

$= \dfrac{b^2}{a^{3/2}}$ Definition of negative exponents   ∎

Radical Notation

The exponent $1/n$ and the **radical sign** $\sqrt[n]{}$ are both used to indicate the nth root.

Definition:
Radical Notation

> If n is a positive integer and a is a number for which $a^{1/n}$ is defined, then the expression $\sqrt[n]{a}$ is called a **radical,** and
> $$\sqrt[n]{a} = a^{1/n}.$$
> If $n = 2$, we write $\sqrt{a}$ rather than $\sqrt[2]{a}$.

The number a is called the **radicand** and n is the **index** of the radical. Expressions such as $\sqrt{-3}$, $\sqrt[4]{-81}$, and $\sqrt[6]{-1}$ do not represent real numbers because each is an even root of a negative number.

Example **7** Evaluating radicals

Evaluate each expression.

a. $\sqrt{49}$ **b.** $\sqrt[3]{-1000}$ **c.** $\sqrt[4]{\dfrac{16}{81}}$ **d.** $\sqrt[3]{125^2}$

Solution

a. The symbol $\sqrt{49}$ indicates the positive square root of 49. So $\sqrt{49} = 49^{1/2} = 7$. Writing $\sqrt{49} = \pm 7$ is incorrect.

b. $\sqrt[3]{-1000} = (-1000)^{1/3} = -10$ Check that

c. $\sqrt[4]{\dfrac{16}{81}} = \left(\dfrac{16}{81}\right)^{1/4} = \dfrac{2}{3}$ Check that $\left(\dfrac{2}{3}\right)^4 = \dfrac{16}{81}$.

d. $\sqrt[3]{125^2} = (125^2)^{1/3} = 125^{2/3} = 5^2 = 25$ ∎

Since $a^{1/n} = \sqrt[n]{a}$, expressions involving rational exponents can be written with radicals.

Rule: Converting $a^{m/n}$ to Radical Notation

If a is a real number and m and n are integers for which $\sqrt[n]{a}$ is real, then
$$a^{m/n} = (\sqrt[n]{a})^m = \sqrt[n]{a^m}.$$

Example **8** **Converting rational exponents to radicals**

Write each expression in radical notation. Assume that all variables represent positive real numbers. Simplify the radicand if possible.

a. $2^{2/3}$ **b.** $(3x)^{3/4}$ **c.** $2(x^2 + 3)^{-1/2}$

Solution

a. $2^{2/3} = \sqrt[3]{2^2} = \sqrt[3]{4}$

b. $(3x)^{3/4} = \sqrt[4]{(3x)^3} = \sqrt[4]{27x^3}$

c. $2(x^2 + 3)^{-1/2} = 2 \cdot \dfrac{1}{(x^2 + 3)^{1/2}} = \dfrac{2}{\sqrt{x^2 + 3}}$ ■

The Product and Quotient Rules for Radicals

Using the power of a product and the power of a quotient rules for rational exponents we can write

$$(ab)^{1/n} = a^{1/n}b^{1/n} \quad \text{and} \quad \left(\frac{a}{b}\right)^{1/n} = \frac{a^{1/n}}{b^{1/n}}.$$

These rules are expressed in radical notation as follows.

Rules: Product and Quotient Rules for Radicals

For any positive integer n and real numbers a and b ($b \neq 0$),

1. $\sqrt[n]{ab} = \sqrt[n]{a} \cdot \sqrt[n]{b}$ **Product Rule for Radicals**

2. $\sqrt[n]{\dfrac{a}{b}} = \dfrac{\sqrt[n]{a}}{\sqrt[n]{b}}$ **Quotient Rule for Radicals**

provided that all of the roots are real.

In words, the nth root of a product (or quotient) is the product (or quotient) of the nth roots.

An expression that is the square of a term that is free of radicals is called a **perfect square.** For example, $9x^6$ is a perfect square because $9x^6 = (3x^3)^2$. Likewise, $27y^{12}$ is a **perfect cube.** In general, an expression that is the nth power of an expression free of radicals is a **perfect nth power.** In the next example, the product and quotient rules for radicals are used to simplify radicals containing perfect squares, cubes, and so on.

Example **9** **Using the product and quotient rules for radicals**

Simplify each radical expression. Assume that all variables represent positive real numbers.

a. $\sqrt[3]{125a^6}$ **b.** $\sqrt{\dfrac{3}{16x^2}}$

Solution

a. Both 125 and a^6 are perfect cubes. So use the product rule to simplify:

$$\sqrt[3]{125a^6} = \sqrt[3]{125} \cdot \sqrt[3]{a^6} = 5a^2 \quad \text{Since } \sqrt[3]{a^6} = a^{6/3} = a^2$$

b. Since 16 is a perfect square, use the quotient rule to simplify the radical:

$$\sqrt{\dfrac{3}{16x^2}} = \dfrac{\sqrt{3}}{\sqrt{16x^2}} = \dfrac{\sqrt{3}}{4x}$$

■

Simplified Form and Rationalizing the Denominator

We have been simplifying radical expressions by just making them look simpler. However, a radical expression is in *simplified form* only if it satisfies the following three specific conditions.

Definition: Simplified Form for Radicals of Index *n*

A radical of index *n* in **simplified form** has

1. *no* perfect *n*th powers as factors of the radicand,
2. *no* fractions inside the radical, and
3. *no* radicals in a denominator.

The process of removing radicals from a denominator is called **rationalizing the denominator.**

Example **10** **Simplified form of a radical expression**

Write each radical expression in simplified form. Assume that all variables represent positive real numbers.

a. $\sqrt{20x^8y^9}$ **b.** $\dfrac{9}{\sqrt{3}}$ **c.** $\sqrt[3]{\dfrac{3}{5a^4}}$

Solution

a. Use the product rule to factor the radical, putting all perfect squares in the first factor:

$$\sqrt{20x^8y^9} = \sqrt{4x^8y^8}\,\sqrt{5y} \qquad \text{Product rule}$$

$$= 2x^4y^4\,\sqrt{5y} \qquad \text{Simplify the first radical.}$$

b. Since $\sqrt{3}$ appears in the denominator, we multiply the numerator and denominator by $\sqrt{3}$ to rationalize the denominator:

$$\frac{9}{\sqrt{3}} = \frac{9\sqrt{3}}{\sqrt{3}\sqrt{3}} = \frac{9\sqrt{3}}{3} = 3\sqrt{3}$$

c. To rationalize this denominator, we must get a perfect cube in the denominator. The radicand $5a^4$ can be made into the perfect cube $125a^6$ by multiplying by $25a^2$.

$$\sqrt[3]{\frac{3}{5a^4}} = \frac{\sqrt[3]{3}}{\sqrt[3]{5a^4}} \qquad \text{Quotient rule for radicals}$$

$$= \frac{\sqrt[3]{3}\,\sqrt[3]{25a^2}}{\sqrt[3]{5a^4}\sqrt[3]{25a^2}} \qquad \text{Multiply numerator and denominator by } \sqrt[3]{25a^2}.$$

$$= \frac{\sqrt[3]{75a^2}}{\sqrt[3]{125a^6}} \qquad \text{Product rule for radicals}$$

$$= \frac{\sqrt[3]{75a^2}}{5a^2} \qquad \text{Since } (5a^2)^3 = 125a^6 \qquad ■$$

Operations with Radical Expressions

We can use the properties of radicals to add, subtract, multiply, and divide radical expressions with the same index.

Example **11** **Operations with radicals of the same index**

Perform each operation and write the answer in simplified form for radicals. Assume that each variable represents a positive real number.

a. $\sqrt{20} + \sqrt{5}$ **b.** $\sqrt[4]{4y^3}\,\sqrt[4]{12y^2}$ **c.** $\sqrt{40} \div \sqrt{5}$

Solution

a. $\sqrt{20} + \sqrt{5} = \sqrt{4}\sqrt{5} + \sqrt{5}$ Product rule for radicals

$\qquad\qquad\qquad = 2\sqrt{5} + \sqrt{5}$ Simplify.

$\qquad\qquad\qquad = (2 + 1)\sqrt{5}$ Distributive property

$\qquad\qquad\qquad = 3\sqrt{5}$ Simplify.

b. $\sqrt[4]{4y^3}\,\sqrt[4]{12y^2} = \sqrt[4]{48y^5}$ Product rule for radicals

$\qquad\qquad\qquad = \sqrt[4]{16y^4}\,\sqrt[4]{3y}$ Factor out the perfect fourth powers.

$\qquad\qquad\qquad = 2y\sqrt[4]{3y}$ Simplify.

c. $\sqrt{40} \div \sqrt{5} = \sqrt{\dfrac{40}{5}}$ Quotient rule for radicals

$\qquad\qquad\qquad = \sqrt{8}$ Divide.

$\qquad\qquad\qquad = \sqrt{4}\sqrt{2}$ Product rule for radicals

$\qquad\qquad\qquad = 2\sqrt{2}$ Simplify.   ■

Radical expressions that can be added or subtracted using the distributive property as in Example 11(a) are called **like radicals.** Note that it is not necessary to write out the distributive step. We can simply write $2\sqrt{5} + \sqrt{5} = 3\sqrt{5}$ just like we write $2x + x = 3x$.

A.1 Exercises

Evaluate each expression.

1. 4^3

2. 3^4

3. -7^2

4. -9^2

5. $(-4)^2$

6. $(-10)^4$

7. 3^{-4}

8. $\dfrac{1}{2^{-3}}$

9. $6^{-1} + 5^{-1}$

10. $2^0 + 2^{-1}$

11. $\dfrac{3^{-2}}{6^{-3}}$

12. $\dfrac{3^{-1}}{2^3}$

13. $\left(\dfrac{1}{2}\right)^{-3}$

14. $\left(-\dfrac{1}{10}\right)^{-4}$

Simplify each expression. Write answers without negative exponents. Assume that all variables represent nonzero real numbers.

15. $(-3x^2y^3)(2x^9y^8)$

16. $(-6a^7b^4)(3a^3b^5)$

17. $y^3y^2 + 2y^4y$

18. $x^2x^5 + x^3x^4$

19. $-1(2x^3)^2$

20. $(-3y^{-1})^{-1}$

21. $\left(\dfrac{-2x^2}{3}\right)^3$

22. $\left(\dfrac{-1}{2a}\right)^{-2}$

23. $\dfrac{6x^7}{2x^3}$

24. $\dfrac{-9x^2y}{3xy^2}$

25. $\left(\dfrac{y^2}{5}\right)^{-2}$

26. $\left(-\dfrac{y^2}{2a}\right)^4$

27. $\left(\dfrac{1}{2}x^{-4}y^3\right)\left(\dfrac{1}{3}x^4y^{-6}\right)$

28. $\left(\dfrac{1}{3}a^{-5}b\right)(a^4b^{-1})$

29. $\left(\dfrac{-3m^{-1}n}{-6m^{-1}n^{-1}}\right)^2$

30. $\left(\dfrac{-p^{-1}q^{-1}}{-3pq^{-3}}\right)^{-2}$

Evaluate each expression.

31. $-9^{1/2}$

32. $27^{1/3}$

33. $64^{1/2}$

34. $-144^{1/2}$

35. $(-64)^{1/3}$

36. $81^{1/4}$

37. $(-27)^{4/3}$

38. $125^{-2/3}$

39. $8^{-4/3}$

40. $4^{-3/2}$

Simplify each expression. Assume that all variables represent positive real numbers. Write your answers without negative exponents.

41. $(x^4y)^{1/2}$

42. $(a^{1/2}b^{1/3})^2$

43. $(2a^{1/2})(3a)$

44. $(-3y^{1/3})(-2y^{1/2})$

45. $\dfrac{6a^{1/2}}{2a^{1/3}}$

46. $\dfrac{-4y}{2y^{2/3}}$

47. $(a^2b^{1/2})(a^{1/3}b^{1/2})$

48. $(4^{3/4}a^2b^3)(4^{3/4}a^{-2}b^{-5})$

49. $\left(\dfrac{x^6y^3}{z^9}\right)^{1/3}$

50. $\left(\dfrac{x^{1/2}y}{y^{1/2}}\right)^3$

Evaluate each radical expression.

51. $\sqrt{900}$　　**52.** $\sqrt{400}$　　**53.** $\sqrt[3]{-8}$　　**54.** $\sqrt[3]{64}$

55. $\sqrt[3]{-\dfrac{8}{1000}}$　**56.** $\sqrt[4]{\dfrac{1}{625}}$　**57.** $\sqrt[4]{16^3}$　　**58.** $\sqrt[3]{8^5}$

Write each expression involving rational exponents in radical notation, and each expression involving radicals in exponential notation.

59. $10^{2/3}$　　　　**60.** $-2^{3/4}$　　　　**61.** $3y^{-3/5}$

62. $a(b^4 + 1)^{-1/2}$　**63.** $\dfrac{1}{\sqrt{x}}$　　**64.** $-4\sqrt{x^3}$

65. $\sqrt[5]{x^3}$　　　　**66.** $\sqrt[3]{x^3 + y^3}$

Simplify each radical expression. Assume that all variables represent positive real numbers.

67. $\sqrt{16x^2}$　**68.** $\sqrt{121y^4}$　**69.** $\sqrt[3]{8y^9}$　**70.** $\sqrt[3]{125x^{18}}$

71. $\sqrt{\dfrac{xy}{100}}$　**72.** $\sqrt{\dfrac{t}{81}}$　**73.** $\sqrt[3]{\dfrac{-8a^3}{b^{15}}}$　**74.** $\sqrt[4]{\dfrac{16t^4}{y^8}}$

Write each radical expression in simplified form. Assume that all variables represent positive real numbers.

75. $\sqrt{28}$　**76.** $\sqrt{50}$　**77.** $\dfrac{1}{\sqrt{5}}$　**78.** $\dfrac{7}{\sqrt{7}}$

79. $\sqrt{\dfrac{x}{8}}$ **80.** $\sqrt{\dfrac{3y}{20}}$ **81.** $\sqrt[3]{40}$ **82.** $\sqrt[3]{54}$

83. $\sqrt[3]{-250x^4}$ **84.** $\sqrt[3]{-24a^5}$ **85.** $\sqrt[3]{\dfrac{1}{2}}$ **86.** $\sqrt[3]{\dfrac{3x}{25}}$

Perform the indicated operations and simplify your answer. Assume that all variables represent positive real numbers.

87. $3\sqrt{6} + 9 - 5\sqrt{6}$

88. $3\sqrt{2} + 8 - 5\sqrt{2}$

89. $\sqrt{8} + \sqrt{20} - \sqrt{12}$

90. $\sqrt{18} - \sqrt{50} + \sqrt{12} - \sqrt{75}$

91. $\left(-2\sqrt{3}\right)\left(5\sqrt{6}\right)$

92. $\left(-3\sqrt{2}\right)\left(-2\sqrt{3}\right)$

93. $\left(3\sqrt{5a}\right)\left(4\sqrt{5a}\right)$

94. $\left(-2\sqrt{6}\right)\left(3\sqrt{6}\right)$

95. $\sqrt{18a} \div \sqrt{2a^4}$ **96.** $\sqrt{21x^7} \div \sqrt{3x^2}$

97. $\sqrt{20x^3} + \sqrt{45x^3}$ **98.** $\sqrt[3]{16a^4} + \sqrt[3]{54a^4}$

A.2   Polynomials

In this section we will review some basic facts about polynomials.

Definitions

A **term** is the product of a number and one or more variables raised to powers. A **polynomial** is simply a single term or a finite sum of terms in which the powers of the variables are whole numbers. A polynomial in one variable is defined as follows.

Definition: Polynomial in One Variable *x*

> If n is a nonnegative integer and $a_0, a_1, a_2, \ldots, a_n$ are real numbers, then
> $$a_n x^n + a_{n-1} x^{n-1} + a_{n-2} x^{n-2} + \cdots + a_1 x + a_0$$
> is a **polynomial** in one variable x.

In algebra a single number is often referred to as a **constant.** The last term a_0 is called the **constant term.**

Polynomials with one, two, and three terms are called **monomials, binomials, and trinomials,** respectively. We usually write the terms of a polynomial in a single variable so that the exponents are in descending order from left to right. When a polynomial is written in this manner, the coefficient of the first term is the **leading coefficient.** The **degree** of a polynomial in one variable is the highest power of the variable in the polynomial. A constant such as 5 is a monomial with zero degree because $5 = 5x^0$. First-, second-, and third-degree polynomials are called **linear, quadratic,** and **cubic polynomials,** respectively.

Example **1** **Using the definitions**

Find the degree and leading coefficient of each polynomial and determine whether the polynomial is a monomial, binomial, or trinomial.

a. $\dfrac{x^3}{2} - \dfrac{1}{8}$ **b.** $5y^2 + y - 9$ **c.** $3w$

Solution

Polynomial (a) is a third-degree binomial, (b) is a second-degree trinomial, and (c) is a first-degree monomial. The leading coefficients are $1/2$, 5, and 3, respectively. We can also describe (a) as a cubic polynomial, (b) as a quadratic polynomial, and (c) as a linear polynomial. ■

Naming and Evaluating Polynomials

Polynomials are often used to model quantities such as profit, revenue, and cost. A profit polynomial might be named P. For example, if the expression $3x - 10$ is a profit polynomial, we write $P = 3x - 10$ or $P(x) = 3x - 10$. P and $P(x)$ (read "P of x") both represent the profit when x units are sold. If $x = 6$, then the value of the polynomial is $3 \cdot 6 - 10$ or 8. If $x = 7$ the value is 11. Using the $P(x)$ notation we write $P(6) = 8$ and $P(7) = 11$. With the $P(x)$ notation it is clear that the profit for 6 units is 8 and the profit for 7 units is 11. The $P(x)$ notation is called **function notation.** Functions are discussed in great detail in Chapter 1.

Example **2** **Evaluating a polynomial**

Let $P(x) = x^2 - 5$ and $C(x) = -x^3 + 5x - 3$. Find the following.

a. $P(3)$ **b.** $C(10)$

Solution

a. $P(3) = 3^2 - 5 = 4$
b. $C(10) = -10^3 + 5(10) - 3 = -1000 + 50 - 3 = -953$ ■

Addition and Subtraction of Polynomials

We add or subtract polynomials by adding or subtracting the like terms. You can arrange the work horizontally, as in Example 3, or vertically, as in Example 4.

Example **3** **Adding and subtracting polynomials horizontally**

Find each sum or difference.

a. $(3x^3 - x + 5) + (-8x^3 + 3x - 9)$ **b.** $(x^2 - 5x) - (3x^2 - 4x - 1)$

Solution

a. We use the commutative and associative properties of addition to rearrange the terms.

$$(3x^3 - x + 5) + (-8x^3 + 3x - 9) = (3x^3 - 8x^3) + (-x + 3x) + (5 - 9)$$

$$= -5x^3 + 2x - 4 \quad \text{Combine like terms.}$$

b. The first step is to distribute the multiplication by -1 over the three terms of the second polynomial, changing the sign of every term.

$$(x^2 - 5x) - (3x^2 - 4x - 1) = x^2 - 5x - 3x^2 + 4x + 1 \quad \text{Distributive property}$$

$$= -2x^2 - x + 1 \quad \text{Combine like terms.}$$

■

Example **4** Adding and subtracting
polynomials vertically

Find each sum or difference.

a. $(3x^3 - x + 5) + (-8x^3 + 3x - 9)$ **b.** $(x^2 - 5x) - (3x^2 - 4x - 1)$

Solution

a. Add:
$$\begin{array}{r} 3x^3 - x + 5 \\ -8x^3 + 3x - 9 \\ \hline -5x^3 + 2x - 4 \end{array}$$

b. Subtract:
$$\begin{array}{r} x^2 - 5x \\ 3x^2 - 4x - 1 \\ \hline -2x^2 - x + 1 \end{array}$$ ■

Multiplication of Polynomials

To multiply two polynomials, we use the distributive property.

Example **5** Multiplying polynomials

Use the distributive property to find each product.

a. $-3x(2x - 3)$ **b.** $(x^2 - 3x + 4)(2x - 3)$ **c.** $(x + 5)(2x - 3)$

Solution

a. $-3x(2x - 3) = -6x^2 + 9x$ Distributive property

b. $(x^2 - 3x + 4)(2x - 3) = x^2(2x - 3) - 3x(2x - 3) + 4(2x - 3)$ Distributive property

$$= 2x^3 - 3x^2 - 6x^2 + 9x + 8x - 12$$ Distributive property

$$= 2x^3 - 9x^2 + 17x - 12$$ Combine like terms.

c. $(x + 5)(2x - 3) = x(2x - 3) + 5(2x - 3)$ Distributive property

$$= 2x^2 - 3x + 10x - 15$$ Distributive property

$$= 2x^2 + 7x - 15$$ Combine like terms. ■

Using FOIL

The product of two binomials (as in Example 5(c)) results in four terms:

the product of the <u>First</u> term of each,

the product of the <u>Outer</u> terms,

the product of the <u>Inner</u> terms, and

the product of the <u>Last</u> term of each.

We use FOIL as a memory aid for these four products. The FOIL method allows us to quickly obtain the product of two binomials.

Example **6** **Multiplying binomials using FOIL**

Find each product using FOIL.

a. $(x + 4)(2x - 3)$ **b.** $(a^2 - 3)(2a - 3)$

Solution

$$\overset{\text{F}\quad\text{O}\quad\text{I}\quad\text{L}}{}$$

a. $(x + 4)(2x - 3) = \overset{\frown}{2x^2} - \overset{\frown}{3x} + \overset{\frown}{8x} - \overset{\frown}{12} = 2x^2 + 5x - 12$

b. $(a^2 - 3)(2a - 3) = 2a^3 - 3a^2 - 6a + 9$ ■

Special Products

The products $(a + b)^2$, $(a - b)^2$, and $(a + b)(a - b)$ are called the **special products.** We could use FOIL to find these products, but it is better to memorize the following rules for finding these products quickly.

The Special Products

$(a + b)^2 = a^2 + 2ab + b^2$	**The Square of a Sum**
$(a - b)^2 = a^2 - 2ab + b^2$	**The Square of a Difference**
$(a + b)(a - b) = a^2 - b^2$	**The Product of a Sum and a Difference**

Example **7** **Finding special products**

Find each product by using the special product rules.

a. $(2x + 3)^2$ **b.** $(x^3 - 9)^2$ **c.** $(3x + 5)(3x - 5)$

Solution

a. To find $(2x + 3)^2$ substitute $2x$ for a and 3 for b in $(a + b)^2 = a^2 + 2ab + b^2$:

$$(2x + 3)^2 = (2x)^2 + 2(2x)(3) + 3^2$$
$$= 4x^2 + 12x + 9$$

b. To find $(x^3 - 9)^2$ substitute x^3 for a and 9 for b in $(a - b)^2 = a^2 - 2ab + b^2$:

$$(x^3 - 9)^2 = (x^3)^2 - 2(x^3)(9) + 9^2$$
$$= x^6 - 18x^3 + 81$$

c. $(3x + 5)(3x - 5) = (3x)^2 - 5^2 = 9x^2 - 25$ ■

The expressions $3 - \sqrt{6}$ and $3 + \sqrt{6}$ are called **conjugates.** Their product is a rational number because of the rule $(a + b)(a - b) = a^2 - b^2$. This fact is used to rationalize a denominator in the next example.

Example **8** **Using conjugates to rationalize a denominator**

Simplify the expression $\dfrac{\sqrt{3}}{3 - \sqrt{6}}$.

Solution

Multiply the numerator and denominator by $3 + \sqrt{6}$, the conjugate of $3 - \sqrt{6}$:

$$\frac{\sqrt{3}}{3 - \sqrt{6}} = \frac{\sqrt{3}(3 + \sqrt{6})}{(3 - \sqrt{6})(3 + \sqrt{6})} = \frac{3\sqrt{3} + \sqrt{18}}{9 - 6}$$

$$= \frac{3\sqrt{3} + 3\sqrt{2}}{3} = \sqrt{3} + \sqrt{2} \qquad ■$$

Division of Polynomials

If 20 is divided by 3, the quotient is 6 and the remainder is 2. So we can write $20 = 6 \cdot 3 + 2$. If the **dividend** $P(x)$ and the **divisor** $D(x)$ are polynomials such that $D(x) \neq 0$ and the degree of $P(x)$ is greater than or equal to the degree of $D(x)$, then there exist unique polynomials, the **quotient** $Q(x)$ and the **remainder** $R(x)$, such that

$$P(x) = Q(x)D(x) + R(x)$$

where $R(x) = 0$ or the degree of $R(x)$ is less than the degree of $D(x)$. To find the quotient and remainder polynomials we use an algorithm that is similar to the well-known algorithm for dividing whole numbers.

Example **9** **Using the division algorithm to divide polynomials**

Find the quotient and remainder when $x^3 - 8$ is divided by $x - 2$.

Solution

To keep the division organized, insert $0x^2$ and $0x$ for the missing x^2 and x terms.

$$
\begin{array}{r}
x^2 + 2x + 4 \\
x - 2\overline{)x^3 + 0x^2 + 0x - 8} \quad {\scriptstyle x^3 \div x = x^2} \\
\underline{x^3 - 2x^2} \qquad\qquad {\scriptstyle x^2(x-2) = x^3 - 2x^2} \\
2x^2 + 0x \qquad {\scriptstyle 0x^2 - (-2x^2) = 2x^2} \\
\underline{2x^2 - 4x} \qquad {\scriptstyle 2x(x-2) = 2x^2 - 4x} \\
4x - 8 \quad {\scriptstyle 4(x-2) = 4x - 8} \\
\underline{4x - 8} \\
0
\end{array}
$$

The quotient is $x^2 + 2x + 4$ and the remainder is 0. To check, find the product $(x^2 + 2x + 4)(x - 2)$.

The dividend is equal to the quotient times the divisor plus the remainder:

$$\text{dividend} = (\text{quotient})(\text{divisor}) + \text{remainder}$$

We can also express this relationship as follows:

$$\frac{\text{dividend}}{\text{divisor}} = \text{quotient} + \frac{\text{remainder}}{\text{divisor}} \qquad ■$$

A.2 Exercises

Find the degree and leading coefficient of each polynomial. Determine whether the polynomial is a monomial, binomial, or trinomial.

1. $x^3 - 4x^2 + \sqrt{5}$ **2.** $-x^7 - 6x^4$ **3.** $x - 3x^2$

4. $x + 5 + x^2$ **5.** 79   **6.** $-\dfrac{x}{\sqrt{2}}$

Let $P(x) = x^2 - 3x + 2$ and $M(x) = -x^3 + 5x^2 - x + 2$. Find the following.

7. $P(-2)$ **8.** $P(-1)$ **9.** $M(-3)$ **10.** $M(50)$

Find each sum or difference.

11. $(3x^2 - 4x) + (5x^2 + 7x - 1)$

12. $(-3x^2 - 4x + 2) + (5x^2 - 8x - 7)$

13. $(4x^2 - 3x) - (9x^2 - 4x + 3)$

14. $(x^2 + 2x + 4) - (x^2 + 4x + 4)$

15. $(4ax^3 - a^2x) - (5a^2x^3 - 3a^2x + 3)$

16. $(x^2y^2 - 3xy + 2x) - (6x^2y^2 + 4y - 6x)$

Perform the indicated operation.

17. Add:
$$\begin{array}{r} 3x - 4 \\ -x + 3 \end{array}$$

18. Add:
$$\begin{array}{r} -2x^2 - 5 \\ 3x^2 - 6 \end{array}$$

19. Subtract:
$$\begin{array}{r} x^2 \qquad - 8 \\ -2x^2 + 3x - 2 \end{array}$$

20. Subtract:
$$\begin{array}{r} -2x^2 - 5x + 9 \\ 4x^2 - 7x \end{array}$$

Use the distributive property to find each product.

21. $-3a^3(6a^2 - 5a + 2)$ **22.** $-2m(m^2 - 3m + 9)$

23. $(3b^2 - 5b + 2)(b - 3)$ **24.** $(-w^2 - 5w + 6)(w + 5)$

25. $(2x - 1)(4x^2 + 2x + 1)$ **26.** $(3x - 2)(9x^2 + 6x + 4)$

27. $(x + 5)(x^2 - 5x + 25)$ **28.** $(a + 3)(a^2 - 3a + 9)$

29. $(x - 4)(z + 3)$ **30.** $(a - 3)(b + c)$

31. $(a - b)(a^2 + ab + b^2)$ **32.** $(a + b)(a^2 - ab + b^2)$

Find each product using FOIL.

33. $(a + 9)(a - 2)$ **34.** $(z - 3)(z - 4)$

35. $(2y - 3)(y + 9)$ **36.** $(2y - 1)(3y + 4)$

37. $(2x - 9)(2x + 9)$ **38.** $(4x - 6y)(4x + 6y)$

39. $(2x^2 + 4)(3x^2 + 5)$ **40.** $(3x^3 - 2)(5x^3 + 6)$

41. $(2x + 5)^2$ **42.** $(5x - 3)^2$

Find each product using the special product rules.

43. $(3x + 5)^2$ **44.** $(x^3 - 2)^2$

45. $(x^2 - 3)(x^2 + 3)$ **46.** $(2z^3 + 1)(2z^3 - 1)$

47. $(\sqrt{2} - 5)(\sqrt{2} + 5)$ **48.** $(6 - \sqrt{3})(6 + \sqrt{3})$

49. $(3x^3 - 4)^2$ **50.** $(2x^2y^3 + 1)^2$

51. $(2xy - 5)^2$ **52.** $(-3x^4 - 2)^2$

Simplify each expression by rationalizing the denominator.

53. $\dfrac{\sqrt{10}}{\sqrt{5} - 2}$ **54.** $\dfrac{\sqrt{3}}{\sqrt{2} - \sqrt{3}}$

55. $\dfrac{\sqrt{6}}{6 + \sqrt{3}}$ **56.** $\dfrac{\sqrt{2}}{\sqrt{8} + \sqrt{3}}$

Find the quotient and remainder when the first polynomial is divided by the second.

57. $x^2 + 6x + 9, x + 3$ **58.** $x^2 - 3x - 54, x - 9$

59. $a^3 - 1, a - 1$ **60.** $b^6 + 8, b^2 + 2$

61. $x^2 + 3x + 3, x - 2$ **62.** $3x^2 - x + 4, x + 2$

63. $2x^2 - 5, x + 3$ **64.** $-4x^2 + 1, x - 1$

65. $x^3 - 2x^2 - 2x - 3, x - 3$

66. $x^3 + 3x^2 - 3x + 4, x + 4$

67. $6x^2 - 7x + 2, 2x + 1$ **68.** $-3x^2 + 4x + 9, 2 - x$

69. $2x^3 + 3x^2 - 7x - 12, x^2 - 4$

70. $x^3 + 2x^2 + x - 3, x^2 - 1$

71. $x^3 + 2x^2 - 7x - 6, x^2 - x - 2$

72. $2x^3 - 3x^2 - 19x - 8, x^2 - 2x - 8$

Perform the indicated operations mentally. Write down only the answer.

73. $(x - 4)(x + 6)$

74. $(z^4 + 5)(z^4 - 4)$

75. $(2a^5 - 9)(a^5 + 3)$

76. $(3b^2 + 1)(2b^2 + 5)$

77. $(y - 3) - (2y + 6)$

78. $(a^2 + 3) - (a^2 - 6)$

79. $(w + 4)^2$

80. $(t - 2)^2$

81. $3y^2(y^3 - 3x)$

82. $a^4b(a^2b^2 + 1)$

83. $(6b^3 - 3b^2) \div (3b^2)$

84. $(3w - 8) \div (8 - 3w)$

85. $(3w^2 - 2n)^2$

86. $(7y^2 - 3x)^2$

A.3	**Factoring Polynomials**

In this section we will factor polynomials. **Factoring** "reverses" multiplication.

Factoring Out the Greatest Common Factor

To factor $6x^2 - 3x$, notice that $3x$ is a monomial that can be divided evenly into each term. We can use the distributive property to write

$$6x^2 - 3x = 3x(2x - 1).$$

We call this process **factoring out** $3x$. Both $3x$ and $2x - 1$ are **factors** of $6x^2 - 3x$. Since 3 is a factor of $6x^2$ and $3x$, 3 is a **common factor** of the terms of the polynomial. The **greatest common factor** (GCF) is a monomial that includes every number or variable that is a factor of all terms of the polynomial. The monomial $3x$ is the greatest common factor of $6x^2 - 3x$. Usually the common factor has a positive coefficient, but at times it is useful to factor out a common factor with a negative coefficient.

Example **1** Factoring out the greatest common factor

Factor out the greatest common factor from each polynomial, first using the GCF with a positive coefficient and then using a negative coefficient.

a. $9x^4 - 6x^3 + 12x^2$

b. $x^2y + 10xy + 25y$

Solution

a. $9x^4 - 6x^3 + 12x^2 = 3x^2(3x^2 - 2x + 4)$

$$= -3x^2(-3x^2 + 2x - 4)$$

b. $x^2y + 10xy + 25y = y(x^2 + 10x + 25)$

$$= -y(-x^2 - 10x - 25)$$ ■

Factoring by Grouping

Some four-term polynomials can be factored by **grouping** the terms in pairs and factoring out a common factor from each pair of terms.

Example **2** Factoring four-term polynomials by grouping

Factor each polynomial by grouping.

a. $x^3 + x^2 + 3x + 3$ **b.** $aw + bc - bw - ac$

Solution

a. Factor the common factor x^2 out of the first two terms and the common factor 3 out of the last two terms:

$$x^3 + x^2 + 3x + 3 = x^2(x + 1) + 3(x + 1) \quad \text{Factor out common factors.}$$

$$= (x + 1)(x^2 + 3) \quad \text{Factor out the common factor } (x + 1).$$

b. We must first arrange the polynomial so that the first group of two terms has a common factor and the last group of two terms also has a common factor.

$$aw + bc - bw - ac = aw - bw - ac + bc \quad \text{Rearrange.}$$

$$= w(a - b) - c(a - b) \quad \text{Factor out common factors.}$$

$$= (w - c)(a - b) \quad \text{Factor out } a - b. \quad ■$$

Factoring $ax^2 + bx + c$

Factoring by grouping can be used also to factor a trinomial that is the product of two binomials.

Example **3** Factoring $ax^2 + bx + c$ with $a = 1$

Factor each trinomial.

a. $x^2 - 5x - 14$ **b.** $x^2 + 4x - 21$

Solution

a. First find two numbers that have a product of -14 and a sum of -5. The numbers are -7 and 2. To get four terms that we can factor by grouping, we replace $-5x$ with $-7x + 2x$.

$$x^2 - 5x - 14 = x^2 - 7x + 2x - 14 \quad \text{Replace } -5x \text{ with } -7x + 2x.$$

$$= (x - 7)x + (x - 7)2 \quad \text{Factor out common factors.}$$

$$= (x - 7)(x + 2) \quad \text{Factor out } (x - 7).$$

Check by using FOIL. Note that once we have found -7 and 2, we can skip the grouping step and simply write the answer $(x - 7)(x + 2)$.

b. Two numbers that have a product of -21 and a sum of 4 are 7 and -3.

$$x^2 + 4x - 21 = (x + 7)(x - 3)$$

Check by using FOIL. ■

To factor $ax^2 + bx + c$ with $a \neq 1$, we can use the **ac-method**.

Procedure: The ac-Method for Factoring $ax^2 + bx + c$ with $a \neq 1$

To factor $ax^2 + bx + c$ with $a \neq 1$:

1. Find two numbers whose sum is b and whose product is ac.
2. Replace b with the sum of these two numbers.
3. Factor the resulting four-term polynomial by grouping.

Example **4** Factoring $ax^2 + bx + c$ with $a \neq 1$

Factor each trinomial using the *ac*-method.

a. $2x^2 + 5x + 2$ **b.** $6x^2 - x - 12$

Solution

a. Since $ac = 2 \cdot 2 = 4$ and $b = 5$, we need two numbers that have a product of 4 and a sum of 5. The numbers are 4 and 1.

$$2x^2 + 5x + 2 = 2x^2 + 4x + x + 2 \qquad \text{Replace } 5x \text{ by } 4x + x.$$
$$= (x + 2)2x + (x + 2)1 \qquad \text{Factor by grouping.}$$
$$= (x + 2)(2x + 1) \qquad \text{Check using FOIL.}$$

b. Since $ac = 6(-12) = -72$ and $b = -1$, we need two numbers that have a product of -72 and a sum of -1. The numbers are 8 and -9.

$$6x^2 - x - 12 = 6x^2 - 9x + 8x - 12 \qquad \text{Replace } -x \text{ by } -9x + 8x.$$
$$= (2x - 3)3x + (2x - 3)4 \qquad \text{Factor by grouping.}$$
$$= (2x - 3)(3x + 4) \qquad \text{Check using FOIL.} \qquad ■$$

Factoring the Special Products

Trinomials of the form $a^2 + 2ab + b^2$ and $a^2 - 2ab + b^2$ are called **perfect square trinomials.** We can factor the perfect square trinomials and the difference of two squares by using the same rules that we used to obtain these special products.

Factoring the Special Products

$a^2 - b^2 = (a + b)(a - b)$	**Difference of Two Squares**
$a^2 + 2ab + b^2 = (a + b)^2$	**Perfect Square Trinomial**
$a^2 - 2ab + b^2 = (a - b)^2$	**Perfect Square Trinomial**

Example **5** Factoring the special products

Factor each polynomial.

a. $4x^2 - 1$ **b.** $x^2 - 6x + 9$ **c.** $9y^2 + 30y + 25$

Solution

a. $4x^2 - 1 = (2x)^2 - 1^2$ $\qquad$ Recognize the difference of two squares.

$\qquad\qquad = (2x + 1)(2x - 1)$

b. $x^2 - 6x + 9 = x^2 - 2 \cdot 3 \cdot x + 3^2$ Recognize the perfect square trinomial.

$\qquad\qquad = (x - 3)^2$

c. $9y^2 + 30y + 25 = (3y)^2 + 2(3y)(5) + 5^2$ Recognize the perfect square trinomial.

$\qquad\qquad\qquad = (3y + 5)^2$ ▪

Factoring the Difference and Sum of Two Cubes

The following formulas are used to factor the difference of two cubes and the sum of two cubes. You should verify these formulas using multiplication.

Factoring the Difference and Sum of Two Cubes

$a^3 - b^3 = (a - b)(a^2 + ab + b^2)$	**Difference of Two Cubes**
$a^3 + b^3 = (a + b)(a^2 - ab + b^2)$	**Sum of Two Cubes**

Example **6** **Factoring differences and sums of two cubes**

Factor each polynomial.

a. $x^3 - 27$ **b.** $8w^6 + 125z^3$

Solution

a. Since $x^3 - 27 = x^3 - 3^3$, we use $a = x$ and $b = 3$ in the formula for factoring the difference of two cubes.

$$x^3 - 27 = (x - 3)(x^2 + 3x + 9)$$

b. Since $8w^6 + 125z^3 = (2w^2)^3 + (5z)^3$, we use $a = 2w^2$ and $b = 5z$ in the formula for factoring the sum of two cubes.

$$8w^6 + 125z^3 = (2w^2 + 5z)(4w^4 - 10w^2z + 25z^2)$$ ▪

Factoring Completely

Polynomials that cannot be factored using integral coefficients are called **prime** or **irreducible over the integers.** For example, $a^2 + 1$, $b^2 + b + 1$, and $x + 5$ are prime polynomials because they cannot be expressed as a product (in a nontrivial manner). A polynomial is **factored completely** when it is written as a product of prime polynomials. When factoring polynomials, we usually do not factor integers that are common factors. For example, $4x^2(2x - 3)$ is factored completely even though the coefficient 4 could be factored.

Example **7** **Factoring completely**

Factor each polynomial completely.

a. $2w^4 - 32$ **b.** $-6x^7 + 6x$

Solution

a. $2w^4 - 32 = 2(w^4 - 16)$ Factor out the greatest common factor.

$\qquad\qquad = 2(w^2 - 4)(w^2 + 4)$ Difference of two squares

$\qquad\qquad = 2(w - 2)(w + 2)(w^2 + 4)$ Difference of two squares

The polynomial is now factored completely because $w^2 + 4$ is prime.

b. $-6x^7 + 6x = -6x(x^6 - 1)$ Greatest common factor

$\qquad\qquad = -6x(x^3 - 1)(x^3 + 1)$ Difference of two squares

$\qquad\qquad = -6x(x - 1)(x^2 + x + 1)(x + 1)(x^2 - x + 1)$ Difference of two cubes; sum of two cubes

The polynomial is factored completely because all of the factors are prime. ■

A.3 Exercises

Factor out the greatest common factor from each polynomial, first using a positive coefficient on the GCF and then using a negative coefficient.

1. $6x^3 - 12x^2$

2. $12x^2 + 18x^3$

3. $-ax^3 + 5ax^2 - 6ax$

4. $-sa^3 + sb^3 - sb$

5. $m - n$

6. $y - x$

Factor each polynomial by grouping.

7. $x^3 + 2x^2 + 5x + 10$

8. $2w^3 - 2w^2 + 3w - 3$

9. $y^3 - y^2 - 3y + 3$

10. $x^3 + x^2 - 7x - 7$

11. $ady - w + d - awy$

12. $xy + ab + by + ax$

13. $x^2y^2 + ab - ay^2 - bx^2$

14. $6yz - 3y - 10z + 5$

Factor each trinomial.

15. $x^2 + 10x + 16$

16. $x^2 + 7x + 12$

17. $x^2 - 4x - 12$

18. $y^2 + 3y - 18$

19. $m^2 - 12m + 20$

20. $n^2 - 8n + 7$

21. $t^2 + 5t - 84$

22. $s^2 - 6s - 27$

23. $2x^2 - 7x - 4$

24. $3x^2 + 5x - 2$

25. $8x^2 - 10x - 3$

26. $18x^2 - 15x + 2$

27. $6y^2 + 7y - 5$

28. $15x^2 - 14x - 8$

29. $12b^2 + 17b + 6$

30. $8h^2 + 22h + 9$

Factor each special product.

31. $t^2 - u^2$

32. $9t^2 - v^2$

33. $t^2 + 2t + 1$

34. $m^2 + 10m + 25$

35. $4w^2 - 4w + 1$

36. $9x^2 - 12xy + 4y^2$

37. $9z^2x^2 + 24zx + 16$

38. $25t^2 - 20tw^3 + 4w^6$

Factor each sum or difference of two cubes.

39. $t^3 - u^3$

40. $m^3 + n^3$

41. $a^3 - 8$

42. $b^3 + 1$

43. $27y^3 + 8$

44. $1 - 8a^6$

45. $27x^3y^6 - 8z^9$

46. $8t^3h^3 + n^9$

Factor each polynomial completely.

47. $-3x^3 + 27x$

48. $a^4b^2 - 16b^2$

49. $16t^4 + 54w^3t$

50. $8a^6 - a^3b^3$

51. $a^3 + a^2 - 4a - 4$

52. $2b^3 + 3b^2 - 18b - 27$

53. $x^4 - 2x^3 - 8x + 16$

54. $a^4 - a^3 + a - 1$

55. $-36x^3 + 18x^2 + 4x$

56. $-6a^4 - a^3 + 15a^2$

57. $a^7 - a^6 - 64a + 64$

58. $a^5 - 4a^4 - 4a + 16$

59. $-6x^2 - x + 15$

60. $-6x^2 - 9x + 42$

A.4 Rational Expressions

In this section we will review the basic operations with rational expressions.

Reducing

A **rational expression** is a ratio of two polynomials in which the denominator is not the zero polynomial. The **domain** of a rational expression is the set of all real numbers that can be used in place of the variable.

Example **1** Domain of a rational expression

Find the domain of each rational expression.

a. $\dfrac{2x-1}{x+3}$ **b.** $\dfrac{2x+4}{(x+2)(x+3)}$ **c.** $\dfrac{1}{x^2+8}$

Solution

a. The domain is the set of all real numbers except those that cause $x+3$ to have a value of 0. So -3 is excluded from the domain, because $x+3$ has a value of 0 for $x=-3$. We write the domain in set notation as $\{x \mid x \neq -3\}$.

b. The domain is the set of all real numbers except -2 and -3, because replacing x by either of these numbers would cause the denominator to be 0. The domain is written in set notation as $\{x \mid x \neq -2 \text{ and } x \neq -3\}$.

c. The value of x^2+8 is positive for any real number x. So the domain is the set of all real numbers, R. ■

In arithmetic we learned that each rational number has infinitely many equivalent forms. This fact is due to the **basic principle of rational numbers.**

Basic Principle of Rational Numbers

If a, b, and c are integers with $b \neq 0$ and $c \neq 0$, then

$$\frac{ac}{bc} = \frac{a}{b}.$$

The basic principle holds true when a, b, and c are real numbers as well as when they are integers, but we use it most often with integers, when we reduce fractions. For example, we reduce $3/6$ as follows:

$$\frac{3}{6} = \frac{1 \cdot 3}{2 \cdot 3} = \frac{1}{2}$$

Note that since 3 and 6 have a common factor of 3, we divide both the numerator and denominator by 3 to reduce the fraction. We reduce rational expressions in the same manner. Factor the numerator and denominator completely, then *divide out* the common factors. A rational expression is in *lowest terms* when all common factors have been divided out.

Example **2** **Reducing to lowest terms**

Reduce each rational expression to lowest terms.

a. $\dfrac{2x + 4}{x^2 + 5x + 6}$ **b.** $\dfrac{b - a}{a^3 - b^3}$ **c.** $\dfrac{x^2 z^3}{x^5 z}$

Solution

a. $\dfrac{2x + 4}{x^2 + 5x + 6} = \dfrac{2(x + 2)}{(x + 2)(x + 3)}$ Factor the numerator and denominator.

$\qquad\qquad\qquad = \dfrac{2}{x + 3}$ Divide out the common factor $x + 2$.

b. $\dfrac{b - a}{a^3 - b^3} = \dfrac{-1(a - b)}{(a - b)(a^2 + ab + b^2)}$ Factor -1 out of $b - a$.

$\qquad\qquad\quad = \dfrac{-1}{a^2 + ab + b^2}$

c. $\dfrac{x^2 z^3}{x^5 z} = \dfrac{(x^2 z)(z^2)}{(x^2 z)(x^3)} = \dfrac{z^2}{x^3}$ (The quotient rule yields the same result.) ▪

Be careful when reducing. The <u>only</u> way to reduce rational expressions is to factor and divide out the common *factors*. Identical terms that are not factors cannot be eliminated from a rational expression. For example,

$$\frac{x + 3}{3} \neq x$$

for all real numbers, because 3 is not a factor of the numerator.

Multiplication and Division

We multiply two rational numbers by multiplying their numerators and their denominators. For example, $\frac{2}{3} \cdot \frac{5}{7} = \frac{10}{21}$.

Definition: Multiplication of Rational Numbers	If a/b and c/d are rational numbers, then $$\frac{a}{b} \cdot \frac{c}{d} = \frac{ac}{bd}.$$

We multiply rational expressions in the same manner as rational numbers. Of course, any common factor can be divided out as we do when reducing rational expressions.

Example **3** **Multiplying rational expressions**

Find each product.

a. $\dfrac{2a - 2b}{6} \cdot \dfrac{9a}{a^2 - b^2}$ **b.** $\dfrac{x - 1}{x^2 + 4x + 4} \cdot \dfrac{x + 2}{x^2 + 2x - 3}$

Solution

a. $\dfrac{2a - 2b}{6} \cdot \dfrac{9a}{a^2 - b^2} = \dfrac{2(a - b)}{2 \cdot 3} \cdot \dfrac{3 \cdot 3a}{(a - b)(a + b)}$ Factor completely.

$\qquad\qquad = \dfrac{3a}{a + b}$ Divide out common factors.

b. $\dfrac{x - 1}{x^2 + 4x + 4} \cdot \dfrac{x + 2}{x^2 + 2x - 3} = \dfrac{x - 1}{(x + 2)^2} \cdot \dfrac{x + 2}{(x + 3)(x - 1)}$ Factor completely.

$\qquad\qquad = \dfrac{1}{x^2 + 5x + 6}$ Divide out common factors. ■

We divide rational numbers by multiplying by the reciprocal of the divisor, or *invert and multiply*. For example, $6 \div \frac{1}{2} = 6 \cdot 2 = 12$.

Definition: Division of Rational Numbers

If a/b and c/d are rational numbers with $c \neq 0$, then

$$\frac{a}{b} \div \frac{c}{d} = \frac{a}{b} \cdot \frac{d}{c}.$$

Rational expressions are divided in the same manner as rational numbers.

Example **4** **Dividing rational expressions**

Perform the indicated operations.

a. $\dfrac{9}{2x} \div \dfrac{3}{x}$ **b.** $\dfrac{4 - x^2}{6} \div \dfrac{x - 2}{2}$

Solution

a. $\dfrac{9}{2x} \div \dfrac{3}{x} = \dfrac{9}{2x} \cdot \dfrac{x}{3}$ Invert and multiply.

$\qquad\qquad = \dfrac{3 \cdot 3}{2x} \cdot \dfrac{x}{3}$ Factor completely.

$\qquad\qquad = \dfrac{3}{2}$ Divide out common factors.

b. $\dfrac{4 - x^2}{6} \div \dfrac{x - 2}{2} = \dfrac{-1(x - 2)(x + 2)}{2 \cdot 3} \cdot \dfrac{2}{x - 2}$ Invert and multiply.

$\qquad\qquad = \dfrac{-x - 2}{3}$ Divide out common factors. ■

Note that the division in Example 4(a) is valid only if $x \neq 0$. The division in Example 4(b) is valid only if $x \neq 2$ because $x - 2$ appears in the denominator after the rational expression is inverted.

Building Up the Denominator

The addition of fractions can be carried out only when their denominators are identical. To get a required denominator, we may **build up** the denominator of a fraction. To build up the denominator we use the basic principle of rational numbers in the reverse of the way we use it for reducing. We multiply the numerator and denominator of a fraction by the same nonzero number to get an equivalent fraction.

Example **5** **Writing equivalent rational expressions**

Convert the first rational expression into an equivalent one that has the indicated denominator.

a. $\dfrac{3}{2a}, \dfrac{?}{6ab}$ **b.** $\dfrac{x-1}{x+2}, \dfrac{?}{x^2+6x+8}$ **c.** $\dfrac{a}{3b-a}, \dfrac{?}{a^2-9b^2}$

Solution

a. Compare the two denominators. Since $6ab = 2a(3b)$, we multiply the numerator and denominator of the first expression by $3b$:

$$\frac{3}{2a} = \frac{3 \cdot 3b}{2a \cdot 3b} = \frac{9b}{6ab}$$

b. Factor the second denominator and compare it to the first. Since $x^2 + 6x + 8 = (x+2)(x+4)$, we multiply the numerator and denominator by $x + 4$:

$$\frac{x-1}{x+2} = \frac{(x-1)(x+4)}{(x+2)(x+4)} = \frac{x^2+3x-4}{x^2+6x+8}$$

c. Factor the second denominator as

$$a^2 - 9b^2 = (a-3b)(a+3b) = -1(3b-a)(a+3b).$$

Since $3b - a$ is a factor of $a^2 - 9b^2$, we multiply the numerator and denominator by $-1(a+3b)$:

$$\frac{a}{3b-a} = \frac{a(-1)(a+3b)}{(3b-a)(-1)(a+3b)} = \frac{-a^2-3ab}{a^2-9b^2}$$ ■

Addition and Subtraction

Fractions can be added or subtracted only if their denominators are identical. For example, $\frac{1}{3} + \frac{1}{3} = \frac{2}{3}$ and $\frac{7}{12} - \frac{2}{12} = \frac{5}{12}$.

Definition: Addition and Subtraction of Rational Numbers

If a/b and c/b are rational numbers, then

$$\frac{a}{b} + \frac{c}{b} = \frac{a+c}{b} \quad \text{and} \quad \frac{a}{b} - \frac{c}{b} = \frac{a-c}{b}.$$

For fractions with different denominators, we build up one or both denominators to get denominators that are equal to the least common multiple (LCM) of the denominators. The **least common denominator (LCD)** is the smallest number that is a multiple of all of the denominators. Use the following steps to find the LCD.

Procedure:
Finding the LCD

1. Factor each denominator completely.
2. Write a product using each factor that appears in a denominator.
3. For each factor, use the highest power of that factor that occurs in the denominators.

For example, to find the LCD for 10 and 12, we write $10 = 2 \cdot 5$ and $12 = 2^2 \cdot 3$. The LCD contains the factors 2, 3, and 5. Using the highest power of each we get $2^2 \cdot 3 \cdot 5 = 60$ for the LCD. So, to add fractions with denominators of 10 and 12, we build up each fraction to a denominator of 60:

$$\frac{1}{10} + \frac{1}{12} = \frac{1 \cdot 6}{10 \cdot 6} + \frac{1 \cdot 5}{12 \cdot 5} = \frac{6}{60} + \frac{5}{60} = \frac{11}{60}$$

We use the same method to add or subtract rational expressions.

Example **6** **Adding and subtracting rational expressions**

Perform the indicated operations.

a. $\dfrac{x}{x - 1} + \dfrac{2x + 3}{x^2 - 1}$ **b.** $\dfrac{x}{x^2 + 6x + 9} - \dfrac{x - 3}{x^2 + 5x + 6}$

Solution

a. $\dfrac{x}{x - 1} + \dfrac{2x + 3}{x^2 - 1} = \dfrac{x}{x - 1} + \dfrac{2x + 3}{(x - 1)(x + 1)}$ Factor denominators completely.

$= \dfrac{x(x + 1)}{(x - 1)(x + 1)} + \dfrac{2x + 3}{(x - 1)(x + 1)}$ Build up using the LCD $(x - 1)(x + 1)$.

$= \dfrac{x^2 + x + 2x + 3}{(x - 1)(x + 1)}$ Add the fractions.

$= \dfrac{x^2 + 3x + 3}{(x - 1)(x + 1)}$ Simplify the numerator.

b. $\dfrac{x}{x^2 + 6x + 9} - \dfrac{x - 3}{x^2 + 5x + 6} = \dfrac{x}{(x + 3)^2} - \dfrac{x - 3}{(x + 2)(x + 3)}$

$= \dfrac{x(x + 2)}{(x + 3)^2(x + 2)} - \dfrac{(x - 3)(x + 3)}{(x + 2)(x + 3)(x + 3)}$

$= \dfrac{x^2 + 2x}{(x + 3)^2(x + 2)} - \dfrac{x^2 - 9}{(x + 3)^2(x + 2)}$

$= \dfrac{2x + 9}{(x + 3)^2(x + 2)}$ ∎

A.4 Exercises

Find the domain of each rational expression.

1. $\dfrac{x-3}{x+2}$

2. $\dfrac{x^2-1}{x-5}$

3. $\dfrac{x^2-9}{(x-4)(x+2)}$

4. $\dfrac{2x-3}{(x+1)(x-3)}$

5. $\dfrac{x+1}{(x-3)(x+3)}$

6. $\dfrac{x+2}{(x+2)(x+1)}$

7. $\dfrac{3x^2-2x+1}{x^2+3}$

8. $\dfrac{-2x^2-7}{3x^2+8}$

Reduce each rational expression to lowest terms.

9. $\dfrac{3x-9}{x^2-x-6}$

10. $\dfrac{-2x-4}{x^2-3x-10}$

11. $\dfrac{10a-8b}{12b-15a}$

12. $\dfrac{a^2-b^2}{b-a}$

13. $\dfrac{a^3b^6}{a^2b^3-a^4b^2}$

14. $\dfrac{18u^6v^5+24u^3v^3}{42u^2v^5}$

15. $\dfrac{x^4y^5z^2}{x^7y^3z}$

16. $\dfrac{t^3u^7}{-t^8u^5}$

17. $\dfrac{a^3-b^3}{a^2-b^2}$

18. $\dfrac{a^3+b^3}{a^2+b^2}$

19. $\dfrac{ab+3a-by-3y}{a^2-2ay+y^2}$

20. $\dfrac{x^4-16}{x^4+8x^2+16}$

Find the products or quotients.

21. $\dfrac{2a}{3b^2}\cdot\dfrac{9b}{14a^2}$

22. $\dfrac{14w}{51y}\cdot\dfrac{3w}{7y}$

23. $\dfrac{12a}{7}\div\dfrac{2a^3}{49}$

24. $\dfrac{20x}{y^3}\div\dfrac{30}{y^5}$

25. $\dfrac{a^2-9}{3a-6}\cdot\dfrac{a^2-4}{a^2-a-6}$

26. $\dfrac{6x^2+x-1}{6x+3}\cdot\dfrac{15}{9x^2-1}$

27. $\dfrac{x^2-y^2}{9}\div\dfrac{x^2+2xy+y^2}{18}$

28. $\dfrac{a^3-b^3}{a^2-2ab+b^2}\div\dfrac{2a^2+2ab+2b^2}{9a^2-9b^2}$

29. $\dfrac{x^2-y^2}{-3xy}\cdot\dfrac{6x^2y^3}{2y-2x}$

30. $\dfrac{a^2-a-2}{2}\cdot\dfrac{1}{4-a^2}$

31. $\dfrac{wx-x}{x^2}\div\dfrac{1-w^2}{2}$

32. $\dfrac{2+x}{2}\div\dfrac{x^2-4}{4}$

Convert the first rational expression into an equivalent one that has the indicated denominator.

33. $\dfrac{4}{3a},\dfrac{?}{12a^2}$

34. $\dfrac{a+2}{4a^2},\dfrac{?}{20a^3b}$

35. $\dfrac{x-5}{x+3},\dfrac{?}{x^2-9}$

36. $\dfrac{x+2}{x-8},\dfrac{?}{16-2x}$

37. $\dfrac{x}{x+5},\dfrac{?}{x^2+6x+5}$

38. $\dfrac{3a-b}{a+b},\dfrac{?}{9b^2-9a^2}$

39. $\dfrac{t}{2t+2},\dfrac{?}{2t^2+4t+2}$

40. $\dfrac{x-1}{2x+4},\dfrac{?}{4x^2-16x-48}$

Find the least common denominator (LCD) for each given pair of rational expressions.

41. $\dfrac{1}{4ab^2},\dfrac{7}{6a^2b^3}$

42. $\dfrac{3}{2x^2y},\dfrac{a}{5xy}$

43. $\dfrac{-7a}{3a+3b},\dfrac{5b}{2a+2b}$

44. $\dfrac{1}{3a-3b},\dfrac{2}{a^2-b^2}$

45. $\dfrac{2x}{x^2+5x+6},\dfrac{3x}{x^2-x-6}$

46. $\dfrac{x+7}{2x^2+7x-15},\dfrac{x-5}{2x^2-5x+3}$

Perform the indicated operations.

47. $\dfrac{3}{2x}+\dfrac{1}{6}$

48. $\dfrac{-7}{3a^2b}+\dfrac{4}{6ab^2}$

49. $\dfrac{x+3}{x-1}-\dfrac{x+4}{x+1}$

50. $\dfrac{x+2}{x-3}-\dfrac{x^2+3x-2}{x^2-9}$

51. $3+\dfrac{1}{a}$

52. $-1-\dfrac{3}{c}$

53. $t-1-\dfrac{1}{t+1}$

54. $w+\dfrac{1}{w-1}$

55. $\dfrac{x}{x^2 + 3x + 2} + \dfrac{x - 1}{x^2 + 5x + 6}$

56. $\dfrac{x - 1}{x^2 + x - 6} - \dfrac{x - 2}{x^2 + 4x + 3}$

57. $\dfrac{1}{x - 3} - \dfrac{5}{6 - 2x}$

58. $\dfrac{5}{4 - x^2} - \dfrac{2x}{x - 2}$

59. $\dfrac{y^2}{x^3 - y^3} + \dfrac{x + y}{x^2 + xy + y^2}$

60. $\dfrac{ab}{a^3 + b^3} + \dfrac{a}{2a^2 - 2ab + 2b^2}$

61. $\dfrac{x - 2}{2x^2 + 7x - 15} - \dfrac{x + 1}{2x^2 - 5x + 3}$

62. $\dfrac{x - 1}{2x^2 - 5x - 3} - \dfrac{x}{4x^2 - 1}$

Chapter 1

Section 1.1

For Thought: **1.** F **2.** F **3.** T **4.** F **5.** F **6.** F
7. F **8.** F **9.** T **10.** T
Exercises: **1.** e, true **3.** h, true **5.** g, true **7.** c, true
9. All **11.** $\{-\sqrt{2}, \sqrt{3}, \pi, 5.090090009\ldots\}$ **13.** $\{0, 1\}$
15. $x + 7$ **17.** $5x + 15$ **19.** $5(x + 1)$ **21.** $(-13 + 4) + x$
23. 8 **25.** $\sqrt{3}$ **27.** $y^2 - x^2$ **29.** 7.2 **31.** $\sqrt{5}$ **33.** 5
35. 22 **37.** 12 **39.** 3/4 **41.** $|x - 7| = 3, \{4, 10\}$
43. $|-1 - x| = 3, \{-4, 2\}$ **45.** $|x - (-9)| = 4, \{-13, -5\}$
47. $\{-9, 9\}$ **49.** $\{4/5\}$ **51.** $\{-2, 5\}$ **53.** $\{-10, 0\}$
55. $\{2/3\}$ **57.** $\varnothing$

59. Smallest to largest: $-\dfrac{1}{2}, -\dfrac{5}{12}, -\dfrac{1}{3}, 0, \dfrac{1}{3}, \dfrac{5}{12}, \dfrac{1}{2}$

Section 1.2

For Thought: **1.** T **2.** F **3.** F **4.** T **5.** F **6.** F
7. F **8.** F **9.** F **10.** T
Exercises: **1.** $x < 12$ **3.** $x \geq -7$
5. $[-8, \infty)$ **7.** $(-\infty, \pi/2)$
9. $(5, \infty)$

11. $[2, \infty)$

13. $(-\infty, 54)$

15. $(-\infty, 13/3]$

17. $(-\infty, 3/2]$

19. $(-\infty, 0]$

21. $(-3, \infty)$ **23.** $(-3, \infty)$ **25.** $(-5, -2)$ **27.** $\varnothing$
29. $(-\infty, 5]$
31. $(3, 6)$

33. $(1/2, \infty)$

35. $(-3, \infty)$

37. $(-\infty, \infty)$

39. $\varnothing$
41. $(2, 4)$

43. $(-3, 1]$

45. $(-1/3, 1)$

47. $[1, 3/2]$

49. $[-1, 9]$

51. $\varnothing$

53. $\varnothing$

55. $(-\infty, 1) \cup (3, \infty)$

57. $(-\infty, 1) \cup (5, \infty)$

59. $|x| < 5$

61. $|x| > 3$ **63.** $|x - 6| < 2$ **65.** $|x - 4| > 1$
67. $|x| \geq 9$ **69.** $|x - 7| \leq 4$ **71.** $|x - 5| > 2$
73. $(93, 115)$ **75.** $(86, 102.5)$ **77.** $[\$0, \$7000]$
79. 96, 79, 68, 56, 47, yes **81.** $|x - 74{,}595| > 25{,}000, x > 99{,}595$
or $x < 49{,}595$

83. $\dfrac{|x - 35|}{35} < 0.01, (34.65°, 35.35°)$ **85.** [2.26 cm, 2.32 cm]

Section 1.3

For Thought: **1.** F **2.** F **3.** F **4.** F **5.** T **6.** F
7. T **8.** T **9.** T **10.** F
Exercises: **1.** $(4, 1)$, I **3.** $(1, 0)$, x-axis **5.** $(5, -1)$, IV
7. $(-4, -2)$, III **9.** $(-2, 4)$, II **11.** $5, (2.5, 5)$
13. $2\sqrt{2}, (0, -1)$ **15.** $6, \left(\dfrac{-2 + 3\sqrt{3}}{2}, \dfrac{5}{2}\right)$
17. $\sqrt{74}, (-1.3, 1.3)$ **19.** $|a - b|, \left(\dfrac{a + b}{2}, 0\right)$
21. $\dfrac{\sqrt{\pi^2 + 4}}{2}, \left(\dfrac{3\pi}{4}, \dfrac{1}{2}\right)$
23. $(0, 0), 4$ **25.** $(-6, 0), 6$

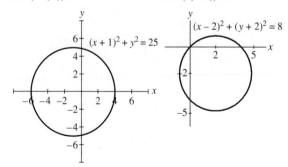

27. $(-1, 0), 5$ **29.** $(2, -2), 2\sqrt{2}$

31. $x^2 + y^2 = 49$ **33.** $(x + 2)^2 + (y - 5)^2 = 1/4$
35. $(x - 3)^2 + (y - 5)^2 = 34$ **37.** $(x - 5)^2 + (y + 1)^2 = 32$

39. (0, 0), 3

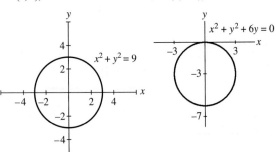

41. (0, −3), 3

57. (0, 30), (−90, 0)

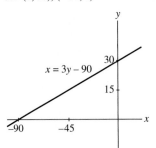

59. (0, 600), (−800, 0)

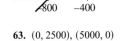

43. (3, 4), 5

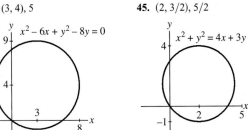

45. (2, 3/2), 5/2

61. (0, 0.0025), (0.005, 0)

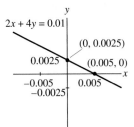

63. (0, 2500), (5000, 0)

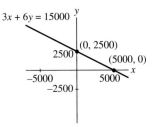

47. (1/4, −1/6), 1/6

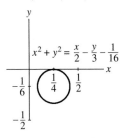

65.

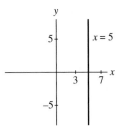

67.

49. a. $x^2 + y^2 = 49$ **b.** $(x − 1)^2 + y^2 = 20$
   c. $(x − 1)^2 + (y − 2)^2 = 13$

51. a. $(x − 2)^2 + (y + 3)^2 = 4$ **b.** $(x + 2)^2 + (y − 1)^2 = 1$
   c. $(x − 3)^2 + (y + 1)^2 = 9$ **d.** $x^2 + y^2 = 1$

53. (0, −4), (4/3, 0)

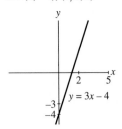

55. (0, −6), (2, 0)

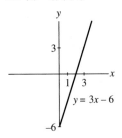

69.

71.

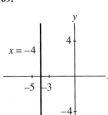

73. {3.6}   **75.** {14}   **77.** {−2.83}   **79.** {116,566.67}
81. {4.91}
83. a. (15, 22.95) The median of age at first marriage in 1985 was 22.95.
   b. 30.3, Because of the units, distance is meaningless.

85. $C = 1.8$

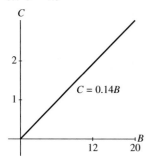

$C = 0.14B$

Section 1.4

For Thought: **1.** F **2.** F **3.** F **4.** F **5.** T **6.** F
7. F **8.** T **9.** F **10.** T

Exercises: **1.** $\dfrac{1}{3}$ **3.** -4 **5.** 0 **7.** 2 **9.** No slope

11. $y = \dfrac{5}{4}x + \dfrac{1}{4}$ **13.** $y = -\dfrac{7}{6}x + \dfrac{11}{3}$ **15.** $y = 5$ **17.** $x = 4$

19. $y = \dfrac{2}{3}x - 1$ **21.** $y = \dfrac{5}{2}x + \dfrac{3}{2}$ **23.** $y = -2x + 4$

25. $y = \dfrac{3}{2}x + \dfrac{5}{2}$ **27.** $y = \dfrac{3}{5}x - 2, \dfrac{3}{5}, (0, -2)$

29. $y = 2x - 5, 2, (0, -5)$ **31.** $y = \dfrac{1}{2}x + \dfrac{1}{2}, \dfrac{1}{2}, \left(0, \dfrac{1}{2}\right)$

33. $y = 4, 0, (0, 4)$ **35.** $y = 0.03x - 2.6, 0.03, (0, -2.6)$

37.

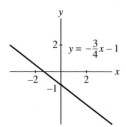

$y = \dfrac{1}{2}x - 2$

39.

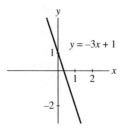

$y = -3x + 1$

41.

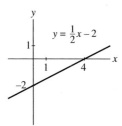

$y = -\dfrac{3}{4}x - 1$

43.

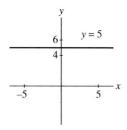

$x - y = 3$

45.

$y = 5$

47. $4x - 3y = 12$ **49.** $4x - 5y = -7$ **51.** $x = -4$
53. 0.5 **55.** -1 **57.** 0 **59.** $2x - y = 4$ **61.** $3x + y = 7$

63. $2x + y = -5$ **65.** $y = 5$ **67.** $F = \dfrac{9}{5}C + 32, 302°F$

69. $c = 50 - n$, \$400

71. $c = -\dfrac{3}{4}p + 30, -\dfrac{3}{4}$, If p increases by 4 then c decreases by 3.

Section 1.5

For Thought: **1.** F **2.** F **3.** T **4.** F **5.** F **6.** F
7. T **8.** T **9.** T **10.** F
Exercises: **1.** Both **3.** a is a function of b **5.** b is a function of a
7. Neither **9.** Both **11.** No **13.** Yes **15.** Yes
17. Yes **19.** No **21.** No **23.** Yes **25.** Yes **27.** Yes
29. No **31.** Yes **33.** Yes **35.** No **37.** $\{-3, 4, 5\}, \{1, 2, 6\}$
39. $(-\infty, \infty), \{4\}$ **41.** $(-\infty, \infty), [5, \infty)$ **43.** $[-3, \infty), (-\infty, \infty)$
45. $[4, \infty), [0, \infty)$ **47.** $(-\infty, 0], (-\infty, \infty)$ **49.** 6 **51.** 11
53. 3 **55.** 7 **57.** 22 **59.** $3a^2 - a$ **61.** $4a + 6$
63. $3x^2 + 5x + 2$ **65.** $4x + 4h - 2$ **67.** $6xh + 3h^2 - h$
69. $-\$2,400$ per yr
71. $-32, -48, -62.4, -63.84$, and -63.984 ft/sec **73.** 4 **75.** 3
77. $2x + h + 1$ **79.** $-2x - h + 1$
81. $\dfrac{3}{\sqrt{x + h} + \sqrt{x}}$ **83.** $\dfrac{1}{\sqrt{x + h + 2} + \sqrt{x + 2}}$
85. $\dfrac{-1}{x(x + h)}$ **87.** $\dfrac{-3}{(x + 2)(x + h + 2)}$

89. a. $A = s^2$ **b.** $s = \sqrt{A}$ **c.** $s = \dfrac{d\sqrt{2}}{2}$ **d.** $d = s\sqrt{2}$

 e. $P = 4s$ **f.** $s = P/4$ **g.** $A = \dfrac{P^2}{16}$ **h.** $d = \sqrt{2A}$

91. a. 4 atm **b.** 130 ft **93.** $h = \left(2\sqrt{3} + 2\right)a$

Section 1.6

For Thought: **1.** T **2.** F **3.** T **4.** T **5.** F **6.** T
7. T **8.** T **9.** F **10.** T
Exercises:
1. $(-\infty, \infty), (-\infty, \infty)$, yes **3.** $(-\infty, \infty), (-\infty, \infty)$, yes

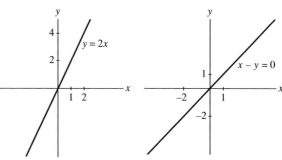

$y = 2x$

$x - y = 0$

5. $(-\infty, \infty), \{5\}$, yes **7.** $(-\infty, \infty), [0, \infty)$, yes

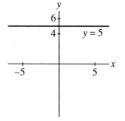

$y = 5$

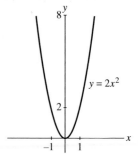

$y = 2x^2$

9. $(-\infty, \infty)$, $(-\infty, 1]$, yes

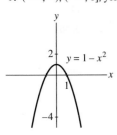

11. $[0, \infty)$, $[1, \infty)$, yes

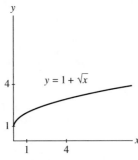

29. $(-\infty, \infty)$, $(-\infty, 0]$, yes

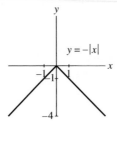

31. $[0, \infty)$, $(-\infty, \infty)$, no

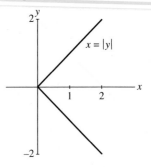

13. $[1, \infty)$, $(-\infty, \infty)$, no

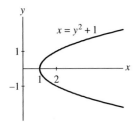

15. $[0, \infty)$, $[0, \infty)$, yes

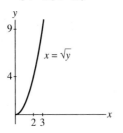

33. $(-\infty, \infty)$, $\{-2, 2\}$

$$f(x) = \begin{cases} 2 & x < -1 \\ -2 & x \ge -1 \end{cases}$$

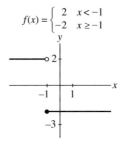

35. $(-\infty, \infty)$, $(-\infty, -2] \cup (2, \infty)$

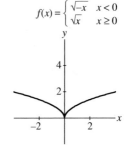

$$f(x) = \begin{cases} x+1 & x > 1 \\ x-3 & x \le 1 \end{cases}$$

17. $(-\infty, \infty)$, $(-\infty, \infty)$, yes

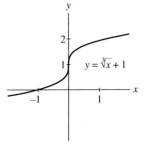

19. $(-\infty, \infty)$, $(-\infty, \infty)$, yes

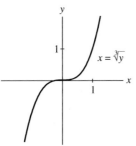

37. $[-2, \infty)$, $(-\infty, 2]$

$$f(x) = \begin{cases} \sqrt{x+2} & -2 \le x \le 2 \\ 4-x & x > 2 \end{cases}$$

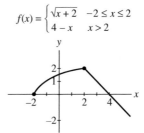

39. $(-\infty, \infty)$, $[0, \infty)$

$$f(x) = \begin{cases} \sqrt{-x} & x < 0 \\ \sqrt{x} & x \ge 0 \end{cases}$$

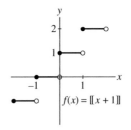

21. $[-1, 1]$, $[-1, 1]$, no

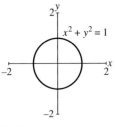

23. $[-1, 1]$, $[0, 1]$, yes

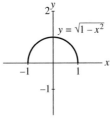

41. $(-\infty, \infty)$, $(-\infty, \infty)$

$$f(x) = \begin{cases} x^2 & x < -1 \\ -x & x \ge -1 \end{cases}$$

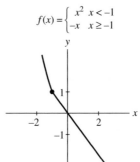

43. $(-\infty, \infty)$, integers

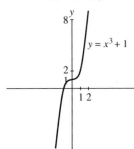

$$f(x) = [\![x+1]\!]$$

25. $(-\infty, \infty)$, $(-\infty, \infty)$, yes

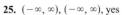

27. $(-\infty, \infty)$, $[0, \infty)$, yes

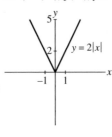

45. [0, 4), {2, 3, 4, 5}

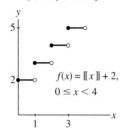

$f(x) = [\![x]\!] + 2,$
$0 \le x < 4$

47. a. $D(-\infty, \infty)$, $R(-\infty, \infty)$, dec $(-\infty, \infty)$
  b. $D(-\infty, \infty)$, $R(-\infty, 4]$, inc $(-\infty, 0)$, dec $(0, \infty)$
49. a. $D[-2, 6]$, $R[3, 7]$, inc $(-2, 2)$, dec $(2, 6)$
  b. $D(-\infty, 2]$, $R(-\infty, 3]$, inc $(-\infty, -2)$, constant $(-2, 2)$
51. a. $D(-\infty, \infty)$, $R[0, \infty)$, dec $(-\infty, 0)$, inc $(0, \infty)$
  b. $D(-\infty, \infty)$, $R(-\infty, \infty)$, dec $(-\infty, -2)$ and $(-2/3, \infty)$, inc $(-2, -2/3)$
53. a. $D(-\infty, \infty)$, $R(-\infty, \infty)$, inc $(-\infty, \infty)$
  b. $D[-2, 5]$, $R[1, 4]$, dec $(-2, 1)$, inc $(1, 2)$, constant $(2, 5)$
55. $(-\infty, \infty)$, $(-\infty, \infty)$, inc $(-\infty, \infty)$

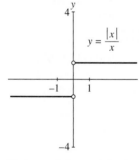

$f(x) = 2x + 1$

57. $(-\infty, \infty)$, $[0, \infty)$, dec $(-\infty, 1)$, inc $(1, \infty)$

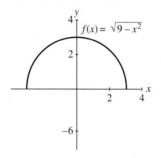

$f(x) = |x - 1|$

59. $(-\infty, 0) \cup (0, \infty)$, {-1, 1}, constant $(-\infty, 0)$, $(0, \infty)$

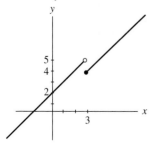

$y = \dfrac{|x|}{x}$

61. $[-3, 3]$, $[0, 3]$, inc $(-3, 0)$, dec $(0, 3)$

$f(x) = \sqrt{9 - x^2}$

63. $(-\infty, \infty)$, $(-\infty, \infty)$, inc $(-\infty, 3)$, $(3, \infty)$

$f(x) = \begin{cases} x + 1 & x \ge 3 \\ x + 2 & x < 3 \end{cases}$

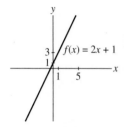

65. $(-\infty, \infty)$, $(-\infty, 2]$,
  inc $(-\infty, -2)$, $(-2, 0)$
  dec $(0, 2)$, $(2, \infty)$

$f(x) = \begin{cases} x + 3 & x \le -2 \\ \sqrt{4 - x^2} & -2 < x < 2 \\ -x + 3 & x \ge 2 \end{cases}$

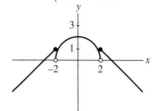

67. $f(x) = \begin{cases} 2 & \text{for } x > -1 \\ -1 & \text{for } x \le -1 \end{cases}$ **69.** $f(x) = \begin{cases} x - 1 & \text{for } x \ge -1 \\ -x & \text{for } x < -1 \end{cases}$

71. $f(x) = \begin{cases} 2x - 2 & \text{for } x \ge 0 \\ -x - 2 & \text{for } x < 0 \end{cases}$

73. Dec $(-\infty, 0.83)$, inc $(0.83, \infty)$
75. Inc $(-\infty, -1)$, $(1, \infty)$, dec $(-1, 1)$
77. Dec $(-\infty, -1.73)$, $(0, 1.73)$, inc $(-1.73, 0)$, $(1.73, \infty)$
79. Inc $(30, 50)$, $(70, \infty)$, dec $(-\infty, 30)$, $(50, 70)$
81. 565 million, 800 million, 14.5 million/yr
83. $[5, \infty)$

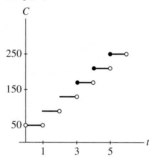

Section 1.7

For Thought: **1.** F   **2.** T   **3.** F   **4.** T   **5.** T   **6.** F
7. T   **8.** F   **9.** T   **10.** T
Exercises:
1.   **3.**

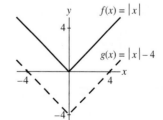

$f(x) = |x|$
$g(x) = |x| - 4$

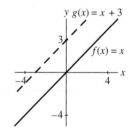

$y\ g(x) = x + 3$
$f(x) = x$

5.

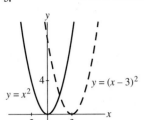

7.

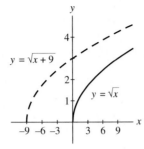

39. $(-\infty, \infty), (-\infty, \infty)$

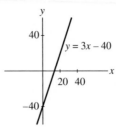

41. $(-\infty, \infty), (-\infty, \infty)$

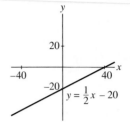

43. $(-\infty, \infty), (-\infty, 40]$

45. $(-\infty, \infty), (-\infty, 0]$

9.

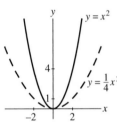

11.

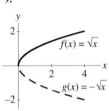

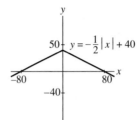

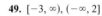

47. $[3, \infty), (-\infty, 1]$

49. $[-3, \infty), (-\infty, 2]$

13.

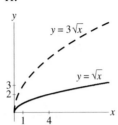

15.

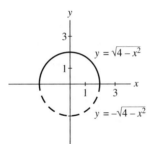

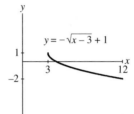

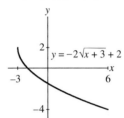

51. y-axis, even **53.** No symmetry, neither
55. $x = -3$, neither **57.** $x = 2$, neither **59.** Origin, odd
61. No symmetry, neither **63.** No symmetry, neither
65. y-axis, even **67.** e **69.** g **71.** b **73.** c
75. $(-\infty, -1] \cup [1, \infty)$ **77.** $(-\infty, -1) \cup (5, \infty)$

17. g **19.** b **21.** c **23.** f **25.** $y = \sqrt{x} + 2$
27. $y = (x - 5)^2$ **29.** $y = (x - 10)^2 + 4$
31. $y = -3\sqrt{x} - 5$ **33.** $y = -3|x - 7| + 9$
35. $(-\infty, \infty), [2, \infty)$ **37.** $(-\infty, \infty), [3, \infty)$

79. $(-2, 4)$ **81.** $[0, 25]$ **83.** $\left(-\infty, 2 - \sqrt{3}\right) \cup \left(2 + \sqrt{3}, \infty\right)$
85. $(-5, 5)$ **87.** $(-3.36, 1.55)$
89. a.

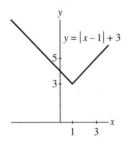

b.

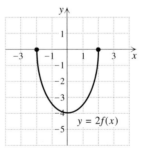
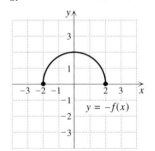

c.

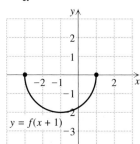

$y = f(x + 1)$

d.

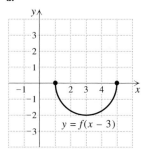

$y = f(x - 3)$

e.

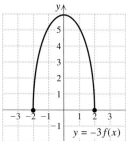

$y = -3f(x)$

f.

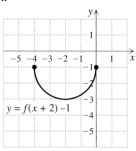

$y = f(x + 2) - 1$

g.

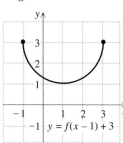

$y = f(x - 1) + 3$

h.

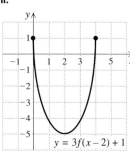

$y = 3f(x - 2) + 1$

91. $N(x) = x + 2000$

Section 1.8

For Thought: **1.** F   **2.** T   **3.** T   **4.** T   **5.** T   **6.** T
7. F   **8.** T   **9.** F   **10.** T
Exercises: **1.** 1   **3.** −11   **5.** −8   **7.** 1/12   **9.** $a^2 - 3$
11. $a^3 - 4a^2 + 3a$ **13.** $\{(-3, 3), (2, 6)\}, \{-3, 2\}$
15. $\{(-3, -1), (2, -6)\}, \{-3, 2\}$ **17.** $\{(-3, 2), (2, 0)\}, \{-3, 2\}$
19. $\{(-3, 2)\}, \{-3\}$ **21.** $(f + g)(x) = \sqrt{x} + x - 4, [0, \infty)$
23. $(f - h)(x) = \sqrt{x} - \dfrac{1}{x - 2}, [0, 2) \cup (2, \infty)$
25. $(g \cdot h)(x) = \dfrac{x - 4}{x - 2}, (-\infty, 2) \cup (2, \infty)$
27. $(g/f)(x) = \dfrac{x - 4}{\sqrt{x}}, (0, \infty)$ **29.** $\{(-3, 0), (1, 0), (4, 4)\}$
31. $\{(1, 4)\}$ **33.** $\{(-3, 4), (1, 4)\}$ **35.** 5   **37.** 5
39. 59.816   **41.** 5   **43.** 5   **45.** a   **47.** $3t^2 + 2$
49. $(f \circ g)(x) = \sqrt{x} - 2, [0, \infty)$
51. $(f \circ h)(x) = \dfrac{1}{x} - 2, (-\infty, 0) \cup (0, \infty)$

53. $(h \circ g)(x) = \dfrac{1}{\sqrt{x}}, (0, \infty)$ **55.** $(f \circ f)(x) = x - 4, (-\infty, \infty)$

57. $(h \circ g \circ f)(x) = \dfrac{1}{\sqrt{x - 2}}, (2, \infty)$

59. $(h \circ f \circ g)(x) = \dfrac{1}{\sqrt{x - 2}}, (0, 4) \cup (4, \infty)$

61. $F = g \circ h$ **63.** $H = h \circ g$ **65.** $N = h \circ g \circ f$
67. $P = g \circ f \circ g$ **69.** $S = g \circ g$ **71.** $y = 6x - 1$
73. $y = x^2 + 6x + 7$ **75.** $y = x$ **77.** $[-1, \infty), [-7, \infty)$
79. $[1, \infty), [0, \infty)$ **81.** $[0, \infty), [4, \infty)$
83. $P(x) = 28x - 200, x \geq 8$ **85.** $A = d^2/2$ **87.** $T(x) = 1.26x$
89. $D = \dfrac{1.16 \times 10^7}{L^3}$ **91.** $W = \dfrac{(8 + \pi)s^2}{8}$

Section 1.9

For Thought: **1.** F   **2.** F   **3.** F   **4.** T   **5.** F   **6.** F   **7.** F
8. F   **9.** F   **10.** T
Exercises: **1.** Yes   **3.** No   **5.** No   **7.** Not one-to-one
9. One-to-one   **11.** Not one-to-one   **13.** One-to-one
15. One-to-one   **17.** Not one-to-one   **19.** Not one-to-one
21. One-to-one   **23.** Invertible, $\{(3, 9), (2, 2)\}$
25. Not invertible   **27.** Invertible, $\{(3, 3), (2, 2), (4, 4), (7, 7)\}$
29. Not invertible   **31.** Not invertible   **33.** Invertible
35. Invertible   **37.** $\{(1, 2), (5, 3)\}, 3, 2$
39. $\{(-3, -3), (5, 0), (-7, 2)\}, 0, 2$ **41.** Not invertible

43. Not invertible   **45.** $f^{-1}(x) = \dfrac{x + 7}{3}$

47. $f^{-1}(x) = (x - 2)^2 + 3$ for $x \geq 2$ **49.** $f^{-1}(x) = -x - 9$

51. $f^{-1}(x) = \dfrac{5x + 3}{x - 1}$ **53.** $f^{-1}(x) = -\dfrac{1}{x}$

55. $f^{-1}(x) = (x - 5)^3 + 9$ **57.** $f^{-1}(x) = \sqrt{x} + 2$
59. $f(g(x)) = x, g(f(x)) = x$, yes   **61.** $f(g(x)) = x, g(f(x)) = |x|$, no
63. $f(g(x)) = x, g(f(x)) = x$, yes
65. $f(g(x)) = x, g(f(x)) = x$, yes
67. The functions y_1 and y_2 are inverses.   **69.** No   **71.** Yes
73.   **75.**

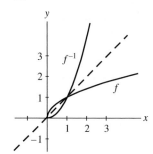

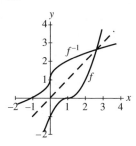

77. $f^{-1}(x) = \dfrac{x - 2}{3}$ **79.** $f^{-1}(x) = \sqrt{x + 4}$

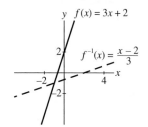

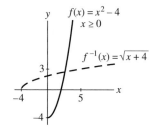

81. $f^{-1}(x) = \sqrt[3]{x}$

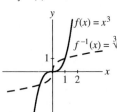

83. $f^{-1}(x) = (x + 3)^2$ for $x \ge -3$

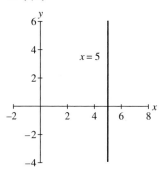

85. a. $f^{-1}(x) = \dfrac{x}{5}$ **b.** $f^{-1}(x) = x + 88$ **c.** $f^{-1}(x) = \dfrac{x + 7}{3}$

d. $f^{-1}(x) = \dfrac{x - 4}{-3}$ **e.** $f^{-1}(x) = 2x + 18$ **f.** $f^{-1}(x) = -x$

g. $f^{-1}(x) = (x + 9)^3$ **h.** $f^{-1}(x) = \sqrt[3]{\dfrac{x + 7}{3}}$

87. $C = 1.08P, P = \dfrac{C}{1.08}$ **89.** Yes, $r = \dfrac{7.89 - t}{0.39}, 6$

91. $w = \dfrac{V^2}{1.496}, 8840$ lb   **93. a.** 10.9%   **b.** $V = 50,000(1 - r)^5$

Chapter 1 Review Exercises

1. F   **3.** F   **5.** F   **7.** F   **9.** F   **11.** F

13. $\left\{\dfrac{2}{3}, 2\right\}$ **15.** $\{3/2\}$ **17.** No solutions

19. $(3, \infty)$

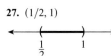

21. $(-\infty, 4)$

23. $(-\infty, -14/3)$

25. $(-1, 13]$

27. $(1/2, 1)$

29. $(-4, \infty)$

31. $(-\infty, 1) \cup (5, \infty)$

33. $\{7/2\}$

35. $(-\infty, \infty)$

37. $(0, 0), 5$

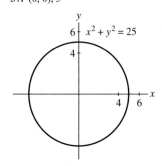

39. $(-2, 0), 2$

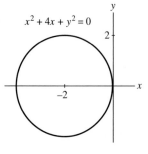

41. $(25, 0), (0, 25)$

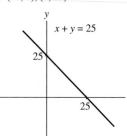

43. $(4/3, 0), (0, -4)$

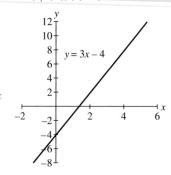

45. $(5, 0)$

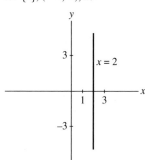

47. $\sqrt{34}$ **49.** $(x + 3)^2 + (y - 5)^2 = 3$ **51.** $(4, 0), (0, -3)$

53. -2 **55.** $y = -\dfrac{4}{7}x + \dfrac{13}{7}$ **57.** $x - 3y = 14$

59. $\{-2, 0, 1\}, \{-2, 0, 1\}$, yes   **61.** $(-\infty, \infty), (-\infty, \infty)$, yes

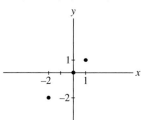

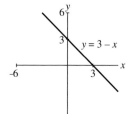

63. $\{2\}, (-\infty, \infty)$, no   **65.** $[-0.1, 0.1], [-0.1, 0.1]$, no

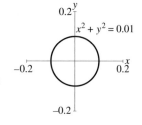

67. $[1, \infty), (-\infty, \infty)$, no

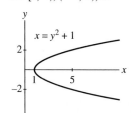

69. $[0, \infty), [-3, \infty)$, yes

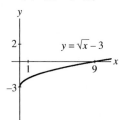

71. 12   **73.** 17   **75.** ± 4 **77.** 17   **79.** 4   **81.** -36
83. 12   **85.** $4x^2 - 28x + 52$ **87.** $x^4 + 6x^2 + 12$
89. $a^2 + 2a + 4$ **91.** $6 + h$ **93.** $2x + h$
95. x **97.** $\dfrac{x + 7}{2}$

99.

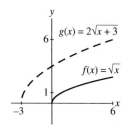

101.

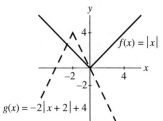

103.

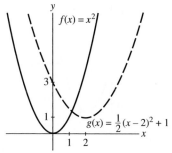

105.

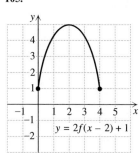

107.

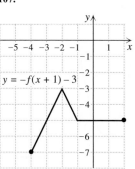

109.

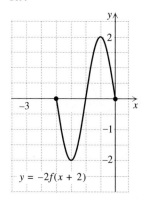

111.

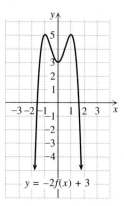

113. $F = f \circ g$ **115.** $H = f \circ h \circ g \circ j$ **117.** $N = h \circ f \circ j$
119. $R = g \circ h \circ j$ **121.** -5 **123.** $\dfrac{-1}{2x(x + h)}$
125. $[-10, 10], [0, 10]$,   **127.** $(-\infty, \infty), (-\infty, \infty)$,
   inc $(-10, 0)$, dec $(0, 10)$ inc $(-\infty, \infty)$

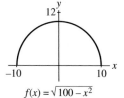
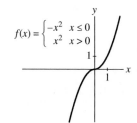

129. $(-\infty, \infty), [-2, \infty)$,
   inc $(-2, 0), (2, \infty)$,
   dec $(-\infty, -2)$ and $(0, 2)$

$$f(x) = \begin{cases} -x - 4 & x \le -2 \\ -|x| & -2 < x < 2 \\ x - 4 & x \ge 2 \end{cases}$$

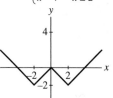

131. $y = |x| - 3, (-\infty, \infty), [-3, \infty)$
133. $y = -2|x| + 4, (-\infty, \infty), (-\infty, 4]$
135. $y = |x + 2| + 1, (-\infty, \infty), [1, \infty)$ **137.** y-axis   **139.** Origin
141. Neither symmetry   **143.** y-axis

145. Inverse functions

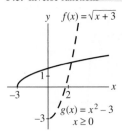

147. Inverse functions

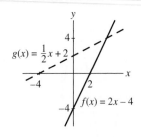

149. Not invertible   **151.** $f^{-1}(x) = \dfrac{x + 21}{3}$, $(-\infty, \infty)$, $(-\infty, \infty)$

153. Not invertible   **155.** $f^{-1}(x) = x^2 + 9$ for $x \geq 0$, $[0, \infty)$, $[9, \infty)$

157. $f^{-1}(x) = \dfrac{5x + 7}{1 - x}$, $(-\infty, 1) \cup (1, \infty)$, $(-\infty, -5) \cup (-5, \infty)$

159. $f^{-1}(x) = -\sqrt{x - 1}$, $[1, \infty)$, $(-\infty, 0]$

161. $C(x) = 1.20x + 40$, $R(x) = 2x$, $P(x) = 0.80x - 40$ where x is the number of roses, 51 or more roses

163. $t = \dfrac{\sqrt{64 - h}}{4}$, domain $[0, 64]$ **165.** $d = 2\sqrt{A/\pi}$

167. 0.5 in./lb

Chapter 2

Section 2.1

For Thought: **1.** F   **2.** F   **3.** T   **4.** T   **5.** T   **6.** T
7. T   **8.** T   **9.** T   **10.** F

Exercises:

1. $y = (x + 2)^2 - 4$ **3.** $y = \left(x - \dfrac{3}{2}\right)^2 - \dfrac{9}{4}$

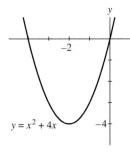

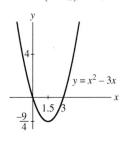

5. $y = 2(x - 3)^2 + 4$ **7.** $y = -3(x - 1)^2$

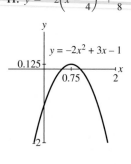

9. $y = \left(x + \dfrac{3}{2}\right)^2 + \dfrac{1}{4}$ **11.** $y = -2\left(x - \dfrac{3}{4}\right)^2 + \dfrac{1}{8}$

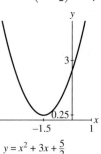

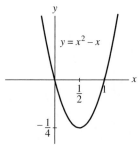

$y = x^2 + 3x + \dfrac{5}{2}$

13. $(2, -11)$ **15.** $(4, 1)$ **17.** $(-1/3, 1/18)$

19. Up, $(1, -4)$, $x = 1$, $[-4, \infty)$, min -4, dec $(-\infty, 1)$, inc $(1, \infty)$

21. $(-\infty, 3]$, max 3, inc $(-\infty, 0)$, dec $(0, \infty)$

23. $[-1, \infty)$, min -1, dec $(-\infty, 1)$, inc $(1, \infty)$

25. $[-18, \infty)$, min value -18, dec $(-\infty, -4)$, inc $(-4, \infty)$

27. $[4, \infty)$, min value 4, dec $(-\infty, 3)$, inc $(3, \infty)$

29. $(-\infty, 27/2]$, max 27/2, inc $(-\infty, 3/2)$, dec $(3/2, \infty)$

31. $(-\infty, 9]$, max 9, inc $(-\infty, 1/2)$, dec $(1/2, \infty)$

33. $(0, -3)$, $x = 0$, $(0, -3)$, $(\pm\sqrt{3}, 0)$, up

35. $\left(\dfrac{1}{2}, -\dfrac{1}{4}\right)$, $x = \dfrac{1}{2}$, $(0, 0)$, $(1, 0)$, up

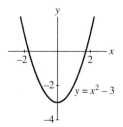

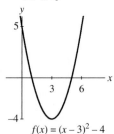

37. $(-3, 0)$, $x = -3$, $(0, 9)$, $(-3, 0)$, up

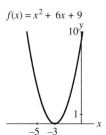

39. $(3, -4)$, $x = 3$, $(0, 5)$, $(1, 0)$, $(5, 0)$, up

41. $(2, 12), x = 2, (0, 0),$
$(4, 0)$, down

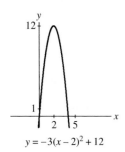

$$y = -3(x - 2)^2 + 12$$

43. $(1, 3), x = 1, (0, 1),$

$$\left(1 \pm \frac{\sqrt{6}}{2}, 0\right), \text{ down}$$

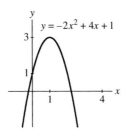

$$y = -2x^2 + 4x + 1$$

h. $(-2, 0), (5, 0), (0, -10), x = \frac{3}{2}, \left(\frac{3}{2}, -\frac{49}{4}\right)$, opens upward, dec on $\left(-\infty, \frac{3}{2}\right)$ and inc on $\left(\frac{3}{2}, \infty\right)$

77. 261 ft   **79. a.** 408 ft   **b.** $(10 + \sqrt{102})/2 \approx 10.05$ sec
81. 50 yd by 50 yd   **83.** 20 ft by 30 ft   **85.** 7.5 ft by 15 ft
87. a. $p = 50 - n$ **b.** $R = 50n - n^2$ **c.** \$625

Section 2.2

For Thought: 1. T   **2.** T   **3.** F   **4.** T   **5.** F   **6.** F
7. T   **8.** T   **9.** T   **10.** F

Exercises: 1. Imaginary, $0 + 6i$ **3.** Imaginary, $\frac{1}{3} + \frac{1}{3}i$

5. Real, $\sqrt{7} + 0i$ **7.** Real, $\frac{\pi}{2} + 0i$ **9.** $7 + 2i$ **11.** $-2 - 3i$
13. $-12 - 18i$ **15.** 26   **17.** 29   **19.** 4   **21.** $-7 + 24i$
23. $1 - 4i\sqrt{5}$ **25.** i **27.** -1 **29.** 1   **31.** $-i$ **33.** 90
35. $17/4$ **37.** 1   **39.** 12   **41.** $\frac{2}{5} + \frac{1}{5}i$ **43.** $\frac{3}{2} - \frac{3}{2}i$
45. $-1 + 3i$ **47.** $3 + 3i$ **49.** $\frac{1}{13} - \frac{5}{13}i$ **51.** $-i$
53. $-4 + 2i$ **55.** -6 **57.** -10 **59.** $-1 + i\sqrt{5}$
61. $-3 + i\sqrt{11}$ **63.** $-4 + 8i$ **65.** $-1 + 2i$
67. $\frac{-2 + i\sqrt{2}}{2}$ **69.** $-3 - 2i\sqrt{2}$ **71.** $\frac{3 + \sqrt{21}}{2}$ **73.** $\{\pm i\}$
75. $\left\{\pm 2i\sqrt{2}\right\}$ **77.** $\left\{\pm\frac{i\sqrt{2}}{2}\right\}$ **79.** $\{1 \pm i\}$ **81.** $\{2 \pm 3i\}$
83. $\left\{1 \pm i\sqrt{3}\right\}$ **85.** $\left\{\frac{1}{2} \pm \frac{3}{2}i\right\}$ **87.** $\left\{1 \pm \frac{\sqrt{3}}{2}i\right\}$

Section 2.3

For Thought: 1. F   **2.** T   **3.** T   **4.** T   **5.** F   **6.** F
7. F   **8.** T   **9.** T   **10.** F
Exercises: 1. $x - 3, 1$ **3.** $-2x^2 + 6x - 14, 33$ **5.** $s^2 + 2, 16$
7. $x + 6, 13$ **9.** $-x^2 + 4x - 16, 57$ **11.** $4x^2 + 2x - 4, 0$
13. $2a^2 - 4a + 6, 0$ **15.** $x^3 + x^2 + x + 1, -2$
17. $x^4 + 2x^3 - 2x^2 - 4x - 4, -13$ **19.** 0   **21.** -33
23. 5   **25.** $\frac{55}{8}$ **27.** 0   **29.** 8   **31.** $(x + 3)(x + 2)(x - 1)$
33. $(x - 4)(x + 3)(x + 5)$ **35.** Yes   **37.** No   **39.** Yes
41. No   **43.** $\pm(1, 2, 3, 4, 6, 8, 12, 24)$ **45.** $\pm(1, 3, 5, 15)$
47. $\pm\left(1, 3, 5, 15, \frac{1}{2}, \frac{1}{4}, \frac{1}{8}, \frac{3}{2}, \frac{3}{4}, \frac{3}{8}, \frac{5}{2}, \frac{5}{4}, \frac{5}{8}, \frac{15}{2}, \frac{15}{4}, \frac{15}{8}\right)$
49. $\pm\left(1, 2, \frac{1}{2}, \frac{1}{3}, \frac{2}{3}, \frac{1}{6}, \frac{1}{9}, \frac{2}{9}, \frac{1}{18}\right)$ **51.** $2, 3, 4$
53. $-3, 2 \pm i$ **55.** $\frac{1}{2}, \frac{3}{2}, \frac{5}{2}$ **57.** $\frac{1}{2}, \frac{1 \pm i}{3}$ **59.** $\pm i, 1, -2$
61. $-1, \pm\sqrt{2}$ **63.** $\frac{1}{4}, \frac{1}{3}, \frac{1}{2}$ **65.** $\frac{1}{16}, 1 \pm 2i$ **67.** $-\frac{6}{7}, \frac{7}{3}, \pm i$
69. $-5, -2, 1, \pm 3i$ **71.** $1, 3, 5, 2 \pm \sqrt{3}$ **73.** $2 + \frac{5}{x - 2}$
75. $a + \frac{5}{a - 3}$ **77.** $1 + \frac{-3c}{c^2 - 4}$ **79.** $2 + \frac{-7}{2t + 1}$
81. a. 6 hr   **b.** ≈ 120 ppm   **c.** ≈ 3 hr   **d.** ≈ 4 hr
83. 5 in. by 9 in. by 14 in.

45. $(-1, 3/2)$

$$\xleftrightarrow{\hspace{1cm}\underset{-1}{(}\hspace{1cm}\underset{\frac{3}{2}}{)}\hspace{1cm}}$$

47. $(-\infty, -3) \cup (5, \infty)$

$$\xleftrightarrow{\hspace{0.8cm}\underset{-3}{)}\hspace{1.5cm}\underset{5}{(}\hspace{0.8cm}}$$

49. $(-\infty, -2] \cup [6, \infty)$

$$\xleftrightarrow{\hspace{0.8cm}\underset{-2}{]}\hspace{1.5cm}\underset{6}{[}\hspace{0.8cm}}$$

51. $[-4, 4]$

$$\xleftrightarrow{\hspace{1cm}\underset{-4}{[}\hspace{1cm}\underset{4}{]}\hspace{1cm}}$$

53. $\{-3\}$

$$\xleftrightarrow{\hspace{1cm}\underset{-3}{\bullet}\hspace{1cm}}$$

55. $(-\infty, 3/2) \cup (3/2, \infty)$

$$\xleftrightarrow{\hspace{1.5cm}\underset{\frac{3}{2}}{\times}\hspace{1.5cm}}$$

57. $\left(2 - \sqrt{2}, 2 + \sqrt{2}\right)$

$$\xleftrightarrow{\hspace{0.8cm}\underset{2 - \sqrt{2}}{(}\hspace{1cm}\underset{2 + \sqrt{2}}{)}\hspace{0.8cm}}$$

59. $\left(-\infty, -\sqrt{10}\right) \cup \left(\sqrt{10}, \infty\right)$

$$\xleftrightarrow{\hspace{0.6cm}\underset{-\sqrt{10}}{)}\hspace{1.5cm}\underset{\sqrt{10}}{(}\hspace{0.6cm}}$$

61. $\left(-\infty, 5 - \sqrt{7}\right) \cup$
$\left(5 + \sqrt{7}, \infty\right)$

$$\xleftrightarrow{\hspace{0.6cm}\underset{5 - \sqrt{7}}{)}\hspace{1.5cm}\underset{5 + \sqrt{7}}{(}\hspace{0.6cm}}$$

63. $(-\infty, \infty)$

$$\xleftrightarrow{\hspace{4cm}}$$

65. No solution

67. $(-\infty, \infty)$

$$\xleftrightarrow{\hspace{4cm}}$$

69. $(-\infty, -1] \cup [3, \infty)$ **71.** $(-3, 1)$ **73.** $[-3, 1]$
75. a. $\{-2, 5\}$ **b.** $\{0, 3\}$ **c.** $(-\infty, -2) \cup (5, \infty)$ **d.** $[-2, 5]$
e. $f(x) = \left(x - \frac{3}{2}\right)^2 - \frac{49}{4}$, Move $y = x^2$ to the right $\frac{3}{2}$ and down $\frac{49}{4}$ to obtain f.
f. $(-\infty, \infty), \left[-\frac{49}{4}, \infty\right)$, minimum $-\frac{49}{4}$

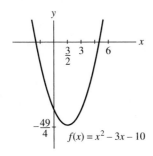

$$f(x) = x^2 - 3x - 10$$

g. The graph of f is above the x-axis when x is in $(-\infty, -2) \cup (5, \infty)$ and on or below the x-axis when x is in $[-2, 5]$.

Section 2.4

For Thought: **1.** F **2.** T **3.** T **4.** F **5.** F **6.** T
7. F **8.** F **9.** T **10.** T

Exercises: **1.** Degree 2, 5 with multiplicity 2
3. Degree 5, ± 3, 0 with multiplicity 3
5. Degree 4, 0, 1 each with multiplicity 2
7. Degree 4, $-\frac{4}{3}, \frac{3}{2}$ each with multiplicity 2 **9.** Degree 3, 0, $2 \pm \sqrt{10}$

11. $x^2 + 9$ **13.** $x^2 - 2x - 1$ **15.** $x^2 - 6x + 13$
17. $x^3 - 8x^2 + 37x - 50$ **19.** $x^2 - 2x - 15 = 0$
21. $x^2 + 16 = 0$ **23.** $x^2 - 6x + 10 = 0$
25. $x^3 + 2x^2 + x + 2 = 0$ **27.** $x^3 + 3x = 0$
29. $x^3 - 5x^2 + 8x - 6 = 0$ **31.** $x^3 - 6x^2 + 11x - 6 = 0$
33. $x^3 - 5x^2 + 17x - 13 = 0$ **35.** $24x^3 - 26x^2 + 9x - 1 = 0$
37. $x^4 - 2x^3 + 3x^2 - 2x + 2 = 0$ **39.** 3 neg; 1 neg, 2 imag
41. 1 pos, 2 neg; 1 pos, 2 imag **43.** 4 imag
45. 4 pos; 2 pos, 2 imag; 4 imag **47.** 4 imag and 0

49. $-2, 1, 5$ **51.** $-3, \dfrac{3 \pm \sqrt{13}}{2}$ **53.** $\pm i, 2, -4$ **55.** $-5, \dfrac{1}{3}, \dfrac{1}{2}$

57. 1, -2 each with multiplicity 2 **59.** 0, 2 with multiplicity 3

61. 0, 1, ± 2, $\pm i\sqrt{3}$ **63.** $-2, -1, 1/4, 1, 3/2$
65. 4 hr and 5 hr **67.** 3 in.

Section 2.5

For Thought: **1.** F **2.** F **3.** F **4.** F **5.** T **6.** F
7. F **8.** T **9.** T **10.** F

Exercises: **1.** $\{\pm 2, -3\}$ **3.** $\left\{-500, \pm \dfrac{\sqrt{2}}{2}\right\}$

5. $\left\{0, \dfrac{15 \pm \sqrt{205}}{2}\right\}$ **7.** $\{0, \pm 2\}$ **9.** $\{\pm 2, \pm 2i\}$ **11.** $\{8\}$

13. $\{25\}$ **15.** $\left\{\dfrac{1}{4}\right\}$ **17.** $\left\{\dfrac{2 + \sqrt{13}}{9}\right\}$ **19.** $\{-4, 6\}$

21. $\{9\}$ **23.** $\{5\}$ **25.** $\{10\}$ **27.** $\{\pm 2\sqrt{2}\}$ **29.** $\left\{\pm \dfrac{1}{8}\right\}$

31. $\left\{\dfrac{1}{49}\right\}$ **33.** $\left\{\dfrac{5}{4}\right\}$ **35.** $\{\pm 3, \pm \sqrt{3}\}$ **37.** $\left\{-\dfrac{17}{2}, \dfrac{13}{2}\right\}$

39. $\left\{\dfrac{3}{20}, \dfrac{4}{15}\right\}$ **41.** $\{-2, -1, 5, 6\}$ **43.** $\{1, 9\}$ **45.** $\{9, 16\}$

47. $\{8, 125\}$ **49.** $\{\pm \sqrt{7}, \pm 1\}$ **51.** $\{0, 8\}$ **53.** $\{-3, 0, 1, 4\}$

55. $\{-2, 4\}$ **57.** $\left\{\dfrac{1}{2}\right\}$ **59.** $\{\pm 2, -1 \pm i\sqrt{3}, 1 \pm i\sqrt{3}\}$

61. $\{\sqrt{3}, 2\}$ **63.** $\{-2, \pm 1\}$ **65.** $\{5 \pm 9i\}$ **67.** $\left\{\dfrac{1 \pm 4\sqrt{2}}{3}\right\}$

69. $\{\pm 2\sqrt{6}, \pm \sqrt{35}\}$ **71.** $\{\pm 3, 2\}$ **73.** $\{-11\}$ **75.** $\{2\}$

77. 279.56 m^2 **79.** $\dfrac{25}{4}$ and $\dfrac{49}{4}$ **81.** 17,419.3 lbs

83. $6 - \sqrt{10}$ ft **85.** 37.5 min, 38.2 min, 1.75 mi or 4.69 mi

Section 2.6

For Thought: **1.** F **2.** T **3.** T **4.** F **5.** T **6.** T
7. F **8.** F **9.** T **10.** F
Exercises: **1.** Symmetric about y-axis **3.** Symmetric about $x = 3/2$
5. Neither symmetry **7.** Symmetric about origin

9. Symmetric about $x = 5$ **11.** Symmetric about origin
13. Does not cross at $(4, 0)$ **15.** Crosses at $(1/2, 0)$
17. Crosses at $(1/4, 0)$ **19.** No x-intercepts
21. Does not cross at $(0, 0)$, crosses at $(3, 0)$
23. Crosses at $(1/2, 0)$, does not cross at $(1, 0)$
25. Does not cross at $(-3, 0)$, crosses at $(2, 0)$
27. $y \to \infty$ **29.** $y \to -\infty$ **31.** $y \to -\infty$ **33.** $y \to \infty$
35. $y \to \infty$
37. Neither symmetry; crosses at $(-2, 0)$; does not cross at $(1, 0)$; $y \to \infty$
as $x \to \infty$; $y \to -\infty$ as $x \to -\infty$
39. Symmetric about y-axis; no x-intercepts; $y \to \infty$ as $x \to \infty$; $y \to \infty$ as
$x \to -\infty$
41. ∞ **43.** $-\infty$ **45.** ∞ **47.** $-\infty$

49.
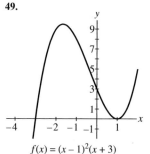

51.
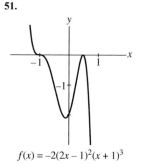

$f(x) = (x - 1)^2(x + 3)$

$f(x) = -2(2x - 1)^2(x + 1)^3$

53. (e) **55.** (g) **57.** (b) **59.** (c)
61.
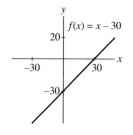

63.
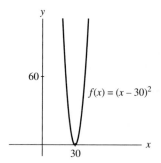

$f(x) = x - 30$

$f(x) = (x - 30)^2$

65.
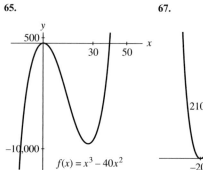

67.
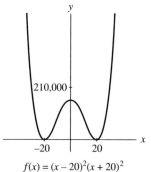

$f(x) = x^3 - 40x^2$

$f(x) = (x - 20)^2(x + 20)^2$

69.

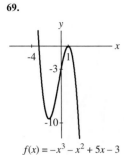

$$f(x) = -x^3 - x^2 + 5x - 3$$

71.

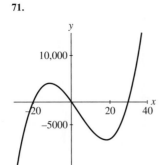

$$f(x) = x^3 - 10x^2 - 600x$$

73.

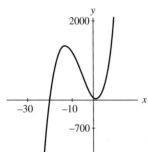

$$f(x) = x^3 + 18x^2 - 37x + 60$$

75.

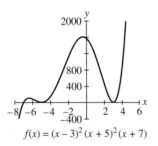

$$f(x) = -x^4 + 196x^2$$

77.

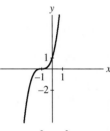

$$f(x) = x^3 + 3x^2 + 3x + 1$$

79.

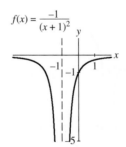

$$f(x) = (x - 3)^2 (x + 5)^2 (x + 7)$$

81. $\left(-\sqrt{3}, 0\right) \cup \left(\sqrt{3}, \infty\right)$ **83.** $\left(-\infty, -\sqrt{2}\right] \cup \{0\} \cup \left[\sqrt{2}, \infty\right)$
85. $(-4, -1) \cup (1, \infty)$ **87.** $[-4, 2] \cup [6, \infty)$ **89.** $(-\infty, 1)$
91. $\left[-\sqrt{10}, -3\right] \cup \left[3, \sqrt{10}\right]$ **93.** d
95. 20.07 in.^{3} **97.** $V = 3x^3 - 24x^2 + 48x$, $4/3$ in. by 4 in. by $16/3$ in.

Section 2.7

For Thought: **1.** F   **2.** F   **3.** F   **4.** F   **5.** T   **6.** F
7. T   **8.** F   **9.** T   **10.** T
Exercises: **1.** $(-\infty, -2) \cup (-2, \infty)$
3. $(-\infty, -2) \cup (-2, 2) \cup (2, \infty)$ **5.** $(-\infty, 3) \cup (3, \infty)$
7. $(-\infty, 0) \cup (0, \infty)$ **9.** $(-\infty, -1) \cup (-1, 0) \cup (0, 1) \cup (1, \infty)$
11. $(-\infty, -3) \cup (-3, -2) \cup (-2, \infty)$
13. $(-\infty, 2) \cup (2, \infty)$, $y = 0$, $x = 2$
15. $(-\infty, 0) \cup (0, \infty)$, $y = x$, $x = 0$ **17.** $x = 2$, $y = 0$

19. $x = \pm 3$, $y = 0$ **21.** $x = 1$, $y = 2$ **23.** $x = 0$, $y = x - 2$
25. $x = -1$, $y = 3x - 3$ **27.** $x = -2$, $y = -x + 6$
29. $x = 0$, $y = 0$ **31.** $x = 2$, $y = 0$, $(0, -1/2)$

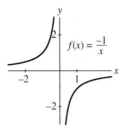

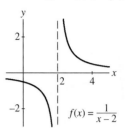

$$f(x) = \frac{-1}{x}$$

$$f(x) = \frac{1}{x - 2}$$

33. $x = \pm 2$, $y = 0$, $(0, -1/4)$

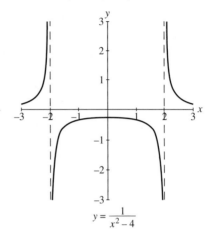

$$y = \frac{1}{x^2 - 4}$$

35. $x = -1$, $y = 0$, $(0, -1)$ **37.** $x = 1$, $y = 2$, $(0, -1)$,
$(-1/2, 0)$

$$f(x) = \frac{-1}{(x + 1)^2}$$

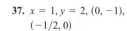

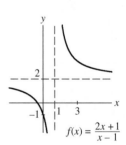

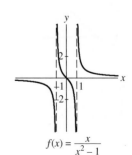

$$f(x) = \frac{2x + 1}{x - 1}$$

39. $x = -2$, $y = 1$, $(3, 0)$,
$(0, -3/2)$

41. $x = \pm 1$, $y = 0$, $(0, 0)$

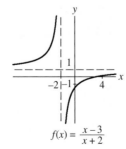

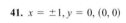

$$f(x) = \frac{x - 3}{x + 2}$$

$$f(x) = \frac{x}{x^2 - 1}$$

43. $x = 1, y = 0, (0, 0)$

45. $x = \pm 3, y = -1, (0, -8/9),$ $\left(\pm \sqrt{8}, 0 \right)$

65. $y = x - 1$

67. $y = -x + 2$

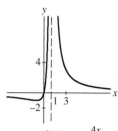

$$f(x) = \frac{4x}{x^2 - 2x + 1}$$

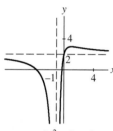

$$f(x) = \frac{8 - x^2}{x^2 - 9}$$

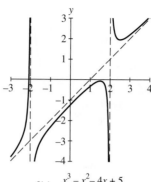

$$f(x) = \frac{x^3 - x^2 - 4x + 5}{x^2 - 4}$$

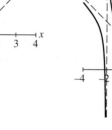

$$f(x) = \frac{-x^3 + x^2 + 5x - 4}{x^2 + x - 2}$$

47. $x = -1, y = 2, (0, 2),$ $\left(-2 \pm \sqrt{3}, 0 \right)$

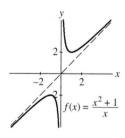

$$f(x) = \frac{2x^2 + 8x + 2}{x^2 + 2x + 1}$$

69. (e)   **71.** (a)   **73.** (b)   **75.** (c)
77. $x \neq \pm 1$ **79.** $x \neq 1$

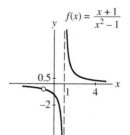

$$f(x) = \frac{x + 1}{x^2 - 1}$$

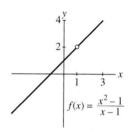

$$f(x) = \frac{x^2 - 1}{x - 1}$$

49. 0   **51.** 2   **53.** ∞ **55.** ∞
57. $y = x$ **59.** $y = x$

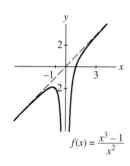

$$f(x) = \frac{x^2 + 1}{x}$$

$$f(x) = \frac{x^3 - 1}{x^2}$$

81. $(-2, 4]$ **83.** $(-\infty, -8) \cup (-3, \infty)$ **85.** $[-2, 3] \cup (6, \infty)$
87. $(-2, 3)$ **89.** $(-\infty, -2) \cup (4, 5)$ **91.** $[-1, 3] \cup (5, \infty)$
93. $\left(-\infty, -\sqrt{7} \right] \cup \left(-\sqrt{2}, \sqrt{2} \right) \cup \left[\sqrt{7}, \infty \right)$
95. $(-\infty, -3) \cup \{-1\} \cup (5, \infty)$ **97.** $(1, \infty)$ **99.** $(3, \infty)$
101. $C = \dfrac{100 + x}{x}, \$2, C \to \$1$ **103.** $S = \dfrac{100}{4 - x}, S \to \infty$ as $x \to 4$
105. a.

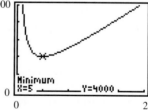

b. 5

61. $y = x - 1$

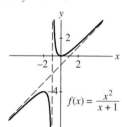

$$f(x) = \frac{x^2}{x + 1}$$

63. $y = 2x + 1$

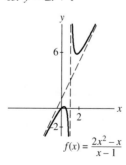

$$f(x) = \frac{2x^2 - x}{x - 1}$$

Chapter 2 Review Exercises

1. $f(x) = 3\left(x - \dfrac{1}{3} \right)^2 + \dfrac{2}{3}$

3. $(1, -3), x = 1, \left(\dfrac{2 \pm \sqrt{6}}{2}, 0 \right), (0, -1)$

5. $y = -2x^2 + 4x + 6$ **7.** $-1 - i$ **9.** $-9 - 40i$

11. 20   **13.** $-3 - 2i$ **15.** $\dfrac{1}{5} - \dfrac{3}{5}i$ **17.** $3 + i\sqrt{2}$

19. $-1 - i$ **21.** $1/3$ **23.** $\pm 2\sqrt{2}$ **25.** $\dfrac{1}{2}, \dfrac{-1 \pm i\sqrt{3}}{4}$

27. $\pm\sqrt{10}, \pm i\sqrt{10}$ **29.** $-\frac{1}{2}, \frac{1}{2}$ with multiplicity 2

31. $0, -1 \pm \sqrt{7}$ **33.** 83   **35.** 5   **37.** $\pm\left(1, 2, \frac{1}{3}, \frac{2}{3}\right)$

39. $\pm\left(1, 3, \frac{1}{2}, \frac{1}{3}, \frac{1}{6}, \frac{3}{2}\right)$ **41.** $2x^2 - 5x - 3 = 0$

43. $x^2 - 6x + 13 = 0$ **45.** $x^3 - 4x^2 + 9x - 10 = 0$

47. $x^2 - 4x + 1 = 0$ **49.** 0 with multiplicity 2, 6 imag

51. 1 pos, 2 imag; 3 pos   **53.** 3 neg; 1 neg, 2 imag   **55.** 1, 2, 3

57. $\frac{1}{2}, \frac{1}{3}, \pm i$ **59.** $3, 3 \pm i$ **61.** $2, 1 \pm i\sqrt{2}$

63. $0, \frac{1}{2}, 1 \pm \sqrt{3}$ **65.** $\{1/5\}$ **67.** $\{\pm\sqrt{2}\}$ **69.** $\{30\}$

71. $\{16\}$ **73.** $\{\pm 2\}$ **75.** $\{-7, 9\}$ **77.** No solution

79. $\{11/4\}$ **81.** $x = 3/4$ **83.** y-axis   **85.** Origin

87. $(-\infty, -2.5) \cup (-2.5, \infty)$

89. $(-\infty, \infty)$

91. $(-1, 0), (2, 0), (0, -2)$ **93.** $(-1, 0), (2, 0), (0, -2)$

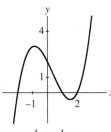

$f(x) = x^2 - x - 2$

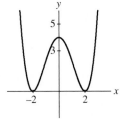

$f(x) = x^3 - 3x - 2$

95. $(\pm 2, 0), (1, 0), (0, 2)$ **97.** $(\pm 2, 0), (0, 4)$

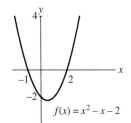

$f(x) = \frac{1}{2}x^3 - \frac{1}{2}x^2 - 2x + 2$ $f(x) = \frac{1}{4}x^4 - 2x^2 + 4$

99. $(0, 2/3), x = -3, y = 0$ **101.** $(0, 0), x = \pm 2, y = 0$

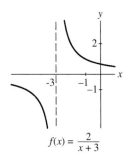

$f(x) = \frac{2}{x+3}$ $f(x) = \frac{2x}{x^2 - 4}$

103. $(1, 0), \left(0, -\frac{1}{2}\right), x = 2,$ **105.** $\left(\frac{1}{2}, 0\right), \left(0, -\frac{1}{2}\right), x = 2,$
$y = x$ $y = -2$

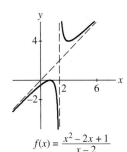

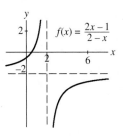

$f(x) = \frac{2x-1}{2-x}$

$f(x) = \frac{x^2 - 2x + 1}{x - 2}$

107. $(-2, 0), (0, 2)$

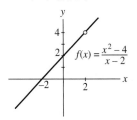

$f(x) = \frac{x^2 - 4}{x - 2}$

109. $(1/4, 1/2)$ **111.** $[-5, 3]$ **113.** $[-1/2, 1/2] \cup [100, \infty)$

115. $(-\infty, -2) \cup (0, \infty)$ **117.** $(-\infty, 0) \cup (0, 3) \cup (4, \infty)$

119. $(-\infty, 1] \cup [2, 3) \cup (4, \infty)$ **121.** $x^2 - 3x, -15$

123. 380.25 ft   **125.** $A = -a^3 + 16a, (2.3, -10.7)$

Chapter 3

Section 3.1

For Thought: **1.** T   **2.** F   **3.** F   **4.** F   **5.** T   **6.** F
7. F   **8.** T   **9.** T   **10.** T

Exercises: **1.** $420°, 780°, -300°, -660°$

3. $344°, 704°, -376°, -736°$ **5.** Yes   **7.** No   **9.** I

11. III   **13.** IV   **15.** I   **17.** $45°$ **19.** $60°$ **21.** $120°$

23. $40°$ **25.** $20°$ **27.** $340°$ **29.** $13.2°$ **31.** $-8.505°$

33. $28.0858°$ **35.** $75°30'$ **37.** $-17°19'48''$ **39.** $18°7'23''$

41. $\pi/6$ **43.** $\pi/10$ **45.** $-3\pi/8$ **47.** $7\pi/2$ **49.** 0.653

51. -0.241 **53.** -0.936 **55.** $75°$ **57.** $315°$ **59.** $-1080°$

61. $136.937°$ **63.** $7\pi/3, 13\pi/3, -5\pi/3, -11\pi/3$

65. $11\pi/6, 23\pi/6, -13\pi/6, -25\pi/6$ **67.** π **69.** $\pi/2$

71. $\pi/3$ **73.** $5\pi/3$ **75.** 2.04   **77.** No   **79.** Yes   **81.** I

83. III   **85.** IV   **87.** IV

89. $30° = \pi/6, 45° = \pi/4, 60° = \pi/3, 90° = \pi/2, 120° = 2\pi/3,$
$135° = 3\pi/4, 150° = 5\pi/6, 180° = \pi, 210° = 7\pi/6,$
$225° = 5\pi/4, 240° = 4\pi/3, 270° = 3\pi/2,$
$300° = 5\pi/3, 315° = 7\pi/4, 330° = 11\pi/6, 360° = 2\pi$

91. 3π ft   **93.** 209.4 mi   **95.** 1 mi   **97.** 3.18 km

99. 1.68 m.   **101.** 41,143 km, 40,074 km

Section 3.2

For Thought: **1.** F   **2.** F   **3.** T   **4.** F   **5.** F   **6.** T
7. F   **8.** F   **9.** F   **10.** T

Exercises: **1.** $(1, 0), \left(\sqrt{2}/2, \sqrt{2}/2\right), (0, 1), \left(-\sqrt{2}/2, \sqrt{2}/2\right), (-1, 0),$
$\left(-\sqrt{2}/2, -\sqrt{2}/2\right), (0, -1), \left(\sqrt{2}/2, -\sqrt{2}/2\right)$

3. 0   **5.** 0   **7.** 0   **9.** 0   **11.** $\sqrt{2}/2$ **13.** $-\sqrt{2}/2$

15. $1/2$ **17.** $1/2$ **19.** $-\sqrt{3}/2$ **21.** $\sqrt{3}/2$ **23.** $1/2$

25. $1/2$ **27.** $\sqrt{3}/2$ **29.** $\sqrt{2}/2$ **31.** -1 **33.** $\sqrt{3}/2$

35. $1/2$ **37.** $\sqrt{3}/3$ **39.** -1 **41.** 1   **43.** $2 + \sqrt{3}$

45. $\sqrt{2}$ **47.** $+$ **49.** $+$ **51.** $-$ **53.** $-$ **55.** 0.9999

57. 0.4035 **59.** -0.7438 **61.** 1.0000 **63.** -0.2588 **65.** 1

67. $1/2$ **69.** $\sqrt{2}/2$ **71.** $\sqrt{3}/2$ **73.** $-12/13$ **75.** $-4/5$

77. $2\sqrt{2}/3$ **79.** $x = 4 \sin t - 3 \cos t, 3.53$

81. 1.708 in., 1.714 in.

Section 3.3

For Thought: **1.** F   **2.** F   **3.** F   **4.** T   **5.** T   **6.** T
7. F   **8.** T   **9.** T   **10.** T
Exercises: **1.** $y = -2 \sin(x), 2$
3. $y = 3 \cos(x), 3$ **5.** $2, 2\pi, 0$
7. $1, 2\pi, \pi/2$ **9.** $2, 2\pi, -\pi/3$
11. $1, 0, (0, 0), (\pi/2, -1), (\pi, 0),$ **13.** $3, 0, (0, 0), (\pi/2, -3),$
$(3\pi/2, 1), (2\pi, 0)$ $(\pi, 0), (3\pi/2, 3), (2\pi, 0)$

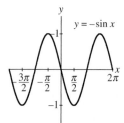

15. $1/2, 0, (0, 1/2), (\pi/2, 0),$ **17.** $1, -\pi, (0, 0), (\pi/2, -1),$
$(\pi, -1/2), (3\pi/2, 0),$ $(\pi, 0), (3\pi/2, 1), (2\pi, 0)$
$(2\pi, 1/2)$

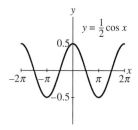

19. $1, \pi/3, (-2\pi/3, -1),$ **21.** $1, 0, (0, 3), (\pi/2, 2),$
$(-\pi/6, 0), (\pi/3, 1),$ $(\pi, 1), (3\pi/2, 2), (2\pi, 3)$
$(5\pi/6, 0), (4\pi/3, -1)$

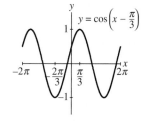

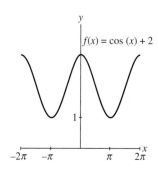

23. $1, 0, (0, -1), (\pi/2, -2), (\pi, -1), (3\pi/2, 0), (2\pi, -1)$

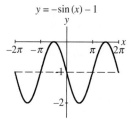

25. $1, -\pi/4, (-\pi/4, 2), (\pi/4, 3), (3\pi/4, 2), (5\pi/4, 1), (7\pi/4, 2)$

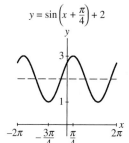

27. $2, -\pi/6, (-\pi/6, 3), (\pi/3, 1), (5\pi/6, -1), (4\pi/3, 1), (11\pi/6, 3)$

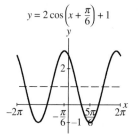

29. $2, \pi/3, (-\pi/6, 3), (\pi/3, 1), (5\pi/6, -1), (4\pi/3, 1), (11\pi/6, 3)$

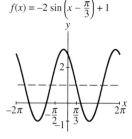

31. $3, \pi/2, 0$ **33.** $1, 4\pi, 0$ **35.** $2, 2\pi, \pi$ **37.** $2, \pi, -\pi/4$
39. $2, 4, -2$ **41.** $y = 2 \sin [2(x + \pi/2)] + 5$
43. $y = 5 \sin [\pi(x - 2)] + 4$ **45.** $y = 6 \sin [4\pi(x + \pi)] - 3$
47. $y = -\sin(x - \pi/4) + 1$ **49.** $y = -3 \cos(x - \pi) + 2$

51. $2\pi/3, 0, [-1, 1], (0, 0),$
$(\pi/6, 1), (\pi/3, 0),$
$(\pi/2, -1), (2\pi/3, 0)$

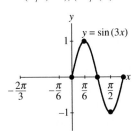

53. $\pi, 0, [-1, 1], (0, 0),$
$(\pi/4, -1), (\pi/2, 0),$
$(3\pi/4, 1), (\pi, 0)$

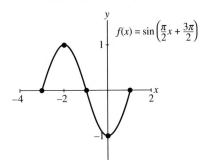

63. $4, -3, [-1, 1], (-3, 0), (-2, 1), (-1, 0), (0, -1), (1, 0)$

55. $\pi/2, 0, [1, 3], (0, 3), (\pi/8, 2),$
$(\pi/4, 1), (3\pi/8, 2), (\pi/2, 3)$

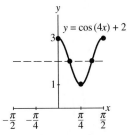

57. $8\pi, 0, [1, 3], (0, 2), (2\pi, 1),$
$(4\pi, 2), (6\pi, 3), (8\pi, 2)$

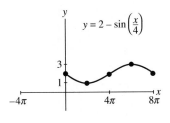

65. $\pi, -\pi/6, [-1, 3], (-\pi/6, 3),$
$(\pi/12, 1), (\pi/3, -1),$
$(7\pi/12, 1), (5\pi/6, 3)$

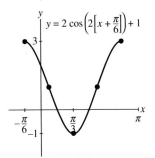

59. $6, 0, [-1, 1], (0, 0), (1.5, 1), (3, 0), (4.5, -1), (6, 0)$

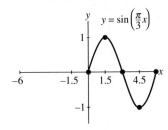

67. $2\pi/3, \pi/6, [-3/2, -1/2], (\pi/6, -1), (\pi/3, -3/2), (\pi/2, -1),$
$(2\pi/3, -1/2), (5\pi/6, -1)$

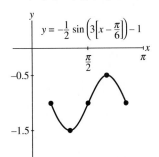

61. $\pi, \pi/2, [-1, 1], (\pi/2, 0),$
$(3\pi/4, 1), (\pi, 0),$
$(5\pi/4, -1), (3\pi/2, 0)$

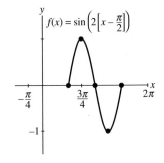

69. $y = 2 \sin\left(2\left[x - \dfrac{\pi}{4}\right]\right)$　　**71.** $y = 3 \sin\left(\dfrac{3}{2}\left[x + \dfrac{\pi}{3}\right]\right) + 3$

73. 100 cycles/sec　　**75.** 40 cycles/hr

77. $x = 3 \sin(2t), 3, \pi$

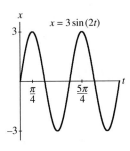

79. 11 yr   **81. a.** 1300 cc, 500 cc   **b.** 30

83. 12, 15,000, −3, 25,000, $y = 15,000 \sin(\pi x/6 + \pi/2) +$ 25,000, $17,500

85. $y = \sin(\pi x/10) + 1$

Section 3.4

For Thought: **1.** T   **2.** F   **3.** T   **4.** F   **5.** F   **6.** T   **7.** F   **8.** T   **9.** T   **10.** T

Exercises:

1. $\tan(0) = 0, \tan(\pi/4) = 1, \tan(\pi/2)$ undefined, $\tan(3\pi/4) = -1$, $\tan(\pi) = 0, \tan(5\pi/4) = 1, \tan(3\pi/2)$ undefined, $\tan(7\pi/4) = -1$

3. $\sqrt{3}$ **5.** −1   **7.** 0

9. $-\sqrt{3}/3$ **11.** $2\sqrt{3}/3$

13. Undefined   **15.** Undefined   **17.** $\sqrt{2}$

19. −1   **21.** $\sqrt{3}$

23. −2   **25.** $-\sqrt{2}$

27. 0   **29.** 48.0785

31. −2.8413   **33.** 500.0003

35. 1.0353   **37.** 636.6192

39. −1.4318   **41.** 71.6221

43. −0.9861   **45.** 4   **47.** $\sqrt{3}/3$ **49.** $-\sqrt{2}$

51. $\pi/3$ **53.** π

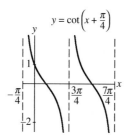

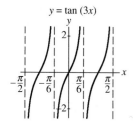

55. 2π **57.** 1

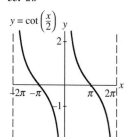

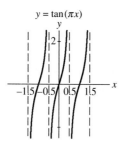

59. π **61.** π

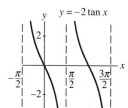

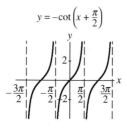

63. $\pi/2$ **65.** 2

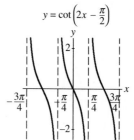

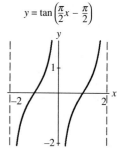

67. $\pi, (-\infty, -1] \cup [1, \infty)$ **69.** $2\pi, (-\infty, -1] \cup [1, \infty)$

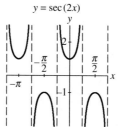

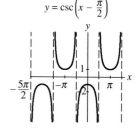

71. $4\pi, (-\infty, -1] \cup [1, \infty)$ **73.** $4, (-\infty, -1] \cup [1, \infty)$

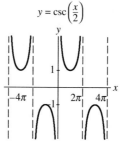

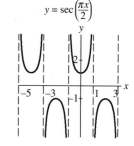

75. $2\pi, (-\infty, -2] \cup [2, \infty)$ **77.** $\pi, (-\infty, -1] \cup [1, \infty)$

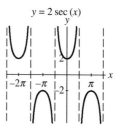

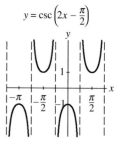

79. 4, $(-\infty, -1] \cup [1, \infty)$

$$y = -\csc\left(\frac{\pi}{2}x + \frac{\pi}{2}\right)$$

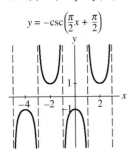

81. π, $(-\infty, 0] \cup [4, \infty)$

$$y = 2 + 2\sec(2x)$$

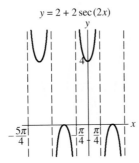

83. $\pi/2$, $(-\infty, \infty)$ **85.** 4π, $(-\infty, -3] \cup [1, \infty)$
87. π, $(-\infty, -7] \cup [-1, \infty)$ **89.** $y = 3\tan(x - \pi/4) + 2$
91. $y = -\sec(x + \pi) + 2$

Section 3.5

For Thought: **1.** T   **2.** T   **3.** F   **4.** F   **5.** F   **6.** T
7. T   **8.** T   **9.** F   **10.** F
Exercises: **1.** $-\pi/6$ **3.** $\pi/6$ **5.** $\pi/4$ **7.** $-45°$ **9.** $30°$
11. $0°$ **13.** $-19.5°$ **15.** $34.6°$ **17.** $3\pi/4$ **19.** $\pi/3$
21. π **23.** $135°$ **25.** $180°$ **27.** $120°$ **29.** $173.2°$
31. $89.9°$ **33.** $-\pi/4$ **35.** $\pi/3$ **37.** $\pi/4$ **39.** $-\pi/6$
41. 0   **43.** $\pi/2$ **45.** $3\pi/4$ **47.** $2\pi/3$ **49.** 0.60
51. 3.02   **53.** -0.14 **55.** 1.87   **57.** 1.15   **59.** -0.36
61. 3.06   **63.** 0.06   **65.** $\sqrt{3}$ **67.** $-\pi/6$ **69.** $\pi/6$
71. $\pi/4$ **73.** 1   **75.** $\pi/2$ **77.** 0   **79.** $\pi/2$
81. $f^{-1}(x) = 0.5\sin^{-1}(x)$, $[-1, 1]$

83. $f^{-1}(x) = \dfrac{1}{\pi}\tan^{-1}(x - 3)$, $(-\infty, \infty)$

85. $f^{-1}(x) = 2\sin(x - 3)$, $[3 - \pi/2, 3 + \pi/2]$ **87.** $67.1°$

Section 3.6

For Thought: **1.** F   **2.** T   **3.** F   **4.** T   **5.** T   **6.** F
7. F   **8.** T   **9.** T   **10.** F
Exercises: **1.** 4/5, 3/5, 4/3, 5/4, 5/3, 3/4
3. $3\sqrt{10}/10, -\sqrt{10}/10, -3, \sqrt{10}/3, -\sqrt{10}, -1/3$
5. $-\sqrt{3}/3, -\sqrt{6}/3, \sqrt{2}/2, -\sqrt{3}, -\sqrt{6}/2, \sqrt{2}$
7. $-1/2, \sqrt{3}/2, -\sqrt{3}/3, -2, 2\sqrt{3}/3, -\sqrt{3}$
9. $\sqrt{5}/5, 2\sqrt{5}/5, 1/2, 2\sqrt{5}/5, \sqrt{5}/5, 2$
11. $3\sqrt{34}/34, 5\sqrt{34}/34, 3/5, 5\sqrt{34}/34, 3\sqrt{34}/34, 5/3$
13. 4/5, 3/5, 4/3, 3/5, 4/5, 3/4   **15.** $80.5°$ **17.** $60°$ **19.** 1.0
21. 0.4   **23.** $\beta = 30°, a = 10\sqrt{3}, b = 10$
25. $c = 10, \alpha = 36.9°, \beta = 53.1°$
27. $a = 5.7, \alpha = 43.7°, \beta = 46.3°$
29. $\beta = 74°, a = 5.5, b = 19.2$ **31.** $\beta = 50°51', b = 11.1, c = 14.3$
33. 50 ft   **35.** 1.7 mi   **37.** 43.2 m   **39.** 25.1 ft
41. 22 m, 57.6°   **43.** 153.1 m   **45.** 4.5 km   **47.** 4391 mi
49. 5.987 ft

Section 3.7

For Thought: **1.** T   **2.** F   **3.** F   **4.** F   **5.** F   **6.** T   **7.** T
8. T   **9.** F   **10.** T
Exercises: **1.** 1   **3.** $-\sin^2 x \cos^3 x$ **5.** $\cos^2 w$ **7.** $\csc x$
9. $\cos x$ **11.** 0   **13.** $\cos(11)$ **15.** $\cos(k)$ **17.** 1
19. $-\cos(\pi/5)$ **21.** $\tan(5\pi/18)$ **23.** $\cos(2k)$ **25.** $\sin(26°)$

27. $\sqrt{3}/6$ **29.** $-\sin(2\pi/9)$
31. $\sin\alpha = \sqrt{5}/5, \cos\alpha = 2\sqrt{5}/5, \csc\alpha = \sqrt{5},$
  $\sec\alpha = \sqrt{5}/2, \cot\alpha = 2$
33. $\sin\alpha = -\sqrt{22}/5, \tan\alpha = \sqrt{66}/3,$
  $\csc\alpha = -5\sqrt{22}/22, \sec\alpha = -5\sqrt{3}/3,$
  $\cot\alpha = \sqrt{66}/22$
35. $\sqrt{5}/5, 2\sqrt{5}/5, 1/2, \sqrt{5}, \sqrt{5}/2, 2$
37. $-\sqrt{15}/8, -7/8, \sqrt{15}/7, -8\sqrt{15}/15, -8/7, 7\sqrt{15}/15$
39. $\dfrac{\sqrt{2} - \sqrt{3}}{2}$ or $\dfrac{\sqrt{6} - \sqrt{2}}{4}$ **41.** $\dfrac{\sqrt{6} - \sqrt{2}}{4}$ **43.** $\dfrac{\sqrt{2} + \sqrt{6}}{4}$

45. $\dfrac{1 + \sqrt{3}}{\sqrt{3} - 1}$ **77.** $0.5(\cos 4° - \cos 22°)$

79. $0.5(\cos(\pi/30) + \cos(11\pi/30))$ **81.** $\dfrac{\sqrt{2} - 1}{4}$

83. $\dfrac{\sqrt{2} + \sqrt{3}}{4}$ **85.** $2\cos(10°)\sin(2°)$

87. $-2\sin(4\pi/15)\sin(\pi/15)$ **89.** $\sqrt{6}/2$ **91.** $\dfrac{\sqrt{2} - \sqrt{2}}{2}$
93. $\sqrt{2}\sin(x - \pi/4)$ **95.** $\sin(x + 2\pi/3)$

97. $\sin(x - \pi/6)$ **99.** $A = \dfrac{d^2}{2}\sin(2\alpha)$

101. $x = 2\sin(t + \pi/6)$, 2 m

Section 3.8

For Thought: **1.** F   **2.** F   **3.** T   **4.** T   **5.** T   **6.** F
7. F   **8.** T   **9.** F   **10.** F
Exercises: **1.** $\{x | x = \pi + 2k\pi\}$ **3.** $\{x | x = k\pi\}$

5. $\left\{x | x = \dfrac{3\pi}{2} + 2k\pi\right\}$

7. $\left\{x | x = \dfrac{\pi}{3} + 2k\pi \text{ or } x = \dfrac{5\pi}{3} + 2k\pi\right\}$

9. $\left\{x | x = \dfrac{\pi}{4} + 2k\pi \text{ or } x = \dfrac{3\pi}{4} + 2k\pi\right\}$ **11.** $\left\{x | x = \dfrac{\pi}{4} + k\pi\right\}$

13. $\left\{x | x = \dfrac{5\pi}{6} + 2k\pi \text{ or } x = \dfrac{7\pi}{6} + 2k\pi\right\}$

15. $\left\{x | x = \dfrac{5\pi}{4} + 2k\pi \text{ or } x = \dfrac{7\pi}{4} + 2k\pi\right\}$

17. $\left\{x | x = \dfrac{3\pi}{4} + k\pi\right\}$ **19.** $\{x | x = 90° + k180°\}$

21. $\{x | x = 90° + k360°\}$ **23.** $\{x | x = k180°\}$
25. $\{x | x = 29.2° + k360° \text{ or } x = 330.8° + k360°\}$
27. $\{x | x = 345.9° + k360° \text{ or } x = 194.1° + k360°\}$
29. $\{x | x = 79.5° + k180°\}$

31. $\left\{x | x = \dfrac{2\pi}{3} + 4k\pi \text{ or } x = \dfrac{10\pi}{3} + 4k\pi\right\}$ **33.** $\left\{x | x = \dfrac{2k\pi}{3}\right\}$

35. $\left\{x | x = \dfrac{\pi}{3} + 4k\pi \text{ or } x = \dfrac{5\pi}{3} + 4k\pi\right\}$

37. $\left\{x | x = \dfrac{5\pi}{8} + k\pi \text{ or } x = \dfrac{7\pi}{8} + k\pi\right\}$ **39.** $\left\{x | x = \dfrac{\pi}{6} + \dfrac{k\pi}{2}\right\}$

41. $\left\{x | x = \dfrac{k\pi}{4}\right\}$ **43.** $\left\{x | x = \dfrac{1}{6} + 2k \text{ or } x = \dfrac{5}{6} + 2k\right\}$

45. $\left\{x | x = \dfrac{1}{4} + \dfrac{k}{2}\right\}$ **47.** $\{240°, 300°\}$

49. $\{22.5°, 157.5°, 202.5°, 337.5°\}$
51. $\{45°, 75°, 165°, 195°, 285°, 315°\}$ **53.** $\{60°\}$
55. $\{\alpha | \alpha = 6.6° + k120° \text{ or } \alpha = 53.4° + k120°\}$

57. $\{\alpha \mid \alpha = 72.3° + k120°$ or $\alpha = 107.7° + k120°\}$
59. $\{\alpha \mid \alpha = 38.6° + k180°$ or $\alpha = 141.4° + k180°\}$
61. $\{\alpha \mid \alpha = 668.5° + k720°$ or $\alpha = 411.5° + k720°\}$
63. $\{0, 0.3, 2.8, \pi\}$ **65.** $\{\pi, 2\pi/3, 4\pi/3\}$
67. $\{0.7, 2.5, 3.4, 6.0\}$ **69.** $\{11\pi/6\}$
71. $\{1.0, 4.9\}$ **73.** $\{0, \pi\}$ **75.** $\{\pi/2, 3\pi/2\}$
77. $\{7\pi/12, 23\pi/12\}$ **79.** $\{7\pi/6, 11\pi/6\}$
81. $\{0°\}$ **83.** $\{26.6°, 206.6°\}$
85. $\{30°, 90°, 150°, 210°, 270°, 330°\}$
87. $\{67.5°, 157.5°, 247.5°, 337.5°\}$ **89.** $\{221.8°, 318.2°\}$
91. $\{120°, 300°\}$ **93.** $\{30°, 45°, 135°, 150°, 210°, 225°, 315°, 330°\}$
95. $\{0°, 60°, 120°, 180°, 240°, 300°\}$
97. $\dfrac{5\pi}{12} + \dfrac{k\pi}{2}$ for k a nonnegative integer
99. $44.4°$ or $45.6°$ **101.** $12.5°$ or $77.5°$, 6.3 sec

Section 3.9

For Thought: **1.** T   **2.** F   **3.** F   **4.** T   **5.** T   **6.** F
7. T   **8.** F   **9.** F   **10.** T
Exercises: **1.** $\gamma = 44°, b = 14.4, c = 10.5$
3. $\beta = 134.2°, a = 5.2, c = 13.6$ **5.** $\beta = 26°, a = 14.6, b = 35.8$
7. $\alpha = 45.7°, b = 587.9, c = 160.8$ **9.** None
11. One: $\alpha = 30°, \beta = 90°, a = 10$
13. One: $\alpha = 26.3°, \gamma = 15.6°, a = 10.3$
15. Two: $\alpha_1 = 134.9°, \gamma_1 = 12.4°, c_1 = 11.4$; or $\alpha_2 = 45.1°$,
     $\gamma_2 = 102.2°, c_2 = 51.7$ **17.** One: $\alpha = 25.4°, \beta = 55.0°, a = 5.4$
19. $\alpha = 30.4°, \beta = 28.3°, c = 5.2$
21. $\alpha = 26.4°, \beta = 131.3°, \gamma = 22.3°$
23. $\alpha = 163.9°, \gamma = 5.6°, b = 4.5$
25. $\alpha = 130.3°, \beta = 30.2°, \gamma = 19.5°$
27. $\beta = 48.3°, \gamma = 101.7°, a = 6.2$
29. $\alpha = 53.9°, \beta = 65.5°, \gamma = 60.6°$
31. $\alpha = 120°, b = 3.5, c = 4.8$ **33.** 0   **35.** 1   **37.** 0   **39.** 1
41. 0   **43.** 18.4 mi   **45.** 159.4 ft   **47.** 28.9 ft, 15.7 ft
49. 9.90 ft^2 **51.** 783.45 ft   **53.** 20.6 mi
55. $\theta_1 = 13.3°, \theta_2 = 90.6°$

Chapter 3 Review Exercises

1. $28°$ **3.** $206°45'33''$ **5.** $180°$ **7.** $108°$ **9.** $300°$
11. $270°$ **13.** $11\pi/6$ **15.** $-5\pi/3$
17.

θ deg	0	30	45	60	90	120	135	150	180
θ rad	0	$\dfrac{\pi}{6}$	$\dfrac{\pi}{4}$	$\dfrac{\pi}{3}$	$\dfrac{\pi}{2}$	$\dfrac{2\pi}{3}$	$\dfrac{3\pi}{4}$	$\dfrac{5\pi}{6}$	π
$\sin\theta$	0	$\dfrac{1}{2}$	$\dfrac{\sqrt{2}}{2}$	$\dfrac{\sqrt{3}}{2}$	1	$\dfrac{\sqrt{3}}{2}$	$\dfrac{\sqrt{2}}{2}$	$\dfrac{1}{2}$	0
$\cos\theta$	1	$\dfrac{\sqrt{3}}{2}$	$\dfrac{\sqrt{2}}{2}$	$\dfrac{1}{2}$	0	$-\dfrac{1}{2}$	$-\dfrac{\sqrt{2}}{2}$	$-\dfrac{\sqrt{3}}{2}$	-1
$\tan\theta$	0	$\dfrac{\sqrt{3}}{3}$	1	$\sqrt{3}$		$-\sqrt{3}$	-1	$-\dfrac{\sqrt{3}}{3}$	0

19. $-\sqrt{2}/2$ **21.** $\sqrt{3}$ **23.** $-2\sqrt{3}/3$ **25.** 0   **27.** 0
29. -1 **31.** $\sqrt{3}/3$ **33.** $-\sqrt{2}/2$ **35.** -2 **37.** $-\sqrt{3}/3$
39. 5/13, 12/13, 5/12, 13/5, 13/12, 12/5   **41.** 0.6947
43. -0.0923 **45.** 0.1869   **47.** 1.0356   **49.** $-\pi/6$
51. $-\pi/4$ **53.** $\pi/4$ **55.** $\pi/6$ **57.** $90°$ **59.** $135°$
61. $30°$ **63.** $90°$ **65.** $c = \sqrt{13}, \alpha = 33.7°, \beta = 56.3°$

67. $\beta = 68.7°, c = 8.8, b = 8.2$
69. $2\pi/3, [-2, 2]$

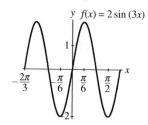

71. $\pi/2, (-\infty, \infty)$

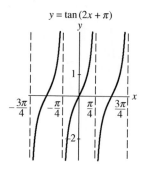

73. $4\pi, (-\infty, -1] \cup [1, \infty)$

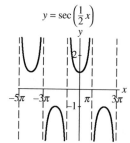

75. $\pi, [-1/2, 1/2]$

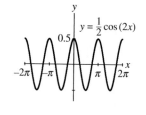

77. $\pi/2, (-\infty, \infty)$

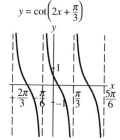

79. $\pi, (-\infty, -1/3] \cup [1/3, \infty)$

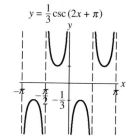

81. $y = 2\sin\left(\dfrac{\pi}{2}[x - 2]\right)$ **83.** $y = 20\sin(x) + 40$
85. $150°$ **87.** $-2\sqrt{6}/5$ **89.** 6.6 ft   **91.** $53.1°$
93. $\cos^2\alpha$ **95.** $-\cot^2 x$ **97.** $\sec^2\alpha$ **99.** $\tan 4s$
101. $-\sin 3\theta$ **103.** $\tan z$
115. $\left\{ x \mid x = \dfrac{\pi}{3} + k\pi \text{ or } x = \dfrac{2\pi}{3} + k\pi \right\}$
117. $\left\{ x \mid x = \dfrac{\pi}{3} + 2k\pi, \dfrac{2\pi}{3} + 2k\pi, \dfrac{\pi}{6} + 2k\pi, \dfrac{5\pi}{6} + 2k\pi \right\}$
119. $\left\{ x \mid x = \dfrac{\pi}{6} + 2k\pi, \dfrac{5\pi}{6} + 2k\pi, \dfrac{\pi}{2} + 2k\pi \right\}$
121. $\left\{ x \mid x = \dfrac{2\pi}{3} + 4k\pi \text{ or } x = \dfrac{4\pi}{3} + 4k\pi \right\}$
123. $\left\{ x \mid x = \pi + 2k\pi, \dfrac{\pi}{3} + 4k\pi, \dfrac{5\pi}{3} + 4k\pi \right\}$
125. $\left\{ x \mid x = \dfrac{\pi}{2} + k\pi \right\}$ **127.** $\alpha = 82.7°, \beta = 49.3°, c = 2.5$
129. $\gamma = 103°, a = 4.6, b = 18.4$ **131.** No triangle
133. $\alpha = 107.7°, \beta = 23.7°, \gamma = 48.6°$

135. $\alpha_1 = 110.8°, \gamma_1 = 47.2°, a_1 = 6.2; \alpha_2 = 25.2°, \gamma_2 = 132.8°,$
$a_2 = 2.8$

137. 1.08×10^{-8} sec   **139.** 6.9813 ft

141.

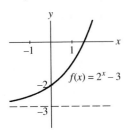

$$y = 2\sin\left(\frac{\pi}{10}[x+5]\right) + 14$$

143. No   **145.** 1117 ft   **147.** 11.6 mi

Chapter 4

Section 4.1

For Thought: **1.** F   **2.** T   **3.** T   **4.** T   **5.** T   **6.** T
7. T   **8.** F   **9.** T   **10.** T

Exercises: **1.** 9   **3.** 1/9   **5.** 1/2   **7.** 8   **9.** 4   **11.** 2
13. $(-\infty, \infty), (0, \infty)$, inc   **15.** $(-\infty, \infty), (0, \infty)$, dec

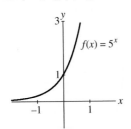

17. $(-\infty, \infty), (0, \infty)$, dec

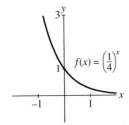

19. ∞ **21.** 0   **23.** 0   **25.** ∞
27. $(-\infty, \infty), (-3, \infty),$ **29.** $(-\infty, \infty), (-5, \infty),$
$y = -3$, increasing $y = -5$, increasing

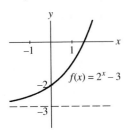

31. $(-\infty, \infty), (-\infty, 0),$
$y = 0$, increasing

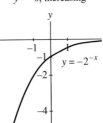

33. $(-\infty, \infty), (-\infty, 1),$
$y = 1$, decreasing

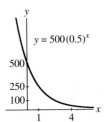

35. $(-\infty, \infty), (0, \infty),$
$y = 0$, increasing

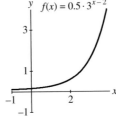

37. $(-\infty, \infty), (0, \infty),$
$y = 0$, decreasing

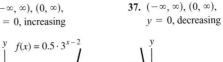

39. $y = 2^{x-5} - 2$ **41.** $y = -(1/4)^{x-1} - 2$
43. $\{6\}$ **45.** $\{-1\}$ **47.** $\{3\}$ **49.** $\{-2\}$ **51.** $\{1/3\}$
53. $\{-2\}$ **55.** $\{-3\}$ **57.** $\{-1\}$ **59.** 2   **61.** -1 **63.** 0
65. -3 **67.** 3   **69.** -1 **71.** 1   **73.** -1
75. 9, 1, 1/3, -2 **77.** 1, -2, 5, 1   **79.** $-16, -2, -1/2, 5$
81. a. \$7934.37, \$2934.37   **b.** \$8042.19, \$3042.19
c. \$8067.51, \$3067.51   **d.** \$8079.95, \$3079.95
83. a. \$8080.37   **b.** \$9671.31   **c.** \$7694.93   **d.** \$26,570.30
85. \$2121.82   **87.** \$3934.30   **89. a.** \$6.85   **b.** \$6.87
91. 200 g, 121.3 g   **93.** $P = 10\left(\frac{1}{2}\right)^n$ **95.** 6

Section 4.2

For Thought: **1.** T   **2.** F   **3.** T   **4.** T   **5.** F   **6.** T
7. T   **8.** F   **9.** T   **10.** T

Exercises: **1.** 6   **3.** -4 **5.** 1/4   **7.** -3 **9.** 6   **11.** -4
13. 1/4   **15.** -3 **17.** -1 **19.** 0   **21.** 1   **23.** -5
25. $(0, \infty), (-\infty, \infty)$ **27.** $(0, \infty), (-\infty, \infty)$

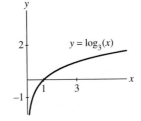

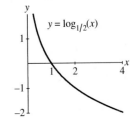

29. $(1, \infty), (-\infty, \infty)$ **31.** $(-2, \infty), (-\infty, \infty)$

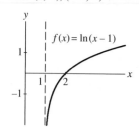

$f(x) = \ln(x - 1)$

$f(x) = -3 + \log(x + 2)$

33. $(1, \infty), (-\infty, \infty)$

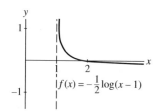

$f(x) = -\frac{1}{2}\log(x - 1)$

35. ∞ **37.** ∞ **39.** $-\infty$ **41.** ∞ **43.** $y = \ln(x - 3) - 4$
45. $y = -\log_2(x - 5) - 1$ **47.** $2^5 = 32$ **49.** $5^y = x$
51. $10^z = 1000$ **53.** $\log_5(125) = 3$ **55.** $\ln(y) = 3$
57. $\log(y) = m$ **59.** $f^{-1}(x) = \log_2(x)$ **61.** $f^{-1}(x) = 7^x$
63. $f^{-1}(x) = e^x + 1$ **65.** $f^{-1}(x) = \log_3(x) - 2$
67. $f^{-1}(x) = \log(2x - 10) + 1$ **69.** 256   **71.** $\sqrt{3}$ **73.** 4
75. $\log_3(77)$ **77.** 6   **79.** $3\sqrt{2}$ **81.** $-1 + \log_3(7)$ **83.** 2
85. 1.3979   **87.** 0.5493   **89.** -0.2231 **91.** -0.3010
93. a. 34.7 yr   **b.** 17.3 yr   **c.** 8.7 yr   **d.** 4.3 yr
95. a. 22.0%   **b.** 11.0%   **c.** 5.5%   **d.** 2.7%
97. 49 yr 125 days   **99.** $r = \ln(A/P)/t$, 23.1%
101. a. 99.2 billion gigabits/yr   **b.** 2005

Section 4.3

For Thought: **1.** F   **2.** T   **3.** T   **4.** T   **5.** F   **6.** F
7. F   **8.** T   **9.** F   **10.** F
Exercises: **1.** $\sqrt{y}$ **3.** $y + 1$ **5.** 999   **7.** $\log(15)$
9. $\log_2(x^2 - x)$ **11.** $\log_4(6)$ **13.** $\ln(x^5)$ **15.** $\log_2(3) + \log_2(x)$
17. $\log(x) - \log(2)$ **19.** $\log(x - 1) + \log(x + 1)$
21. $3 \cdot \log_a(5)$ **23.** $\frac{1}{2} \cdot \log_a(5)$ **25.** $-1 \cdot \log_a(5)$
27. $\log_a(2) + \log_a(5)$ **29.** $\log_a(5) - \log_a(2)$
31. $\log_a(2) + \frac{1}{2} \cdot \log_a(5)$ **33.** $\log_3(5) + \log_3(x)$
35. $\log_2(5) - \log_2(2) - \log_2(y)$ **37.** $\log(3) + \frac{1}{2}\log(x)$
39. $\log(3) + (x - 1)\log(2)$ **41.** $\frac{1}{3}\ln(x) + \frac{1}{3}\ln(y) - \frac{4}{3}\ln(t)$
43. $\log_2(5x^3)$ **45.** $\log_7(x^{-3})$ **47.** $\log\left(\frac{2xy}{z}\right)$ **49.** $\log\left(\frac{z\sqrt{x}}{y\sqrt[3]{w}}\right)$
51. 3.1699   **53.** -3.5864 **55.** 11.8957   **57.** 1.5850
59. 0.3772   **61.** -3.5850 **63.** 13.8695   **65.** 34.3240
67. 2.0172   **69.** 1.5928   **71.** 11 yr 166 days   **73.** 44 quarters
75. 3.58%   **77.** $\log(I/I_0), 3$ **79.** $t = \frac{1}{r}\ln(P) - \frac{1}{r}\ln(P_0)$

Section 4.4

For Thought: **1.** T   **2.** T   **3.** F   **4.** T   **5.** F   **6.** F
7. T   **8.** T   **9.** T   **10.** T

Exercises: **1.** 8   **3.** 80   **5.** ± 5 **7.** 3   **9.** 1/2   **11.** $\sqrt[3]{10}$
13. 1/4   **15.** 6   **17.** 20   **19.** 2   **21.** $\dfrac{\sqrt{5}}{2}$ **23.** $\dfrac{1}{2}$
25. $\dfrac{2}{\ln(6)}$ **27.** 3.8074   **29.** 3.5502   **31.** -3.0959
33. No solution   **35.** 1.5850   **37.** 0.7677   **39.** -0.2
41. 2.5850   **43.** 1/3   **45.** 1, 100   **47.** 29.4872   **49.** 2
51. $-5/4$ **53.** 0.194, 2.70   **55.** -49.73 **57.** $-0.767, 2, 4$
59. -6.93×10^{-5} **61.** 19,035 yr ago   **63.** 1507 yr
65. 24,850 yr   **67.** 30.5%   **69.** 1 hr 11 min, forever
71. 5:20 A.M.   **73.** 10 yr 3 mo   **75.** 19,328 yr 307 days
77. a. 2211   **b.** 99%   **79.** 10^{-3} watts/m^2 **81.** $6791.91

Chapter 4 Review Exercises

1. 64   **3.** 6   **5.** 0   **7.** 17   **9.** 32   **11.** 3   **13.** 9
15. 3   **17.** -3 **19.** 3   **21.** $\log(x^2 - 3x)$ **23.** $\ln(3x^2y)$
25. $\log(3) + 4 \cdot \log(x)$ **27.** $\log_3(5) + \frac{1}{2}\log_3(x) - 4 \cdot \log_3(y)$
29. $\ln(2) + \ln(5)$ **31.** $\ln(2) + 2 \cdot \ln(5)$ **33.** 10^{10} **35.** 3
37. -5 **39.** -4 **41.** $2 + \ln(9)$ **43.** -2 **45.** $100\sqrt{5}$
47. 8   **49.** 6   **51.** 3   **53.** $3, -3, \sqrt{3}, 0$ **55.** (c)   **57.** (b)
59. (d)   **61.** (e)
63. $(-\infty, \infty), (0, \infty)$, inc, $y = 0$ **65.** $(-\infty, \infty), (0, \infty)$, dec, $y = 0$

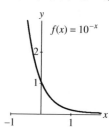

$f(x) = 5^x$

$f(x) = 10^{-x}$

67. $(0, \infty), (-\infty, \infty)$, inc, $x = 0$ **69.** $(-3, \infty), (-\infty, \infty)$, inc, $x = -3$

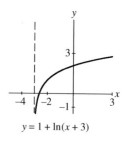

$y = \log_3(x)$

$y = 1 + \ln(x + 3)$

71. $(-\infty, \infty), (1, \infty)$, inc, $y = 1$ **73.** $(-\infty, 2), (-\infty, \infty)$, dec, $x = 2$

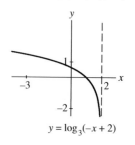

$f(x) = 1 + 2^{x-1}$

$y = \log_3(-x + 2)$

75. $f^{-1}(x) = \log_7(x)$ **77.** $f^{-1}(x) = 5^x$ **79.** $f^{-1}(x) = 10^{x/3} + 1$
81. $f^{-1}(x) = -2 + \ln(x + 3)$ **83.** 2.0959   **85.** 7.8538
87. -4.4243 **89.** T   **91.** F   **93.** T   **95.** F   **97.** F

99. F   **101.** T   **103.** $r = \dfrac{\ln(A/P)}{t}$ **105.** \$122,296.01

107. 56 quarters   **109.** 25 g, 18.15 g, 2166 yr   **111.** 2,877 hr
113. $-2/3, -3, -1/3$

Chapter 5

Section 5.1

For Thought: **1.** F   **2.** T   **3.** T   **4.** F   **5.** T   **6.** F
7. F   **8.** T   **9.** F   **10.** F
Exercises: **1.** $(0, 0), (0, 1), y = -1$
3. $(1, 2), (1, 3/2), y = 5/2$ **5.** $(3, 1), (15/4, 1), x = 9/4$
7. $y = \dfrac{1}{8}x^2$ **9.** $y = -\dfrac{1}{12}x^2$ **11.** $y = \dfrac{1}{6}(x - 3)^2 + \dfrac{7}{2}$
13. $y = -\dfrac{1}{10}(x - 1)^2 - \dfrac{1}{2}$ **15.** $y = 1.25(x + 2)^2 + 1$
17. $y = \dfrac{1}{4}x^2$ **19.** $y = -x^2$ **21.** $(1, 0), (1, 1/4), y = -1/4$
23. $(3, 0), (3, 1), y = -1$ **25.** $(3, 4), (3, 31/8), y = 33/8$
27. $y = (x - 4)^2 - 13, (4, -13), (4, -51/4), y = -53/4$
29. $y = 2(x + 3)^2 - 13, (-3, -13), (-3, -103/8), y = -105/8$
31. $y = -2(x - 3/2)^2 + 11/2, (3/2, 11/2), (3/2, 43/8), y = 45/8$
33. $y = 5(x + 3)^2 - 45, (-3, -45), (-3, -44.95), y = -45.05$
35. $y = \dfrac{1}{8}(x - 2)^2 + 4, (2, 4), (2, 6), y = 2$

37. $(2, -1), (2, -3/4), y = -5/4, \text{up}$
39. $(1, -4), (1, -17/4), y = -15/4, \text{down}$
41. $(1, -2), (1, -23/12), y = -25/12, \text{up}$
43. $(-3, 13/2), (-3, 6), y = 7, \text{down}$ **45.** $(0, 5), (0, 6), y = 4, \text{up}$
47. $(1/2, -9/4), x = 1/2, (-1, 0), (2, 0), (0, -2)$

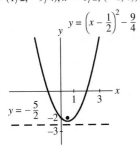

49. $(-1/2, 25/4), x = -1/2, (-3, 0), (2, 0), (0, 6)$

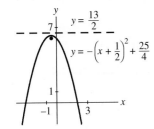

51. $(-2, 2), x = -2, (0, 4), (-2, 5/2), y = 3/2$

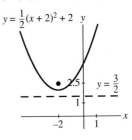

53. $(-4, 2), x = -4, (-4 \pm 2\sqrt{2}, 0), (0, -2), (-4, 1), y = 3$

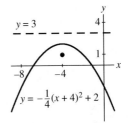

55. $(0, -2), x = 0, (\pm 2, 0), (0, -2), (0, -3/2), y = -5/2$

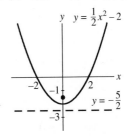

57. $(2, 0), x = 2, (2, 0), (0, 4), (2, 1/4), y = -1/4$

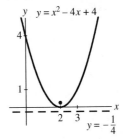

59. $(3/2, -3/4), x = 3/2, (0, 0), (3, 0), (0, 0), (3/2, 0), y = -3/2$

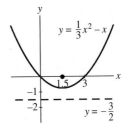

61. $(0, 0), y = 0, (0, 0), (0, 0), (-1/4, 0), x = 1/4$

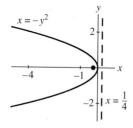

63. $(1, 0), y = 0, (1, 0), (0, \pm 2), (0, 0), x = 2$

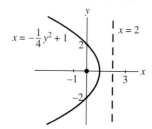

65. $(-25/4, -1/2), y = -1/2, (-6, 0), (0, -3), (0, 2), (-6, -1/2),$
$x = -13/2$

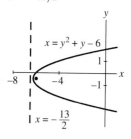

67. $(-7/2, -1), y = -1, (-4, 0), (-4, -1), x = -3$

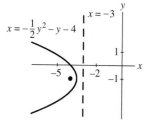

69. $(3, 1), y = 1, (5, 0), (25/8, 1), x = 23/8$

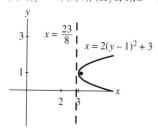

71. $(1, -2), y = -2, (-1, 0), (0, -2 \pm \sqrt{2}), (1/2, -2), x = 3/2$

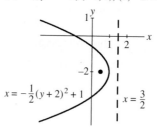

73. $y = \dfrac{1}{4}(x - 1)^2 + 4$ **75.** $x = \dfrac{1}{8}y^2$ **77.** $y = \dfrac{1}{2640}x^2$, 26.8 in.
79. $y = 6.3 \times 10^{-5}x^2$, $(0, 3946)$

Section 5.2

For Thought: **1.** F   **2.** T   **3.** T   **4.** T   **5.** T   **6.** F
7. T   **8.** F   **9.** F   **10.** F
Exercises: **1.** Foci $(\pm \sqrt{5}, 0)$, vertices $(\pm 3, 0)$, center $(0, 0)$
3. Foci $(2, 1 \pm \sqrt{5})$, vertices $(2, 4)$ and $(2, -2)$, center $(2, 1)$
5.   **7.**

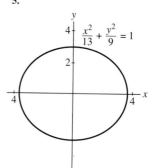   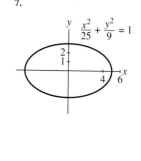

9.   **11.**

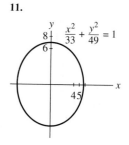   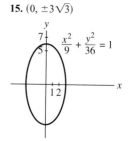

13. $(\pm 2\sqrt{3}, 0)$ **15.** $(0, \pm 3\sqrt{3})$

17. $(\pm 2\sqrt{6}, 0)$

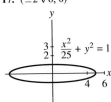

19. $(0, \pm 4)$

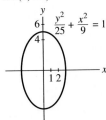

33. $(3, -2 \pm \sqrt{5})$

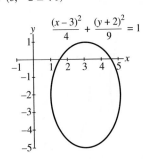

21. $(0, \pm 2\sqrt{2})$

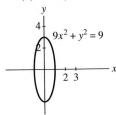

23. $(\pm \sqrt{5}, 0)$

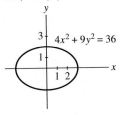

35. $\dfrac{x^2}{16} + \dfrac{y^2}{4} = 1, (\pm 2\sqrt{3}, 0)$

37. $\dfrac{(x+1)^2}{4} + \dfrac{(y+2)^2}{16} = 1, (-1, -2 \pm 2\sqrt{3})$

39. $x^2 + y^2 = 41$ **41.** $(x-2)^2 + (y+3)^2 = 20$

43. $(x-1)^2 + (y-3)^2 = 5$

45. $(0, 0), 10$

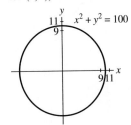

47. $(1, 2), 2$

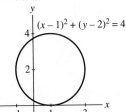

25. $(1 \pm \sqrt{7}, -3)$

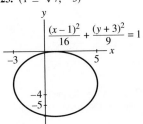

27. $(3, 2), (3, -6)$

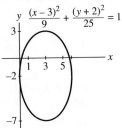

49. $(0, -1), 3$ **51.** $(-4, 5), \sqrt{41}$ **53.** $(-2, 0), 3$
55. $(0.5, -0.5), 1$ **57.** Circle   **59.** Ellipse   **61.** Parabola
63. Parabola   **65.** Ellipse   **67.** Circle   **69.** $\dfrac{x^2}{16} + \dfrac{y^2}{12} = 1$
71. $(12, 0)$ **73.** $\dfrac{x^2}{260.5^2} + \dfrac{y^2}{520} = 1, 0.996$ **75.** 5.25×10^9 km

29. $(-4 \pm \sqrt{35}, -3)$

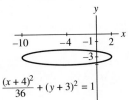

Section 5.3
For Thought: **1.** F   **2.** F   **3.** T   **4.** T   **5.** F   **6.** T
7. T   **8.** T   **9.** F   **10.** F
Exercises: **1.** Vertices $(\pm 1, 0)$, foci $(\pm \sqrt{2}, 0)$, asymptotes $y = \pm x$
3. Vertices $(1, \pm 3)$, foci $(1, \pm \sqrt{10})$, asymptotes $y = 3x - 3$ and
$y = -3x + 3$

5. $(\pm \sqrt{13}, 0), y = \pm \dfrac{3}{2}x$ **7.** $(0, \pm \sqrt{29}), y = \pm \dfrac{2}{5}x$

31. $(1, -2 \pm \sqrt{5})$

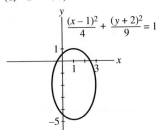

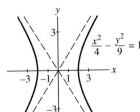

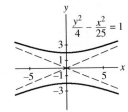

9. $(\pm\sqrt{5}, 0), y = \pm\frac{1}{2}x$ **11.** $(\pm\sqrt{10}, 0), y = \pm3x$

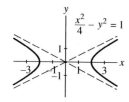

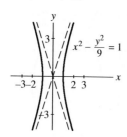

13. $(\pm5, 0), y = \pm\frac{4}{3}x$ **15.** $(\pm\sqrt{2}, 0), y = \pm x$

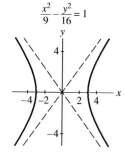

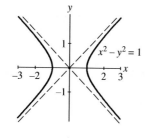

17. $(-1 \pm \sqrt{13}, 2), y = \frac{3}{2}x + \frac{7}{2}, y = -\frac{3}{2}x + \frac{1}{2}$

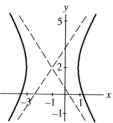

19. $(-2, 1 \pm \sqrt{5}), y = 2x + 5, y = -2x - 3$

21. $(3, 3), (-7, 3), y = \frac{3}{4}x + \frac{9}{2}, y = -\frac{3}{4}x + \frac{3}{2}$

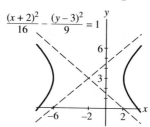

23. $(3, 3 \pm \sqrt{2}), y = x, y = -x + 6$

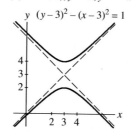

25. $\frac{x^2}{36} - \frac{y^2}{9} = 1$ **27.** $\frac{x^2}{9} - \frac{y^2}{16} = 1$ **29.** $\frac{x^2}{9} - \frac{y^2}{25} = 1$

31. $\frac{x^2}{9} - \frac{y^2}{16} = 1$ **33.** $\frac{y^2}{9} - \frac{(x-1)^2}{9} = 1$

35. $y^2 - (x - 1)^2 = 1$, hyperbola   **37.** $y = x^2 + 2x$, parabola

39. $x^2 + y^2 = 100$, circle   **41.** $x = -4y^2 + 100$, parabola

43. $(x - 1)^2 + (y - 2)^2 = \frac{1}{2}$, circle

45. $\frac{(x + 1)^2}{2} + \frac{(y + 3)^2}{4} = 1$, ellipse

47. $\frac{(x - 3)^2}{4} - \frac{(y + 1)^2}{25} = 1$, hyperbola   **49.** Ellipse

51. Parabola   **53.** Hyperbola   **55.** $\frac{y^2}{64} - \frac{x^2}{36} = 1$

57. $\left(\frac{4\sqrt{14}}{7}, \frac{3\sqrt{7}}{7}\right)$ **59.** $\left(\pm\frac{3\sqrt{2}}{2}, \frac{\sqrt{14}}{2}\right)\left(\pm\frac{3\sqrt{2}}{2}, -\frac{\sqrt{14}}{2}\right)$

61. 0.01

Section 5.4

For Thought: **1.** T   **2.** F   **3.** F   **4.** F   **5.** T   **6.** T
7. T   **8.** F   **9.** F   **10.** F
Exercises: **1.** $(3, \pi/2)$ **3.** $\left(3\sqrt{2}, \pi/4\right)$
5–19 odd

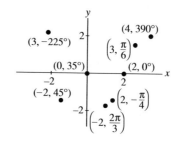

21. $(4, 0)$　**23.** $(0, 0)$　**25.** $\left(\dfrac{\sqrt{3}}{2}, \dfrac{1}{2}\right)$　**27.** $(0, 3)$

29. $(-1, 1)$　**31.** $\left(-\dfrac{\sqrt{6}}{2}, \dfrac{3\sqrt{2}}{2}\right)$　**33.** $\left(2\sqrt{3}, 60°\right)$

35. $\left(2\sqrt{2}, 135°\right)$　**37.** $(2, 90°)$　**39.** $\left(3\sqrt{2}, 225°\right)$

41. $\left(\sqrt{17}, 75.96°\right)$　**43.** $\left(\sqrt{6}, -54.7°\right)$　**45.** $(3.60, 1.75)$

47. $(1.80, 0.87)$　**49.** $(-0.91, -1.78)$　**51.** $(6.4, 51.3°)$

53. $(7.3, 254.1°)$

55.

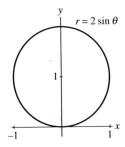

57.

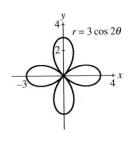

59.

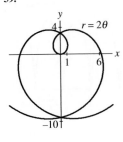

61.

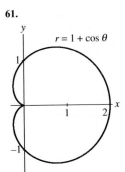

63.

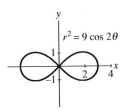

65.

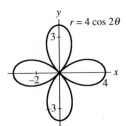

67.

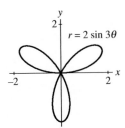

69.

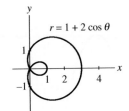

71.

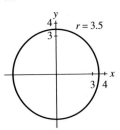

73.

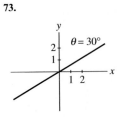

75. $x^2 - 4x + y^2 = 0$　**77.** $y = 3$　**79.** $x = 3$

81. $x^2 + y^2 = 25$　**83.** $y = x$　**85.** $x^2 - 4y = 4$

87. $r\cos\theta = 4$　**89.** $\theta = -\pi/4$　**91.** $r = 4\tan\theta\sec\theta$

93. $r = 2$　**95.** $r = \dfrac{1}{2\cos\theta - \sin\theta}$　**97.** $r = 2\sin\theta$

Section 5.5

For Thought: **1.** T　**2.** F　**3.** T　**4.** T　**5.** T　**6.** F

7. T　**8.** T　**9.** T　**10.** T

Exercises: **1.** 2, hyperbola, 3　**3.** 1, parabola, $\dfrac{3}{4}$　**5.** $\dfrac{4}{3}$, hyperbola, $\dfrac{3}{4}$

7. Parabola, $y = -2$

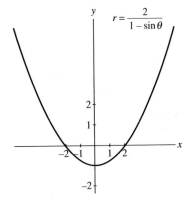

9. Ellipse, $x = \dfrac{5}{2}$

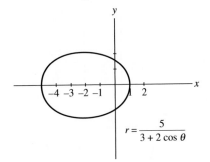

11. Hyperbola, $x = -\dfrac{1}{6}$

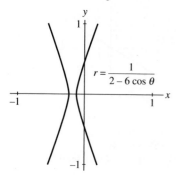

$r = \dfrac{1}{2 - 6\cos\theta}$

13. Ellipse, $y = 6$

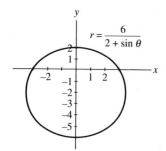

$r = \dfrac{6}{2 + \sin\theta}$

15. $r = \dfrac{2}{1 + \sin\theta}$ **17.** $r = \dfrac{10}{1 + 2\cos\theta}$ **19.** $r = \dfrac{4}{2 + \sin\theta}$

21. $r = \dfrac{24}{4 - 3\cos\theta}$ **23.** $x^2 + 6y - 9 = 0$, parabola

25. $15x^2 + 16y^2 - 6x - 9 = 0$, ellipse

27. $72x^2 - 9y^2 - 18x + 1 = 0$, hyperbola

29. $36x^2 + 35y^2 - 4y - 4 = 0$, ellipse   **31.** $(0, 0), (3, \pi)$

33. $e = \dfrac{1}{2}$

Section 5.6

For Thought: **1.** F   **2.** T   **3.** T   **4.** F   **5.** T   **6.** F
7. T   **8.** T   **9.** F   **10.** T

Exercises:

1.

t	x	y
0	1	-2
1	5	-1
1.5	7	-0.5
3	13	1

3.

t	x	y
1	1	2
2.5	6.25	6.5
$\sqrt{5}$	5	$3\sqrt{5} - 1$
4	16	11
5	25	14

5. $[-2, 10], [3, 7]$

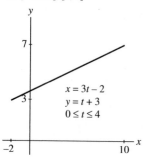

$x = 3t - 2$
$y = t + 3$
$0 \le t \le 4$

7. $(-\infty, \infty), [0, \infty)$

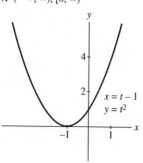

$x = t - 1$
$y = t^2$

9. $(0, 1), (0, 1)$

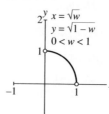

$x = \sqrt{w}$
$y = \sqrt{1 - w}$
$0 < w < 1$

11. $[-1, 1], [-1, 1]$

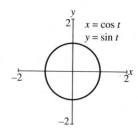

$x = \cos t$
$y = \sin t$

13. $x + y = -2$, line, $(-\infty, \infty), (-\infty, \infty)$
15. $x^2 + y^2 = 16$, circle, $[-4, 4], [-4, 4]$
17. $y = e^{4x}$, exponential, $(-\infty, \infty), (0, \infty)$
19. $y = 2x + 3$, line, $(-\infty, \infty), (-\infty, \infty)$
21. $x = \dfrac{3}{2}t + 2, y = 3t + 3, 0 \le t \le 2$
23. $x = 2\cos t, y = 2\sin t, \pi < t < 3\pi/2$
25. $x = 3, y = t, -\infty < t < \infty$ **27.** $x = \sin(2t), y = 2\sin^2(t)$
29.   **31.**

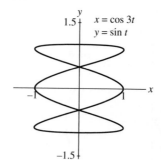

$x = \cos 3t$
$y = \sin t$

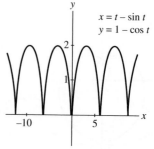

$x = t - \sin t$
$y = 1 - \cos t$

33.

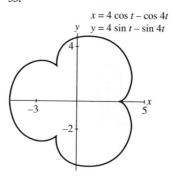

$x = 4\cos t - \cos 4t$
$y = 4\sin t - \sin 4t$

35.

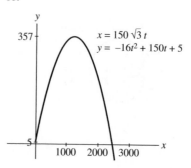

$x = 150\sqrt{3}\, t$
$y = -16t^2 + 150t + 5$

37. 9.4 sec

Chapter 5 Review Exercises

1. $(-6, 0), (2, 0), (0, -12), (-2, -16), x = -2, (-2, -63/4),$
$y = -65/4$

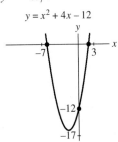

$$y = x^2 + 4x - 12$$

3. $(0, 0), (3, 0), (3/2, 9/2), x = 3/2,$
$(3/2, 35/8), y = 37/8$

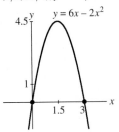

$$y = 6x - 2x^2$$

5. $(-6, 0), (0, -2 \pm \sqrt{10}), (-10, -2), y = -2,$
$(-39/4, -2), x = -41/4$

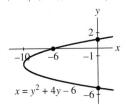

$$x = y^2 + 4y - 6$$

7. $(0, \pm 2\sqrt{5})$

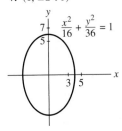

$$\frac{x^2}{16} + \frac{y^2}{36} = 1$$

9. $(1, 5), (1, -3)$

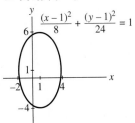

$$\frac{(x-1)^2}{8} + \frac{(y-1)^2}{24} = 1$$

11. $(1, -3 \pm \sqrt{2})$

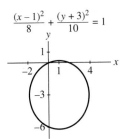

$$\frac{(x-1)^2}{8} + \frac{(y+3)^2}{10} = 1$$

13. $(0, 0), 9$

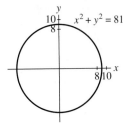

$$x^2 + y^2 = 81$$

15. $(-1, 0), 2$

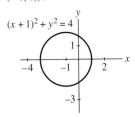

$$(x+1)^2 + y^2 = 4$$

17. $(-5/2, 0), \sqrt{6}$

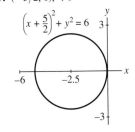

$$\left(x + \frac{5}{2}\right)^2 + y^2 = 6$$

19. $x^2 + (y + 4)^2 = 9$ **21.** $(x + 2)^2 + (y + 7)^2 = 6$

23. $(\pm 10, 0), y = \pm\frac{3}{4}x$

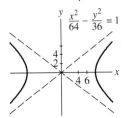

$$\frac{x^2}{64} - \frac{y^2}{36} = 1$$

25. $(4, 2 \pm 4\sqrt{5}), y = 2x - 6, y = -2x + 10$

$$\frac{(y-2)^2}{64} - \frac{(x-4)^2}{16} = 1$$

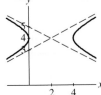

27. $(2 \pm \sqrt{5}, 4), y = \frac{1}{2}x + 3, y = -\frac{1}{2}x + 5$

$$\frac{(x-2)^2}{4} - (y - 4)^2 = 1$$

29. Hyperbola   **31.** Ellipse   **33.** Parabola   **35.** Hyperbola

37.

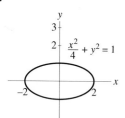

39.

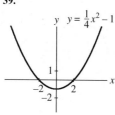

75. $x + y = 1$ **77.** $x^2 + y^2 = 25$ **79.** $r = \dfrac{3}{\sin \theta}$ **81.** $r = 7$

83. Parabola, $y = 3$

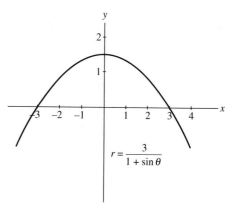

41.

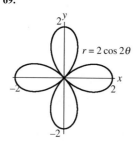

43.

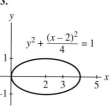

85. Ellipse, $x = 4$

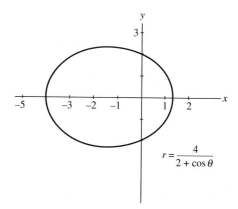

45. $x = (y - 3)^2 + \dfrac{3}{4}$ **47.** $\dfrac{x^2}{36} + \dfrac{y^2}{20} = 1$

49. $(x - 1)^2 + (y - 3)^2 = 20$

51. $\dfrac{x^2}{4} - \dfrac{y^2}{5} = 1$

53. $(x + 2)^2 + (y - 3)^2 = 9$

55. $\dfrac{(x + 2)^2}{9} + (y - 1)^2 = 1$

57. $\dfrac{(y - 1)^2}{9} - \dfrac{(x - 2)^2}{4} = 1$

59. $\left(2.5, 5\sqrt{3}/2\right)$ **61.** $(-0.3, 1.7)$

63. $(4, 4\pi/3)$

65. $\left(\sqrt{13}, -0.98\right)$

67.

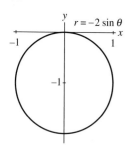

69.

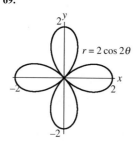

87. Hyperbola, $x = -\dfrac{1}{6}$

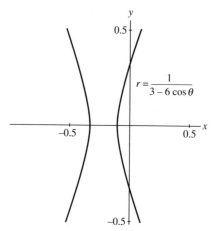

71.

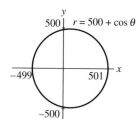

73.

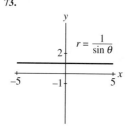

89. $r = \dfrac{3}{1 + \sin \theta}$ **91.** $r = \dfrac{18}{1 - 3 \cos \theta}$ **93.** $r = \dfrac{9}{3 + \sin \theta}$

95.

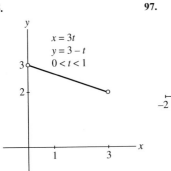

97.

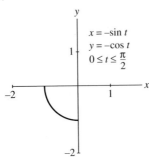

99. $\dfrac{x^2}{100^2} - \dfrac{y^2}{120^2} = 1$

101. $\dfrac{x^2}{34^2} + \dfrac{y^2}{16^2} = 1$, 10.81 ft

Appendix A

Section A.1

Exercises: **1.** 64   **3.** -49 **5.** 16   **7.** $\dfrac{1}{81}$ **9.** 11/30   **11.** 24

13. 8   **15.** $-6x^{11}y^{11}$ **17.** $3y^5$ **19.** $-4x^6$ **21.** $-\dfrac{8}{27}x^6$

23. $3x^4$ **25.** $\dfrac{25}{y^4}$ **27.** $\dfrac{1}{6y^3}$ **29.** $\dfrac{n^4}{4}$ **31.** -3 **33.** 8

35. -4 **37.** 81   **39.** 1/16   **41.** $x^2y^{1/2}$ **43.** $6a^{3/2}$ **45.** $3a^{1/6}$

47. $a^{7/3}b$ **49.** $\dfrac{x^2y}{z^3}$ **51.** 30   **53.** -2 **55.** $-\dfrac{1}{5}$ **57.** 8

59. $\sqrt[3]{10^2}$ **61.** $\dfrac{3}{\sqrt[5]{y^3}}$ **63.** $x^{-1/2}$ **65.** $x^{3/5}$ **67.** $4x$ **69.** $2y^3$

71. $\dfrac{\sqrt{xy}}{10}$ **73.** $\dfrac{-2a}{b^5}$ **75.** $2\sqrt{7}$ **77.** $\dfrac{\sqrt{5}}{5}$ **79.** $\dfrac{\sqrt{2x}}{4}$ **81.** $2\sqrt[3]{5}$

83. $-5x\sqrt[3]{2x}$ **85.** $\dfrac{\sqrt[3]{4}}{2}$ **87.** $9 - 2\sqrt{6}$

89. $2\sqrt{2} + 2\sqrt{5} - 2\sqrt{3}$ **91.** $-30\sqrt{2}$ **93.** $60a$

95. $\dfrac{3\sqrt{a}}{a^2}$ **97.** $5x\sqrt{5x}$

Section A.2

Exercises: **1.** 3, 1, trinomial   **3.** 2, -3, binomial

5. 0, 79, monomial   **7.** 12   **9.** 77   **11.** $8x^2 + 3x - 1$
13. $-5x^2 + x - 3$ **15.** $(-5a^2 + 4a)x^3 + 2a^2x - 3$
17. $2x - 1$ **19.** $3x^2 - 3x - 6$ **21.** $-18a^5 + 15a^4 - 6a^3$

23. $3b^3 - 14b^2 + 17b - 6$ **25.** $8x^3 - 1$ **27.** $x^3 + 125$
29. $xz - 4z + 3x - 12$ **31.** $a^3 - b^3$ **33.** $a^2 + 7a - 18$
35. $2y^2 + 15y - 27$ **37.** $4x^2 - 81$ **39.** $6x^4 + 22x^2 + 20$
41. $4x^2 + 20x + 25$ **43.** $9x^2 + 30x + 25$ **45.** $x^4 - 9$ **47.** -23
49. $9x^6 - 24x^3 + 16$ **51.** $4x^2y^2 - 20xy + 25$ **53.** $5\sqrt{2} + 2\sqrt{10}$

55. $\dfrac{2\sqrt{6} - \sqrt{2}}{11}$ **57.** $x + 3, 0$ **59.** $a^2 + a + 1, 0$

61. $x + 5, 13$ **63.** $2x - 6, 13$ **65.** $x^2 + x + 1, 0$ **67.** $3x - 5, 7$
69. $2x + 3, x$ **71.** $x + 3, -2x$ **73.** $x^2 + 2x - 24$
75. $2a^{10} - 3a^5 - 27$ **77.** $-y - 9$ **79.** $w^2 + 8w + 16$
81. $3y^5 - 9xy^2$ **83.** $2b - 1$ **85.** $9w^4 - 12w^2n + 4n^2$

Section A.3

Exercises: **1.** $6x^2(x - 2), -6x^2(-x + 2)$
3. $ax(-x^2 + 5x - 6), -ax(x^2 - 5x + 6)$
5. $1(m - n), -1(-m + n)$ **7.** $(x^2 + 5)(x + 2)$
9. $(y^2 - 3)(y - 1)$ **11.** $(d - w)(ay + 1)$ **13.** $(y^2 - b)(x^2 - a)$
15. $(x + 2)(x + 8)$ **17.** $(x - 6)(x + 2)$ **19.** $(m - 2)(m - 10)$
21. $(t - 7)(t + 12)$ **23.** $(2x + 1)(x - 4)$ **25.** $(4x + 1)(2x - 3)$
27. $(3y + 5)(2y - 1)$ **29.** $(3b + 2)(4b + 3)$
31. $(t - u)(t + u)$ **33.** $(t + 1)^2$ **35.** $(2w - 1)^2$ **37.** $(3zx + 4)^2$
39. $(t - u)(t^2 + tu + u^2)$ **41.** $(a - 2)(a^2 + 2a + 4)$
43. $(3y + 2)(9y^2 - 6y + 4)$
45. $(3xy^2 - 2z^3)(9x^2y^4 + 6xy^2z^3 + 4z^6)$ **47.** $-3x(x - 3)(x + 3)$
49. $2t(2t + 3w)(4t^2 - 6tw + 9w^2)$ **51.** $(a - 2)(a + 2)(a + 1)$
53. $(x - 2)^2(x^2 + 2x + 4)$ **55.** $-2x(6x + 1)(3x - 2)$
57. $(a - 2)(a^2 + 2a + 4)(a + 2)(a^2 - 2a + 4)(a - 1)$
59. $-(3x + 5)(2x - 3)$

Section A.4

Exercises: **1.** $\{x \mid x \neq -2\}$ **3.** $\{x \mid x \neq 4 \text{ and } x \neq -2\}$

5. $\{x \mid x \neq 3 \text{ and } x \neq -3\}$ **7.** All real numbers   **9.** $\dfrac{3}{x + 2}$

11. $-\dfrac{2}{3}$ **13.** $\dfrac{ab^4}{b - a^2}$ **15.** $\dfrac{y^2z}{x^3}$ **17.** $\dfrac{a^2 + ab + b^2}{a + b}$ **19.** $\dfrac{b + 3}{a - y}$

21. $\dfrac{3}{7ab}$ **23.** $\dfrac{42}{a^2}$ **25.** $\dfrac{a + 3}{3}$ **27.** $\dfrac{2x - 2y}{x + y}$ **29.** $x^2y^2 + xy^3$

31. $\dfrac{-2}{x + wx}$ **33.** $\dfrac{16a}{12a^2}$ **35.** $\dfrac{x^2 - 8x + 15}{x^2 - 9}$ **37.** $\dfrac{x^2 + x}{x^2 + 6x + 5}$

39. $\dfrac{t^2 + t}{2t^2 + 4t + 2}$ **41.** $12a^2b^3$ **43.** $6(a + b)$

45. $(x + 2)(x + 3)(x - 3)$ **47.** $\dfrac{9 + x}{6x}$ **49.** $\dfrac{x + 7}{(x - 1)(x + 1)}$

51. $\dfrac{3a + 1}{a}$ **53.** $\dfrac{t^2 - 2}{t + 1}$ **55.** $\dfrac{2x^2 + 3x - 1}{(x + 1)(x + 2)(x + 3)}$

57. $\dfrac{7}{2x - 6}$ **59.** $\dfrac{x^2}{x^3 - y^3}$ **61.** $\dfrac{-9x - 3}{(2x - 3)(x + 5)(x - 1)}$

CREDITS

Chapter 1

P.1, © PhotoDisc Red; p. 21, © PhotoDisc; p. 22, © PhotoDisc; p. 34, © PhotoDisc; p. 35, © Corbis RF; p. 42, © Digital Vision; p. 55, © Chuck Savage/Corbis Stock Market; p. 56, © PhotoDisc; p. 57, (Table for Exercise 91) © PhotoDisc, (Exercise 92) © DigitalVision; p. 68, © PhotoDisc; p. 91, © Corbis RF; p. 92, © PhotoDisc; p. 95, © PhotoDisc; p. 103, © PhotoDisc; p. 104, (Exercise 93) © DigitalVision (PP), (Exercise 94) © PhotoDisc.

Chapter 2

P. 113, © Getty RF; p. 142, © PhotoDisc; p. 194, © PhotoDisc (PP).

Chapter 3

P. 196, © Corbis RM; p. 217, © Digital Vision; p. 230, (Exercise 79) © PhotoDisc, (Exercise 80) © PhotoDisc; p. 231, © PhotoDisc; p. 300, Courtesy of the U.S. Department of Defense.

Chapter 4

P. 311, © Getty/Stone; p. 314, © PhotoDisc; p. 324, (Exercise 95) Courtesy of NASA, (Exercise 96) © PhotoDisc Red (PP); p. 334, © PhotoDisc; p. 350, © PhotoDisc (PP); p. 353, © PhotoDisc; p. 354, © PhotoDisc; p. 358, (Table for Exercise 113) © PhotoDisc, (Exercise 114) © PhotoDisc Red (PP).

Chapter 5

P. 361, © Getty Creative; p. 368, Courtesy of NASA; p. 370, © Roger Ressmeyer/Corbis.

INDEX OF APPLICATIONS

Archeology

Age of Deinonychus, 349
Carbon-14 Dating, 353
Dating a Bone, 353
Finding the Half-Life, 353, 357
Radioactive Decay, 323

Astronomy

Comet Hale-Bopp, 382
Eccentricity of an Orbit, 379
First Pulsar, 230
Hale Telescope, 370
Halley's Comet, 382
Hubble Telescope, 367
Orbit of the Moon, 382
Sun Spots, 230

Biology/Health/Life Sciences

Blood Velocity, 230
Bungee Jumping, 56
Drug Testing, 142
Growth Rate for Bacteria, 149
Half-Life of a Drug, 353
Learning Curve, 358
Lithotripter, 422
Lung Capacity, 230
Pediatric Tuberculosis, 358
Poiseuille's Law, 103
Positioning a Human Arm, 300

Business

Across-the-Board Raise, 82
Balancing the Costs, 189
Big Debt, 323
Billboard Advertising, 189
Carpenters and Helpers, 44
Computers and Printers, 44
Concert Tickets, 125
Cost of Printing Handbooks, 185
Depreciation Rate, 103
Filing a Tax Return, 69
Laying Sod, 91
Maximizing Revenue, 124
Packing Billiard Balls, 150
Paying up, 10
Periodic Cost, 230
Periodic Revenue, 230
Photography from a Spy Plane, 258
Profit, 91
Profitable Business, 90
Retail Store Profit, 149
Sales Tax Function, 111
Seven-Eleven, 335
Shipping Machinery, 69
Turning a Profit, 110
Working by the Hour, 323

Chemistry

Celsius to Fahrenheit Formula, 44
Controlling Temperature, 21
Drug Build-Up, 353
Lorazepam, 353
Temperature, 103

Construction

Area of a Window, 91
Best Fitting Pipe, 69
Blocking a Pipe, 262
Bonus Room, 194
Buckling Bridge, 309
Cellular One, 299
Constructing an Elliptical Arch, 382
Cross Fenced, 124
Designing an Addition, 299
Hanging a Pipe, 262
Heating and Air, 150
Height of Buildings, 263, 299, 308
Height of a Crosswalk, 261
Height of a Skyscraper, 262
Height of a Tower, 257, 296, 299
Height of an Antenna, 257
Installing a Guy Wire, 261
Laying Pipe, 405
Length of a Tunnel, 261
Oscillating Depth, 308
Oscillating Temperature, 308
Painting Problem, 160
Storing Supplies, 160
Tall Antenna, 261

Consumer

Admission to the Zoo, 188
Becoming a Millionaire, 334
Cell Phone Costs, 42
Change in Cost, 57
Computer Spending, 57
Cost of Business Cards, 44
Cost of Food, 92, 94
Cost of Gravel, 56
Cost of a Parking Ticket, 324
Cost-of-Living Raise, 82
Costly Computers, 104
Depreciation and Inflation, 354
Depreciation of a Mustang, 55
Difference in Prices, 21
Doubling Your Money, 334
Expensive Models, 21
Hamburgers, 91
Paying off a Loan, 354
Price Range for a Car, 21
Price of a Burger, 21
Price of a Car, 103
Renting a Car, 189
Ticket Pricing, 44
Traffic Jam, 334
Volume Discount, 44

Design

Bridge Cables, 371
Capsize Control, 35
Capsize Screening Value, 160
Designing Fireworks, 150
Displacement-Length Ratio, 91
Giant Teepee, 174
Limiting the Beam, 35
Making a Gas Tank, 189
Mirror Mirror, 124
Packing Cheese, 173
Pentagon, 300
Robotic Arm, 295
Rocket Propelled Grenade, 124
Sail Area-Displacement Ratio, 91, 159
Spacing Between Teeth, 216
Wing of the F-106 Delta Dart, 294

Environment

Dating a Tree, 353
Extinction of Species, 354
Giant Redwood, 260
Global Warming, 335
Habitat Destruction, 354
Noise Pollution, 354
Nuclear Power, 421
Radioactive Waste, 353

Geometry

Adjacent Circles, 383
Angle Bisectors, 57
Angle of Depression, 261
Angles, 251, 252
Big Barn, 124
Cartridge Box, 143
Circle Inscribed in a Square, 111
Computer Case, 143
Crescent City, 358
Crooked Man, 308
Filling a Triangle, 189
Height of a Balloon, 261
Height of a Rock, 261
Height of a Tower, 257
Hexagon, 300
Lakefront Property, 416
Length of a Chord, 300
Length of a Diagonal of a Square, 53
Length of an Arc, 205, 208
Maximizing Volume, 173
Maximum Area, 122, 124, 194
Open-Top Box, 142
Overlapping Region, 125

Radius of a Pipe, 160
Right Triangle, 159
Shortcut to Snyder, 261
Straight Man, 308
Tangent Circles, 274
Three Circles, 383
Tin Can, 110
Triangles and Circles, 208
Twin Kennels, 124
Two Common Triangles, 287
Viewing Area, 273
Volume of a Cube, 103

Investment

Annual Growth Rate, 104
Ben's Gift to Boston, 344
Ben's Gift to Philadelphia, 344
Comparing Investments, 358
Compounded Daily, 320, 342
Doubling Time with Continuous
     Compounding, 357
Doubling Time with Quarterly
     Compounding, 357
Equality of Investments, 354
Finding Rate, 334
Finding Time, 334, 351
Future Value, 357
Higher Yields, 323
Interest Compounded Continuously, 321, 323,
     324, 331, 335
Interest Compounded Quarterly, 324
Periodic Compounding, 323
Present Value Compounding Continuously, 323
Present Value Compounding Daily, 323
Present Value of a Bond, 354
Present Value of a CD, 354
Saving for Retirement, 323

Miscellaneous

Army of Ants, 44
Attack of the Grizzly, 300
Avoiding a Swamp, 260
Counting Votes, 242
Distance Between Towers, 207
Distance to the Helper, 208
Double-Boxed, 410
Doubling the Bet, 358
Falling Painter, 422
Firing an M-16, 286
Leaning Ladder, 174
Methodical Mower, 35
Moving a Refrigerator, 143
Muzzle Velocity, 287
Observing Traffic, 299

Pile of Pipes, 57, 354
Robin and Marion, 261
Rollover Time, 344
Shot Down, 300
Stacking Pipes, 371
Student's Salary, 89
Telling Time, 217
The Survivor, 231
Watering the Lawn, 301

Navigation

Air Navigation, 395
Average Speed of an Auto Trip, 189
Boating, 300
Course of a Bush Pilot, 297
Course of a Fighter Plane, 299
Flying, 300
Hiking, 300
Increasing Visibility, 309
Instantaneous Rate of Change, 195
Marine Navigation, 395
Passing in the Night, 261
Radar Range, 309

Physics

Load on a Spring, 111
Maximum Height of a Ball, 121, 124
Maximum Height of a Football, 124
Motion of a Spring, 214, 216, 230, 273, 286
Path of a Projectile, 283

Production

A Little Lagniappe, 395
Acceptable Bearings, 21
Acceptable Targets, 21
Cooling Hot Steel, 353
Manufacturing Cost, 324

Science

AM Radio, 308
Aerial Photography, 260
Altitude of a Rocket, 194
Below Sea Level, 56
Broadcasting the Oldies, 308
Challenger Disaster, 324
Cloud Height, 308
Colombian Earthquake of 1906, 344
Communicating Via Satellite, 262
Cooking a Roast, 353
Distance to the North Pole, 204
Doubling the Sound Level, 354
Dropping a Watermelon, 56
Dropping the Ball, 111

Eavesdropping, 371
Eratosthenes Measures the Earth,
     208
Focus of an Elliptical Reflector, 382
Landing Speed, 103
Large Ocean Waves, 231
Newton's Law of Cooling, 350
Ocean Waves, 230
Parabolic Mirror, 395
Richter Scale, 344
Room Temperature, 353
Searchlight, 421
Sonic Boom, 384
Telephoto lens, 395
Testing a Scale, 18
Time of Death, 353
View from Landsat, 262
Whispering Gallery, 421

Sports

Archer Shooting an Arrow, 416
Bicycle Gear Ratio, 21
Changing Speed of a Dragster, 111
Choosing the Right Angle, 286
Clear Sailing, 252
Kicking a Field Goal, 263
Laying Out a Track, 21
Limiting Velocity, 194
Maximum Sail Area, 159
Minimum Displacement for a Yacht,
     159
Position of a Football, 323
Rowers and Speed, 103
Sail Area-Displacement Ratio, 159
Sea Worthiness, 358
Selecting the Cogs, 21
Shooting a Target, 308
Shooting an Arrow, 124
Shooting Baskets, 414
Swim Meet, 324
Throwing a Javelin, 217
Time Swimming and Running, 160

Statistics/Demographics

Bringing Up Your Average, 21
Final Exam Scores, 18, 20
First Marriage, 34
Motor Vehicle Ownership, 68, 69
Population Growth, 314, 323
Population of California, 52
Time for Growth, 344
Unmarried Couples, 34
Weighted Average with Fractions, 21
Weighted Average with Whole Numbers, 21

INDEX

A

Abscissa, 22
Absolute error, 18
Absolute value, 6–9
  defined, 6–7
  equations involving, 8–9
  equations with, solving, 156–157
  properties of, 7
  uses of, 7–8
Absolute value inequalities, 16–18
  defined, 16–17
  modeling with, 18–19
  solving, 17–18
Absolute-value functions, 61–62
  defined, 61
  graphing, 62
  properties of, 79
Acute angle, measure of, 198
Additive inverse property
  defined, 4
  examples of, 5
Adjacent, right triangle, 254
Algebra
  fundamental theorem of, 137
  interrelation to geometry, 58
  review of concepts, Appendix A
Algebraic functions
  defined, 312
  with transformations, gallery of, 359
Ambiguous case (SSA)
  defined, 290
  oblique triangles, solving, 290–292
  SSA with no triangle, 291
  SSA with one triangle, 291
  SSA with two triangles, 291–292
Amplitude
  defined, 220
  of sine wave, finding, 220
Angle(s)
  acute angle, 198
  coterminal, 198–199
  defined, 197
  degree measure of, 197–198
  of depression, 256
  of elevation, 256–258
  obtuse angle, 198
  parts of, 197
  quadrant of, determining, 199–200
  quadrantal angle, 198
  radian measure of, 201–204
  right angle, 198
  straight angle, 198
Angle of elevation, 256–258
  defined, 256
  height of object, finding, 257
  height of object from distance, finding,
    257–258, 296–297
Appolonius of Perga, 378

Arc, sine and cosine of, 212–213
Arc length, 204–205
  defined, 204
  radian measure to find, 204–205
  spy plane, photography from, 258–259
Arccos, 245
Arccot, 246
Arccsc, 246
Archimedes of Syracuse, 295
Arcsec, 246
Arcsin, 243
Arctan, 246
Associative property, 4
Asymptotes
  horizontal and vertical, 175–178
  of hyperbola, 386, 388
  oblique, 178–179
  for rational function, finding, 179
Average rate of change, 51–53
  defined, 51
  finding, 52
Axis of ellipse
  conjugate axis, 386
  in equation of ellipse, 372–374
  major and minor, 362
Axis of hyperbola
  conjugate axis, 386
  defined, 384
  in graph of hyperbola, 386
  transverse axis, 384
Axis of symmetry, parabola, 362

B

Base a, exponential functions with, 312
Base-change formula
  compound interest model, 341–342
  defined, 341
Base-e exponential function, continuous compounding, 320–321
Bearing, to solve triangles, 297–298
Bounded intervals, compound inequalities, 14

C

Calculator use
  absolute value, 7
  circle, graphing, 26–27
  degree-radian conversion, 202–203
  domain and range, determining, 49
  exponential family of functions, 317
  exponential functions, evaluating, 312–313
  function notation, 51
  functions, 47
  fundamental identity, 214
  graph to solve equation, 30
  graphs showing intercepts, 30
  greatest integer functions, 64
  inverse trigonometric functions, 248–249

irrational numbers, approximation, 3
logarithmic functions, 325–326, 328
parabolas, graphing, 116–117
periodic function, 224
perpendicular lines, equation for, 41
piecewise functions, 63
polar equations, graphing, 400–401
polynomials, factoring, 136
radian measure of angles, 202
rational zeros, finding, 140
rule of signs, 147
sine and cosine, 213–214, 218–219, 222–224
synthetic division, 134
trigonometric functions, evaluating, 233
vertical and horizontal asymptotes, 177–178,
  180–181
Calculus
  circle, area of and π, 310
  instantaneous rate of change, 195
  limits, 112
  parabola, reflecting properties of, 423
  transcendental functions, evaluating, 360
Cartesian coordinate system
  elements of, 22–23, 396
  See also Rectangular coordinate system
Center of circle
  defined, 26, 377
  in equation of circle, 378
  finding, 378
Center of ellipse
  defined, 372–373
  ellipse, graphing from, 376–377
  in equation of ellipse, 372–374
Center of hyperbola
  defined, 384–385
  graphing hyperbola from, 389–391
Central angle, 197
Circle(s), 25–28, 377–379
  arc length, 204–205
  completing the square, 28
  defined, 26, 377
  equation of, 26–28, 378
  graphing, 26–27
  unit circles, 201
Closed intervals, 11–12, 14
Closure property, 4
Cofunction identities
  defined, 266
  odd/even, using, 266
Common logarithms
  defined, 326
  equations involving, 331
Commutative property, 4
Completing the square
  conic sections, identifying by, 392–393
  quadratic functions, 114–116
  rule for, 28

Complex conjugates
  defined, 128
  imaginary numbers, dividing, 128–129
  product, finding, 128
Complex numbers, 125–130
  adding/subtracting/multiplying, 127
  complex conjugates, 128
  defined, 126
  dividing, 128–129
  equivalence, 126
  negative numbers, square roots of, 129–130
  quadratic equations, imaginary solutions, 130
  real and imaginary parts of, 126
  simplifying, 127–128
  standard form of, 126
Composition of functions, 85–89
  defined, 86
  defined by equations, 86–88
  defined by sets, 86
  defined with formulas, 88–89
  evaluating, 249–250
  function, writing as composition, 88
  with function notation, 89
  inverse functions, verifying with, 98
  trigonometric functions, 249–250
Compound inequalities, 14–16
  bounded intervals, 14
  defined, 14
  solving, 15–16
Compound interest
  base-change formula, use of, 341–342
  base-*e* exponential function, 320–321
  continuous compounding, 320–321, 331–332
  exponential functions, 319–321
  formula for, 319–320
Conditional equations
  defined, 274
  See also Trigonometric equations
Conic sections
  circles, 377–379
  defined, 409
  degenerate conics, 362
  ellipse, 371–377
  hyperbola, 383–393
  identifying, complete the squares method,
    392–393
  parabola, 362–368
  polar equations of. *See* Polar equations of the
    conics
Conjugate axis, 386
Conjugate pairs theorem
  polynomial equations with real coefficients,
    finding, 145
  theorem, 145
Constant functions
  defined, 65, 75
  on an interval, 65
  properties of, 79
Continuous compounding
  procedure, 320–321
  time, finding, 331–332

Coordinate(s)
  Cartesian coordinate system, 22–23
  defined, 22
  points on number line, 2
Coordinate plane, 23
Cosecant, 232–233
  defined, 232
  domain of, 232
  graphs of, 238–239
  inverse of, 246
  sign and quadrant, 232
  transformations, 239
Cosine, 209–227
  amplitude, 221
  of arc, 212–213
  for common angles, memory tip, 212
  cosine of sum, 265
  defined, 209
  domain of, 209
  evaluating at multiple of 30°, 211–212
  evaluating at multiple of 45°, 210–211
  evaluating at multiple of 90°, 209–210
  evaluating with calculator, 213
  finding angle from, 246
  fundamental cycle, 220–221
  fundamental identity, 213–214
  graphs of, 220–221
  horizontal translations, 222–223
  inverse of, 244–246
  law of. *See* Law of cosines
  motion of spring, modeling, 214–215
  periodic function, 221
  phase shift, 221
  sign and quadrant, 210, 232
  transformations, 221–224
  vertical translation, 222–223
Cosine equations, 274–276
  solving, 275–276
Cotangent, 232–238
  defined, 232
  domain of, 232
  graphs of, 235–236
  inverse of, 246
  sign and quadrant, 232
  transformations, 237
Coterminal angles, 198–199
  defined, 198
  determining from measures, 199
  finding, 199
  measures of, 198
  radian measure, finding with, 204
Cube functions, properties of, 79
Cube root, graphing, 60
Cube-root functions
  graphing, 60
  properties of, 79

D

Decimal degrees
  converting to degrees/minutes/seconds, 201
  degrees/minutes/seconds, converting to,
    200–201

Decreasing functions
  defined, 65
  on an interval, 65
Degenerate conics, 362
Degree(s), measure of angles, 197–198
Degree measure of angles
  decimal degrees, converting to degrees/minutes/
    seconds, 201
  defined, 197
  degree-radian conversion, 202–203
  degrees/minutes/seconds, converting to deci-
    mal degrees, 200–201
  examples of, 198
  minutes and seconds, 200
Dependent variables, 46
Descartes, René, 22, 146
  rule of signs, 146–148
Difference(s)
  complex numbers, 127
  defined, 5
  functions, 83–85
Difference quotient, 52–53
  defined, 52
  finding, 52–53
Directed length, radian measure of angles
    as, 201
Directrix of parabola
  defined, 362
  in equation of parabola, 363
  equation of parabola from, 363–364
  finding, 364–365
  graphing parabola from, 365–366
  vertical, 367
Distance, absolute value equations, 8
Distance formula, 23–24
  circle, standard equation for, 26
  distance between points, finding, 24
  theorem, 24
Distributive property, 4
Division
  synthetic division, 133–135
  See also Quotient(s)
Domain
  cosecant, 232
  cosine, 209
  cotangent, 232
  defined, 48
  determining, 49
  exponential functions, 313
  polynomial functions, 175
  rational functions, 175, 182
  secant, 232
  sine, 209
  tangent, 232
Double-angle, sine equation with, 278–279
Double-angle identities, 266–267
  defined, 266
  using, 267
Downward translations, 70–71
Dummy variables, 50

E

Eccentricity
    of ellipse, 378–379
    of orbit around earth, 379
Ellipse, 371–377
    defined, 372
    eccentricity of, 378–379
    equation of, 372–374
    graphing, 374–376
    parts of, 372–373
    reflecting properties, 372
    translations of, 376–377
Equality, properties of, 6
Equation(s)
    with absolute value, solving, 156–157
    of circles, 378
    composition of functions defined by, 86–88
    defined, 11
    of ellipse, 372–374
    exponential equations, 318–319
    factoring, 151–152
    functions, identifying from, 48
    of hyperbola, 384–385
    logarithmic equations, 329–331
    of parabolas, 362–363
    parametric equations, 411–415
    polar equations, 401–403
    polar equations of the conics, 406–409
    of quadratic type. *See* quadratic type,
        equations of
    with rational exponents, solving, 153–154
    rectangular equations, 398
    solving with graphs, 31
    with square roots, solving, 152–153
    trigonometric equations, 274–283
Equations in two variables, 35–42
    linear equation in two variables, 29
    solution set, 25–26
Equilibrium, 214
Equivalence, complex numbers, 126
Erathosthenes, 205
Euclid, 28
Euler, Leonhard, 320
Even functions, symmetric about the *y*-axis, 76
Exponential equations, 318–319
    first coordinate, finding given second, 319
    one-to-one property, 318
    rewriting as logarithmic equations, 329
    with single exponential expression, 347
    solving, 318–319
    solving, procedure in, 348
    with two exponential expressions, 347–348
Exponential family of functions, 316–318
    defined, 316
    graphing members of family, 316–317
    graphing with reflection, 317–318
Exponential functions, 312–318
    with base *a*, 312
    base-*e* exponential function, 320–321
    compound interest model, 319–321
    defined, 312

domain, 313
evaluating, 312–313
graphing, 313–316
inverse of, 329–330
one-to-one property of, 318
properties of, 316
types/properties, gallery of, 332
Extraneous roots, and equations with square
    roots, 152–153

F

Factor theorem, 136–137
    polynomials, factoring, 136
    theorem, 136
Factoring
    equations, solving by, 151–152
    quadratic type, equations of, 281
    trigonometric equations, 280
Families of functions
    defined, 69
    linear family, 75
    square family, 70
Focal length, parabola, 362–363
Focus(ii) of ellipse
    defined, 372–373
    ellipse, graphing from, 375–376
    in equation of ellipse, 372–374
Focus of hyperbola, in equation of
    hyperbola, 385
Focus of parabola
    defined, 362
    in equation of parabola, 363
    equation of parabola from, 363–364
    finding, 364–365
    graphing parabola from, 365–366
Formula(s)
    composition of functions defined by, 88–89
    functions as, 48
Four-leaf rose, 400
Frequency
    defined, 226
    of sine wave, 226–227
Function(s), 45–53
    absolute value functions, 61–62
    algebraic functions, 312
    average rate of change, 51–53
    composition of, 85–89
    constant functions, 75
    constructing, 53
    cube functions, 60
    cube-root functions, 60
    defined, 46
    difference quotient, 52–53
    evaluating, 84
    even functions, 76
    exponential functions, 312–318
    families of. *See* Families of functions
    as formulas, 48
    greatest integer functions, 64
    identifying from equation, 48
    identifying from graph, 47

identifying from table/list, 47–48
increasing/decreasing/constant functions, 65
inverse functions, 94–100
"is a function of", 45–46
linear functions, 75
logarithmic functions, 325–329
minimum and maximum value, 118
odd functions, 77
one-to-one functions, 92–94
piecewise functions, 61–65
polynomial functions, 114, 132–143
quadratic functions, 114–122
rational functions, 175–182
square functions, 58
square-root function, 59
sum, difference, product, quotient functions,
    83–85
transcendental functions, 312
transformations of, 69
types/properties, gallery of, 79
vertical line test, 46–47
Function notation, 50–51
    composition of functions defined by, 89
    defined, 50
    inverse function notation, 95–96
    with variables, using, 50–51
Fundamental cycle
    of cosine, 220–221
    of sine, 219
    of tangent, 234–235
Fundamental identity
    defined, 213
    using, 214
Fundamental rectangle, 386
Fundamental theorem of algebra, 137

G

Galois, Evariste, 146
Galois theory, 146
Gauss, Carl Friedrich, 137
Geometry, interrelationship to algebra, 58
Graph(s)
    absolute value functions, 62
    axis, 23
    circle, 26–27
    cosecant, 238–239
    cosine, 220–221
    cotangent, 235–236
    cube functions, 60
    cube-root functions, 60
    ellipse, 374–376
    exponential functions, 313–316
    function and its inverse, 98
    functions, identifying from, 47
    greatest integer functions, 64
    horizontal lines, 30
    hyperbola, 386–391
    increasing/decreasing/constant functions, 65
    inequalities, solving with, 78
    linear functions, graphing with transforma-
        tions, 75

Graph(s) (*continued*)
  parabolas, 59, 365–367
  piecewise functions, 62–64
  point of intersection, 22–23
  quadratic functions, 115
  quadratic inequalities, 120–121
  rational function graphs, 179–182
  reflection, 72–73
  and rigid versus nonrigid transformations, 69
  rule of signs, 147
  secant, 237–238
  semicircles, 61
  showing intercepts 29–30
  sign graphs, 119–120
  sine, 217–220
  slope of line, 36–39
  slope of line, finding, 36–37
  to solve equations, 31
  square functions, 58
  square-root functions, 59
  stretching and shrinking, 73–74
  symmetry, 76–77, 162–163
  tangent, 233–235
  translations, 70–72
  vertical line test, functions, 46–47
  vertical lines, 30
Graphs in two variables, 22–31
  circle, 25–28
  distance formula, 23–24
  lines, 28–31
  midpoint formula, 24–25
  parallel lines, 40
  perpendicular lines, 40–41
Greatest integer functions
  defined, 64
  graphing, 64
  properties of, 79

H

Half-angle identities, 267–268
  defined, 267
  using, 267
Horizontal asymptotes, 175–178
  defined, 176
  identifying, 177–178
  rational function crosses asymptote, 181–182
  rational function, graphing, 180–182
  x-axis, 176–178, 185
Horizontal line(s), graphing, 30–31
Horizontal line test
  defined, 93–94
  one-to-one functions, finding, 93–94
Horizontal translations, 71–72
  sine and cosine, 222–223
Hyperbola, 383–393
  asymptotes of, 386, 388
  defined, 384
  equation, finding, 391–392
  equation for, 384–385
  graphing, 386–391
  opening left and right, 384, 387–388
  opening up and down, 384, 389–390

  parts of, 384–385
  reflecting properties, 383–384
Hypotenuse, 254

I

i
  and complex numbers, 126
  defined, 126
  power of, simplifying, 127–128
  whole-number powers, finding, 127
Identities, trigonometric. *See* Trigonometric
    identities
Identity functions, properties of, 79
Identity properties, of real numbers, 4
Imaginary numbers
  dividing, 128–129
  purpose of, 125
Increasing functions
  defined, 65
  on an interval, 65
Independent variables, 46
Inequality(ies)
  absolute value inequalities, 16–18
  compound inequalities, 14–16
  defined, 11
  linear inequalities, 12–14
  modeling with, 18–19
  polynomial inequalities, 169
  quadratic inequalities, 119–121
  rational inequalities, 183–185
  solving with graphs, 78
Initial side of angle, 197
Instantaneous rate of change, 195
Intercepted arc of angle, 197
Intermediate value theorem, polynomial function
    graphs, drawing, 161
Intersection of sets, 14–15
Interval notation, 11–12
  defined, 11
  open/closed types, 11–12
  unbounded intervals, 11–12
Inverse functions, 94–100, 97–98
  composition to verify, 98
  cosecant, 246
  cosine, 244–246
  cotangent, 246
  defined, 95
  of exponential functions, 329–330
  finding, 95
  finding mentally, 98–99
  graphing, 98
  inverse function notation, 95–96
  secant, 246
  sine, 242–244, 250
  switch-and-solve method, 96–97
  tangent, 246
  trigonometric, 242–249
  types/properties, gallery of, 100
Inverse rules
  logarithms, 335–336
  using, 336

Invertible functions, one-to-one functions, 95, 98
Irrational numbers, 2–3
  defined, 2
  examples of, 2–3
  purpose of, 125

K

Kepler, Johannes, 365

L

Lambert, Johann Heinrich, 3
Law of cosines, 292–294
  defined, 292
  proof, 292–293
  three sides of triangle (SSS), solving, 293–294
  two sides and included angle (SAS),
    solving, 294
Law of sines, 288–292
  ambiguous case (SSA), 290–292
  defined, 288
  proof, 288–289
  two angles and included side (ASA), solving,
    289–290
Leading coefficient test, 164–166
  behaviors of graph, examples of, 165–166
  defined, 166
Limits, in calculus, 112
Line(s), 28–31
  defined, 28
  equation for, standard form, 29, 39–40
  graphing, 29–31
  parallel lines, 40
  perpendicular lines, 40–41
  slope of line, 36–39
Line segment
  graphing, 411–412
  parametric equation, writing for, 413
Linear equation in two variables, equation,
    finding, 42
Linear functions
  defined, 75
  graphs using transformations, 75
  properties of, 79
Linear inequalities, 12–14
  defined, 12
  solving, 13–14
Lists, functions, identifying from, 47–48
Logarithm(s)
  common logarithms, 326
  defined, 325
  logarithmic equations, 329–331
  logarithmic functions, 325–329
  Naperian logarithms, 326
  natural logarithms, 326
  rules of. *See* Logarithmic rules
Logarithmic equations, 329–331
  base-change formula, 341–342
  with common and natural logarithms, 331
  loan payoff time application, 351
  with multiple logarithms, 346–347
  Newton's law of cooling application, 350–351

radioactive dating application, 349–350
rational approximation, 341
rewriting as exponential equations, 329
with single logarithm, 345
solving, 330–331
solving, procedure in, 348
Logarithmic family of functions
defined, 328
graphing members of family, 328–329
Logarithmic functions, 325–329
with base *a*, 325
base-*a* function, graphing, 326–327
defined, 325
evaluating, 325–326
one-to-one property, 330–331
properties of, 327
types/properties, gallery of, 332
Logarithmic rules
with base *a*, 339
inverse rules, 335–336
logarithmic expressions, rewriting as single
logarithm, 340
of natural logarithms, 339–340
power rule, 338–339
product rule, 336–337
quotient rule, 337–338

M

Major axis of ellipse, 372
Maximum
height, finding, 121
rectangle, maximizing area, 122
Maximum value, functions, 118
Midpoint formula, 24–25
finding midpoint, 25
theorem, 24–25
Minimum, height, finding, 121
Minimum value, functions, 118
Minor axis of ellipse, 372
Minutes (of degrees)
decimal degrees, converting to degrees/
minutes/seconds, 201
defined, 200
degrees/minutes/seconds, converting to decimal degrees, 200–201
Models/modeling
basketball, flight of, 414–415
compound interest, 319–321
eccentricity of orbit, 379
Hubble telescope mirror thickness, 367–368
with inequalities, 18–19
loan payoff time, 351
motion of spring, 214–215
Newton's law of cooling, 350–351
projectile motion, 283–284
radioactive dating, 349–350
Multiplication property of zero, 4
Multiplicative inverse property, 4
Multiplicity, 143–145
defined, 144
n-root theorem, 144

real and imaginary roots of polynomial equations, finding, 144–145

N

Naperian logarithms, 326
Napier, John, 326
Natural logarithms
defined, 326
equations involving, 331
rules of, 339
Negative numbers
purpose of, 125
square root of, 129–130
Negative sign, meaning of, 5
Newton's law of cooling, 350–351
Nonrigid transformations
defined, 69
stretching and shrinking, 73–74
n-root theorem
defined, 137
theorem, 144
Number(s)
complex numbers, 125–130
imaginary numbers, 125
irrational numbers, 2–3
rational numbers, 2–3
real numbers, 2–8
relations, 6
sets, 2–3
Number line
defined, 2
distance between points, finding, 8
sign graphs, 119–120

O

Oblique asymptotes, 178–179
defined, 178
rational function, graphing, 178–179, 182
Oblique triangles
ambiguous case (SSA), 290–292
defined, 287
law of cosines, 292–294
law of sines, 288–292
three sides of triangle (SSS), solving, 293–294
two angles and included side (ASA), solving,
289–290
two sides and included angle (SAS),
solving, 294
Obtuse angle, measure of, 198
Odd and even identities, 264–265
defined, 264
using, 265
Odd functions, symmetric about the origin, 77
One-to-one functions, 92–94
determining, 92–93
horizontal line test, 93–94
inverse of. *See* Inverse functions
as invertible, 95, 98
One-to-one property
exponential functions, 318
logarithmic functions, 330–331
Open intervals, 11–12, 14

Opposite, right triangle, 254
Opposites, properties of, 5
Orbit around earth, eccentricity of, 379
Ordered pairs
equations in two variables, 25–26
functions, 46–53
one-to-one functions, 92–93
Origin of number line, 22–23

P

Parabola(s), 116–119, 362–368
characteristics, identifying, 118
defined, 362
equation, writing from focus and directrix,
363–364
equation of, 362–363
graphing, 365–367
Hubble telescope application, 367–368
intercepts, 118–119
opening to left and right, graphs, 366–367
opening to right, graph, 59
opening upward and downward, graphs,
116–117
parts of, 362
reflecting properties of, 367–368, 423
and square family, 70
vertex, focus, directrix, finding, 364–365
vertex of, 116–118, 362
with vertical directrix, graphing, 367
Parallel lines
defined, 40
equations, writing, 40
theorem, 40
Parametric equations, 411–415
basketball flight application, 414–415
defined, 411
for line segment, 413
line segment, graphing, 411–412
parameter, eliminating, 412–413
polar equation converted to, 413–414
Period of function, 219
Periodic function
for cosine, 221
graphing, 219–220, 221
period of, 219
for sine, 218–220
Perpendicular lines
defined, 40
equations, writing, 41
theorem, 41
Phase shift, sine and cosine, 221
Piecewise functions, 61–65
absolute value functions, 61–62
defined, 61
graphing, 62–64
Points, graphs, 23
Point-slope form, 37
theorem, 37
Polar axis, polar coordinate system, 396
Polar coordinates 396–403
defined, 396

Polar coordinates (*continued*)
   equations. *See* Polar equations
   parts of, 396
   points, plotting, 396–397
   polar-rectangular conversions, 397–398
Polar equations, 398–401
   of the conics. *See* Polar equations of the conics
   converting to parametric equations, 413–414
   converting to rectangular equation, 402
   four-leaf rose graph, 400
   rectangular equations converted to, 402–403
   spiral of Archimedes graph, 400–401
Polar equations of the conics, 406–409
   definitions, equivalency of, 408–409
   finding, 408
   identifying/graphing conic from polar equation, 407–408
   polar conic, converting to rectangular coordinates, 409
   theorem, 407
Pole, 396
Polynomial equations
   conjugate pairs theorem, 145
   fourth-degree, solving, 155
   multiplicity, 143–145
   n-root theorem, 144
   rule of signs, 146–148
Polynomial function(s), 132–143
   cubic, 170
   defined, 114
   domain, 175
   factor theorem, 136–137
   fundamental theorem of algebra, 137
   graphing. *See* Polynomial function graphs
   linear, 170
   quadratic, 170
   quartic, 170
   rational zero theorem, 137–140
   remainder theorem, 132–133
   synthetic division, 133–136
Polynomial function graphs, 161–169
   behavior at *x*-intercepts, 163–164
   drawing, 161
   graphing procedure, 167–168
   intermediate value theorem, 161
   leading coefficient test, 164–166
   second-degree. *See* Parabola(s)
   symmetry, 161–163
   types, gallery of, 170
   0 or 1 degree. *See* Line(s)
Polynomial inequalities, test-point method, 169
Potassium-argon dating, 349–350
Power rule
   defined, 338
   logarithms, 338–339
   using, 339
Principal square root, 129
Product(s)
   complex conjugates, 128
   complex numbers, 127
   functions, 83–85

Product rule
   logarithms, 336–337
   proof, 336
   using, 337
Product-to-sum identities, 268–270
   defined, 268
   product-to-sum, expressing, 268–269
   sum or difference, expressing as product, 269–270
Projectile, maximum height, finding, 121–122
Projectile motion, modeling with trigonometric equations, 283–284
Pythagoras of Samos, 23
Pythagorean identities, 263–264
   producing, 263–264
   using, 264
Pythagorean theorem, right triangles, solving, 256

Q

Quadrant(s), 23
Quadrant of angles, determining, 199–200
Quadrantal angle, measure of, 198
Quadratic formula
   quadratic equations, imaginary solutions, 130
   theorem, 116
   trigonometric equations, solving with, 281–282
Quadratic functions, 114–122
   completing the square, 114–116
   maximum and minimum, 121–122
   quadratic formula, 116
   quadratic inequalities, 119–121
   theorem, 115
Quadratic inequalities, 119–121
   defined, 119
   solving with sign graphs, 119–120
   solving with test-point method, 120–121
Quadratic type, equations of, 154–156
   form for, 154
   and rational exponents, 154
   trigonometric equation, 281
Quotient(s)
   functions, 83–85
   imaginary numbers, 128–129
   synthetic division, 133–135
Quotient rule
   defined, 338
   logarithms, 337–338
   using, 338

R

Radian measure of angles, 201–204
   arc length, finding, 204–205
   coterminal angles, finding, 204
   degree-radian conversion, 202–203
   as directed length, 201
Radical(s)
   and equations with square roots, 152–153
Radioactive dating, 349–350
   and half-life, 349
   potassium-argon dating, 349–350

Radius of circle
   defined, 26, 377
   in equation of circle, 378
   finding, 378
Range
   defined, 48
   determining, 49
Rational approximation, logarithmic equations, 341
   equations with, solving, 153–154
   and quadratic type equations, 154
Rational function(s), 175–179
   asymptotes, finding for, 179
   defined, 175
   domain, 175, 182
   graphing. *See* Rational function graphs
   oblique asymptotes, 178–179
   project cost, application, 184–185
   vertical and horizontal asymptotes, 175–178
Rational function graphs, 179–182
   crossing asymptotes, 181–182
   graph with hole, 182
   graphing procedure, 179
   oblique asymptotes, 178–179, 182
   vertical and horizontal asymptotes, 180–182
Rational inequalities, 183–185
   defined, 183
   sign graph of factors, 183–184
   test-point method, 184
Rational numbers, 2–3
   defined, 2
   set of, 2
Rational zero theorem, 137–140
   polynomial functions, finding rational zeros, 138–139
   polynomial functions, finding real and imaginary zeros, 139–140
   theorem, 138
Ray of angle, 197
Real numbers, 2–8
   defined, 3
   examples of, 3
   properties of, 4–6
   set of, 2
   trichotomy property, 6
Real part, complex numbers, 126
Rectangle(s)
   area, maximizing, 122
   fundamental rectangle, 386
Rectangular coordinate system
   polar conic converted to, 409
   polar-rectangular conversions, 397–398, 401
   rectangular equations, 398
   See also Cartesian coordinate system
Rectangular equations, converting to polar equation, 402–403
Reduction formula, 270–271
   defined, 270
   using, 270–271
Reflection, 72–73
   defined, 72

exponential functions, graphing with, 317–318
graphing, 72–73
symmetry, 76–77
Reflexive property of equality, 6
Relations
defined, 48
domain and range of, 48–49
equality, properties of, 6
symbols for, 6
Relative error, 18
Remainder theorem, 132–133
polynomials, evaluating with, 133
theorem, 133
Right angle, measure of, 198
Right triangles, 254–259
angle of depression, 256
angle of elevation, 256–258, 296–297
parts of, 254
robotic arm, positioning, 295–296
solving, 255–256
trigonometric functions in, 254–255
Rigid transformations
defined, 69
reflection, 72–73
translations, 69–70
Rise, 36
Robotic arm, positioning, 295–296
Rule of signs, 146–148
defined, 146
graphing, 147
using, 146–147
variation of sign, 146
Run, 36

S

Secant, 232–233, 237–238
defined, 232
domain of, 232
graphs of, 237–238
inverse of, 246
sign and quadrant, 232
transformations, 238
Seconds (of degrees)
decimal degrees, converting to degrees/
minutes/seconds, 201
defined, 200
degrees/minutes/seconds, converting to decimal degrees, 200–201
Semicircle(s), graphing, 61
Sets, 2–3
composition of functions defined by, 86
defined, 2
intersection of, 14–15
member of set, symbol for, 3
of real numbers, 2
union of, 14–15
Sign graphs, 119–120
defined, 119
quadratic inequalities, solving with,
119–120
rational inequalities, solving with, 183–184

Sine, 209–227
of arc, 212–213
for common angles, memory tip, 212
defined, 209
domain of, 209
evaluating at multiple of 30°, 211–212
evaluating at multiple of 45°, 210–211
evaluating at multiple of 90°, 209–210
evaluating with calculator, 213
finding angle from, 244
fundamental identity, 213–214
general sine wave, 224–225
graphs of, 217–220
horizontal translation, 222–223
inverse of, 242–244, 250
law of. *See* Law of sines
motion of spring, modeling, 214–215
periodic function, 218–220
phase shift, 221
sign and quadrant, 210, 232
sinusoid. *See* Sine wave
transformations, 221–224
Sine equations, 276–277
with double angle, 278–279
solving, 277
Sine wave, 218–227
amplitude, 220
defined, 218
frequency of, 226–227
fundamental cycle of, 219
graphing, 224–227
Slant asymptote. *See* Oblique asymptotes
Slope of line, 36–39
average rate of change, 51–52
defined, 36
finding slope, 36
to graph line, 38–39
point-slope form, 37
positive/negative, 39
run and rise, 36
slope-intercept form, 37–38
Slope-intercept form, 37–38
theorem, 38
Solution set, equations in two variables,
25–26
Spiral of Archimedes, 400–401
Spring, motion of, sine and cosine for modeling,
214–215
Square family, and parabola, 70
Square functions
graphing, 58
properties of, 79
Square root
equations with, solving, 152–153
of negative numbers, 129–130
principal square root, 129
Square-root functions
graphing, 59
properties of, 79
Squaring sides, trigonometric equations, 282–283
Standard form, equation for circle, 26–27
Standard position of angle, 197

Straight angle, measure of, 198
Stretching and shrinking, 73–74
defined, 73
graphing, 73–74
Substitution property of equality, 6
Sum(s)
complex numbers, 127
functions, 83–85
Sum and difference identities
cosine of sum, 265
defined, 265
Switch-and-solve method, inverse functions,
finding, 96–97
Symmetry, 76–77, 161–163
about the origin, 77
y-axis, 76, 162
in graphs, determining, 77, 162–163
property of equality, 6
Synthetic division, 133–136
compared to ordinary division, 134
defined, 134
polynomials, evaluating with, 135
procedure in, 134

T

Tables, functions, identifying from, 47–48
Tangent, 232–236
defined, 232
domain of, 232
fundamental cycle, 234–235
graphs of, 233–235
inverse of, 246
sign and quadrant, 232
transformations, 234–235
Tangent equations, 277–278
with multiple angles, 279–280
solving, 278
Terminal side of angle, 197
Test-point method
polynomial inequalities, solving with, 169
quadratic inequalities, solving with, 120–121
rational inequalities, solving with, 183–184
Transcendental functions
defined, 312
evaluating, 360
Transformations
cosecant, 239
cotangent, 237
families of functions, 69
linear functions, graphing with, 75
multiple, graphing, 74–75
quadratic functions, 115
reflection, 72–73
rigid and nonrigid, 69
secant, 238
sine and cosine, 221–224
stretching and shrinking, 73–74
tangent, 234–235
translations, 69–70
Transitive property of equality, 6

Translations
    defined, 70
    of ellipse, 376–377
    horizontal, 71–72
    to right or left, 71
    sine and cosine, 222–223
    upward or downward, 70–71
Transverse axis, 384
Trends, limits, 112
Triangles
    oblique triangles, 287–294
    right triangles, 254–259
Trichotomy property, 6
Trigonometric equations
    cosine equations, 274–276
    factoring, solving by, 280
    with multiple angles, 278–280
    projectile motion, modeling, 283–284
    quadratic formula, solving with, 281–282
    quadratic type equation, 281
    sine equations, 276–277
    solving, procedure in, 283
    squaring sides of equation, 282–283
    tangent equations, 277–278
Trigonometric functions
    angles, 197–209
    composition of functions, 249–250
    cosecant, 232–233, 239
    cotangent, 232–238
    evaluating, 232–233
    inverse functions, 242–249
    reduction formula, 270–271
    in right triangles, 254–255
    secant, 232–233, 237–238
    signs of, 210, 232
    sine and cosine, 209–227
    tangent, 232–236
Trigonometric identities, 263–270
    cofunction identities, 266
    defined, 263
    double-angle identities, 266–267
    half-angle identities, 267–268
    odd and even identities, 264–265

    product-to-sum identities, 268–270
    Pythagorean identities, 263–264
    sum and difference identities, 265
    using, 268
Trigonometric ratios, 252–253
    theorem, 253
    trigonometric functions, finding, 253
Trigonometry
    fundamental identity, 213–214
    history of, 197
    law of cosines, 292–294
    law of sines, 288–292
    right triangles, 254–259
    trigonometric equations, 274–283
    trigonometric identities, 263–270
    trigonometric ratios, 252–253

U

Unbounded intervals, interval notation, 11–12
Union of sets, 14–15
Unit circles, defined, 201
Units of number line, 2
Upward translations, 70–71

V

Variables
    defined, 2
    dependent and independent, 46
    dummy variables, 50
    function notation with, 50–51
    two, relationship between. *See* Equations in
        two variables; Graphs in two variables
Vertex of angle, 197
Vertex(ices) of ellipse
    defined, 373
    in equation of ellipse, 372–374
Vertex(ices) of hyperbola, defined, 384
Vertex of parabola, 116–118
    defined, 116
    in equation of parabola, 363
    finding, 117–118, 364–365

Vertical asymptotes, 175–178
    defined, 176
    identifying, 177–178
    rational function crosses asymptote, 181–182
    rational function, graphing, 180–182
    y-axis, 176, 185–186
Vertical line test, 46–47
    theorem, 46
Vertical lines, graphing, 30–31
Vertical translation, cosine, 222–223

X

x-axis
    defined, 23
    horizontal asymptotes, 176–178, 185
x-intercept
    behavior at, theorem, 163
    crossing by functions, 164
    defined, 29
    parabolas, finding, 119
    polynomial function graphs, 163–164
xy-plane, 23

Y

y-axis
    defined, 23
    symmetry about the, 76, 162
    vertical asymptotes, 176, 185–186
y-intercept
    defined, 29
    graphing, 29, 38–39
    parabolas, finding, 119

Z

Zero
    multiplication property of, 4
    n-root theorem, 137
    of polynomial functions, 132, 136
    rational zero theorem, 137–140

 **Geometry**

Rectangle

Area = LW

Perimeter = $2L + 2W$

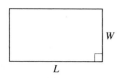

Square

Area = s^2

Perimeter = $4s$

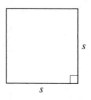

Triangle

Area = $\frac{1}{2}bh$

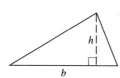

Right Triangle

Area = $\frac{1}{2}ab$

Pythagorean theorem:
$c^2 = a^2 + b^2$

Parallelogram

Area = bh

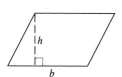

Trapezoid

Area = $\frac{1}{2}h(b_1 + b_2)$

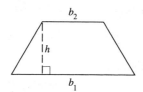

Circle

Area = πr^2

Circumference = $2\pi r$

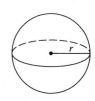

Right Circular Cone

Volume = $\frac{1}{3}\pi r^2 h$

Lateral surface area = $\pi r\sqrt{r^2 + h^2}$

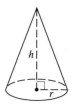

Right Circular Cylinder

Volume = $\pi r^2 h$

Lateral surface area = $2\pi rh$

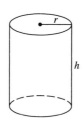

Sphere

Volume = $\frac{4}{3}\pi r^3$

Surface area = $4\pi r^2$

Metric Abbreviations

Length		Volume		Weight	
mm	millimeter	mL	milliliter	mg	milligram
cm	centimeter	cL	centiliter	cg	centigram
dm	decimeter	dL	deciliter	dg	decigram
m	meter	L	liter	g	gram
dam	dekameter	daL	dekaliter	dag	dekagram
hm	hectometer	hL	hectoliter	hg	hectogram
km	kilometer	kL	kiloliter	kg	kilogram

English-Metric Conversion

Length	Volume (U.S.)	Weight
1 in. = 2.540 cm	1 pt = 0.4732 L	1 oz = 28.35 g
1 ft = 30.48 cm	1 qt = 0.9464 L	1 lb = 453.6 g
1 yd = 0.9144 m	1 gal = 3.785 L	1 lb = 0.4536 kg
1 mi = 1.609 km		

Length	Volume (U.S.)	Weight
1 cm = 0.3937 in.	1 L = 2.2233 pt	1 g = 0.0353 oz
1 cm = 0.03281 ft	1 L = 1.0567 qt	1 g = 0.002205 lb
1 m = 1.0936 yd	1 L = 0.2642 gal	1 kg = 2.205 lb
1 km = 0.6215 mi		

# Algebra

Subsets of the Real Numbers

Natural numbers = $\{1, 2, 3, \ldots\}$
Whole numbers = $\{0, 1, 2, 3, \ldots\}$
Integers = $\{\ldots -3, -2, -1, 0, 1, 2, 3, \ldots\}$
Rational = $\left\{ \dfrac{a}{b} \middle| a \text{ and } b \text{ are integers with } b \neq 0 \right\}$
Irrational = $\{x | x \text{ is not rational}\}$

Properties of the Real Numbers

For all real numbers a, b, and c

$a + b$ and ab are real numbers.	Closure
$a + b = b + a$; $a \cdot b = b \cdot a$	Commutative
$(a + b) + c = a + (b + c)$; $(ab)c = a(bc)$	Associative
$a(b + c) = ab + ac$; $a(b - c) = ab - ac$	Distributive
$a + 0 = a$; $1 \cdot a = a$	Identity
$a + (-a) = 0$; $a \cdot \dfrac{1}{a} = 1 \quad (a \neq 0)$	Inverse
$a \cdot 0 = 0$	Multiplication property of 0

Absolute Value

$$|a| = \begin{cases} a & \text{for } a \geq 0 \\ -a & \text{for } a < 0 \end{cases}$$

$\sqrt{x^2} = |x|$ for any real x

$|x| = k \Leftrightarrow x = k \text{ or } x = -k \qquad (k > 0)$

$|x| < k \Leftrightarrow -k < x < k \qquad (k > 0)$

$|x| > k \Leftrightarrow x < -k \text{ or } x > k \qquad (k > 0)$

(The symbol $\Leftrightarrow$ means "if and only if.")

Interval Notation

$(a, b) = \{x | a < x < b\}$

$(a, b] = \{x | a < x \leq b\}$

$(-\infty, a) = \{x | x < a\}$

$(-\infty, a] = \{x | x \leq a\}$

$[a, b] = \{x | a \leq x \leq b\}$

$[a, b) = \{x | a \leq x < b\}$

$(a, \infty) = \{x | x > a\}$

$[a, \infty) = \{x | x \geq a\}$

Exponents

$a^n = a \cdot a \cdot \cdots \cdot a$ (n factors of a)

$a^0 = 1 \qquad\qquad a^{-n} = \dfrac{1}{a^n}$

$a^r a^s = a^{r+s} \qquad\qquad \dfrac{a^r}{a^s} = a^{r-s}$

$(a^r)^s = a^{rs} \qquad\qquad (ab)^r = a^r b^r$

$\left(\dfrac{a}{b}\right)^r = \dfrac{a^r}{b^r} \qquad\qquad \left(\dfrac{a}{b}\right)^{-r} = \left(\dfrac{b}{a}\right)^r$

Radicals

$a^{1/n} = \sqrt[n]{a} \qquad\qquad a^{m/n} = \left(\sqrt[n]{a}\right)^m = \sqrt[n]{a^m}$

$\sqrt[n]{ab} = \sqrt[n]{a} \cdot \sqrt[n]{b} \qquad\qquad \sqrt[n]{\dfrac{a}{b}} = \dfrac{\sqrt[n]{a}}{\sqrt[n]{b}}$

Factoring

$a^2 + 2ab + b^2 = (a + b)^2$

$a^2 - 2ab + b^2 = (a - b)^2$

$a^2 - b^2 = (a + b)(a - b)$

$a^3 - b^3 = (a - b)(a^2 + ab + b^2)$

$a^3 + b^3 = (a + b)(a^2 - ab + b^2)$

Rational Expressions

$\dfrac{ac}{bc} = \dfrac{a}{b} \qquad\qquad \dfrac{a}{b} + \dfrac{c}{d} = \dfrac{ad + bc}{bd}$

$\dfrac{a}{b} \cdot \dfrac{c}{d} = \dfrac{ac}{bd} \qquad\qquad \dfrac{a}{b} \div \dfrac{c}{d} = \dfrac{a}{b} \cdot \dfrac{d}{c}$

Quadratic Formula

The solutions to $ax^2 + bx + c = 0$ with $a \neq 0$ are

$$x = \dfrac{-b \pm \sqrt{b^2 - 4ac}}{2a}.$$

Distance Formula

The distance from (x_1, y_1) to (x_2, y_2), is

$$\sqrt{(x_2 - x_1)^2 + (y_2 - y_1)^2}.$$

# Algebra

Midpoint Formula

The midpoint of the line segment with endpoints (x_1, y_1) and (x_2, y_2) is

$$\left(\frac{x_1 + x_2}{2}, \frac{y_1 + y_2}{2} \right).$$

Slope Formula

The slope of the line through (x_1, y_1) and (x_2, y_2) is

$$\frac{y_2 - y_1}{x_2 - x_1} \quad \text{(for } x_1 \neq x_2\text{)}.$$

Linear Function

$f(x) = mx + b$ with $m \neq 0$

Graph is a line with slope m.

Quadratic Function

$f(x) = ax^2 + bx + c$ with $a \neq 0$

Graph is a parabola.

Polynomial Function

$f(x) = a_n x^n + a_{n-1} x^{n-1} + \cdots + a_1 x + a_0$ for n a nonnegative integer

Rational Function

$f(x) = \dfrac{p(x)}{q(x)}$, where p and q are polynomial functions with $q(x) \neq 0$

Exponential and Logarithmic Functions

$f(x) = a^x$ for $a > 0$ and $a \neq 1$

$f(x) = \log_a(x)$ for $a > 0$ and $a \neq 1$

Properties of Logarithms

Base-a logarithm:	$y = \log_a(x) \Leftrightarrow a^y = x$
Natural logarithm:	$y = \ln(x) \Leftrightarrow e^y = x$
Common logarithm:	$y = \log(x) \Leftrightarrow 10^y = x$
One-to-one:	$a^{x_1} = a^{x_2} \Leftrightarrow x_1 = x_2$
	$\log_a(x_1) = \log_a(x_2) \Leftrightarrow x_1 = x_2$

$\log_a(a) = 1 \qquad \log_a(1) = 0$

$\log_a(a^x) = x \qquad a^{\log_a(N)} = N$

$\log_a(MN) = \log_a(M) + \log_a(N)$

$\log_a(M/N) = \log_a(M) - \log_a(N)$

$\log_a(M^x) = x \cdot \log_a(M)$

$\log_a(1/N) = -\log_a(N)$

$$\log_a(M) = \frac{\log_b(M)}{\log_b(a)} = \frac{\ln(M)}{\ln(a)} = \frac{\log(M)}{\log(a)}$$

Compound Interest

$P = $ principal, $t = $ time in years, $r = $ annual interest rate, and $A = $ amount:

$$A = P\left(1 + \frac{r}{n}\right)^{nt} \text{ (compounded } n \text{ times/year)}$$

$$A = Pe^{rt} \text{ (compounded continuously)}$$

Variation

Direct: $y = kx$ $(k \neq 0)$

Inverse: $y = k/x$ $(k \neq 0)$

Joint: $y = kxz$ $(k \neq 0)$

Straight Line

Slope-intercept form: $y = mx + b$

Slope: m y-intercept: $(0, b)$

Point-slope form: $y - y_1 = m(x - x_1)$

Standard form: $Ax + By = C$

Horizontal: $y = k$ Vertical: $x = k$

 Algebra

Parabola

$y = a(x - h)^2 + k$ $\quad (a \neq 0)$

Vertex: (h, k)

Axis of symmetry: $x = h$

Focus: $(h, k + p)$, where $a = \dfrac{1}{4p}$

Directrix: $y = k - p$

Circle

$(x - h)^2 + (y - k)^2 = r^2$ $\quad (r > 0)$

Center: (h, k) $\quad$ Radius: r

$x^2 + y^2 = r^2$

Center $(0, 0)$ $\quad$ Radius: r

Ellipse

$\dfrac{x^2}{a^2} + \dfrac{y^2}{b^2} = 1$ $\quad (a > b > 0)$

Center: $(0, 0)$ $\qquad\qquad$ Major axis: horizontal

Foci: $(\pm c, 0)$, where $c^2 = a^2 - b^2$

$\dfrac{x^2}{b^2} + \dfrac{y^2}{a^2} = 1$ $\quad (a > b > 0)$

Center: $(0, 0)$ $\qquad\qquad$ Major axis: vertical

Foci: $(0, \pm c)$, where $c^2 = a^2 - b^2$

Hyperbola

$\dfrac{x^2}{a^2} - \dfrac{y^2}{b^2} = 1$

Center: $(0, 0)$ $\qquad\qquad$ Vertices: $(\pm a, 0)$

Foci: $(\pm c, 0)$, where $c^2 = a^2 + b^2$

Asymptotes: $y = \pm \dfrac{b}{a} x$

$\dfrac{y^2}{a^2} - \dfrac{x^2}{b^2} = 1$

Center: $(0, 0)$ $\qquad\qquad$ Vertices: $(0, \pm a)$

Foci: $(0, \pm c)$, where $c^2 = a^2 + b^2$

Asymptotes: $y = \pm \dfrac{a}{b} x$

Arithmetic Sequence

$a_1, a_1 + d, a_1 + 2d, a_1 + 3d, \ldots$

Formula for nth term: $a_n = a_1 + (n - 1)d$

Sum of n terms:

$$S_n = \sum_{i=1}^{n} [a_1 + (i - 1)d] = \dfrac{n}{2}(a_1 + a_n)$$

Geometric Sequence

$a_1, a_1 r, a_1 r^2, a_1 r^3, \ldots$

Formula for nth term: $a_n = a_1 r^{n-1}$

Sum of n terms when $r \neq 1$:

$$S_n = \sum_{i=1}^{n} a_1 r^{i-1} = \dfrac{a_1 - a_1 r^n}{1 - r}$$

Sum of all terms when $|r| < 1$:

$$S = \sum_{i=1}^{\infty} a_1 r^{i-1} = \dfrac{a_1}{1 - r}$$

Counting Formulas

Factorial notation: $n! = 1 \cdot 2 \cdot 3 \cdot \cdots \cdot (n - 1) \cdot n$

Permutation: $P(n, r) = \dfrac{n!}{(n - r)!}$ for $0 \leq r \leq n$

Combination: $C(n, r) = \dbinom{n}{r} = \dfrac{n!}{(n - r)!r!}$ for $0 \leq r \leq n$

Binomial Expansion

$(a + b)^2 = a^2 + 2ab + b^2$

$(a + b)^3 = a^3 + 3a^2 b + 3ab^2 + b^3$

$(a + b)^4 = a^4 + 4a^3 b + 6a^2 b^2 + 4ab^3 + b^4$

$(a + b)^n = \displaystyle\sum_{r=0}^{n} \binom{n}{r} a^{n-r} b^r$, where $\dbinom{n}{r} = \dfrac{n!}{(n - r)!r!}$

Trigonometry

Trigonometric Functions

If the angle α
(in standard position)
intersects the unit circle
at (x, y), then

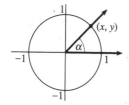

$$\sin \alpha = y \qquad \cos \alpha = x \qquad \tan \alpha = \frac{y}{x}$$

$$\csc \alpha = \frac{1}{y} \qquad \sec \alpha = \frac{1}{x} \qquad \cot \alpha = \frac{x}{y}$$

Trigonometric Ratios

If (x, y) is any point
other than the origin
on the terminal side of α
and $r = \sqrt{x^2 + y^2}$, then

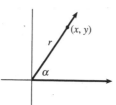

$$\sin \alpha = \frac{y}{r} \qquad \cos \alpha = \frac{x}{r} \qquad \tan \alpha = \frac{y}{x}$$

$$\csc \alpha = \frac{r}{y} \qquad \sec \alpha = \frac{r}{x} \qquad \cot \alpha = \frac{x}{y}$$

Right Triangle Trigonometry

If α is an acute angle
of a right triangle, then

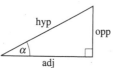

$$\sin \alpha = \frac{\text{opp}}{\text{hyp}} \qquad \cos \alpha = \frac{\text{adj}}{\text{hyp}} \qquad \tan \alpha = \frac{\text{opp}}{\text{adj}}$$

$$\csc \alpha = \frac{\text{hyp}}{\text{opp}} \qquad \sec \alpha = \frac{\text{hyp}}{\text{adj}} \qquad \cot \alpha = \frac{\text{adj}}{\text{opp}}$$

Special Right Triangles

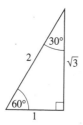

Exact Values of Trigonometric Functions

x degrees	x radians	$\sin x$	$\cos x$	$\tan x$
$0°$	0	0	1	0
$30°$	$\dfrac{\pi}{6}$	$\dfrac{1}{2}$	$\dfrac{\sqrt{3}}{2}$	$\dfrac{\sqrt{3}}{3}$
$45°$	$\dfrac{\pi}{4}$	$\dfrac{\sqrt{2}}{2}$	$\dfrac{\sqrt{2}}{2}$	1
$60°$	$\dfrac{\pi}{3}$	$\dfrac{\sqrt{3}}{2}$	$\dfrac{1}{2}$	$\sqrt{3}$
$90°$	$\dfrac{\pi}{2}$	1	0	$—$

Basic Identities

$$\tan x = \frac{\sin x}{\cos x} = \frac{1}{\cot x} \qquad \cot x = \frac{\cos x}{\sin x} = \frac{1}{\tan x}$$

$$\sin x = \frac{1}{\csc x} \qquad \csc x = \frac{1}{\sin x}$$

$$\cos x = \frac{1}{\sec x} \qquad \sec x = \frac{1}{\cos x}$$

Pythagorean Identities

$$\sin^2 x + \cos^2 x = 1 \qquad 1 + \cot^2 x = \csc^2 x$$

$$\tan^2 x + 1 = \sec^2 x$$

Odd Identities

$$\sin(-x) = -\sin(x) \qquad \csc(-x) = -\csc(x)$$

$$\tan(-x) = -\tan(x) \qquad \cot(-x) = -\cot(x)$$

Even Identities

$$\cos(-x) = \cos(x) \qquad \sec(-x) = \sec(x)$$